Ventricular/Vascular Coupling

Ventricular/Vascular Coupling

Clinical, Physiological, and Engineering Aspects

Edited by Frank C.P. Yin

With Contributions by

A.P. Avolio K.P. Brin K.B. Campbell C.R. Conti
J.W. Covell R.L. Engler R.A. Hopkins W.C. Hunter
J.S. Janicki T.W. Latson Z. Liu W.L. Maughan
J.P. Murgo W.W. Nichols M.F. O'Rourke C.J. Pepine
S. Permutt N.S. Peterson H. Piene J.A. Ringo
K. Sagawa S.G. Shroff K. Sunagawa G.J. van den Horn
L.K. Waldman K.T. Weber N. Westerhof R.A. Wise
T. Yaginuma F.C.P. Yin

With 195 Figures

Springer-Verlag New York Berlin Heidelberg
London Paris Tokyo

Frank C.P. Yin
Department of Medicine
Johns Hopkins Medical Institutions
Baltimore, Maryland 21205
U.S.A.

Library of Congress Cataloging in Publication Data
Ventricular/vascular coupling.
Includes bibliographies and index.
1. Cardiovascular system—Diseases. 2. Physiology, Pathological. 3. Heart—Ventricles. 4. Arteries. 5. Blood—Circulation—Regulation. I. Yin, F.C.P. (Frank C.P.) II. Avolio, A.P. (Albert P.) [DNLM: 1. Cardiovascular System—physiology. WG 102 V467]
RC669.9.V46 1986 616.1 85-30421

Softcover reprint of the hardcover 1st edition 1987

Typeset by David Seham Associates, Inc., Metuchen, New Jersey.

9 8 7 6 5 4 3 2 1

ISBN-13: 978-1-4613-8636-0 e-ISBN-13: 978-1-4613-8634-6
DOI: 10.1007/978-1-4613-8634-6

Preface

Physiologists have long been interested in the interaction, or coupling, between the heart and the vasculature. The early literature consists mainly of phenomenological descriptions of cardiac alterations resulting from specific interventions in the vasculature. Hundreds of studies, for example, describe functional aspects of hypertrophied myocardium associated with the excessive vascular loading produced by various types of experimental hypertension. Recently, the concepts of ventricular/vascular interaction have found important clinical application. The widespread use of vasodilators and of intraaortic counterpulsation balloons for unloading an overburdened, diseased heart is a prime example.

Despite the interest in this field, until as recently as 20 years ago we were not able to describe ventricular or vascular function in a framework suitable for quantitatively expressing the interaction between these two complex systems. Three major developments—description of ventricular function in terms of both the time-varying elastance and the pump function graph and quantification of vascular function in terms of impedance—have changed this. These functional descriptions now enable systems engineers, bioengineers, physiologists, and clinicians to address very specific quantitative aspects of ventricular/vascular interaction and have resulted in a flurry of papers and symposia devoted to this subject.

This book is an attempt to bring together the viewpoints and interests of both the clinical and the basic sciences. Among the contributors are many whose work has had major impact in this field in recent years. It is hoped that through synthesis of these different approaches a better understanding will arise. Specifically, I hope that the insight provided by these basic investigations will stimulate those with clinical interests to seek newer and better focused diagnostic and therapeutic modalities based on firm fundamental principles. By the same token, I hope that the very challenging clinical applications and unique aspects of disease in humans will stimulate and challenge basic scientists to further refine their approach toward areas of urgently needed applied research.

The book is divided into three major sections. The first section, composed of the first six chapters, addresses the more clinically oriented aspects of ventricular/vascular coupling of the venous, pulmonary, and arterial systems. The effects of aging, atherosclerosis, hypertension, exercise, and congenital heart disease on ventricular/vascular interaction are discussed, and areas needing further research are pointed out. In the following section, comprising Chapters 7 through 13, coupling of these same systems is examined from a more systems-oriented viewpoint. Most of the conceptual approaches in this section are based on recently developed functional descriptions of ventricular function (time-varying elastance, pump function graph, internal resistance) together with descriptions of vascular function in terms of vascular impedance. The section concludes with a chapter that identifies, using sensitivity analysis, the most important variables in left ventricular/arterial coupling. The third section, comprising Chapters 14 through 16, presents engineering-oriented approaches toward solving the extremely complicated time-varying interaction problem. The first is an approach for bridging the gap between isolated ventricular muscle to the intact heart and for accounting for the nonlinear effects of the aortic valve. The second is a time-domain approach utilizing the impulse response function to account for reflections as well as for the other components of the vascular load. The final chapter presents an approach for estimating vascular compliance. Although compliance estimates are a key descriptor of vascular properties, the means to estimate compliance have not received as much careful attention as the means to determine its counterpart, vascular impedance. Most of the work represents initial attempts, which are necessarily limited in scope. Nevertheless, the approaches outlined should provide fertile foundations for future research.

I have purposefully not provided a unified view for each topic. Rather, wherever possible, differing or at least complementary viewpoints are presented. For example, one of the chapters dealing with the effects of age on vascular impedance finds no effect in dogs (Ch. 5) whereas the other finds a significant effect in humans (Ch. 4). There are two differing but complementary views of the effect of the venous system on ventricular/vascular interaction, and two chapters deal with coupling of the pulmonary arterial system to the right ventricle. Most of the chapters avoid the complexity of the aortic valve by using time-averaged values, whereas Chapter 14 presents a method to account explicitly for the complex effects of the valve. While the approach adopted in this book may be somewhat disquieting for the reader, I think this approach is the reality of our state of knowledge. The hope is that readers will be so stimulated by the ideas presented that they will pursue related areas of research in the coming years and so add to our knowledge in this challenging, rewarding field.

The impetus for this book arose from a symposium dealing with ventricular/vascular interaction sponsored by the Biomedical Engineering Society in the spring of 1983. I am grateful to those original participants and to the other contributors, for this book is a reflection of their ideas and efforts. I would also like to acknowledge the invaluable editorial assistance of Susan Edmunds and the secretarial support of Linda Stasky and Deborah M. Cotton, whose efforts have greatly expedited the preparation of the book. Finally, I express my thanks to my wife and children for their understanding and support throughout this project.

Baltimore 1986 Frank C.P. Yin

Contents

Contributors

Albert P. Avolio Department of Medicine, University of New South Wales, St. Vincent's Hospital, Sydney, Australia

Kenneth P. Brin Department of Medicine, Johns Hopkins Medical Institutions, Baltimore, Maryland 21205, U.S.A.

Kenneth B. Campbell Department of Veterinary and Comparative Anatomy, Physiology, and Pharmacology, Washington State University, Pullman, Washington 99164, U.S.A.

C. Richard Conti Department of Medicine, University of Florida, Gainesville, Florida 32611, U.S.A.

James W. Covell Department of Medicine, University of California San Diego School of Medicine, La Jolla, California 92093, U.S.A.

Robert L. Engler Department of Medicine, University of California San Diego School of Medicine, La Jolla, California 92093, U.S.A.

Richard A. Hopkins Department of Surgery, Washington University School of Medicine, St. Louis, Missouri 63110, U.S.A.

William C. Hunter Department of Biomedical Engineering, Johns Hopkins Medical Institutions, Baltimore, Maryland 21205, U.S.A.

Joseph S. Janicki Division of Cardiology, Michael Reese Hospital, University of Chicago, Chicago, Illinois 60616, U.S.A.

Terry W. Latson Department of Anesthesiology, Johns Hopkins Medical Institutions, Baltimore, Maryland 21205, U.S.A.

Zhaorong Liu Department of Physiology, Johns Hopkins Medical Institutions, Baltimore, Maryland 21205, U.S.A. (Visiting Scientist, Fudan University, Shanghai, China)

W. Lowell Maughan Department of Medicine, Johns Hopkins Medical Institutions, Baltimore, Maryland 21205, U.S.A.

Joseph P. Murgo Department of Medicine, Brooke Army Medical Center, Fort Sam Houston, Texas 78234 U.S.A.

Wilmer W. Nichols Department of Medicine, University of Florida, Gainesville, Florida, 32611, U.S.A.

Michael F. O'Rourke Department of Medicine, University of New South Wales, St. Vincent's Hospital, Sydney, Australia

Carl J. Pepine Department of Medicine, University of Florida, Gainesville, Florida, 32611, U.S.A.

Solbert Permutt Department of Medicine, Francis Scott Key Medical Center, Baltimore, Maryland 21224, U.S.A.

N.S. Peterson Department of Veterinary and Comparative Anatomy, Physiology, and Pharmacology, Washington State University, Pullman, Washington 99164, U.S.A.

Hroar Piene Department of Biomedical Engineering, University of Trondheim, Trondheim, Norway

John A. Ringo Department of Electrical Engineering, Washington State University, Pullman, Washington 99164, U.S.A.

Kiichi Sagawa Department of Biomedical Engineering, Johns Hopkins Medical Institutions, Baltimore, Maryland 21205, U.S.A.

Sanjeev G. Shroff Division of Cardiology, Michael Reese Hospital, University of Chicago, Chicago, Illinois 60616, U.S.A.

Kenji Sunagawa Research Institute of Angiocardiology and Cardiovascular Clinic, Kyushu University School of Medicine, Fukuoka, Japan 812

Gerardus J. van den Horn Academic Medical Centre, University of Amsterdam, Amsterdam, The Netherlands

Lewis K. Waldman Department of Medicine, University of California San Diego School of Medicine, La Jolla, California 92093, U.S.A.

Karl T. Weber Division of Cardiology, Michael Reese Hospital, University of Chicago, Chicago, Illinois 60616, U.S.A.

Nicolaas Westerhof Laboratory for Physiology, Free University, Amsterdam, The Netherlands

Robert A. Wise Department of Medicine, Francis Scott Key Medical Center, Baltimore, Maryland 21224, U.S.A.

Toshio Yaginuma Department of Medicine, University of Florida and Veterans Administration Medical Center, Gainesville, Florida 32611, U.S.A.

Frank C.P. Yin Department of Medicine, Johns Hopkins Medical Institutions, Baltimore, Maryland 21205, U.S.A.

1
Left Ventricular-Systemic Arterial Coupling in Humans and Strategies to Improve Coupling in Disease States

MICHAEL F. O'ROURKE, ALBERT P. AVOLIO,
AND WILMER W. NICHOLS

Introduction

It is paradoxical that this book on ventricular/vascular coupling will occupy space in medical libraries alongside clinical texts on hypertension, cardiac failure, and ischemic heart disease. Paradoxical, because the approaches described here are rarely if ever addressed in these clinical texts, even though they are of paramount relevance and importance. Neglect of these approaches is all the more surprising when one considers that hypertension, cardiac failure, and ischemic heart disease are among the most common serious diseases handled regularly by practicing physicians.

Why are these approaches ignored? Not because they are so complex and esoteric that physicians lack the intellectual abilities to cope with them. In all countries, for years young persons with intelligent enquiring minds have competed more eagerly to enter medicine than virtually any other graduate course. No, rather the medical profession, worldwide, seems satisfied with the current approaches to description of cardiac load. Yet these approaches are firmly based on the sphygmomanometric technique of arterial pressure measurement, which is now almost 100 years old and has long been surpassed by new technology and theory. It is as though the medical profession has become uncritically secure with sphygmomanometry as a method of recording arterial pressure and expressing cardiac load.

Sphygmomanometry is of course an unsurpassed noninvasive and rapid method for determining brachial artery pressure in individuals and for collecting epidemiologic data from groups of human subjects. The real problem is that it has led to simplistic and erroneous views of arterial hemodynamics and cardiac function. These simplistic views are widespread today, despite the work described and referred to in this book.

Perhaps our profession has been deceived because sphygmomanometry provides numbers. The great British physicist Lord Kelvin stressed over 100 years ago that unless one can measure what one is describing and express the measurement in numbers, one's knowledge is meager and unsatisfactory. The numbers from the sphygmomanometer can be misleading, however, and then our knowledge can be even more unsatisfactory than if no numbers were available in the first place.

Misconceptions

There are a number of widespread and serious misconceptions in the present conventional approach to arterial function and left ventricular load.

The main misconception arises from reliance on the sphygmomanometer for measurement of arterial pressure. Description of pressure in terms of systolic and diastolic values implies that the arterial system behaves as a "Windkessel"—a distensible chamber—connected at one end to the heart and at the other to the resistance vessels of the body. This model was first suggested by Stephen Hales in 1769, but has been found oversimplified in more recent studies, even those conducted in Kelvin's time during the nineteenth century. This conceptual model of the arterial system does not consider nonuniform elasticity of arteries nor pressure wave reflection. It does not allow for wave travel or pressure wave amplification, but considers pressure wave amplitude to be everywhere the same, with systolic and diastolic pressures identical in the brachial and all other arteries, including the ascending aorta, under all conditions. This is incorrect. Arterial

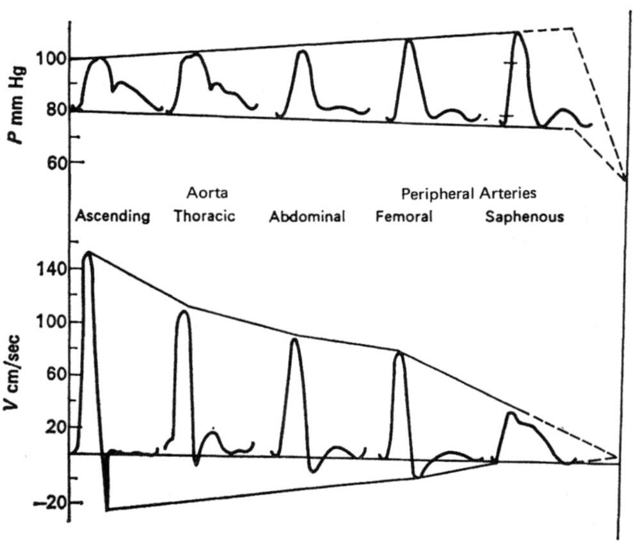

FIGURE 1.1. Diagrammatic representation of change in pressure *(P)* and flow waves *(V)* between the ascending aorta and peripheral arteries. There is a progressive rise in amplitude of the pressure pulse and decrease in amplitude of the flow pulse. (From McDonald, 1974.)

pressure is *not* the same in all arteries. There is a progressive rise in systolic pressure and fall in diastolic pressure between the ascending aorta and the peripheral arteries (Fig. 1.1). In the ascending aorta, systolic pressure is lower and diastolic pressure higher than in the brachial artery (McDonald 1974, O'Rourke 1982, Rowell et al. 1968). The relationship between brachial and aortic pressure is influenced by a variety of conditions including aging, hypertension, valvular heart disease, posture, exercise, heart rate, and vasoactive agents (Kroeker and Wood 1955, 1956; O'Rourke 1970, 1982; Rowell et al. 1968; Fig. 1.2). To describe alteration in left ventricular afterload as a function of brachial artery systolic pressure during exercise is apt to be erroneous.

Perhaps the worst misconception that arises from reliance on the sphygmomanometer is that systolic pressure is a manifestation of cardiac strength and diastolic pressure a measure of arteriolar tone and peripheral resistance. This concept, first stated by Mackenzie (1926), is the basis for (until recently) considering elevated systolic pressure unimportant and for basing a diagnosis of hypertension on diastolic pressure alone. This notion still persists in recent books and reviews (Frohlich 1977), despite abundant

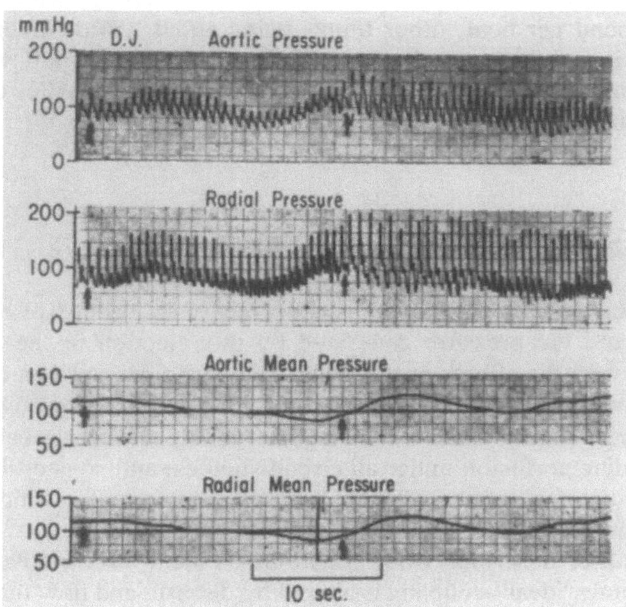

FIGURE 1.2. Pulsatile (upper two tracings) and mean (lower two tracings) pressure in the proximal aorta and radial artery of a human subject during transition from the seated to standing position, at rest (arrow at left) and at the beginning of exercise (arrow at right) on a treadmill. (From Rowell et al. 1968. Reproduced by permission of the American Heart Association, Inc.)

evidence to the contrary (O'Rourke 1983). Diastolic pressure is neither a measure of arteriolar tone nor an index of peripheral resistance. A more logical approach is to consider the systolic and diastolic pressures in any artery as representing the extremes of pressure fluctuation in that artery about an unknown mean. This approach permits consideration of the arterial pressure wave in terms of separate mean and pulsatile components (McDonald 1960, O'Rourke 1982). It is the basis of the concept of steady-state analysis of the pulse, popularized by McDonald in 1960 and extended by many since.

There are other misconceptions that are not directly associated with the sphygmomanometer. One of the most common is the idea that peripheral resistance is the only component of ventricular afterload. This excludes the contribution of arterial elasticity and the pulsatile component of ventricular load—factors that were under study long ago before the sphygmomanometer was being developed and popularized. Roy in 1880 and Bramwell and Hill in 1922 were more concerned with arterial distensibility than with resistance in describing left ventricular load. In describing changes due to age, Roy stated, "Only in the case of young children do we find that the elasticity of arteries is so perfectly adapted to the requirements of the organism as in the case of the lower animals," and Bramwell and Hill remarked, ". . . the amount of energy which the heart has to expend per beat, other things being equal, varies inversely with the elasticity of the arterial system." In accounting for left ventricular load, one must consider both resistive and reactive properties of the systemic circulation.

Importance of Ventricular/Vascular Coupling

Left ventricular/vascular coupling is concerned with ejection of blood from the heart and the pressure generated by this ejection in the ascending aorta, and with the effects of this generated aortic pressure on continuing ejection throughout systole and on ventricular perfusion through the coronary arteries during diastole. The subject is relevant to cardiac function and to cardiac perfusion under all circumstances, and so is crucial in any consideration of the effects of such diseases as hypertension and coronary atherosclerosis and the treatments of these.

This chapter describes what constitutes "ideal" ventricular/vascular coupling, how "ideal" coupling is altered by disease, and how the ill effects of coupling alterations may be offset by therapeutic manipulations. The aim is to provide a framework for interested physicians to use in order to advance beyond existing conventional approaches; they may then gain more from what is written elsewhere in this book.

Basic Principles of Ventricular/Vascular Coupling

The systemic circulation exists to permit constant steady flow of blood through the body's capillaries, where diffusion of nutrients and waste products occurs. The heart's role is to maintain this constant flow. Unlike flow in the smallest vessels, however, the heart's action is intermittent, systoles, when blood is ejected into the ascending aorta, alternating with diastoles, when the aortic valve shuts and blood flows into the ventricular cavity from the atrium and into the left ventricular wall through the coronary arteries.

Coupling concerns two matters, action of the heart as a pump connected to a hydraulic load and perfusion of the heart as an organ.

In considering ventricular function as a pumping action, one must consider the pressure generated by the left ventricle during systole. It is this pressure that opposes ejection of blood from the ventricle, and that determines myocardial oxygen requirements (Sarnoff et al. 1958), the degree of hypertrophy the ventricle might undergo (Wikman-Coffelt, Parmley, and Mason 1979), and under pathologic conditions, whether the heart might fail (Koch-Weser 1973, Sagawa 1978, Weber et al. 1974). Pressure in the ventricle during systole is approximated by aortic pressure during systole, the only differences being in (1) the short period between onset of ventricular contraction and opening of the aortic valve, (2) the short interval after closure of the valve and complete relaxation of the ventricle, and (3) the slight relative fall in lateral pressure as blood gains kinetic energy in the aorta. For practical purposes, ventricular function as pumping action can be described in terms of mean pressure generated in the ascending aorta during systole (O'Rourke 1982, O'Rourke, Yaginuma, and Avolio 1984). This is the approach taken here, though others in this book and elsewhere stress peak systolic pressure (Weber et al. 1974) or pressure at end systole (Sagawa 1978); both are closely related to mean systolic pressure and influenced by the same factors.

In considering the ventricle as an organ requiring adequate blood flow for preservation of normal function and viability, one must consider the pressure maintained in the aorta during diastole. The left ventricular wall cannot be perfused during systole. At this time the ventricle contracts and blood is ejected from its cavity; the coronary arteries within the left ventricular wall are squeezed shut. Coronary flow to the left ventricular wall can occur only during diastole; hence the coronary perfusion pressure head is the pressure maintained in the aorta during diastole (Gregg, Khouri, and Rayford 1965).

In considering the heart as pump and organ, one can usefully think of the ascending aortic pressure wave in terms of two components: mean pressure during systole, which is relevant to left ventricular performance as a pump, and mean pressure during diastole, which is relevant to left

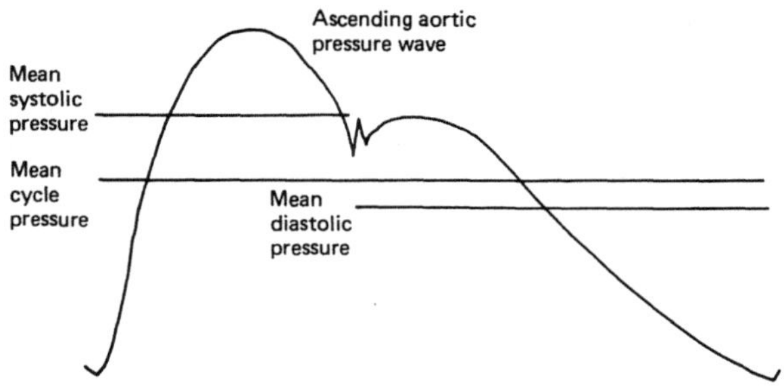

FIGURE 1.3. Diagram of ascending aortic pressure wave and mean systolic pressure, mean diastolic pressure, and mean cycle pressure. (From O'Rourke, Yaginuma, and Avolio 1984).

ventricular perfusion (Fig. 1.3). As far as other bodily organs are concerned, mean pressure during the whole cardiac cycle forces blood through resistance vessels, and thus is the value used for calculation of peripheral resistance.

From the viewpoint of ventricular/vascular interaction, the systemic circulation can be viewed as having three major components, with two interfaces between them: the left ventricle, the arterial system, and the arterioles and capillaries (Fig. 1.4). The function of the left ventricle is to pump blood; output is necessarily intermittent. The function of the arterioles and capillaries is to distribute blood according to need and to permit diffusion of nutrients, waste products, and other elements. Flow through these vessels is continuous, or nearly so. The arterial system has two functions: to convey blood from the left ventricle to the different bodily organs and tissues and to smooth flow oscillations so that pulsatile input from the left ventricle becomes nonpulsatile input into arterioles. The arterial system thus acts as both conduit and cushion at the same time (McDonald 1974, O'Rourke 1982).

Since the arteries are efficient conduits, mean pressure decreases very little along their length. At arterial terminations, however, mean pressure plummets over a very short distance as blood flows through arterioles and capillaries (Fig. 1.4). The sudden change of resistance between artery and arteriole results in strong pressure pulse wave reflection. This reflection is primarily responsible for the increase in pressure pulse wave amplitude between ascending aorta and peripheral arteries.

The strongest evidence for the fact that pressure pulse wave reflection arises predominantly from the interface between low-resistance arteries and high-resistance arterioles is that such reflection can be virtually abolished in peripheral vascular beds by intraarterial administration of arteriolar vasodilator drugs, which cause minimal change in arterial caliber and no change in vascular architecture and branching patterns (O'Rourke and

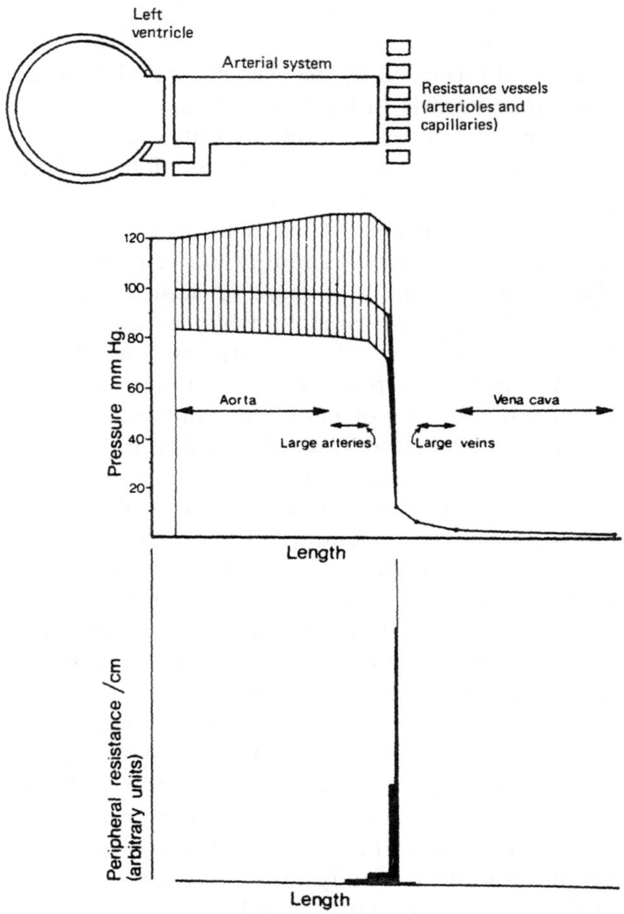

FIGURE 1.4. *Top,* Diagrammatic representation of the left ventricle, arterial system, and resistance vessels. *Center,* change in mean pressure and pulsatile pressure between the central aorta and vena cava; and *bottom,* resistance in corresponding segments of the vascular bed. (Adapted from O'Rourke 1982; by permission of Churchill Livingstone, Edinburgh.)

Taylor 1966). Some wave reflection in the aorta and major arteries also appears to result from nonuniform arterial elasticity (McDonald 1974, O'Rourke 1982) and change in vascular caliber, but this reflection is minor in comparison with arteriolar reflection.

When viewed from the ascending aorta, the systemic circulation appears to present two functionally discrete reflecting sites, one representing the resultant of all individual reflecting sites in the lower part of the body and the other the resultant of all reflecting sites in the upper part of the body (O'Rourke and Taylor 1967).

The hydraulic load presented by the systemic circulation comprises both the major resistance vessels (arterioles and capillaries) and the arteries themselves. Since the arteries are low-resistance conduits, their contribution to peripheral resistance is sufficiently small in conditions of health as to be ignored completely. Unless arteries are obstructed, as by atheroma, the peripheral resistance is taken to represent the properties of the smallest vessels exclusively; changes in peripheral resistance thus represent changes in the caliber of these vessels. We use the term "peripheral" because resistance is predominantly peripheral.

These arguments relate to steady pressure and flow. While arteries can be ignored in discussions of peripheral resistance and the steady flow component of load, they are of paramount importance in the pulsatile component. The arteries convert pulsatile flow at their origin of the systemic circulation (the ascending aorta) to steady flow at their multiple terminations. In considering pulsatile flow and arterial cushioning function, one perceives a second component of left ventricular load—one that is a consequence of the intermittency of the heart. This second, or pulsatile, component adds to the resistive component of load.

The pulsatile component of ventricular load is normally far less than the resistive component. The two can be separated from measurements of pulsatile pressure and flow in the ascending aorta (O'Rourke 1967, 1982). When expressed in terms of energy, the pulsatile component of energy lost in the systemic circulation of healthy young experimental animals is about 10% and the steady component is about 90% of the total.

A number of studies have been conducted to determine what properties of arteries are responsible for the pulsatile component of ventricular load being so low in relation to steady load (O'Rourke 1967, Taylor 1964, 1969, O'Rourke, Yaginuma, and Avolio 1984). These properties include the high distensibility of the proximal aorta (with decreasing distal distensibility), arterial branching patterns, peripheral wave reflection, and spatial localization of reflecting sites such that the timing of wave reflection is optimized in relation to ventricular ejection.

The steady component of ventricular hydraulic load can be viewed as necessary—a value that must be maintained to enable perfusion of all bodily organs. The pulsatile component is a consequence both of the heart's intermittency and the inability of the arterial system to cushion pulsations completely. "Ideal" arterial function entails maintenance of as low as practicable a pulsation around the mean value in the ascending aorta.

Left ventricular/vascular interaction may be assessed from measurement of ventricular pressure, ascending aortic flow and pressure, and coronary artery flow (Fig. 1.5). Hydraulic load presented by the systemic circulation is gauged from recordings of ascending aortic pressure and flow. Rela-

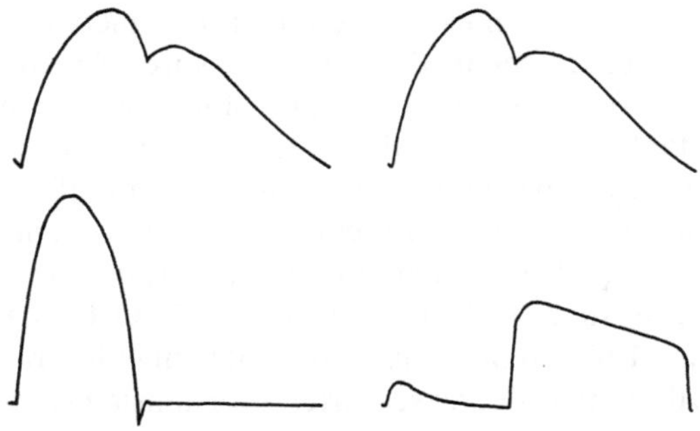

FIGURE 1.5. Diagrammatic representation of pressure in the ascending aorta (*top, right and left*) and of flow from the left ventricle into the ascending aorta (*bottom left*) and down a coronary artery into the left ventricular wall (*bottom right*). Flow from the ventricular cavity is a systolic event. flow into the ventricular wall is a diastolic event. (From O'Rourke, Yaginuma, and Avolio 1984.)

tionships may be analyzed in the time or frequency domain (McDonald 1974, O'Rourke 1982). When viewed in the time domain, the ascending aortic pressure wave is normally of low amplitude, with the incisura (caused by aortic valve closure) immediately followed by a secondary pressure wave (caused by peripheral reflection); this reflected wave augments pressure in early diastole. Contour of the aortic pressure wave is such that mean pressure during cardiac systole is only slightly greater than (by 5–10 mm Hg), and mean pressure during diastole slightly less than (by about 5 mm Hg) mean pressure throughout the cardiac cycle. Arterial function is optimal when the fluctuation around mean pressure is minimal. "Ideal" ventricular/vascular coupling entails as low a mean pressure as practicable for adequate organ flow, with as low a mean systolic pressure and as high a mean diastolic pressure as possible (O'Rourke 1982, 1984). Low mean systolic pressure allows adequate ventricular ejection with low oxygen demands by myocardium and little stimulus to cardiac hypertrophy. High mean diastolic pressure allows adequate coronary perfusion.

Vascular/ventricular coupling will deviate from the ideal if arteriolar resistance is inappropriately high (so that mean aortic and left ventricular pressures are elevated) or if arterial properties are so altered that mean aortic systolic pressure is increased and mean diastolic pressure is reduced in relation to mean pressure throughout the cardiac cycle. Changes in arteriolar properties that increase mean arterial pressure in hypertension are well known; emphasis is given here to the less well-understood changes in arterial properties and their effects on ventricular/vascular coupling.

The two major factors that affect pulsatile pressure are arterial distensibility and the timing and intensity of wave reflection. Increasing arterial stiffness increases pressure pulse wave amplitude, increases mean systolic pressure, and decreases mean diastolic pressure. An increase in wave reflection intensity from its (high) normal value would be expected under normal circumstances to have little effect on the aortic pressure wave, but an alteration in timing would have a substantial effect. Early return of wave reflection would cause reflection to be apparent during systole rather than diastole, and so to augment mean systolic pressure and reduce mean diastolic pressure (O'Rourke 1967, 1982, O'Rourke, Yaginuma, Avolio 1984; Fig. 1.6). Hence vascular/ventricular interaction would be impaired by anything that decreased arterial distensibility or caused earlier wave reflection. Since decreased arterial distensibility is associated with increased pulse wave velocity, both effects are likely to occur together.

Aortic pressure/flow relationships can be viewed in the frequency domain as ascending aortic input impedance (McDonald 1974, O'Rourke 1982, O'Rourke and Taylor 1967). Impedance is described elsewhere in this book. The aortic impedance modulus falls steeply from its value at zero frequency (the peripheral resistance) to very low values at frequencies that correspond to the first three harmonics of the left ventricular ejection wave. This is the manifestation in the frequency domain of ideal arterial function and near-optimal ventricular/vascular interaction. The value of

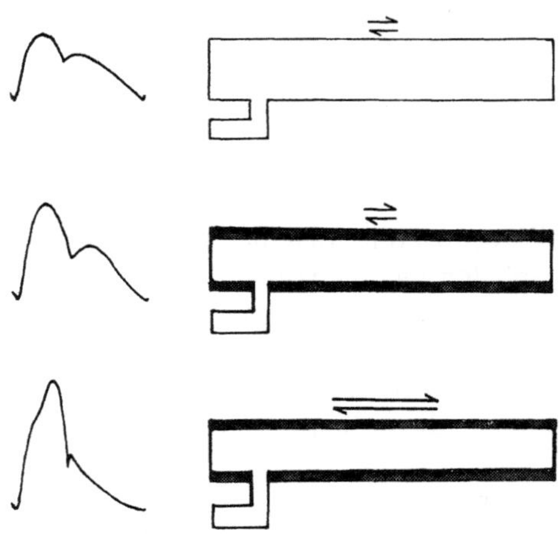

FIGURE 1.6. Simple tubular models of the systemic arterial system. *Top,* normal distensibility and normal pulse wave velocity; *center,* decreased distensibility but normal pulse wave velocity; and *bottom,* decreased distensibility with increased pulse wave velocity. At left are the amplitude and contour of pressure waves that would be generated at the origin of these models by the same ventricular ejection. Decreased distensibility per se increases pressure wave amplitude with increased wave velocity causing the reflected wave to return during ventricular ejection rather than during ventricular diastole.

modulus about which fluctuations occur (the characteristic impedance) depends on aortic stiffness (McDonald 1974, O'Rourke 1982). The frequency of the first minimum depends on the timing of wave reflection. An increase in arterial stiffness elevates the value of characteristic impedance, while early wave reflection increases the frequency of the first minimum of impedance and so shifts the impedance curve to the right. Both increase the impedance modulus over the frequency band that contains most of the energy of the left ventricular ejection wave (Fig. 1.7). This indicates impairment of ideal ventricular/vascular coupling. The changes in Figure 1.6, with the pressure/flow relationship expressed in the time domain, correspond to the changes in Figure 1.7, with the relationship expressed in the frequency domain.

Ventricular/Vascular Coupling in Aging and in Disease States

Ideal coupling in health entails ascending aortic impedance such that mean aortic pressure is as low as possible for perfusion of bodily organs (especially the brain) and pulse pressure is as low as possible so that mean systolic pressure is only slightly greater than, and mean diastolic pressure

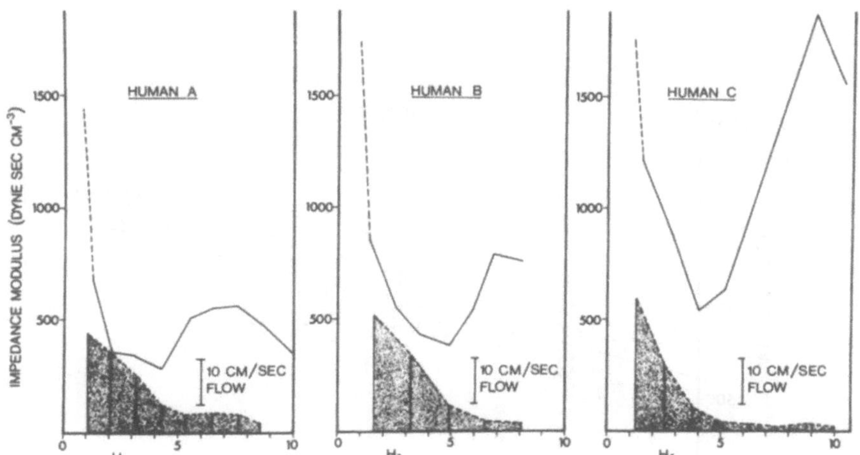

FIGURE 1.7. Impedance modulus in the ascending aorta and harmonics of ascending aortic flow waves (shaded areas) in three groups of human subjects: A, no known cardiac or vascular disease; B, known arterial degenerative disease; and C, hypertension and known arterial degenerative disease. Rearranged from data of W. W. Nichols. (From O'Rourke 1982; by permission of Churchill Livingstone, Edinburgh.)

slightly less than, mean pressure through the complete cardiac cycle. Low mean pressure is achieved by low arteriolar tone. Low pulse pressure results from high arterial distensibility and wave reflection so delayed that the foot of the dicrotic waves follows immediately after aortic valve closure. This indeed is seen in children (O'Rourke 1982, Roy 1880). It is the therapeutic goal in disease states. Aging and different diseases disturb ideal ventricular/vascular coupling by elevating mean pressure through an increase in arteriolar tone, or by increasing pulse pressure through a decrease in aortic distensibility, an increase in arterial pulse wave velocity, or creation of abnormal reflecting sites close to the heart.

Aging

With aging there is progressive increase in arterial stiffening associated with substantial increase in pulse wave velocity (Fig. 1.8). These effects are independent of atherosclerosis and are seen in both Western and Eastern cultures (Avolio et al. 1984, 1985; Ho 1982), though they are exaggerated in societies with high prevalence of hypertension and with high sodium intake. Changes in arterial distensibility and pulse wave velocity with age cause an increase in amplitude of the ascending aortic pressure wave and a change in pressure wave contour, with a late systolic peak and absence of the diastolic pressure wave (O'Rourke 1982, Nichols et al. 1985). These changes and those in ascending aortic impedance (Fig.

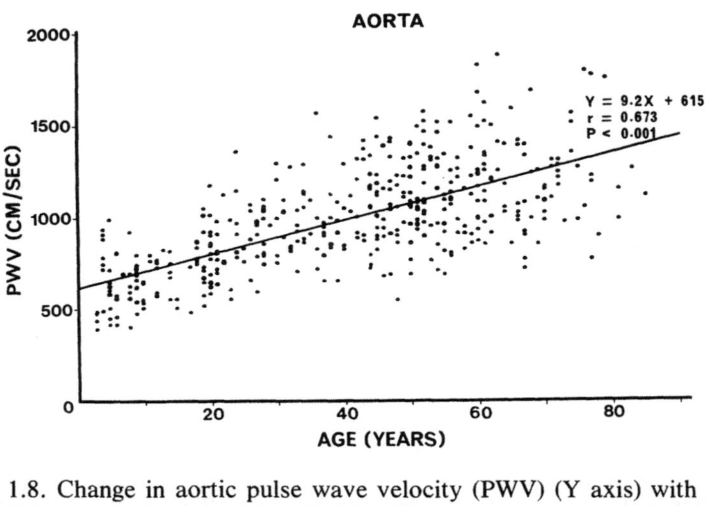

FIGURE 1.8. Change in aortic pulse wave velocity (PWV) (Y axis) with age (X axis) and the regression equation relating PWV to age in a group of apparently normal Chinese subjects in urban Beijing. (From Avolio et al. 1984; by permission of the American Heart Association, Inc.)

1.7), discussed further in Chapter 4, are manifestations of impaired ventricular/vascular coupling.

Arteriosclerosis and Atherosclerosis

Arterial medial degeneration is exaggerated in arteriosclerosis, and changes are greater than would be expected for any given age. The ill effects are the same as, but more marked than, those seen with aging.

Atherosclerosis is a different condition. It is not an inevitable consequence of aging, but occurs very commonly with increasing age in Western societies; it is uncommon in the East and in underdeveloped societies. Atherosclerosis affects primarily and predominantly the arterial intima and causes obstruction of small and medium-sized arteries. Atheroma is usually so localized as to cause little change in arterial distensibility and pulse wave velocity (Avolio et al. 1984, 1985) and little change overall in wave reflection. Its predominant ill effect on ventricular/vascular coupling is through causing coronary artery narrowing and thus myocardial ischemia and impaired ventricular contractility.

Hypertension

In systemic hypertension, mean arterial pressure is abnormally elevated, owing to increased arteriolar tone or to arteriolar rarefaction (Harper et al. 1978, O'Rourke 1983). This causes increased pulse wave velocity and decreased arterial distensibility as a result of greater arterial tension per se and of accelerated arterial degeneration (O'Rourke 1976, 1982), that is, of a reversible and an irreversible mechanism, respectively. In the ascending aorta, mean systolic pressure is also increased. Ascending aortic impedance and the arterial pressure wave are changed in a fashion similar to that seen with aging (Fig. 1.9). Ventricular/vascular coupling is impaired as a result of both increased mean pressure and the accompanying decrease in arterial distensibility.

Aortic Coarctation

In aortic coarctation not only is mean arterial pressure in the upper part of the body abnormally high, as in hypertension, but also there is an abnormal reflection site close to the heart. Arteries in the upper part of the body undergo premature degeneration (Keith 1978, O'Rourke 1982, O'Rourke and Cartmill 1971). Ventricular/vascular interaction is markedly impaired, with high ascending aortic systolic pressure and abnormal ascending aortic impedance (Fig. 1.10). The impairment is more marked than with a corresponding degree of hypertension without coarctation, explaining the more malignant course of this disease without treatment (Keith 1978, O'Rourke 1982).

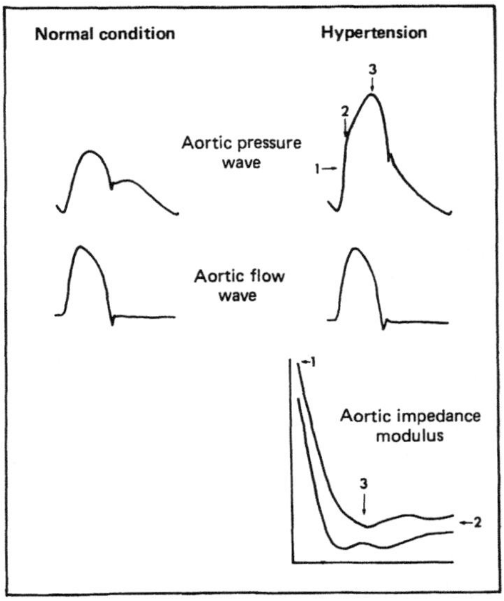

FIGURE 1.9. Effects of hypertension on aortic pressure wave and aortic impedance modulus with the same ascending aortic flow wave. 1, Increased peripheral resistance (increased mean pressure, increased zero frequency component of impedance); 2, Decreased arterial distensibility (increased amplitude of the initial pressure peak, increased characteristic impedance); 3, Earlier return of reflected waves from arterial terminations (late systolic pressure peak, impedance curve shifted to right). (From O'Rourke 1976).

Cardiac Failure and Ischemic Heart Disease

Cardiac failure and ischemic heart disease are not associated with any specific vascular disturbances that alter ventricular/vascular coupling. Their manifestations are, however, very sensitive to alterations in ventricular/vascular coupling—ventricular ejection will be reduced and myocardial ischemia intensified by elevation in mean systolic pressure (O'-Rourke 1982, Sarnoff et al. 1958). Hence a major therapeutic goal is to alter coupling to the heart's advantage. Cardiac failure and ischemic heart disease usually occur at an age when the ideal ventricular/vascular coupling of youth has been distorted by aging changes, causing decreased distensibility of arteries. At age 55, aortic pulse wave velocity is some 60% greater than at age 15, and the arterial pressure wave usually shows a late systolic peak and no diastolic wave (O'Rourke 1982). In most patients with these conditions, therapeutic changes in vascular properties might be expected to improve coupling and alleviate manifestations of disease.

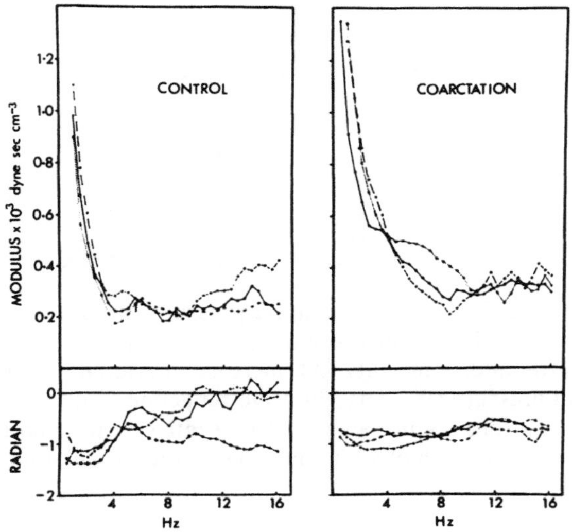

FIGURE 1.10. Imput impedance to the systemic circulation in three normal dogs (left) and in three littermates with chronic surgically induced aortic coarctation. (Phase Radians) (From O'Rourke and Cartmill 1971; by permission of the American Heart Association, Inc.)

Strategies for Improving Ventricular/Vascular Coupling

Abnormal ventricular/vascular coupling is improved by a variety of measures (Fig. 1.11). The most obvious is reduction in mean pressure through reduction in arteriolar tone. Besides reducing mean pressure, a reduction in arteriolar tone secondarily improves arterial distensibility (O'Rourke 1982) and delays pressure wave reflection, leading to an even greater decrease in mean systolic pressure and a lesser reduction in mean diastolic pressure. Arteriolar dilation with reduction in mean pressure has proved advantageous not only in hypertensive heart failure and angina associated with hypertension, but also in normotensive heart failure (Smith and Braunwald 1984). The limiting factor, apart from too great a reduction in mean pressure (and risk of cerebral and other organ ischemia), is too great a reduction in coronary perfusion pressure and the possibility of myocardial ischemia.

Arterial counterpulsation has the most clear cut beneficial effect on ventricular/vascular coupling. Counterpulsation is achieved through cyclic inflation and deflation of a balloon in the aorta in time with cardiac diastole and systole, respectively. Counterpulsation artificially increases mean diastolic pressure and decreases mean systolic pressure, thereby improving

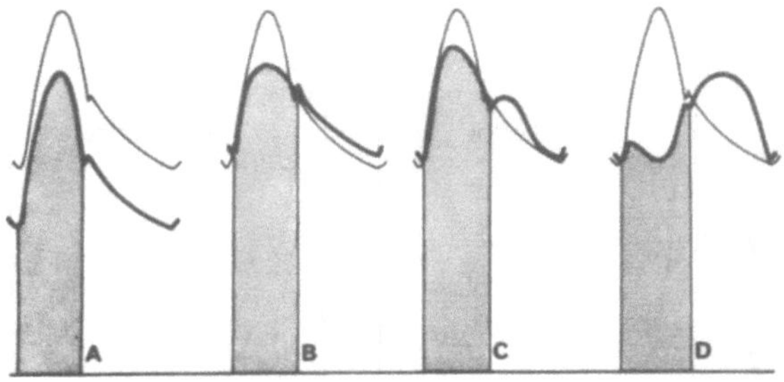

FIGURE 1.11. Reduction in aortic systolic pressure (heavy line), with variable change in mean diastolic pressure, as induced by reduction in mean pressure without change in arterial distensibility (A), increase in arterial distensibility without change in mean pressure (B), delay in wave reflection (C), and arterial counterpulsation (D). Control pressure wave is represented by the thin line. (From O'-Rourke 1982; by permission of Churchill Livingstone, Edinburgh.)

myocardial perfusion and concurrently decreasing myocardial oxygen requirement and improving cardiac output (O'Rourke 1982). Counterpulsation is an invasive technique of mechanical heart assistance of proven value in coronary insufficiency, myocardial infarction, or following cardiac surgery (Smith and Braunwald 1984).

Ventricular/vascular coupling in adult humans theoretically could be improved by measures that decrease aortic distensibility or delay wave reflection (Fig. 1.11). No specific therapies to achieve either are presently known. No known drugs decrease arterial stiffness or decrease wave velocity as a primary effect. Both of these effects, however, follow reduction of mean arterial pressure with antihypertensive therapy. Drugs that could influence primarily large arteries would be of obvious advantage, since coupling could be improved without alteration of peripheral organ perfusion. The production of such drugs is under investigation, as is the possibility that prosthetic arterial grafts could be used to decrease arterial distensibility and delay peripheral wave reflection.

A further possible mechanism for improving ventricular/vascular coupling is currently under scrutiny (Yaginuma et al. 1986). In human adults, wave reflection is disadvantageous to the heart since its inappropriate timing causes increase in systolic and decrease in diastolic pressure in the proximal aorta. Peripheral wave reflection can be decreased by drugs such as nitroglycerin that cause disproportionate dilation of peripheral arteries, even without reduction in mean pressure or alteration in aortic properties. Such effects of nitroglycerin have been reported by Fitchett (1982), Simon and colleagues (1982), and Westling and co-workers (1984). These workers have described nitroglycerin as increasing the distensibility of peripheral arteries. Since aortic properties are unchanged but peripheral reflection is reduced by dilation of small to medium-sized arteries, the functional effects of nitroglycerin on ventricular/vascular coupling are most readily explained on the basis of decreased peripheral reflection (Yaginuma et al. 1986).

Conclusion

The ideal ventricular/vascular interaction of youth is disturbed by the aging process. This disturbance is aggravated by hypertension and by disorders of large arteries. The disturbance can be redressed predictably by therapies that actively or passively decrease arteriolar tone, increase arterial distensibility, delay wave reflection, or artificially alter aortic pressure pulse wave contour.

Bibliography

Avolio, A. P., Chen, Shang-gong, Wang, Ruo-ping, Zhang, Cuhn-lai, Li, Mei-feng, and O'Rourke, M. F. 1984. Effects of aging on changing arterial compliance and left ventricular load in a northern Chinese urban community. *Circulation* 68:50–58.

Avolio, A. P., Deng, Fa-Quan, Li, Wei-Wuiang, Luo, Yao-Fai, Huang, Zhen-Dong, Zing, Lian-Fen, and O'Rourke, M. F. 1985. Effects of aging on arterial distensibility in populations with high and low prevalence of hypertension: comparison between urban and rural communities in China. *Circulation* 71(2):202–210.

Bramwell, J. V., and Hill. A. V. 1922. Velocity of transmission of the pulse wave and elasticity of arteries. *Lancet* 1:891–892.

Fitchett, D. H. 1982. The effects of nitroglycerin on forearm arterial compliance. *J. Clin. Invest. Med.* 5(Suppl.1):31.

Frohlich, E. D. Hemodynamics of hypertension. In Genest, J., Koin, E., and Kuchel, O., (eds.). 1977. *Hypertension: Physiopathology and Treatment*. New York: McGraw-Hill, pp. 15–49.

Gregg, D. E., Khouri, E. M., and Rayford, C. R. 1965. Systemic and coronary energetics in the resting dog. *Circ. Res.* 16:102–113.

Hales, S. 1769. *Statical Essays*. London: Wilson and Nichol.

Harper, R. N., Moore, M. A., Marr, M. C., Watts, L. E. and Hutchings, P. M. 1978. Arteriolar rarefaction in the conjunctiva of human essential hypertensives. *Microvasc. Res.* 16:369–372.

Ho, K. 1982. Effects of aging on arterial distensibility and left ventricular load in an Australian population. B. Sc. (Med) thesis, University of New South Wales, Sydney.

Kannel, W. B., Castelli, W. P., McNamara, P. M., McKee, P. A., and Feinleib, M. 1972. Role of blood pressure in the development of congestive heart failure. The Framingham Study. *N. Engl. J. Med.* 287:781–787.

Keith, J. D. 1978. Coarctation of the aorta. In Keith, J. D., Rowe, R. D., and Vlad, P. (eds.). *Heart Disease in Infancy and Childhood*. New York: Macmillan, pp. 736–760.

Koch-Weser, J. 1973. The therapeutic challenge of systolic hypertension. *N. Engl. J. Med.* 289:481–483.

Kroeker, E. J., and Wood, E. H. 1955. Comparison of simultaneously recorded central and peripheral arterial pressure pulses during rest, exercise and tilted position in man. *Circ. Res.* 3:623–632.

Kroeker, E. J., and Wood, E. H. 1956. Beat to beat alterations in the relationship of simultaneously recorded central and peripheral arterial pressure pulses during Valsalva maneuver and prolonged expiration in man. *J. Appl. Physiol.* 8:483–494.

Mackenzie, J. *Principles of Diagnosis and Treatment in Heart Afflictions*. London: Oxford University Press.

McDonald D. A. 1960. Blood-Flow in Arteries. London: Arnold.

McDonald, D. A. 1974. *Blood Flow in Arteries*. London: Arnold, 2nd Edition.

Nichols, W. W. O'Rourke, M. F., Avolio, A. P., Yaginuma, T., Murgo, J., Pepine, C. J. and Conti, C. R. 1985. Effects of age on ventricular/vascular coupling. *Amer J Cardiol* 55:1179–1184.

O'Rourke, M. F. 1967. Steady and pulsatile energy losses in the systemic circulation under normal conditions and in simulated arterial disease. *Cardiovasc.* Res. 1:313–326.

O'Rourke, M. F. 1970. Influence of ventricular ejection on the relationship between central aortic and brachial pressure pulse in man. *Cardiovasc. Res.* 4:291–300.

O'Rourke, M. F. 1976. Pulsatile arterial hemodynamics in hypertension. *Aust. N. Z. J. Med.* 6 (suppl 2):40–48.

O'Rourke, M. F. 1982. *Arterial Function in Health and Disease*. Edinburgh: Churchill Livingstone, pp. 153–169.

O'Rourke, M. F. 1983. Hypertension is a myth. *Aust. N. Z. J. Med.* 13:83–90.

O'Rourke, M. F., and Cartmill, T. B. 1971. Influence of aortic coarctation on pulsatile hemodynamics in the provisional aorta. Circulation 44:281–292.

O'Rourke, M. F., and Taylor, M. G. 1966. Vascular impedance of the femoral bed. *Circ. Res.* 18:126–139.

O'Rourke, M. F., and Taylor, M. G. 1967. Input impedance of the systemic circulation. *Circ. Res.* 20:365–380.

O'Rourke, M. F., Yaginuma, T., and Avolio, A. P. 1984. Physiological and pathophysiological implications of ventricular/vascular coupling. *Ann. Biomed. Eng.* 12:119.

Rowell, L. V., Brengelmann G. L., Blackmon, J. R., Bruce, R. A., and Murray, J. A. 1968. Disparities between aortic and peripheral pulse pressures induced by upright exercise and vasomotor changes in man. *Circulation* 37:954.

Roy, C. S. 1880. The elastic properties of the arterial wall. *J. Physiol. (London)* 3:125–159.

Sagawa, K. 1978. The ventricular pressure-volume diagram revisited. *Circ. Res.* 43:678–687.

Sarnoff, S. J., Braunwald, E., Welch, G. H., Stainsby, W. N., and Macruz, R. 1958. Hemodynamic determinants of oxygen consumption of the heart with special reference to the tension-time index. *Am. J. Physiol.* 192:148–156.

Simon, A. Ch., Levenson, J. A., Bouthier, J. E., Peronneau, P. P. and Safar, M. E. 1982. Effect of nitroglycerin on peripheral large arteries in hypertension. *Brit. J. Clin. Pharm.* 14:241–246.

Smith, T. W., and Braunwald, E. 1984. The management of heart failure. In Braunwald, E. (ed.). *Heart Disease*. Philadelphia: Saunders, pp. 503–559.

Taylor, M. G. 1964. Wave travel in arteries and the design of the cardiovascular system. In Attinger, E. D. (ed.). *Pulsatile Blood Flow*. New York: McGraw-Hill, pp. 343–372.

Taylor, M. G. 1969. The optimum elastic properties of arteries. In Wolstenholme, G. E. W., and Knight, J. (eds.). *Ciba Foundation Symposium on Circulatory and Respiratory Mass Transport*. London: Churchill Livingstone, pp. 136–147.

Weber, K. T., Janicki, J. S., Reeves, R. C., Hefner, L. L., and Reeves, T. J. 1974. Determinants of stroke volume in the isolated canine heart. *J. Appl. Physiol.* 37:742–747.

Westling, H., Jansson, L., Johnson, B., and Nilsen, R. 1984. Vasoactive drugs and elastic properties of human arteries in vivo, with special reference to the action of nitroglycerin. *Eur. Heart J.* 5:609–616.

Wikman-Coffelt, J., Parmley, W. W., and Mason, D. T. 1979. The cardiac hypertrophy process: analysis of factors determining pathological vs. physiological development. *Circ. Res.* 45:697–707.

Yaginuma, T., Avolio, A. P., O'Rourke, M. F., Nichols, W. W., Morgan, J., Roy, P., Baron, D., Branson, J., and Fengley, M. 1986. Action of nitroglycerin on left ventricular hydraulic load in man: a new hypothesis. *Cardiovasc. Res.* 20:153–160.

2
Influence of the Venous System on Ventricular/Arterial Coupling

ROBERT L. ENGLER AND JAMES W. COVELL

Introduction

This chapter summarizes our understanding of factors that control local pressure and volume in the venous system, and how these factors interact with the circulation in regulating total cardiac output and arterial pressure. Indeed, "the subject of cardiac output regulation is so important that all possible analytical approaches to its understanding deserve widespread support and exploration" (Guyton 1979). A more detailed fundamental analysis can be found in the *Handbook of Physiology* (Rothe 1983).

Much of the material presented here is obtained from open-loop experiments in which feedback control is interrupted. In an open-loop experiment, changes in the dependent variable are prevented from influencing the independent variables and boundary conditions. The use of open-loop physiology in the intact closed-loop circulation may result in circular reasoning, however. To explain a change in cardiac output by a change in venous return is circular reasoning, since both are equal to the net flow around the circuit. In analyzing the factors responsible for changes in venous return and cardiac output, the selection of independent and dependent variables is quite arbitrary. The goal is to select specific independent variables, perhaps specifying other variables to be fixed, and thus gain insight into the effects of these changes in isolation. Only through an integrated modeling technique can we achieve overall understanding of the interplays of data obtained from various open-loop experiments. Figure 2.1 shows a simplified, integrated model. Values of pressure and flow will result from solution of the simultaneous Equations (2) and (3) within this figure (Levy 1979). The boundary conditions for these solutions include the mean systemic pressure, values of resistances and capacitances for the system, blood volume in the system, and venous tone. When the boundary conditions change, the parameters change and the solutions will be different (Levy 1979).

As pointed out by Guyton (1979), the real independent variables in the determination of cardiac output are the sums of the local resistances and

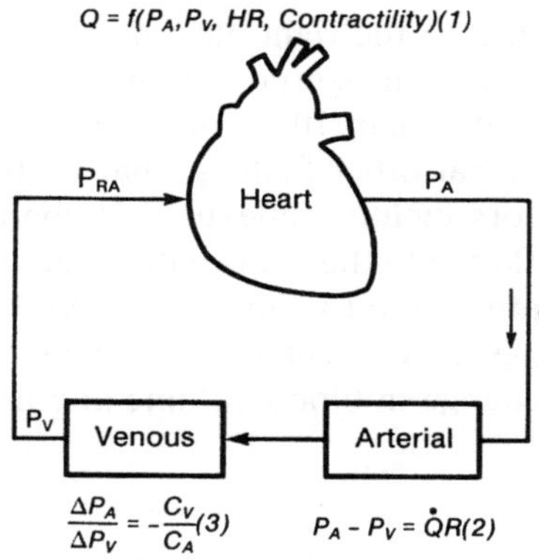

FIGURE 2.1. Diagram of the circulation. $\dot{Q}$ = flow, P_A = arterial pressure, P_v = venous pressure, P_{RA} = right arterial pressure, C_v = venous compliance, C_A = arterial compliance, R = lumped arterial and microvascular resistance.

capacitances in the circulation and the cardiac contractility. Some of the studies discussed in this chapter deal with highly complex preparations in which the dependent variable is one of the usual boundary conditions. For example, several experiments deal with the effect of a baroreceptor reflex on blood volume at constant atrial pressure and cardiac output. The implication of the experiment is that the effect on venous tone can be deduced. Application of this type of information to the venous return curve in a *quantitative* fashion is quite difficult, but one can infer net directional changes. In most of this chapter, the venous return curve is the dependent parameter. Variations in venous capacitance, venous resistance, and blood volume in the system are studied to see what effects they have on this curve. We also examine the effects of changes in the arterial circulation and of reflexes on these various boundary conditions in determining the venous return curve.

The Venous Return Curve

Guyton has defined the venous return curve as follows: The venous return is plotted as the dependent variable; changes in mean right atrial pressure are the chief independent variable, and the boundary conditions are the venous resistance, the venous capacitance, arterial pressure and resistance, and blood volume (Guyton et al. 1973). Thus the venous return is determined at a variety of atrial pressures, with all other parameters (boundary conditions) fixed. Experimentally, this curve can be determined by maintaining the circulation in a stable state and causing an abrupt step change in the atrial pressure. The immediate effect of this change in pressure on the venous return to the right atrium determines a new point on the curve. The flow must be measured within 7 seconds of the change in atrial pressure. When more than 10 seconds have elapsed following the change in pressure, several other factors, that is, boundary conditions, are altered. These factors include reflexes that change the venous tone, and continued arterial flow into the veins with an altered venous drainage, which changes the relative blood volume in the veins and thus the mean systemic pressure. In this experimentally determined venous return curve, the assumption of no change in blood volume and no reflexes can be approximated.

Mean Circulatory, Systemic, and Pulmonary Pressures

The mean circulatory pressure is the pressure in the circulation when, in the absence of changes in arterial or venous tone, the pressure is equal everywhere. Guyton, Jones, and Coleman (1973) initially measured the mean circulatory pressure by fibrillating the heart and rapidly pumping volume from the arteries into the veins so that pressures were equalized everywhere within 7 to 10 seconds. In the dog they found the mean circulatory pressure to be 7 mm Hg, a remarkably constant value. If the same experiment was carried out and a longer period elapsed (i.e., greater than 10–15 sec), the mean circulatory pressure showed a dramatic increase, owing to intense activation of the sympathetic nervous system and arterial and venous constriction. This intense sympathetic activity maintained a high mean circulatory pressure, which gradually returned toward normal over 30 to 45 minutes. During this time many of the control systems are dying and the measurements are meaningless.

The mean systemic pressure is defined slightly differently. This pressure is obtained in the same way as the mean circulatory pressure except that the pulmonary circuit is excluded. The mean pulmonary pressure, likewise, is the equilibration pressure with the systemic circulation excluded. In most clinical and experimental conditions it is better to consider the systemic and pulmonary circuits as separate systems. The mean pulmonary pressure is approximately 5 mm Hg, somewhat lower than the mean systemic pressure. However, since 84% of the blood volume is in the systemic circulation, and the systemic capacitance is greater, most of the mean circulatory pressure is determined by the mean systemic pressure. For this reason many investigators use the terms "mean systemic pressure" and "mean circulatory pressure" interchangeably. We will not do so. The importance of separate consideration of the two circuits is illustrated by the following analysis. The capacitance of the systemic circuit is seven times greater than that of the pulmonary circuit. Therefore, the effect on venous pressure of a 100-cc blood volume shift between the two circuits is seven times greater in the pulmonary circuit than in the systemic circuit.

Thus when a sudden change results in a disparity between the pumping function of the two sides of the heart, the effects of the volume shift on the mean systemic pressure and mean pulmonary pressure will be different not only in direction but also in magnitude. The mean systemic pressure and mean circulatory pressure, however, will remain close, because seven eighths of the mean circulatory pressure is determined by the mean systemic pressure.

The Venous Return Curve and Mean Systemic Pressure

Figure 2.2 shows a typical venous return curve. The atrial pressure intercept is the pressure at which venous return stops (pressure at zero flow) and is approximately the mean (systemic or pulmonary) pressure. Guyton and colleagues found these two pressures (the atrial pressure intercept and the mean systemic or pulmonary pressure) to be equal (Guyton 1955, Lindsey et al. 1957, Lindsey and Guyton 1959). The way they are defined, however, is *not* the same. Lowering the arterial pressure to equal the venous pressure as required by the definition of mean systemic pressure requires a shifting of some small volume, calculated as $\Delta V = (P_A - P_V)C_A$, from arteries into the veins. This small volume does not significantly affect the venous return curve, and thus the intercept at zero flow and the mean systemic pressure are approximately equal.

Table 2.1 lists the various factors that alter the venous return curve. In the following discussions, in each case we take the venous return curve

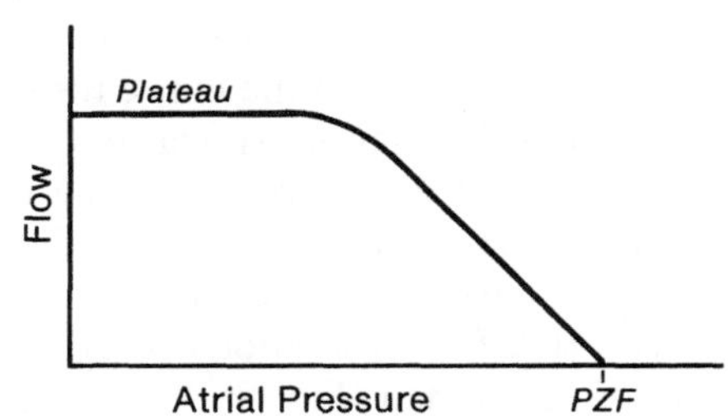

$$VR = (PZF - P_{RA}) / \frac{C_V R_V + C_A(R_V + R_A)}{C_V + C_A}$$

$$\text{If } C_V/C_A = 18, \text{ Then } V_R = (PZF - P_{RA})/(R_V + \frac{R_A}{19})$$

FIGURE 2.2. Typical venous return curve. The equations refer to the sloping portion of the curve. PZF = pressure at zero flow, VR = venous return, P = pressure, R = resistance, V = venous, A = arterial, C = capacitance.

TABLE 2.1. Factors affecting venous return.

Vasomotor tone	Regional distribution of blood volume and vasomotor tone
Blood volume	Hepatic sphincter function
Interstitial fluid volume	Blood viscosity
Intraabdominal pressure	Changes in arterial resistance and arterial capacitance
Muscular compression	

as the dependent function, alter one parameter, and keep all other boundary conditions constant. We do this only to gain insight into the effects of individual factors; many of the factors listed in Table 1 are interdependent variables.

Venous Motor Tone

The venous motor tone can adjust the normal circulatory system mean systemic pressure between 5 mm Hg (areflexic dogs) and 19 mm Hg (vasomotor stimulation induced by either catecholamine infusion or circulatory arrest). Massive catecholamine infusions can raise the mean systemic pressure somewhat higher, but such levels of catecholamine stimulation probably never occur in vivo. Ross, Frahm, and Braunwald (1961) used another approach to study the effect of venous tone. They assessed change in intravascular volume at constant atrial pressure during a constant epinephrine and norepinephrine infusion. They found that intravascular volume decreased 19 ml/kg of body weight during infusion. Complete ganglion blockade in this preparation resulted in an increase in vascular volume of 17.3 ml/kg from normal baseline tone. The effect of baroreceptor reflexes, catecholamine stimulation, and acetylcholine administration on venous return curves was then measured. Venoconstriction was seen to flatten the slope of the curve (or increase resistance) and elevate the plateau. These findings are similar to those of Rothe, Johns, and Bennett (1978), who found that norepinephrine decreased vascular volume by 24% in the intestine. It can be seen that the venous resistance is the slope of the curve between the plateau and the pressure at zero flow. The pressure drop [(PZF − P)/venous return] is the slope of the line and has the units of resistance (Fig. 2.2). Once the plateau level is reached, this parameter is meaningless, as venous return has reached a maximum. We conclude from these studies that venoconstriction in the *whole* circulation increases the venous resistance but elevates the plateau. This has been confirmed in conscious dogs by Appleton and co-workers (1985).

Caldini and colleagues (1974) first proposed that the venous circulation might consist of more than one compartment, with capacitance and resistance parameters that could not be added in a linear fashion. They proposed a two-compartment model, each compartment having different capacitances, resistances, and mean systemic pressures. Their data provide convincing evidence that the systemic venous circulation cannot be treated as a single circulation with a unique capacitance and resistance. There is one minor problem, however, in applying their results in a quantitative fashion to those of subsequent studies. This has to do with selection of boundary conditions. In solving the differential equations for the separate circuits, they assumed that the venous return becomes constant (reaches its plateau) at a right atrial pressure of 0 mm Hg. Several investigations (Guyton, Jones, and Coleman 1973; Guyton et al. 1957) have shown that the plateau of the venous return curve begins at a somewhat lower pressure, between −2 and −4 mm Hg. Thus calculation of the mean systemic pressures from their data may lead to slight errors. Caldini and colleagues also lumped the pulmonary circulation with the systemic circulation in their right-heart bypass preparation. From a theoretical standpoint, however, and from some preliminary work that they reported, it appears that any errors introduced by including the pulmonary circuit were quite small (Caldini et al. 1974).

Regional Differences in Venous Return

The splanchnic circulation contains approximately 30% of the total blood volume (Carneiro and Donald 1977, Greenway and Lister 1974). During moderate or severe hemorrhage in the dog, blood in the spleen and liver accounts for a significant percentage of the total blood volume mobilized. There are probably significant species differences, however. Greenway and Lister (1974) showed that very little volume is mobilized from the splanchnic circulation in the cat during hemorrhage, whereas Carneiro and Donald (1977) demonstrated that 35% of the blood in the spleen and 14% in the liver are mobilized in the dog during moderate hemorrhage. This "reservoir" function may be quite important, as illustrated by the following analysis. The pressure in the venous system (P_{MS}) can be calculated as the total volume (V_T) minus the unstressed volume (V^1) divided by the capacitance (C) for each compartment.

$$P_{MS1} = (V_{T1} - V'_1)/C_1 \qquad \text{and} \qquad P_{MSi} = (V_{Ti} - V'_i)/C_i \qquad (2.1)$$

Mobilizing volume from the liver or spleen to all V_{Ti} compartments will significantly affect P_{MS}.

There are several studies of blood volume shifts from the splanchnic circuit. Donald and Aarhus (1974) studied the effect of active and passive blood release from the dog spleen and intestine. During splanchnic nerve stimulation, significant blood was released from the spleen but very little from the intestine. Most of the blood release from the small intestine was passive (secondary to changes in flow). Johnson and Hanson (1963) have shown that the compliance of the intestine is 0.34 ml/mm Hg/100 gm of tissue, and that the veins in the intestine undergo considerable creep with change in pressure. Brooksby and Donald (1972) demonstrated that an increase in splanchnic blood flow results in an increase in splanchnic blood volume. During hemorrhage, 21% of the splanchnic blood volume is released, which amounts to 65% of the total volume lost during hemorrhage. Thus the splanchnic circulation in the dog provides this important reservoir function during stress. Green and colleagues (Green 1975; Green 1977; Green, Jackman, and Krohn 1978) studied pressure flow and volume flow relations in the splanchnic circulation. Their data support the conclusion of Caldini and co-workers (1974) that the separate venous compartments appear to have different resistances and compliances. The work of Green suggests that the two compartments predicted by Caldini consist in part of the splanchnic and the systemic circulation separately. No work on other subdivisions of the circulation has been forthcoming.

Whether further analytical subdivisions of the venous circulation will improve our understanding is unknown; however, several potentially important regional differences should be kept in mind. One example is the muscular system. Skalak (1984) has shown that skeletal muscle venous compliance in the resting state is extremely low, and Guyton and colleagues (1962) demonstrated that intense muscle contraction can raise the mean systemic pressure by 15 to 20 mm Hg. In considering the mechanism of high venous return during muscular exercise, Riley (1982) hypothesized the disappearance of the vascular waterfall in skeletal muscle during exercise. When one considers an increased flow volume and low compliance in the muscle compartment in light of Equation 2.1, a very high P_{MS} for the muscle compartment during exercise would be predicted. Since $Q_{MAX} = (P_{MS} - P_{RA})/R_V$ (where P_{RA} = right atrial pressure, R_V = venous resistance and Q = flow) the mechanism of high *compartmental* Q_{MAX} can be understood. How this high P_{MS} in the muscle compartment affects P_{MS}, R_V, and Q in other compartments has not been analyzed. A theory for multicompartmental analysis of venous return has not been developed. Thus different organs clearly provide quite dissimilar venous systems. The net effect of these organ systems on the venous return needs further study. At the present time, approximation of the net venous return by separation of the splanchnic from the extrasplanchnic circulation is the best available method.

Regional analysis of drug effects has been studied by Green and coworkers (Green 1977; Green, Jackman, and Krohn 1978; Green, Jackman, and Parsons 1978) and in our laboratory (Engler, Freeman, and Covell 1983). Green studied the effect of isoproterenol on the splanchnic and extrasplanchnic venous capacitance and resistance. He found that venous compliance was 25 ml/mm Hg in the splanchnic circulation and 24 ml/mm Hg in the systemic circulation and was not significantly changed by beta-adrenergic stimulation. The venous resistance in the systemic circulation was 5 mm Hg/1/min and did not change; however, the splanchnic venous resistance decreased significantly, from 16.3 to 9.4 mm Hg/1/min, during beta stimulation. These studies suggest regional differences in beta-adrenergic control of venous parameters. An example of the consequences of such regional variations was demonstrated by Green and colleagues in a study on the effect of morphine on the venous circulation (Green, Jackman, and Parsons 1978). These investigators found that splanchnic venous resistance increased from 13.8 to 20.6 mm Hg/1/min after morphine administration. Morphine also decreased the percentage of cardiac output to the splanchnic circulation from 33% to 24% These findings indicate the potential of the splanchnic circulation as a large reservoir for blood volume separate from the systemic circulation. For example, the well-recognized effect of morphine in relieving acute pulmonary edema in acute left ventricular failure may be the trapping of blood volume in the splanchnic circulation in the "unstressed volume," or V^1 in Equation 2.1. Some of the effects observed by Green and co-workers could be due to the hepatic sphincter effect in the dog. Green showed that morphine increased hepatic sphincter tone and liver volume at constant flow and vena cava pressure and even resulted in the formation of ascites (Green, Jackson, and Parsons 1978). To directly confirm these observations and to test the importance of regional effects, we measured venous return curves in surgically isolated splanchnic and systemic circulations in the dog (Engler, Freeman, and Covell 1983). A diagram of our preparation is shown in Figure 2.3. The preparation was designed in such a way that all venous drainage from the intestine, liver, and spleen, which passed through the hepatic veins, was collected in one cannula leading to a Starling resistor. The extrasplanchnic circulation (which included drainage from the kidneys and lower extremities and from above the diaphragm) was collected in separate venous cannulas and drained through a separate Starling resistor. It is important to note that the renal circulation was treated with the systemic circulation. Several previous studies have included the renal circulation along with all drainage below the diaphragm as the splanchnic compartment. This is a potential problem in compartmental analysis, because the hepatic sphincter affects only some of the "lower" compartment, and arterial blood flow and venous tone may be controlled differently within different vascular beds below the diaphragm.

In our experiment we obtained sudden step changes in atrial pressure by changing the height of the Starling resistors. This is the same technique used by Guyton et al. (1957) He used the pressure at zero flow (the atrial pressure at which venous return is zero) as an estimate of the mean systemic pressure. (Guyton showed that this pressure is essentially the same as the mean circulatory pressure.) Typical results from our studies are shown in Figure 2.4. We administered sodium nitroprusside and found that the entire net change in mean systemic pressure could be attributed to the splanchnic circulation and that there were no significant changes

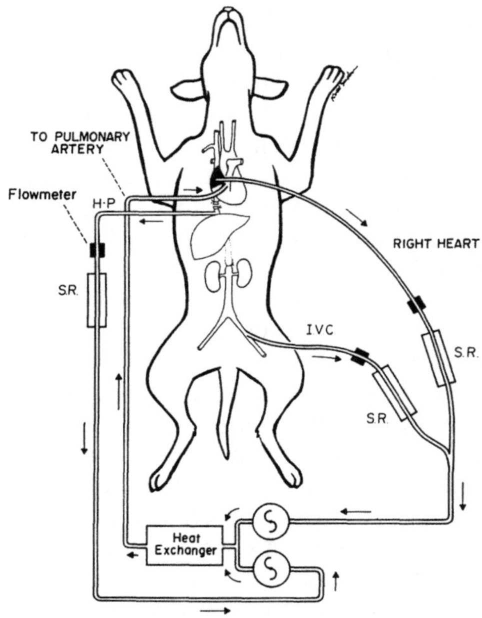

FIGURE 2.3. Diagram of the preparation used to determine separate venous return curves for the splanchnic and extrasplanchnic systemic circulations. The right heart venous drainage included all return to the right atrium and right ventricle. SR = Starling resistor, HP = hepatic portal, IVC = inferior vena cava.

in resistance to venous return or mean systemic pressure in the extrasplanchnic compartment. Since reflexes were, at least in part, intact in our preparation, we cannot differentiate between a regionally specific direct action of nitroprusside and a regionally different reflex change in venous tone in response to the hypotension induced by nitroprusside. Nonetheless, it is of significance that the entire change was confined to the splanchnic bed. In our experiments, the boundary conditions of venous tone and reflex changes in the arterial resistance and capacitance were allowed to change. The implications for using nitroprusside in heart failure, intraoperatively, and in ascites are significant.

To further study how responses to nitroprusside might be altered in the presence of chronic venous congestion, we studied dogs in which tricuspid insufficiency and pulmonic stenosis had been created. Ascites developed in these animals and they had high venous pressures. The most striking finding in the basal state was alteration in the venous resistance in the splanchnic compartment. In their textbook, Guyton, Jones, and Coleman (1973) predicted that such a change might favorably affect cardiac output and that, in fact, it might be a major way of controlling cardiac output. Our findings confirm that in chronic venous congestion, significant changes

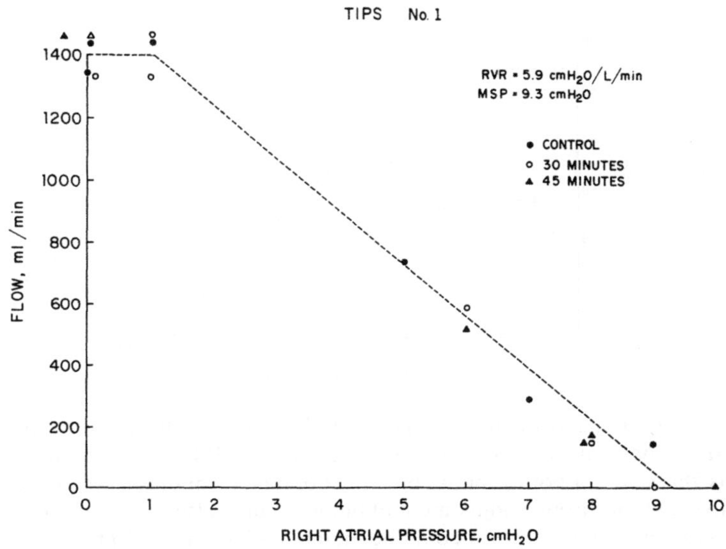

FIGURE 2.4. Venous return curve for the extrasplanchnic (systemic) circulation obtained from a dog with chronic venous congestion. Stability over time is demonstrated by repeated determinations over 45 minutes. The pressure at zero flow may be higher than normal owing to chronic venous congestion or a high initial blood volume in the surgical preparation. TIPS = tricuspid regurgitation and pulmonic stenosis, RVR = resistance to venous return, MSP = mean systemic pressure.

in the venous resistance do occur, and suggest that these changes are important in maintenance of venous return. Furthermore, the regionally specific site of action of nitroprusside for the splanchnic circulation was maintained in the face of chronic venous congestion. In two animals that also received nitroglycerin, the effects appeared to parallel those of nitroprusside. Figure 2.5 provides additional insight into the importance of this decrease in venous resistance. A normal systemic venous return, which existed at atrial pressures of 0 to 3 cm H_2O in controls, could be maintained at an atrial pressure of 9 cm H_2O in the congested animals at the same pressure at zero flow because of the change in slope or venous resistance. (This is not to say that the P_{MS} was not elevated in these animals, which would have further augmented venous return.) One problem in applying our studies to chronic heart failure in humans is that we did not measure blood volume, so the "starting" P_{MS} in the congested dogs was higher in vivo before the preparatory surgery. In patients dying with congestive heart failure, P_{MS} is probably quite high (Starr and Rawson 1940).

Studies by Bell, Zaret, and Rutlen (1983) were performed with extra-

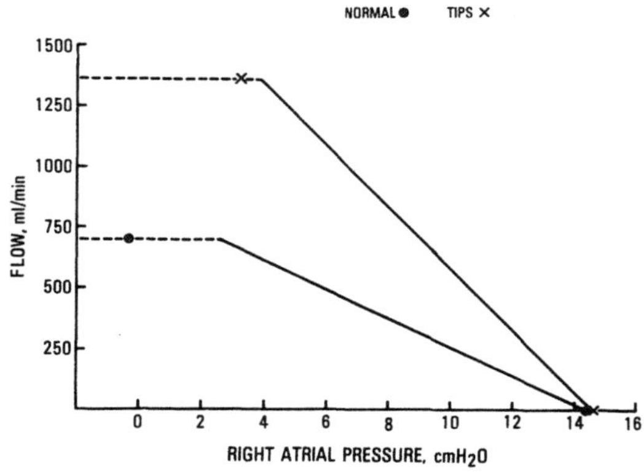

FIGURE 2.5. Regional systemic (extrasplanchnic) venous return curves in control dogs and in dogs with chronic venous congestion (TIPS). The pressure at zero flow was the same in both groups, but this boundary condition was determined by the investigator through adjustment of blood volume after surgical preparation. Normal systemic venous return could be maintained at an atrial pressure of 9 cm H_2O due entirely to a change in venous resistance. TIPS = tricuspid regurgitation and pulmonary stenosis.

corporeal circulation in addition to a blood pool radiolabel to allow measurement of blood volume by nuclear imaging techniques. These investigators found that alpha-adrenergic stimulation with phenylephrine resulted in decreased splenic and intestinal counts, implying a decrease in blood volume due to venoconstriction (active) or decreased flow (passive). Somewhat surprising were their findings after nitroglycerin administration. They found a decrease in counts to the spleen and an increase in intestinal counts, but a net decrease in apparent splanchnic volume. The implication is that hepatic sphincter tone was decreased, allowing blood to leave the splanchnic venous compartment at the same time that the pressure at zero flow decreased. These findings, combined with the decrease we found in mean systemic pressure, leave unclear the exact effect of nitroglycerin on venous tone in the splanchnic vasculature. We found no change in hepatic resistance to blood flow (measured as portal pressure minus vena cava pressure divided by hepatic flow), indicating no change in hepatic sphincter tone (Engler, Freeman, and Covell 1983). Thus our studies would be interpreted as showing a decrease in venous tone resulting in a decrease in mean systemic pressure. The studies of Bergquist, Bell, and Rutlen (1984) and of Rutlen, Wachers, and Zaret (1980) would suggest a change in hepatic sphincter tone resulting in a net decrease in splanchnic volume. However, these investigators did not measure hepatic resistance directly.

Blood Volume

Guyton's original studies demonstrated the important effect of blood volume on the venous return (Guyton 1955; Lindsey and Guyton 1959; Lindsey et al. 1957). Guyton and also Mitzner and Goldberg (1975) demonstrated a linear relationship between mean systemic pressure and blood volume in the absence of reflexes and at constant venous tone. The slope of this relationship is quite steep. For example, a 15% decrease in blood volume causes the mean systemic pressure to be zero. This rather dramatic effect is undoubtedly responsible for syncope in patients with poor vascular reflexes. Patients with diminished ability to maintain venous return reflexively during blood pooling in the lower extremities from orthostasis require such drastic treatment as use of potent salt-retaining steroids, to maintain increased intravascular volume, and of tight-fitting lower-extremity suits, to prevent hydrostatic venodilation and creep.

When blood volume is suddenly increased, mean systemic pressure shows a parallel linear increase. However, stress relaxation occurs in the veins within about 5 to 10 minutes, after which the mean systemic pressure returns toward baseline. Obviously, regional shifts in blood volume due, for example, to changes in hepatic sphincter tone or regionally different reflexes, can affect regional mean systemic pressure without causing a net change in total blood volume. Furthermore, in the dog the splenic reservoir is quite significant and can affect blood volume rather suddenly. The splenic capacity in other experimental animals and in humans is considerably less. Therefore, in studies in dogs on the effect of blood volume, careful consideration of reflexes that affect splenic contracture is required. Sarelius and Sinclair (1981) studied the effect of small changes in blood volume on oxygen delivery in closed-loop experiments. Increases or decreases of 10% of total blood volume were tested for their effects on the response of cardiac output to exercise. A 10% increase did not change flow or oxygen delivery compared with the control condition. However, a 10% decrease in blood volume resulted in a larger than normal increase in cardiac output at comparable oxygen consumption and a 10% augmentation in cardiac output at the same external workload. These changes in blood volume occurred over 20 minutes, so that creep and full reflex effects were allowed to occur. Thus it seems clear that small changes in blood volume do modify response to exercise, although the precise causative mechanisms have not been defined by Sarelius and Sinclair.

Changes in interstitial fluid volume significantly affect vascular compliance. Magrini and Niarchos (1983) demonstrated that venous compliance was decreased by edema and later compliance was improved by diuretic therapy. Under resting or steady-state conditions, changes in venous compliance have little effect on venous return as an independent variable (Guyton, Jones, and Coleman 1973). This is true in the analysis of the systemic venous return with lumped parameters. Guyton's graphic analysis of this situation is reproduced in Figure 2.6. At atrial pressures of −2 to +4 mm Hg, a fourfold increase in compliance has only small effects on venous return. However, when changes are over a large range, such as a high right atrial pressure occurring during exercise in a patient with inadequate cardiac reserve (i.e., congestive heart failure), high venous compliance would buffer increases in the venous return curve and have a definite effect on cardiac output. On the other hand, a decrease in venous compliance from normal has a relatively large effect on venous resistance. A decrease in venous resistance can allow a significantly augmented venous return at comparable right atrial and mean systemic pressures when the set point of the circulation is on the sloping portion of the curve (Fig. 2.6).

Guyton has analyzed the effect of venous resistance on the venous return curve (first equation in Fig. 2.2). Assuming that the relative capacitances of the arterial and venous systems remain constant (which would be approximately true over small ranges, since the normal ratio is 18), the venous return is seen to be affected mostly by venous resistance and in a small way by arterial resistance. Theoretically, then, resistance to venous return can have a profound effect on cardiac output. This was borne out in our studies of chronic venous congestion: The decreased venous resistance allowed the dog to maintain a normal cardiac output at rest despite a very high right atrial pressure. A similar analysis of Equation 2.1 for constant resistance shows that changes in capacitance have a relatively small effect on the venous return when venous capacitance is increased, but that a decrease in venous capacitance relative to arterial capacitance would have a larger relative effect. In our recent study of the properties of the venous

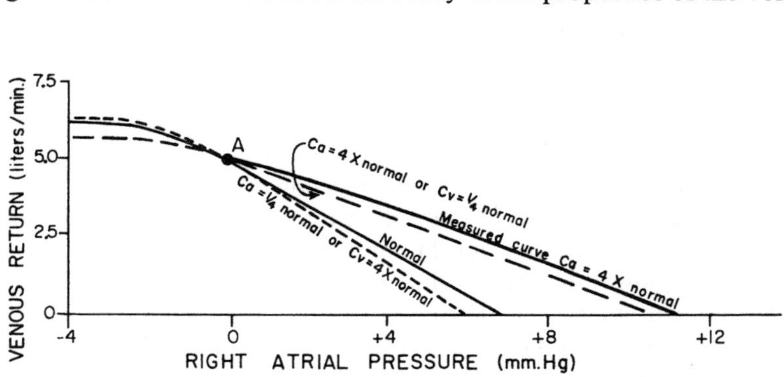

FIGURE 2.6. Graphic analysis of the effect of compliance changes on total systemic venous return. The dashed curves represent calculated effects of changes in venous capacitance C_V and the solid curves represent the measured effects of increasing the arterial capacitance (C_A) to four times normal. (The flow values for the measured curve have been extrapolated from dogs to human beings.) (From Guyton, Jones and Coleman 1973.)

circulation in animals with chronic congestion (Engler, Freeman, and Covell 1983), the resistance to venous return was reduced following long-standing venous congestion. Moreover, the pressure at zero flow in the splanchnic bed was increased. Both these changes would tend to increase cardiac output at high right atrial pressures independent of alterations in cardiac function.

Guyton, Jones, and Coleman (1973) demonstrated that changes in blood viscosity have a significant effect on the venous return curve (Fig. 2.7). This is an expected finding, since venous resistance is proportional to the hindrance times the viscosity. Changes in venous return of 40% were seen with profound anemia and polycythemia. Guyton confirmed the effect of changes in arterial resistance on the venous return curve by opening and closing an arterial venous shunt. In this situation, dramatic changes in arterial resistance could occur suddenly without changes in blood volume or venous tone.

Reflex Effects on Venous Return

Analysis of the effects of reflexes on venous return is quite difficult by the very nature of these experiments. Reflex changes induced by a change in one independent variable result in changes in the other independent

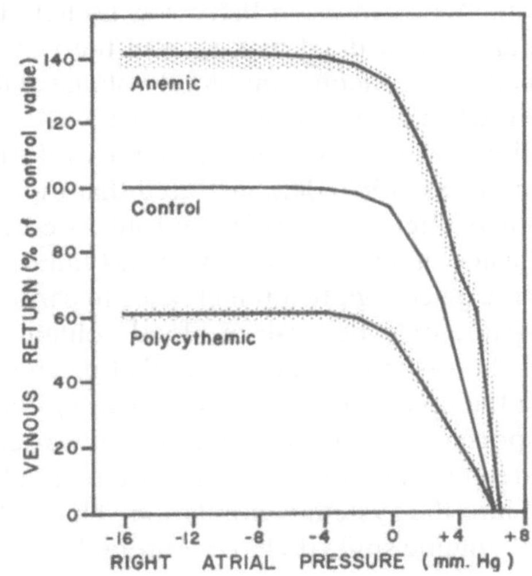

FIGURE 2.7. Average effects on the venous return curve of altering the total peripheral resistance by creating anemia (10 dogs) and polycythemia (8 dogs). Shaded areas indicate probable errors of the means. (From Guyton, Jones, and Coleman 1973.)

variables so that true open-loop criteria are difficult to achieve. Thus in a particular experiment, if the baroreceptors are isolated and stimulated with a step change in pressure, reflex outflow will change the venous capacitance, the venous resistance, and importantly, the distribution of blood volume and flow (arterial effects) between different venous return compartments. If, for example, the investigator then measures the net effect on venous return or on blood volume at a constant atrial pressure, he or she may not be able to analyze the effects of regionally different reflexes or regional shifts in blood volume. With these caveats in mind, considerable insight into the control of cardiac output in the intact situation can be obtained from the literature. In a study of the adrenergic regulation of total systemic distensibility, Rutlen, Supple, and Powell (1981) showed that both alpha and beta stimulation affect venous distensibility. In a study of isolated splanchnic nerve stimulation in cats, Greenway and Innes (1980) demonstrated that such stimulation can have quite significant effects on cardiac preload, afterload, and output. It is clear that changes in alpha-adrenergic tone can affect both resistance to venous return and the mean circulatory pressure (Appleton et al. 1984, 1985). Pagani, Vatner, and Braunwald (1978) demonstrated that the hemodynamic responses to vasodilators such as intravenous nitroprusside are profoundly affected by these regionally different reflexes. For example, they showed that mesenteric, iliac, and coronary beds demonstrate net vasoconstriction after reflex adjustments to nitroprusside. There was an initial increase of cardiac output when nitroprusside lowered the peripheral vascular resistance. However, after 7 minutes of infusion there was no net change in cardiac output, since venodilation resulted in a drop in the venous return curve that was not matched by a concomitant blood volume shift. These effects can be quite different when blood volume shifts are available to compensate for the venodilatation and to maintain the venous return. For example, Pouleur, Covell, and Ross (1980) demonstrated that the administration of nitroprusside failed to increase cardiac output in control conditions, whereas in the presence of acute congestive heart failure induced by myocardial infarction it caused significant increases in cardiac output. These increases were possible only because of blood volume shifts out of the pulmonary and into the systemic circuit, such that mean systemic pressure was maintained. In the control experiments, mean systemic pressure decreased with nitroprusside administration, whereas after myocardial infarction it did not decrease, owing to a 7 ml/kg blood volume shift out of the pulmonary and into the systemic circulation. Thus reflex adjustments resulting in regional volume shifts do have profound effects on the venous return (Fig. 2.8).

Shoukas and colleagues (Brunner, Shoukas, and MacAnespie 1981; Shoukas and Brunner 1980; Shoukas et al. 1981) studied the effect of epinephrine and of carotid sinus baroreceptor reflexes on the total systemic

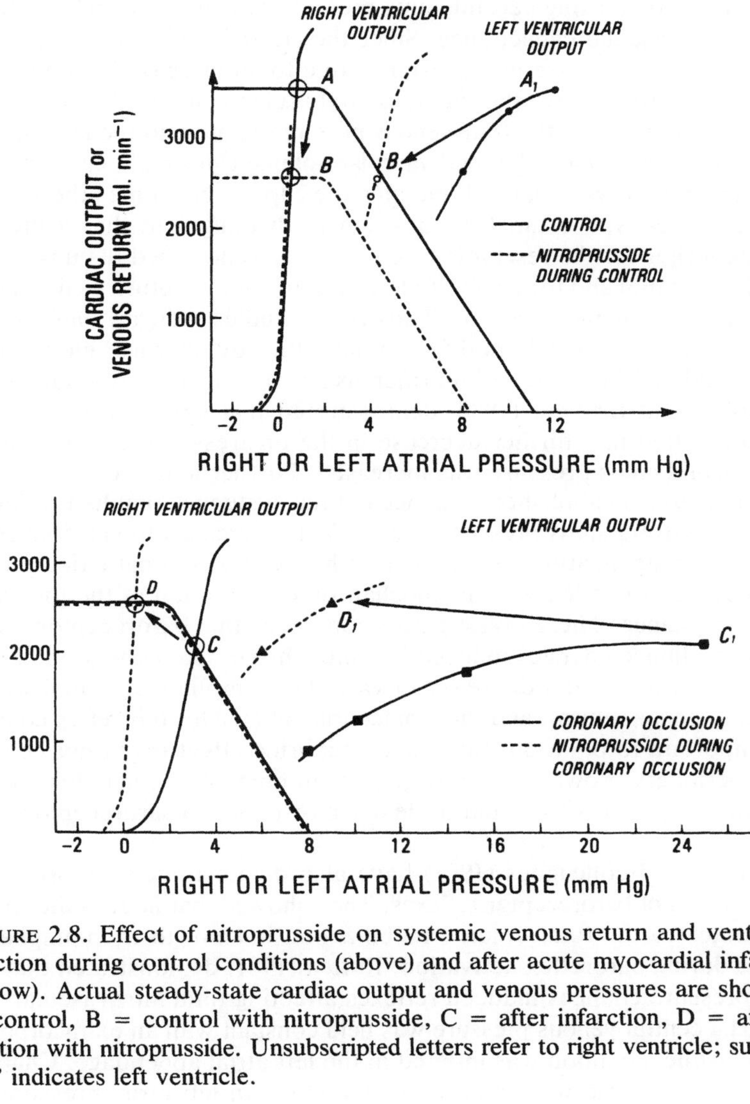

FIGURE 2.8. Effect of nitroprusside on systemic venous return and ventricular function during control conditions (above) and after acute myocardial infarction (below). Actual steady-state cardiac output and venous pressures are shown. A = control, B = control with nitroprusside, C = after infarction, D = after infarction with nitroprusside. Unsubscripted letters refer to right ventricle; subscript "1" indicates left ventricle.

capacitance. Increasing carotid sinus pressure resulted in a 20% increase in systemic vascular capacitance. Since the arterial capacitance remained unchanged, the increase was presumed due to an increase in venous capacitance. Furthermore, the unstressed vascular volume, defined as the vascular volume when the transvenous pressure relative to the atmosphere is zero, decreased. Thus the total vascular volume can simply be calculated as the unstressed vascular volume plus the capacitance times the venous pressure. This type of analysis has some advantages in that it gives a separate feeling for the reservoir capacity of the venous bed (its unstressed vascular volume) and for its ability to generate venous return as measured by its mean systemic pressure. Thus these studies indicate that baroreceptor input changes will modify venous return by altering venous compliance and shifting volume into (increasing) the unstressed volume.

Shoukas and Brunner (1980) demonstrated that epinephrine administration resulted in a further decrease in the unstressed vascular volume when carotid sinus pressure was increased. Furthermore, continuous epinephrine infusion abolished a change in total systemic compliance during increased carotid sinus pressure and, in fact, decreased the arterial compliance. The implication here is that withdrawal of sympathetic tone by the baroreceptor reflex was the mechanism of alteration of the total vascular compliance. These investigators concluded that baroreceptors modulate compliance changes induced by epinephrine, that activation of baroreceptors results in a decrease in vascular compliance (which would augment venous return), and that epinephrine attenuates this reflex control. It is important to point out that the epinephrine effect might not be only in the periphery, however; central effects of epinephrine may have modified the integration of vascular reflexes in response to baroreceptor stimulation.

Shoukas and colleagues (1981) have also demonstrated important regional effects of baroreceptor reflexes. They showed that in dogs the spleen contributes to the total volume mobilization but not to the compliance change during baroreflex activation. Rutlen (1981) demonstrated that left atrial pressure changes induced reflexes affecting total intravascular volume. The central venous pressure was held constant with an extracorporeal circuit, while a balloon was inflated in the left atrial appendage. Changes in reservoir volume were measured. Increases in left atrial pressure resulted in a decrease in intravascular volume, implying one of the following mechanisms: (1) increase in sympathetic venous tone resulting in venoconstriction, (2) decrease in hepatic sphincter tone allowing passive mobilization of blood from the splanchnic circulation, or (3) regional redistribution of cardiac output, again resulting in passive mobilization of blood from various regional circulations. Whichever mechanism was involved, this atrial reflex was shown to have an afferent limb in the vagus nerve and an efferent limb that is probably beta adrenergic in character.

Pharmacologic Effects

No discussion of the influence of the venous system on ventricular/arterial coupling would be complete without some consideration of the direct effects of drugs on venous and arterial tone, although the details of this pharmacology are beyond the scope of this chapter.

Some insight can be gained into integrated responses and the relative open-loop effects of drugs by measurement of the relative arterial dilating and venodilating capacities of each drug. Miller and co-workers (1976, 1982) have suggested taking the ratio of the change in forearm vascular resistance to change in forearm venous tone as an index of arterial-to-venous effects of a particular drug. Although such an analysis is clearly difficult in a closed-loop system, the resulting data are likely at least to help in classifying drugs as major arterial or venous actors. For example, Miller and colleagues found that the ratio of arterial vascular resistance to venous tone changes for sodium nitroprusside is 0.96, indicating approximately equal arterial and venous effects. For nitroglycerin the ratio is 0.53, and for phentolamine it is 1.6. Prazosin has an early ratio of 0.49, which increases to 0.67 one hour after the dose. Hydralazine has a ratio of 10, indicating little if any venous effects. Use of these numbers is appropriate in closed-loop analysis of the circulation. However, in open-loop analysis, the data may not apply. For example, hydralazine might in fact be a significant venodilator, but the potent arterial dilating effect results in baroreceptor activation and resultant intense venoconstriction, overcoming the venous but not the arterial effects. Another limitation of this analysis is the regional specificity of some drugs. For example, indomethacin has different effects on peripheral and coronary arterial resistance, and nitroprusside and nitroglycerin may dilate splanchnic veins more than systemic veins.

Other Effects

In certain disease states, generalized alteration in venous properties could affect venous parameters. For example, in the spontaneously hypertensive rat model of hypertension, significant venous hypertrophy has been demonstrated (Guyton 1981). These hypertrophied veins have augmented responses to catecholamines and thus may have more intense venoconstriction in response to reflexes and are also stiffer when undergoing passive distension. Thus the unstressed vascular volume and the compliance could both be altered. Some information suggests that salt content and venous stiffness are altered in chronic congestive heart failure. These factors have to be considered when using open-loop calculations to form an integrated circulatory model, or the effects of venous congestion, hypertension, and edema on venous capacitance and resistance have to be measured.

Circulatory Analysis With Venous Return as the Independent Variable

Guyton (1981) has taken the admittedly simplified overview that under normal circumstances each tissue controls its local blood flow on an autoregulatory basis with modifying influences from higher centers. Maintenance of adequate pressure in the arteries then allows independent control of local blood flow without flow to other areas in the circulation being affected. For example, if a certain vascular bed has a sudden increase in metabolic demand, it can increase flow without flow to other organs being altered, as long as constant blood pressure is maintained. Thus the major goal of the central nervous system is to control blood pressure, not cardiac output. This situation can change quite dramatically, since the capacity of the heart to pump blood is not unlimited. A large increase in metabolic needs or a decreased ability of the heart to respond to an increase in local blood flow can result in cardiac output that is inadequate for maintaining arterial pressure, secondary failure of local organ blood flow to meet metabolic needs, and circulatory failure. Under these circumstances venous return can no longer be considered a unique independent variable.

Conclusion

The venous bed plays an important role in the regulation of the circulation and is particularly important in the control of cardiac output. The insight gained by the open-loop analysis of venous return is illustrated by the studies of Pouleur, Covell, and Ross (1980) on the effects of nitroprusside on the normal circulation and on the circulation with reduced pumping capacity. These studies clearly show that agents that influence systemic veins, such as nitroprusside, can dramatically alter cardiac output, and perhaps more important, that shifts in blood volume between venous circulations play an important role in circulatory control.

Bibliography

Appleton, C., Martin, G. V., Algeo, S., Olajos, M., and Goldman, S. 1984. $Alpha_1$ and $alpha_2$ adrenergic venoconstriction in intact dogs. *Circulation* 70:II–232.

Appleton, C., Olajos, M., Morkin, E., and Goldman, S. 1985. Alpha-adrenergic control of the venous circulation in conscious dogs. *J. Pharmacol. Exp. Ther.* 233:729–735.

Bell, L., Zaret, B. L., and Rutlen, D. L. 1983. Quantitative radionuclide assessment of the splanchnic capacitance vasculature: validation of a new method. Circulation 68:III–83.

Bergquist, B. D., Bell, L., and Rutlen, D. 1984. Influence of nitroglycerin on the total capacitance vasculature. *Clin. Res.* 32:150A.

Brooksby, G. A., and Donald, D. E. 1972. Release of blood from the splanchnic circulation in dogs. *Circ. Res.* 31:105–118.

Brunner, M. J., Shoukas, A. A., and MacAnespie, C. L. 1981. The effect of the carotid sinus baroreceptor reflex on blood flow and volume redistribution in the total systemic vascular bed of the dog. *Circ. Res.* 48:274–285.

Caldini, P., Permutt, S., Wadell, J. A., and Riley, R. L. 1974. Effect of epinephrine on pressure, flow, and volume relationships in the systemic circulation of dogs. *Circ. Res.* 24:606–623.

Carneiro, J. J., and Donald, D. E. 1977. Blood reservoir function of dog spleen, liver, and intestine. *Am. J. Physiol.* 232:H67–H72.

Donald, D. E., and Aarhus, L. L. 1974. Active and passive release of blood from canine spleen and small intestine. *Am. J. Physiol.* 227:1166–1172.

Engler, R. L., Freeman, G. L., and Covell, J. W. 1983. Regional venous return: nitroprusside effect in normal and chronically congested dogs. *Am. J. Physiol.* 245:H814–H823.

Green, J. F. 1975. Pressure-flow and volume-flow relationships of the systemic circulation of the dog. *Am. J. Physiol.* 229:761–769.

Green, J. F. 1977. Mechanism of action of isoproterenol on venous return. *Am. J. Physiol.* 232:H152–H156.

Green, J. F., Jackman, A. P., and Krohn, K. A. 1978. Mechanism of morphine-induced shifts in blood volume between extracorporeal reservoir and the systemic circulation of the dog under conditions of constant blood flow and vena caval pressures. *Circ. Res.* 42:479–486.

Green, J. F., Jackman, A. P., and Parsons, G. 1978. The effects of morphine on the mechanical properties of the systemic circulation in the dog. Circ. Res. 42:471–478.

Greenway, C. V., and Innes, I. R. 1980. Effects of splanchnic nerve stimulation on cardiac preload, afterload, and output in cats. *Circ. Res.* 46:181–189.

Greenway, C. V., and Lister, G. E. 1974. Capacitance effects and blood reservoir function in the splanchnic vascular bed during non-hypotensive hemorrhage A., and MacAnespie, C. L. 1981. The effect of the carotid sinus baroreceptor reflex on blood flow and volume redistribution in the total systemic vascular bed of the dog. *Circ. Res.* 48:274–285.

Caldini, P., Permutt, S., Wadell, J. A., and Riley, R. L. 1974. Effect of epinephrine on pressure, flow, and volume relationships in the systemic circulation of dogs. *Circ. Res.* 24:606–623.

Carneiro, J. J., and Donald, D. E. 1977. Blood reservoir function of dog spleen, liver, and intestine. *Am. J. Physiol.* 232:H67–H72.

Donald, D. E., and Aarhus, L. L. 1974. Active and passive release of blood from canine spleen and small intestine. *Am. J. Physiol.* 227:1166–1172.

Engler, R. L., Freeman, G. L., and Covell, J. W. 1983. Regional venous return: nitroprusside effect in normal and chronically congested dogs. *Am. J. Physiol.* 245:H814–H823.

Green, J. F. 1975. Pressure-flow and volume-flow relationships of and blood volume expansion in anaesthetized cats. *J. Physiol. (Lond.)* 237:279–294.

Guyton, A. C. 1955. Determination of cardiac output by equating venous return curves with cardiac response curves. *Physiol. Res.* 35:123.

Guyton, A. 1979. Comment on "The cardiac and vascular factors that determine systemic blood flow" by Levy. *Circ. Res.* 44:746–747.
Guyton, A. C. 1981. The relationship of cardiac output and arterial pressure control. *Circulation* 64:1079–1088.
Guyton, A. C., Douglas, B. H., Langston, J. B., and Richardson, T. Q. 1962. Instantaneous increase in mean circulatory pressure and cardiac output at onset of muscular activity. *Circ. Res.* 11:431.
Guyton, A. C., Jones, C. E., and Coleman, T. G. 1973. *Circulatory Physiology: Cardiac Output and Its Regulation*. Philadelphia: Saunders, pp. 205–252.
Guyton, A. C., Lindsey, A. W., Abernathy, B., and Richardson, T. 1957. Venous return at various right atrial pressures and the normal venous return curve. *Am. J. Physiol.* 189:609–615.
Johnson, P. C., and Hanson, K. M. 1963. Relation between venous pressure and blood volume in the intestine. *Am. J. Physiol.* 204:31–34.
Levy, M. N. 1979. The cardiac and vascular factors that determine systemic blood flow. *Circ. Res.* 44:739–746.
Lindsey, A. W., Banahan, B. F., Cannon, R. N., and Guyton, A. C. 1957. Pulmonary blood volume of the dog and its changes in acute heart failure. *Am. J. Physiol.* 190:45.
Lindsey, A. W., and Guyton, A. C. 1959. Continuous recording of pulmonary blood volume, and pulmonary pressure and volume changes during acute right or left ventricular failure. *Am. J. Physiol.* 197:959.
Magrini, F., and Niarchos, A. P. 1983. Hemodynamic effects of massive peripheral edema. *Am. Heart. J.* 105:90–97.
Miller, R. R., Fennell, W. H., Young, J. B., Palomo, A. R., and Quinones, M. A. 1982. Differential systemic arterial and venous actions and consequent cardiac effects on vasodilator drugs. *Prog. Cardiovasc. Dis.* 24:353–374.
Miller, R. R., Vismara, L. A., Williams, D. O., Amsterdam, E. A., and Mason, D. T. 1976. Pharmacological mechanisms of left ventricular unloading in clinical congestive heart failure. *Circ. Res.* 39:127–133.
Mitzner, W., and Goldberg, H. 1975. Effects of epinephrine on resistive and compliant properties of the canine vasculature. *J. Appl. Physiol.* 39:272–280.
Pagani, M., Vatner, S. F., and Braunwald, E. 1978. Hemodynamic effects of intravenous sodium nitroprusside in the conscious dog. *Circulation* 57:144–151.
Pouleur, H., Covell, J. W., and Ross, J. Jr. 1980. Effects of nitroprusside on venous return and central blood volume in the absence and presence of acute heart failure. *Circulation* 61:328–337.
Riley, R. L. 1982. A postscript to *Circulation of the Blood: Men and Ideas* (Dickinson Richards Memorial Lecture, 1982). *Circulation* 66:638–688.
Ross, J. Jr., Frahm, C. J., and Braunwald, E. 1961. Influence of carotid baroreceptors and vasoactive drugs on systemic vascular volume and venous distensibility. *Circ. Res.* 9:75–82.
Rothe, C. F. 1983. Venous system: physiology of the capacitance vessels. In Shepherd, J. T., and Abboud, F. M. (eds.). *The Handboook of Physiology–The Cardiovascular System III*. Bethesda: American Physiological Society, pp. 397–452.
Rothe, C. F., Johns, B. L., and Bennett, T. D. 1978. Vascular capacitance of dog intestine using mean transit time indicator. *Am. J. Physiol.* 234:H7–H13.

Rutlen, D. L. 1981. Reflex effects of left atrial pressure elevation on total intravascular volume. *Am. J. Physiol.* 241:H760–H765.

Rutlen, D. L., Supple, E. W., and Powell, J. W. Jr. 1981. Adrenergic regulation of total systemic distensibility. *Am. J. Cardiol.* 47:579–588.

Rutlen, D. L., Wackers, F. J., and Zaret, B. L. 1980. Effect of dobutamine on systemic capacity in the dog. *Circ. Res.* 46:133–138.

Sarelius, I. H., and Sinclair, J. D. 1981. Effects of small changes of blood volume on oxygen delivery and tissue oxygenation. *Am. J. Physiol.* 240:H177–H184.

Shoukas, A. A., and Brunner, M. C. 1980. Epinephrine and the carotid sinus baroreceptor reflex. *Circ. Res.* 47:249–257.

Shoukas, A. A., MacAnespie, C. L., Brunner, M.J., and Watermeier, L. 1981. The importance of the spleen in blood volume shifts of the systemic vascular bed caused by the carotid sinus baroreceptor reflex in the dog. *Circ. Res.* 49:759–766.

Skalak, T. 1984. A mathematical hemodynamic network model of the microcirculation in skeletal muscle, using measured blood vessel distensibility and topology. University Microfilms International #DA8418303. University of California–San Diego, CA.

Starr, I., and Rawson, A. J. 1940. Role of the "static blood pressure" in abnormal increments of venous pressure, especially in heart failure. I. Theoretical studies on an improved circulation schema whose pumps obey Starling's law of the heart. *Am. J. Med. Sci.* 199:27.

3
The Pathophysiology of Pulmonary Vascular Disease Associated With Congenital Cardiac Lesions

RICHARD A. HOPKINS

Introduction

Many types of congenital cardiac lesions are associated with the development of pulmonary vascular obstructive disease (PVOD). This connection was first reported by Eisenmenger (1897) in reference to a ventricular septal defect. Abbott and Dawson (1924) created some order in the anatomic classifications, and various lists of lesions predisposing to the Eisenmenger reaction have been compiled, but in these, lesions tend to be grouped according to the level of the shunt (Abbott and Dawson 1924, Baumgartner and Abbott 1929, Graham 1979, Moschowitz 1927, O'Rourke 1967, Rosedale 1935, Wood 1958). The lumping of congenital defects according to anatomy is predictive in only the most general sense; such a grouping does not necessarily correlate with rate or severity of pulmonary vascular disease. For example, total anomalous pulmonary venous drainage and atrial septal defect are not similar in their pulmonary vascular effects, although both have atrial-level shunts, nor is atrioventricular septal defect similar to typical large VSD in tendency to cause vascular disease. These different defects at the same "shunt level" do have differing hemodynamics, but the relationship of the differences to the progression of pulmonary vascular disease has not been clarified. Furthermore, factors such as living at high altitude and Down's syndrome (Chi and Krovetz 1975; Yamaki, Horiuchi, and Sekino 1983) can exacerbate a tendency for PVOD, but predicting the precise effect of such factors has been difficult.

Pulmonary vascular disease is one of the primary factors governing morbidity and mortality in many congenital cardiac diseases, and it significantly affects the timing and success of surgery. The reasons for its development have not been delineated, however, and measurement techniques currently in use are inadequate for assessing susceptibility to it, its very early changes, or onset of irreversibility (Brammell et al. 1971; Clarkson et al. 1968; Dexter 1979; DiSesa, Cohn, and Grossman 1983; Hoffman and Rudolph 1965). Pathophysiologic mechanisms remain unexplained by traditional methodology, on the basis of mean hemodynamic measurements (Blank, Muller, and Dammann 1961). A unified conceptual approach to the physiology of pulmonary vascular disease has been lacking.

The premise of this chapter is that the initial stimulus for pulmonary arterial injury should be defined in terms of hydraulic energy dissipation. The concept is based on an analysis of pulsatile hemodynamics and the changes in energy input and dissipation caused by the major stresses associated with development of pulmonary vascular diseases (increased pulmonary blood flow, elevated pulmonary arterial pressure, elevated pulmonary venous pressure, and hypoxemia). This analysis utilizes animal laboratory data, standard principles of hydraulic engineering, and published studies of humans to assess the effects of abnormal hemodynamics typical for various congenital lesions on the distribution and rate of hydraulic energy dissipation. Since clinicians have tended to think in terms of pathologic and morphologic features, this physiologic analysis is related to known pathologic features and clinical tendencies of specific defects.

Pathology

Classical Pathology

Heath and Edwards (1958) described six grades of structural changes in pulmonary arteries associated with congenital cardiac septal defects, on the basis of a study of 67 cases (Fig. 3.1). Their scheme brought some order to the assessment of pulmonary hypertensive changes and, together with Wood's physiologic summation, determined the conceptional approach for the ensuing 25 years (Wood 1958). Although others have disagreed with some aspects of the Heath–Edwards classification, their arguments have not altered the essential concepts (Wagenvoort et al. 1967). The Heath–Edwards disease grades, based on light microscopy, are classified according to a time-dependent progression, beginning with medial thickening and progressing to intimal changes. This approach has validity, but more recent studies, including electron microscopic investigations, have demonstrated that the anatomic responses are actually more rapid than originally perceived, and that very early changes and adaptations can be missed (Esterly, Glagov, and Ferguson 1968). Because the last three Heath–Edwards categories are really just morphologic variations on a theme of severe PVOD, other authors have differentiated only three categories: medial thickening alone, medial thickening plus intimal changes, and the more severe type of fibrotic, fibroelastic, and plexiform lesions (Virmani and Roberts 1979).

Traditionally, increasing thickness of the media has been explained in physiologic terms as a response to increased mean pressure within the

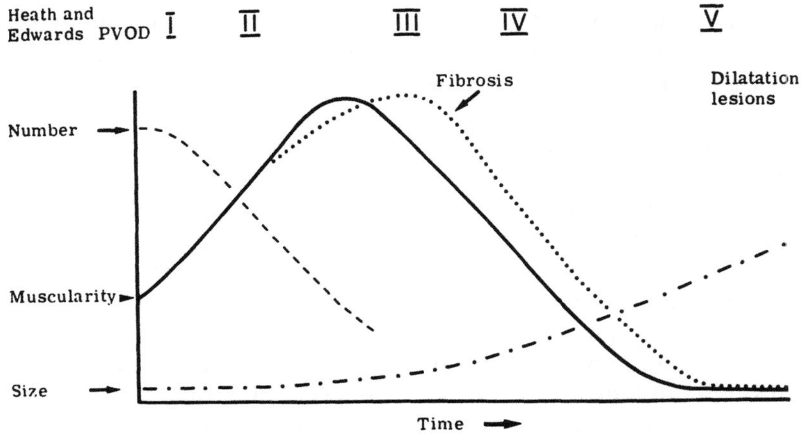

FIGURE 3.1. Generalized relationship between the Heath–Edwards classification for pulmonary vascular obstructive disease (PVOD) and the findings of quantitative morphometrics. (From Haworth 1981; with permission from Churchill Livingstone, Edinburgh).

vessels and not necessarily as a response to increasing flow unaccompanied by pressure elevation (Folkow 1971, Virmani and Roberts 1979, Wagenvoort 1960, Wagenvoort et al. 1967). Intimal thickening, whether cellular or acellular, implies a greater degree of injury. It has been thought to be a result of focal energy dissipation and may have a multiplicity of pathogenetic features (Esterly 1968). Electron microscopy has revealed that endothelial alterations actually occur very rapidly in a high-flow high-pressure shunt model (Van Benthuysen et al. 1977). Despite the contributions of classical pathologic techniques, great variations have been observed, however, which have been attributed to individual, disease-related, and species variability in pulmonary vascular "reactivity"—an undefined but persistent concept (Grover et al. 1963; Pietro et al. 1984; Rich, Ganz, and Levy 1983).

Quantitative Morphometrics

Understanding of the pathology of PVOD associated with congenital cardiac defects has become much more sophisticated and precise in the last 15 years as a number of investigators have differentiated effects of various hemodynamic stresses on the normal growth and maturation of pulmonary vessels, using quantitative morphometric studies in both experimental and clinical material (Davies and Reid 1970, Haworth 1983a, Haworth and Hislop 1983, Rabinovitch et al. 1981.) This work has generated a new understanding of the pathologic anatomic responses of the pulmonary vascular bed. Specifically, the size of distal arteries, the number of arteries associated with each alveolar unit, and the degree of muscularization were found to be altered by congenital cardiac lesions and analogous stresses (Haworth 1981a,b; Haworth 1983a,b; Haworth and Hislop 1983; Hislop et al. 1975). Rabinovitch and co-workers (Rabinovitch and Reid 1981, Rabinovitch et al. 1978, Rabinovitch et al. 1984) have created a structural classification with three categories:

Grade A. Peripheral extension of muscle correlating with high-flow lesions without pulmonary hypertension.

Grade B. Lesions with not only abnormal distal extension of muscle but also medial muscular hypertrophy (increased thickness) in the 50- to 100- μm arteries.

Grade C. Reduced number of arteries and a definite correlation with elevated pulmonary vascular pressure.

Assessment of these changes relative to the patient's age is important, since the pulmonary circulation undergoes dramatic remodeling following birth. Rabinovitch and colleagues have defined a normal progression of distal muscularization with increasing age and have quantitatively related wall thickness to pressure and arteriolar numbers and the rate of arterial vessel taper to the severity of pulmonary vascular disease. Figure 3.1 shows the general relationship between quantitative morphometric analysis of arteriolar number, muscularity, and diameter and the classical Heath–Edwards changes of PVOD (Haworth 1981, Rabinovitch et al. 1978). The increases in muscularity throughout grades I, II, and III are caused not only by increase in medial muscle thickness but also by distal progression of muscularization. As muscularity increases, more proliferation and fibrosis appear, beginning with grade II.

Pulmonary Vascular Physiology

Standard Hemodynamics

Understanding of pulmonary vascular hypertension has traditionally been based on pulmonary artery pressures (mean, systolic, and diastolic) obtained by long fluid-filled catheters, and on mean blood flow measured as cardiac output assessed over a relatively long period (e.g., 1 minute). Pulmonary vascular resistance (PVR) is calculated by dividing the pressure drop across the pulmonary vascular bed by the mean flow (Q) through the pulmonary vascular bed:

$$\mathrm{PVR} = \frac{\mathrm{PAP} - \mathrm{LAP}}{Q} \tag{3.1}$$

where PAP is pulmonary artery pressure and LAP is left atrial pressure.

This calculation of resistance, using mean flow and pressure, is based on a model of the pulmonary vascular bed as a passive complex of conduits with static pressure–volume relationships. Pulmonary vascular resistance is a valuable expression, since it can be correlated with clinical data. It rests, however, on a physiologic model that, while not necessarily wrong, is incomplete. Pulmonary vascular resistance calculated in this way is based upon the Poiseuille relationship, which relates flow directly to the length of a rigid tube:

$$Q = \frac{Kr^4}{8\mu L} \tag{3.2}$$

where Q = mean flow
K = Poiseuille's constant
r = radius
μ = blood viscosity
L = length of system

Poiseuille's law is valid only when certain conditions are met, and there are significant differences between the pulmonary vascular bed and a Poiseuille tube, or even a network of Poiseuille tubes (see Table 3.1). In the first place, this model does not indicate where in the vascular bed changes are occurring. Recruitment and dilatation of the many vessels at the arteriolar–capillary level lower calculated resistance more than any other factor, and it is their effect that is assessed. Vasodilation is not distinguished, however, from vascular recruitment. The proximal pulmonary vessels and their viscoelastic properties are totally ignored. Furthermore, only the net effect of mean pressure and flow is measured. Pulsatile energy is not accounted for, nor is there any differentiation between active and passive vasomotion.

One of the persisting mysteries of pulmonary vascular disease is the interval of time between the onset of hemodynamic stress and the pressure changes that accompany pathologic developments. It may be possible to

TABLE 3.1. Differences between Poiseuille model and pulmonary circulation.

Characteristic	Poiseuille tube	Pulmonary circulation
Flow	Steady, laminar	Pulsatile
Turbulence	None	Flow disturbance present
Fluid	Newtonian	Nonnewtonian
Wall	Rigid	Viscoelastic
Taper	None	Present
Proximal geometry	Circular	Elliptical
Branching	None	Extensive
Total tube length	Long	Short
Curvature	None	Present
Internal surface	Smooth	Imperfect

illuminate what happens during this interval by modeling the vascular bed in terms of pulsatile pressure and flow.

Pulsatile Physiology

Pulmonary impedance expresses the quantitative relationships between pulsatile pressure and flow and allows precise calculation of ventricular power output (and hence input to pulmonary circulation) in both pulsatile and mean terms (Bergel and Milnor 1965; McDonald 1974; Milnor 1982; Milnor, Bergel, and Bargainer 1966; Milnor and Bertram 1978; Milnor et al. 1969; Patel, Schilder, and Mallas 1960). Impedance is a measure of the opposition to flow. Its modulus has the same units as resistance, but in addition to having a resistance term (or opposition to mean-steady flow) it has terms that describe the opposition to and energy cost of pulsations in the pulmonary vascular bed. Impedance relates not only to the magnitude of the relationship between the pressure and flow waveforms but also to their time relationship (phase angle). Impedance is frequency dependent and therefore modulated by heart rate as well as vessel dimensions, stiffness or viscoelastic properties of the vessels, and wave reflections. Impedance is measured using high-fidelity pressure and flow signals, and each waveform can be described as a Fourier series of oscillatory waves. Thus total pulmonary blood flow (Qt) is expressed as

$$Q_{(t)} = Q_m + \sum_{n-1}^{10} Q_n \sin(n\omega t + \theta_n) \tag{3.3}$$

where Q_m = mean flow
Q_n = amplitude of nth harmonic
ω = fundamental angular frequency
t = length of sequence
θ_n = phase angle of nth harmonic

Pressure waves are expressed similarly:

$$P_{(t)} = P_m + \sum_{n=1}^{10} P_n \sin(n\omega t + \beta_n) \tag{3.4}$$

where P_m = mean pressure
P_n = amplitude of nth harmonic
β_n = phase angle of nth harmonic

Division of the mean terms (P_m/Q_m) yields the input impedance to mean flow (Z_m), which is analogous to pulmonary vascular resistance. Similarly, division of the sinusoidal terms (P_n/Q_n) gives the input impedance (Z_n) for the nth harmonic. The corresponding phase angle is derived by subtracting the flow phase from the pressure phase ($\phi_n = \beta_n - \theta_n$). The first harmonic term is at the frequency of the heart rate. The oscillatory terms are the pulsatile components which, when added together, reform the original recorded wave (Bergel and Milnor 1965). Typical impedance spectra are shown in Figure 3.2. Note that increasing cardiac output with isoproterenol or pacing of the heart does not alter the impedance spectra, indicating biological linearity of flow and minimal direct effect of isoproterenol on the vessels. Conversely, serotonin does increase impedance by causing significant increases in pulmonary vascular resistance, stiffening of the proximal pulmonary vessels, and increased wave reflections. Canine and human impedance spectra are similar in shape, although the moduli are somewhat lower in humans.

A given flow wave generated by the right heart at a specific heart rate will generate a pressure wave defined in amplitude, timing, and form by the specific impedance relationship present at the time. Fourier analysis assumes linearity in a distributed system. The pulmonary vasculature behaves as a distributed system (Caro and Harrison 1962, Taylor 1966). Biological linearity has been demonstrated in a number of ways. One con-

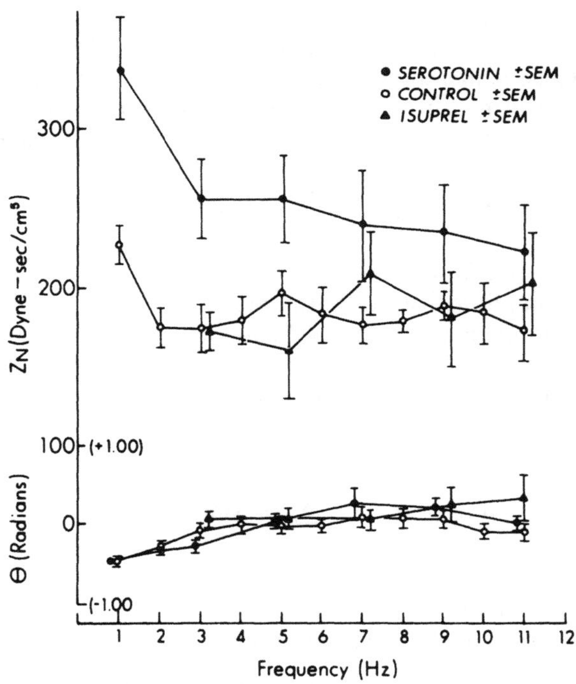

FIGURE 3.2. Pooled impedance (Z_N) and phase angle data from 12 normal, awake, chronically instrumented dogs. These impedance spectra compare data from resting control animals with data obtained during serotonin and isoproterenol infusions. The pulmonary vascular resistance with standard deviation (SEM) for each of the three groups (in dyne sec cm^{-5}) is: control = 320 ± 40; serotonin = 633 ± 65; and isoproterenol = 331 ± 48. (From Hammon et al. 1981).

vincing demonstration involves analysis of low-flow waveforms generated by atrial systoles along with the subsequent ventricular systoles that generate much greater flows. This has been accomplished in awake dogs. The impedance spectra for the atrial and ventricular contractions are equivalent (Fig. 3.3) (Hammon et al. 1981). In addition, many experiments over a wide range of paced fundamental heart rates have demonstrated biological linearity within the noise level of the measurement systems (Milnor, Bergel, and Bargainer 1966).

The characteristic impedance (Z_0) is the impedance in the absence of reflected waves. In blood vessels it therefore is determined only by geometry and viscoelastic properties of the walls and is measured as the average impedance in the higher frequencies (Attinger 1963; Hammon et

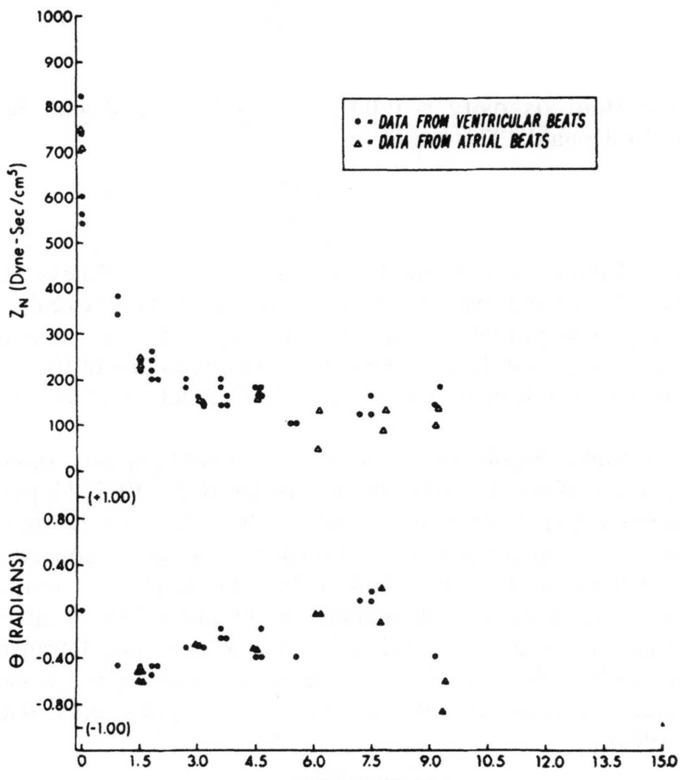

FIGURE 3.3. Comparison of impedance modulus (Z_N) and phase of normal heartbeats from quietly resting, awake, chronically instrumented dogs in which a low fundamental heart rate allows analysis of pulmonary waveforms generated by atrial contraction alone and by the following ventricular systolic contraction. The data are virtually superimposable, indicating biological linearity within the accuracy of the measurement systems. (From Hammon et al. 1981).

al. 1981; Hopkins et al. 1979; McDonald 1974; Milnor 1972a,b; Milnor 1982; Milnor, Bergel, and Bargainer 1966; Westerhof et al. 1972).

Womersley has defined the relationship between wave velocity (C_0), characteristic impedance (Z_0), and vessel radius *(r)* of a strongly tethered elastic tube (Bargainer 1967, Womersley 1957):

$$Z_o = \frac{\rho C_o}{\pi r^2 \sqrt{1 - \sigma^2}} \cdot \frac{1}{M_{10}^1} \cdot e^{-j\varepsilon/2} \tag{3.5}$$

where ρ = density of blood = 1.055 g/ml
σ = Poisson's ratio = 0.5
$j = \sqrt{-1}$

M_{10} and ε are functions of Womersley's nondimensional parameter

$$\alpha = r\sqrt{\frac{\omega\rho}{\mu}}$$

where μ = fluid viscosity = 0.04 poise. Using the Moens–Korteweg equation (McDonald 1974) we have

$$C_o = \sqrt{\frac{Eh}{\rho 2r'}} \tag{3.6}$$

where E is Young's elastic modulus and h is wall thickness. The relationship of Z_0 to E can be defined by substitution. This reveals, in a relatively complex way, that the characteristic impedance increases as vessel stiffness increases and decreases as the effective radius of the larger and more proximal vessels increases (Li et al. 1981; Patel, Schilder, and Mallos 1960).

Both acute and chronic vasomotor regulation of large pulmonary artery geometry and stiffness has been demonstrated (Gow 1972, Hopkins et al. 1979, Hopkins et al. 1980, Somlyo and Somlyo 1964). There are certainly nonlinear effects on the actual relationship between viscoelasticity and impedance (Gow and Taylor 1968, Milnor et al. 1978). Characteristic impedance is more strongly dependent on the properties of the proximal arteries than on the distal vessels (Esterly, Glagov, and Ferguson 1968; Milnor 1972a,b). Mechanical properties of the pulmonary vessels must have a large role in determining the characteristics of energy distribution (Dobrin 1978).

Hydraulic Energy Distribution in the Pulmonary Vascular Bed

Basic hydraulic engineering principles suggest that increased impedance increases the total input energy at the origin of the pulmonary circulation, thereby increasing the total amount of energy to be dissipated. Once vessel walls begin to stiffen in larger arteries there is less attenuation of the pulsatile pressure energy in the proximal vessels and hence greater dissipation distally. The dissipation of both mean and pulsatile energy is primarily dependent on pressure changes, while flow is a less significant factor, contributing directly to energy dissipation only in the kinetic terms and in the modest loss of flow pulsatility in the precapillary region. The more important flow effects are modulated by the relationship of impedance to pressure.

Fry (1968) demonstrated that increased blood velocity gradients can be associated with alterations in vascular endothelial morphology, that is, that there is stress-induced injury and repair. There is a critical value of shearing stress to which endothelial cells "yield" and develop altered structural and chemical characteristics, while intimal changes are induced in regions of turbulence (Fry 1968). Turbulence is associated with marked elevations in local energy dissipation (Roach 1972). Thus Fry was one of the first to demonstrate elegantly that focal energy dissipation can cause anatomic alterations in vascular histology.

Calculation of total hydraulic energy requires measurement of the pulsatile pressure and flow terms (Shepherd, Simpson, and Sharp 1966). How do velocity or flow and pressure changes result in altered energy dissipation? The Bernoulli principle states that in a "narrowing" tube, pressure is converted to kinetic energy as velocity increases (Fry 1968, McDonald 1974). Normally the taper of pulmonary blood vessels is very gradual, and hence this effect is dispersed over a great length. Nevertheless, a drop in pressure represents conversion of this pressure to kinetic energy, which must be dissipated (e.g., as heat or sound) in that region where the pressure drop occurs. Any resistor, such as a precapillary vasoconstrictor region or a point of acute narrowing due to anatomic obstructive changes of the arterial wall (Vogel, McNamara, and Blount 1967), would be a site of such a pressure drop. Rapid tapering (e.g., pruning) would result in conversion of significant amounts of hydraulic energy into heat and sound. Energy is also dissipated in areas of turbulent flow, which are transitions between regions of markedly differing impedance characteristics. Milnor (1972a,b) defined a precapillary region in the pulmonary vascular bed where the majority of pressure drop occurs prior to establishment of the typical intracapillary pressure profile. This is also the region where velocity changes and the most marked damping of pulsations occur (Fig. 3.4). Skalak and colleagues (1966a,b) demonstrated that about 80% of input energy is dissipated in the arteriolar–capillary region, the same region that comprises the primary target vessels of pulmonary vascular disease. Other studies have confirmed this large percentage of input energy dissipation within this critical distal arteriolar region (Brody, Stemmler, and DuBois 1968).

Milnor and colleagues (Milnor 1982; Milnor, Bergel, and Bargainer 1966) described in power terms the relevant hydraulic input energy equations. Mean potential energy obtained per unit time (or work) is mean pressure times mean flow:

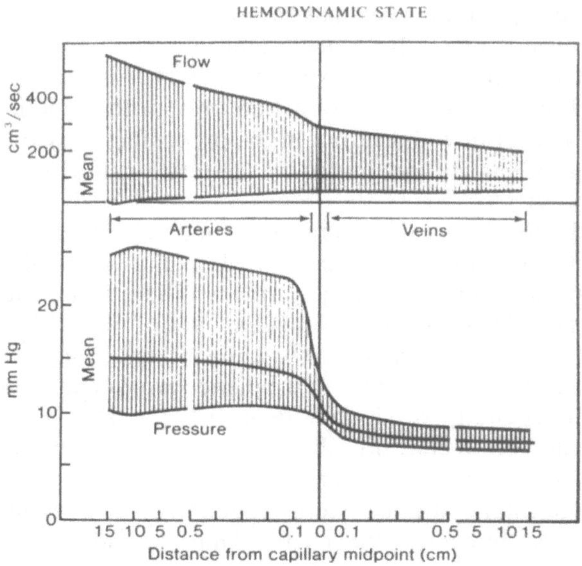

FIGURE 3.4. Pulsatile pressure and flow along an average circulatory pathway through the human pulmonary vascular bed. (From Milnor 1972a).

$$W_m = P_m Q_m \tag{3.7}$$

Utilizing the impedance relationship, the pulsatile energy components are defined as

$$W_o = \frac{1}{2} \sum_{n=1}^{10} (Qn)^2 Z_n \cos \phi_n \tag{3.8}$$

Total power (W_t) is a combination of the mean and pulsatile terms:

$$W_t = W_m + W_o \tag{3.9}$$

The kinetic energy (mV^2) associated with blood flow can also be divided into steady and pulsatile components (Milnor, Bergel, and Bargainer 1966):

$$K_s = \frac{pQ^3}{2A^2} \tag{3.10}$$

$$K_t = \frac{P}{2A^2 J} \sum_{j=0}^{J} (Q_j)^3 \tag{3.11}$$

with

$$K_o = K_t - K_s \tag{3.12}$$

All of these equations define the input power. The actual energy dissipated per unit time can be defined by measuring the output power on the pulmonary venous side of the circulation and subtracting it from the input power.

$$\text{energy in} - \text{energy out} = \text{dissipated energy} \tag{3.13}$$

This can be visualized by calculating the mean energy dissipation:

$$\text{mean energy dissipated} = (P_m * Q_m) - (P_{la} * Q_{la}) \tag{3.14}$$

where P_{la} equals mean pressure in the left atrium (i.e., pulmonary veins) and Q_{la} equals mean total pulmonary venous flow into the left atrium.

It is reasonable to assume that total mean flow into the pulmonary arterial bed equals the flow out. Therefore the dissipated mean hydraulic power is

$$\begin{array}{l}\text{dissipated mean energy}\\ \text{unit time across pulmonary}\\ \text{arterial vasculature}\end{array} = W_D = (P_m - P_c)Q_m \tag{3.15}$$

where P_c equals the mean capillary pressure.

The consensus is that the majority of pressure pulsatility is lost in the arterial precapillary region and only minimally transmitted to the pulmonary veins and left atrium (Bergel and Milnor 1965, Caro and McDonald 1961, Maloney et al. 1968, Milnor 1972a). Although some flow pulsatility is transmitted through the capillaries (Guntheroth et al. 1974, Karatzas and Lee 1970) even with normal arterioles, the flow pulsations are markedly damped on the arterial side, so that at least 75% of pulsatile energy is dissipated. Using impedance measurements, Milnor, Bergel, and Bargainer (1966) demonstrated in dogs that 78% of total input energy was dissipated during transit through the pulmonary circulation. Guntheroth and colleagues (1974) noted that arterial vasoconstriction attenuates the transmitted flow pulsations even more; although this is an acute response, a similar occurrence may be seen in chronically elevated resistance. Reuben and co-workers (1971) demonstrated decreased transmission of flow pulsatility to precapillary regions in patients with chronic pulmonary hypertension. O'Rourke (1967) has shown an increase in the ratio of pulsatile to total work in the systemic circulation when arterial distensibility is reduced. Elkins, Peyton, and Greenfield (1974) have demonstrated a doubling of the pulsatile power in the pulmonary circulation of dogs with elevated characteristic impedance due to chronic pulmonary hypertension, and Milnor (1972a) has measured similar effects in humans.

Total energy dissipation on the arterial side of the circulation is the pulsatile terms plus the mean pressure drop multiplied by the mean flow:

$$\begin{array}{l}\text{total energy dissipation}\\ \text{in pulmonary arteries}\end{array} = W_{DT}$$
$$= (Q_m) * (P_m - P_c) + W_o - W_{oc} \tag{3.16}$$

where W_{oc} is the oscillatory energy remaining in the capillaries after the midpoint and is probably less than 20% of W_o in the arteries. An "effective" pulmonary vascular resistance should be defined on the basis of the rate of energy dissipation within the pulmonary circulation (Skalak et al. 1966a).

Consequences of This Energy Dissipation

The fundamental stresses related to development of hypertensive vascular disease are the amount, rate, duration, and *location* of hydraulic energy dissipation in the vessels. Note that in Figure 3.4, although there is a gradual loss of pulsations in the proximal vessels, the greatest decrease tends to be in the immediate precapillary region, which could be called the resistance region. In the normal pulmonary vascular bed, the pulsatile components of flow and pressure are gradually damped and attenuated by the compliance characteristics of the proximal pulmonary arteries. In the pulmonary circulation, between 30% and 50% of the energy is pulsatile, in contradistinction to the aortic or systemic circulation, in which less than 5% is pulsatile (Mills et al. 1970; Milnor 1972a,b; Nichols et al. 1977). Thus in the normal pulmonary circulation, not only is the actual input energy almost double that estimated by mean pressure flow data, but most of this energy is dissipated proximal to the mid-capillary region.

Proximal Arterial Stiffness and Hydraulic Energy Dissipation

As demonstrated in Womersley's and Moens–Korteweg equations (Eq. 3.5 and 3.6), impedance increases as stiffness increases. Taper and increasing stiffness in vessels also increase the amplitude of the pressure pulse waveform (Fishman 1977). The rate of energy transmission by a sine wave is equal to the oscillatory pressure multiplied by the oscillatory flow (Caro, Pedley, Schroter, and Seed 1978). Since the pressure amplitude is related to impedance such that

$$|P|_n \propto K\sqrt{Z_n} \tag{3.17}$$

then increasing impedance increases the oscillatory pressure value. Factors that increase the stiffness of the proximal pulmonary vessels will decrease the proximal attenuation of the downstream pressure wave (hence pulsatile energy) and enhance transmission to the same precapillary "resistance" region where the majority of energy dissipation occurs. This amplifies the amount of *focal* energy dissipation. Increased viscous effects increase pressure wave damping (Taylor 1966a,b). Increased proximal stiffness tends to attenuate pulsatile flow waves, but this results in small decreases in the pulsatile flow energy dissipated in the distal region, since a greater proportion of the flow is in the mean terms. Increased stiffness of proximal pulmonary arteries is an active response to adrenergic stimulation; this has been clearly demonstrated experimentally in dogs (doubling of elastic modulus), and appears to be blocked by alpha blockade (Ingram et al. 1968). Whether active or passive processes play a role in the stress–strain relationship in humans has not been defined. Chronic increases in proximal stiffness, which apparently are not under adrenergic control, have been demonstrated as a response to chronic pulmonary venous hypertension in dogs (Hopkins et al. 1980) and in humans (Milnor et al. 1969).

Primary Stresses Associated With the Development of Pulmonary Vascular Disease Analyzed With Analogous Clinical Entities

Four factors frequently are associated with the development of pulmonary vascular disease:

1. Elevated pulmonary blood flow.
2. Elevated pulmonary artery pressures.
3. Elevated pulmonary venous pressures.
4. Hypoxemia.

Although it is difficult to isolate each of these factors in the clinical setting, examination of data from experimental and clinical studies can lead to formulation of a unifying theory of the pathogenesis of PVOD.

Elevated Pulmonary Blood Flow; Atrial Septal Defect

Elevated pulmonary blood flow is one of the most frequently studied factors in the etiology of pulmonary vascular disease, although by itself it seems to be a relatively well-tolerated hemodynamic stress which causes no pathologic hemodynamic changes in most individuals for long periods. The preceding section has demonstrated that the pulmonary vascular bed is, in effect, designed to tolerate elevated pulsatile blood flow (Fowler 1969, Fung and Sobin 1972, Marshall et al. 1961, Maseri et al. 1972, Milnor 1972a,b), although neonatal circulations may have reduced tolerance (vide infra) (Friedli, Kent, and Kidd 1975).

But how does the pulmonary vascular bed adapt to persistently and chronically elevated flow, and does it truly tolerate this stress over time? The congenital heart disease model for elevated blood flow is atrial septal defect.

The salient clinical characteristics of atrial septal defect are (Nadas and Fyler 1972):

1. Elevated pulmonary blood flow.
2. Initially normal pulmonary artery pressures.
3. Usually, low pulmonary vascular resistance.

4. No arterial desaturation of oxygen.
5. Pulmonary hypertension in ostium secundum defect (very rare in children, 10% in adults younger than age 40, and 30% in adults over age 40) (Haworth 1983b; Nadas and Fyler 1972).
6. Mitral regurgitation in ostium primum defect; seems to enhance the development of pulmonary vascular disease.
7. PVOD occurring in perhaps 6% of children and occasionally even in infancy (Bull et al. 1981).

EXPERIMENTAL PHYSIOLOGIC STUDIES

Measurement of pulmonary vascular impedance in awake dogs with chronically elevated pulmonary blood flow has revealed progressive reductions of the characteristic impedance and impedance at zero frequency (Z_m) or pulmonary vascular resistance (Hopkins et al. 1979). That the mechanical properties of the larger pulmonary arteries can change under physiologic stress has been repeatedly demonstrated (Ingram et al. 1968; Patel, DeFreitas, and Mallos 1962). The dynamic transmission characteristics of the proximal pulmonary vessels have a contributary role in the control of regional pulmonary flow patterns (Maloney et al. 1968). Chronic flow stress induces alterations in the physical properties of the pulmonary vascular system that are different from those seen in acute elevations of flow (Hammon et al. 1981, Hopkins et al. 1979). The impedance spectra for dogs with chronically elevated blood flow are illustrated in Figure 3.5. The hydraulic input power increases significantly for each increase in chronic pulmonary blood flow, but this increase is far greater in the mean than in the pulsatile terms. Hydraulic power returns to relatively normal after closure of the shunts. The energy consequences of this increased flow in terms of mean and pulsatile components are illustrated in Figure 3.6. Normalizing the work to flow reveals no increase in work per milliliter of blood flow in the pulsatile terms but a modest increase in the mean work (Fig. 3.7). The energy cost is higher for each milliliter of blood moved axially as chronic pulmonary blood flow increases. The right ventricular power required for radial arterial pulsations remains the same, however, for each milliliter of total flow. The energy consequences would be great if this reduction in impedance did not occur. Since mean work normalized to flow increases, this distal mechanism is not perfect. Increased arterial compliance and dilatation also reduce impedance to pulsatile flow. Stiffer vessels would augment the pressure and energy consequences of increased flow (Linde et al. 1972). If the mechanisms altering impedance did not function as flow increased, then the pressure flow energy used in creating pulsations would increase. The magnitude of this hypothetical increase is demonstrated in Figure 3.8, in which the increased potential work of pulsations is demonstrated for the three groups of dogs illustrated in Figure 3.5 with calculations utilizing the original impedance spectra (i.e., no impedance adaptation). Thus this proximal mechanism

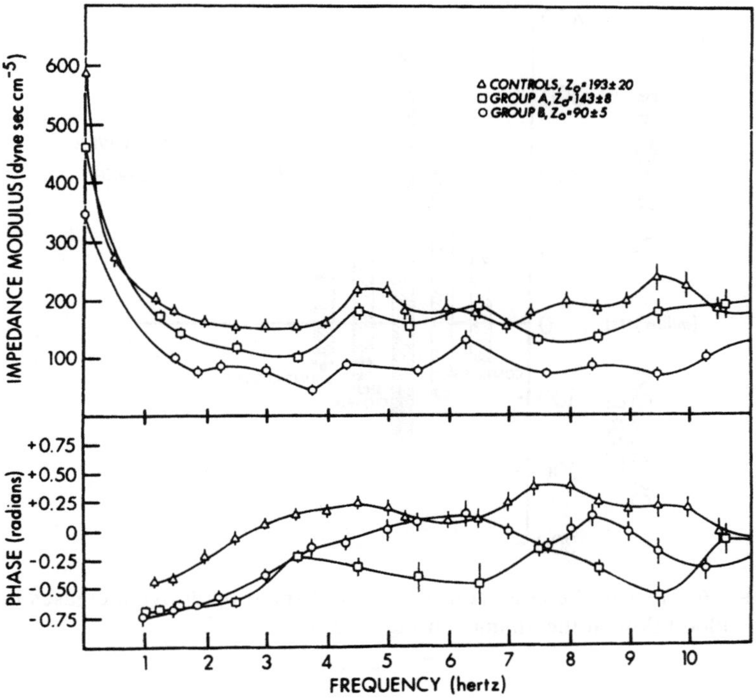

FIGURE 3.5. Impedance spectra for dogs with increased resting pulmonary blood flow as a consequence of surgically created peripheral arterial-venous fistulas. Control animals had average pulmonary blood flow of 2.0 ± 0.15 l/min, while the value for group A was 3.8 ± 0.09 l/min and for group B 5.87 ± 0.16 l/min. The decrease in characteristic impedance (Z_o) from control was significant for both Group A ($p<0.05$) and Group B ($p<0.0005$). Pulmonary vascular resistance also fell in these groups, from 416 ± 32 to 249 ± 6 (Group A) and to 126 ± 14 (Group B) dyne sec cm^{-5}. (From Hopkins et al. 1979; by permission of the American Heart Association, Inc.)

is of major significance in terms of reducing the energy transmitted from the right ventricle to the pulmonary vascular bed.

HUMAN PHYSIOLOGIC DATA

Impedance adaptation to flow correlates with the clinical fact that elevated pulmonary blood flow *alone* is usually a weak stimulus for pulmonary vascular disease (Oakley et al. 1962). Increased stiffness of the pulmonary artery wall in established hypertension has been demonstrated in humans (Jarmakani et al. 1971; Milnor et al. 1969; Patel, DeFreitas, and Mallos 1962). Patients with atrial septal defect with normal pulmonary arterial pressures have been shown to have normal elastic moduli, while patients

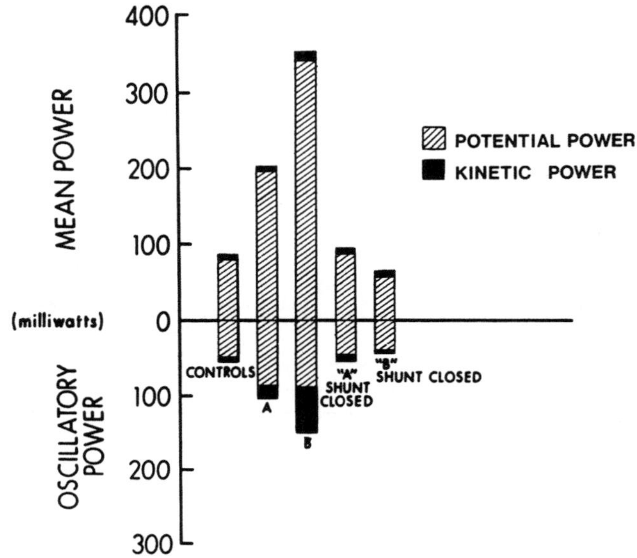

FIGURE 3.6. Measured energy consequences of the progressive increases in pulmonary blood flow in the animals in Figure 3.5.

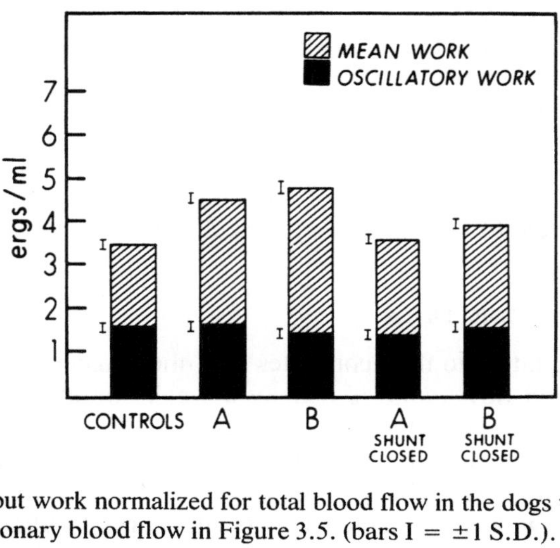

FIGURE 3.7. Input work normalized for total blood flow in the dogs with chronically increased pulmonary blood flow in Figure 3.5. (bars I = ±1 S.D.). (From Hopkins et al. 1979; by permission of the American Heart Association, Inc.)

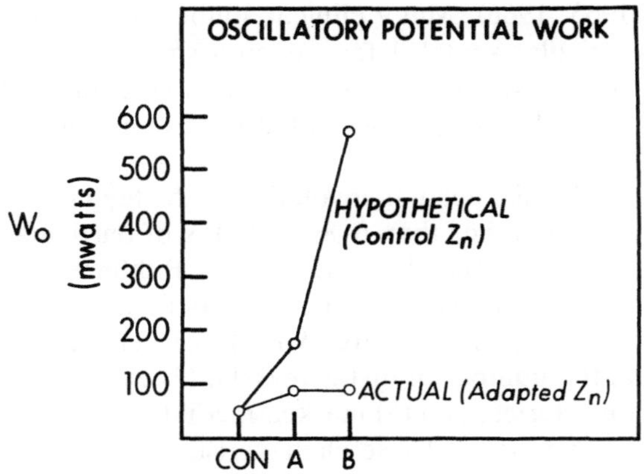

FIGURE 3.8. The calculated potential pulsatile energy (W_o) at increased chronic pulmonary blood flows if the impedance adaptation represented by the fall of the characteristic impedances in Figure 3.5. had not occurred.

with pulmonary hypertension and atrial septal defect have stiffer pulmonary vessels. An extremely interesting study was reported by Boughner and Roach (1971), in which *increased* compliance of the main pulmonary arteries was demonstrated in patients with normotensive atrial septal defect. This human response is analogous to the viscoelastic adaptation of the proximal pulmonary vessels demonstrated in canine impedance studies.

Reasons for Flow Tolerance

There are thus three mechanisms by which the pulmonary vasculature tolerates high blood flow:

1. Minimization of mean pressure elevation due to elevated flow by recruitment and distention of arteriolar-capillary vessels leading to decreased pulmonary vascular resistance (Z_m).
2. Minimization of total pulsatile energy input by a decrease in total impedance.
3. Minimization of pulsatile energy transmission to the distal target vessels by gradual attenuation of pulsations throughout compliant proximal pulmonary arteries by a reduction of characteristic impedance (Z_0).

Elevated Pulmonary Arterial Pressure; Ventricular Septal Defect

Ventricular septal defect causes pulmonary artery pressures to be elevated from the beginning. Natural history studies have documented a much more rapid development of PVOD in ventricular septal defect than in atrial septal defect (Clarkson et al. 1968; Collins et al. 1972; Friedli et al. 1974; Hoffman and Rudolph 1965, 1966). Many clinical studies have confirmed that patients with ventricular septal defect, even a small one, have a tendency toward pulmonary vascular disease (Bisse and Hirschfield 1983). Usually PVOD is prevented by early surgery, but this is not always true (Samaan 1970).

That there is a definite delay in involution of the fetal pulmonary vascular pattern (6–8 months rather than 6–8 weeks) is of particular interest (Haworth and Hislop 1983; Haworth, Sauer, and Bühlmeyer 1977; Hislop et al. 1975). Quantitative morphometrics have defined three characteristics of the pulmonary vessels in children with large ventricular septal defect: (1) reduction in the number of distal vessels, (2) extension of muscle into small peripheral arteries, and (3) marked medial hypertrophy of the small muscular preacinar vessels. These characteristic pathologic changes would be reflected physiologically as increased impedance to pulsatile flow in both the mean and oscillatory terms. Unopposed passive hemodynamic effects of ventricular septal defect would tend to increase both arterial pressure and radius by vessel dilatation. This is important as it results in an increase in wall tension as expressed by the Lame approximation for thin-walled tubes of the relationship of Laplace [tension = (pressure × radius)/height]. As the wall tension of blood vessels increases, the elastic modulus increases and the wall becomes stiffer (Bergel 1961; McDonald 1974; Patel, DeFreitas, and Mallos 1962; Patel, Schilder, and Mallos 1960). Thus, the active and passive responses would all lead to an increase in elastic modulus (E_y) and in characteristic impedance.

The changes noted in morphometric studies suggest a pulmonary arterial circulation that is less distensible and that has reduced recruitable vascular perfusion "radius" (i.e., arteriolar-capillary channels). This effectively decreases the attenuation characteristics of the more proximal vessels and increases the chronic distal hydraulic energy dissipation. There is thus a *positively reinforcing* mechanism by which greater energy dissipation causes increased morphologic changes, which in turn result in stiffer vessels and higher impedances.

Human studies have demonstrated that decreased proximal vascular compliance and pulmonary hypertension coexist, particularly in patients with ventricular septal defect and elevated pressures (Jarmakani et al. 1971). These alterations are independent of flow (Boughner 1971).

In large ventricular septal defect, both mechanisms that reduce energy dissipation in atrial septal defect are at a disadvantage in preventing increased energy dissipation across the target vessels. Both mean pressures and pulsatile pressures are elevated in ventricular septal defect hemodynamics. Although measurements of total (mean + pulsatile) energy dissipation in patients with ventricular septal defect have not been made, the type of calculations demonstrated in the preceding sections reveal that with elevated impedance, the hydraulic energy dissipation per unit time must be greatly increased. In summary:

1. Total input energy is much greater as a consequence of elevated flow and pressure (both mean and pulsatile).
2. Impedance is increased.
3. Total energy dissipation increases, since the pressure drop (both pulsatile and mean) at the resistance region is increased.
4. Increasing pressure and radius increase wall tension and hence wall stiffness (E).
5. Pressure elevations induce medial hypertrophy, which ultimately decreases compliance of larger pulmonary arteries (increased E and increased Z_o).
6. Stiffer proximal vessels increase transmission of pulsatile pressure energy distally.

Elevated Pulmonary Venous Pressure; Anomalous Pulmonary Venous Drainage

An understanding of the role of pulmonary venous hypertension is critical to an understanding of this theoretical approach. Pulmonary venous hypertension is clinically associated with such lesions as mitral stenosis, cor triatriatum, and anomalous pulmonary venous drainage—all three of which have a marked propensity for the development of pulmonary vascular disease (Dexter 1979, Hutchins and Ostrow 1976, Nadas and Fyler 1972). In addition, it is frequently associated with other hemodynamic stresses that seem to lead to rapid progression of pulmonary vascular disease (Babb et al. 1981, Plauth et al. 1970, Saalouke et al. 1977). Results of experimental open-chest *acute* studies of elevated pulmonary venous pressure have varied, but most have shown a fall in pulmonary vascular resistance (Borst et al. 1956, Lloyd and Schneider 1969). Thus, although an acute rise in left atrial pressure also causes a rise in pulmonary arterial pressure, the transpulmonary vascular pressure gradient usually narrows, implying decreased vascular resistance. The chronic situation is fundamentally different (Silove, Tavernor, and Berry 1972).

In such entities as total anomalous pulmonary venous drainage, pathologic studies have demonstrated a marked association with pulmonary vascular disease; the more severe the venous drainage obstruction, the more malignant and rapid the development of the vascular changes (Levy et al. 1965). Quantitative morphometric studies have demonstrated abnormal muscularization of the small arteries beginning shortly after birth, and in addition, marked medial hypertrophy of the more proximal vessels (Haworth and Reid 1977).

Pulsatile Studies

Chronic animal studies have demonstrated a rise in pulmonary vascular resistance in response to venous hypertension, but have usually limited

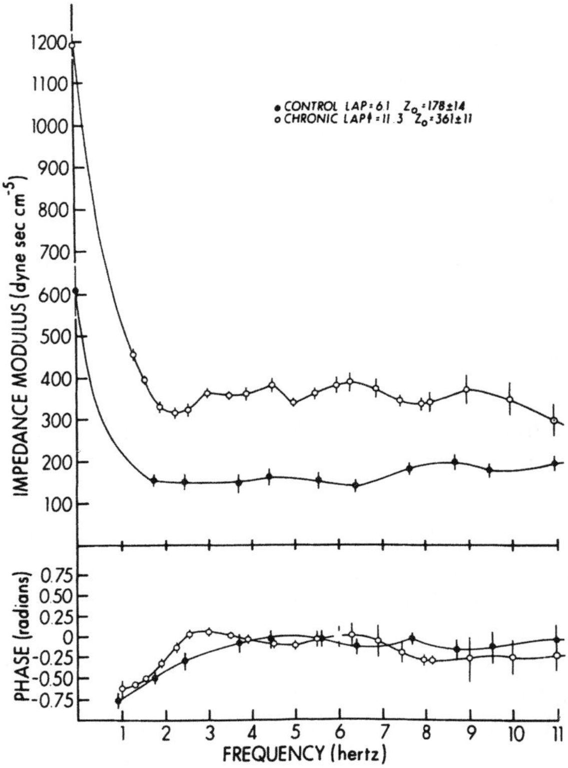

A

FIGURE 3.9A. The effect of chronically elevated left atrial pressure (LAP) in chronically instrumented dogs with an aorta-to-left-artium fistula. Shunts were created two months prior to measurements of pulmonary vascular impedance obtained in the awake, quietly resting state.

the hemodynamic analysis to mean pressure/flow and therefore have identified only "back-pressure" effects and precapillary arteriolar vasoconstriction (Silove, Tavernor, and Berry 1972). Experimental pulsatile studies of the response of the pulmonary vascular bed to *chronically* elevated left atrial pressure have demonstrated a markedly increased characteristic impedance, implying stiffer arteries (Hopkins et al. 1980, Figs. 3.9A,B). The markedly increased characteristic impedance ($P < 0.001$) correlates with a change in Young's elastic modulus (E_y) from 3.17 ± 0.02 to 13.00 ± 0.01 (dyne/cm^2 × 10^6). In contrast, acute elevations in pulmonary venous pressures do not alter characteristic impedance. Milnor and colleagues (1969) in an elegant study measured pulse wave velocity and pulsatile pressure/flow in patients with pulmonary hypertension as a consequence of mitral stenosis. They demonstrated elevations in wave velocity, a doubling of the characteristic impedance, and related energy

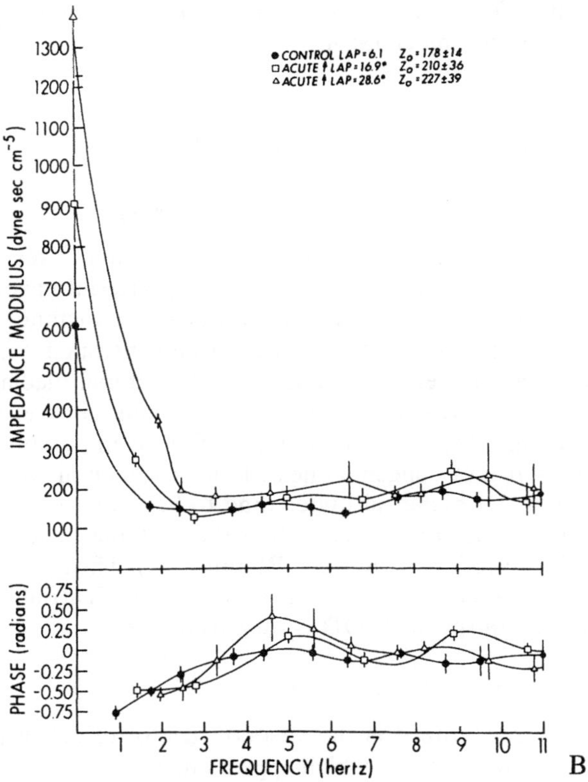

FIGURE 3.9B. Characteristic impedance (Z_o) in acute left atrial pressure elevations in awake, chronically instrumented dogs. (From Hopkins et al. 1980; by permission of the American Heart Association, Inc.)

consequences—a doubling of pulsatile power input and a tripling of the total power input. They described one patient with elevated characteristic impedance and normal pulmonary vascular resistance but with elevated pulsatile input energy, suggesting a stage of proximal arterial stiffening preceding elevated resistance. Reuben (1971) has also shown reduced compliance of pulmonary arteries in patients with mitral valve disease and chronic pulmonary hypertension.

The potent effect of chronic pulmonary venous hypertension on pulmonary arterial impedance and arterial wall stiffness may explain why associated venous hypertension seems to potentiate the rate and degree of progression of PVOD. If, as shown in experimental and human data, even modest elevations of left atrial pressure cause marked elevations in characteristic impedance and stiffening of proximal arterial walls, then mixed congenital lesions, which include mitral regurgitation, mitral stenosis, or left ventricular failure with elevated end-diastolic pressure, might demonstrate this effect, causing a loss of the proximal pulmonary arterial compliance adaptation mechanism.

Hypoxemia and Hypoxia; Shunts and Altitude

Pulmonary Vascular Resistance and Hypoxia

Hypoxemia is frequently present in congenital cardiac disease. Hypoxia has long been known to be a potent pulmonary arterial vasoconstrictor, characterized by elevations in pulmonary vascular resistance (i.e., distal vasoconstriction). Chronic hypoxia has been demonstrated experimentally to induce pulmonary hypertension and muscular arterial medial thickening (Abraham et al. 1971, Hislop and Reid 1976, Rabinovitch et al. 1979).

The potentiating effect of hypoxia has in fact been demonstrated clinically in the presence of congenital heart lesions. Children with ventricular septal defect living at higher altitudes tend to have higher resistance and earlier pulmonary vascular disease than those living at lower altitudes (Blount 1977; Vogel, McNamara, and Blount 1967). The influence of moderate altitude-induced chronic hypoxia in patients with atrial septal defect is to stimulate the onset of PVOD at an earlier age.

Clinical Relevance of Hypoxemia

Less universally recognized is the vasoconstrictor effect of precapillary pulmonary arterial hypoxemia, which appears to be independent of alveolar oxygen concentration. This has been shown experimentally in dogs (Bergofsky et al. 1963).

Fishman (1977) has summarized experimental and clinical data indicating that hypoxemia, either systemic or pulmonary, can contribute to pulmonary artery vasoconstriction. Systemic hypoxemia has been shown to produce pulmonary arterial vasoconstriction through reflex mechanisms (Bergofsky and Holtzman 1967, Bergofsky et al. 1963). A resistance site sensitive to hypoxia and arterial hypoxemia was shown to be closely associated with the precapillary arteriolar region (the same region of preacinar vessels where the earliest changes of pulmonary vascular disease occur).

Systemic-pulmonary shunts have been utilized to treat infants with cyanotic congenital heart disease to improve oxygen saturation in the systemic circulation. Since Blalock's first experiments, it has been thought that this shunt resulted in pulmonary vascular disease in only a very small percentage of patients (Blalock and Taussig 1945; Cole et al. 1971; Ross, Taussig, and Evans 1958). Quantitative morphometric analyses have in fact demonstrated early changes missed by classical techniques (Rendas, Branthwaite, and Reid 1978; Rendas, Lennox, and Reid 1979; Rendas and Reid 1983). The larger shunts, such as the Potts and Waterston, are known to cause a greater incidence of pulmonary vascular changes, but this has been attributed to the larger shunt size and hence greater input of hydraulic energy (Kaplan et al. 1968). Newfeld and colleagues (1977) have found that adequate relief of cyanosis diminishes the occurrence of vascular changes, whatever the type of shunt, which suggests that improved systemic oxygenation modifies the effects of increased pressure and flow within the pulmonary vascular bed. This is an interesting concept, that inadequate relief of hypoxemia could result in a hypoxemic proximal and distal vasoconstrictor response (increased impedance), which would exacerbate the tendency for a systemic shunt to provoke vascular disease.

Cyanosis can elicit pulmonary vascular constriction of the precapillary constrictor site in a response similar to that induced by alveolar hypoxia, but due to decreased oxygen saturations on the right side of the pulmonary circulation. Patients with transposition have wedged pulmonary arterial oxygen saturations similar to the lower systemic saturations, indicating systemic-to-pulmonary anastomoses (Aziz, Paul, and Rowe 1977). Other congenital heart lesions have bronchopulmonary and aortopulmonary collateral circulation pathways, which could establish the systemic level of hypoxemia in this precapillary region of oxygen sensitivity and vasoconstriction (Haworth et al. 1981, Rabinovitch et al. 1981).

Proximal and Distal "Vasoconstriction" Induced by Hypoxia–Hypoxemia

There is evidence that both proximal and distal vasoconstriction in pulmonary arteries is due to contraction of the vascular smooth muscle which, with chronic exposure, becomes increasingly hypertrophied (Fishman 1977, Rabinovitch et al. 1981, Szidon and Fishman 1969). A number of experimenters have reported that in addition to precapillary hypoxic vasoconstriction, there is markedly increased *proximal pulmonary arterial stiffness* induced in intact dogs exposed to hypoxia (Friedman and Wanner 1977; Morgan, Church, and Guntheroth 1968). Reuben and colleagues (1970, 1971) demonstrated a 30% fall in pulmonary arterial compliance in dogs breathing a 10% oxygen mixture, together with major alterations in the pulmonary arterial impedance spectrum. Thus the vasoconstrictor hypoxic response occurs both distally and proximally. Distal arteriolar vasoconstriction causes increased pulmonary resistance, while in the more proximal vessels stiffness increases. The result is an increase in impedance to both mean and pulsatile flow.

These experimental studies and the clinical data all suggest that not only alveolar hypoxia but also mixed venous and arterial hypoxemia are capable of increasing the impedance in the pulmonary circulation to both mean and pulsatile flow (Aziz, Paul, and Rowe 1977). This would have the effect previously described of increasing total energy input and transmission to the precapillary region and hence dissipation in the target vessel region.

Additive Effects: Complex Congenital Lesions

One of the complex congenital defects most closely associated with development of PVOD is transposition of the great arteries, both with and without ventricular septal defect (Clarkson et al. 1976, Macartney et al. 1981). Pulmonary vascular disease typically develops early and progresses rapidly. Discouragingly, even an early operation (within 3 months) cannot always prevent it. Pulmonary vascular resistance that increases prior to surgery has a devastating effect on perioperative mortality (Lindesmith et al. 1975, Macartney et al. 1981). The hemodynamic characteristics of transposition relevant to this discussion are:

1. Elevated pulmonary blood flow.
2. Systemic pressures in the pulmonary artery.
3. Hypoxemia due to shunting.
4. Elevated left atrial pressure.

Thus transposition of the great vessels with ventricular septal defect is associated with all four factors important in the development of pulmonary vascular disease, and all of these factors would tend to override the normal compensatory mechanisms for attenuating the increased energy input into the distal pulmonary arterial bed.

Other complex congenital lesions associated with early onset of pulmonary vascular disease include atrioventricular septal defect (Artman et al. 1984), aortopulmonary window, septal defect with patent ductus arteriosus (Reid, Coleman, and Stevenson 1979) and truncus arteriosus (Ebert et al. 1976). These lesions are associated with elevated pulmonary arterial pressures and flow as well as modest increases in left atrial pressure. However, as was seen in experimental data, even mild elevations in left atrial pressure can cause markedly increased proximal arterial stiffness. Evaluation of complex lesions predisposing to PVOD reveals complex interactions of various hemodynamic factors, all having the effect of increasing the energy dissipation in the distal arteriolar region. Whether increased susceptibility plays a role has not been defined, but there is suggestive evidence in the case of Down's syndrome.

Developing Circulation—Neonatal Versus Later Stress

In understanding the development of pulmonary vascular disease associated with congenital heart disease, it must be recognized that the stresses are usually present at a time when the pulmonary vascular bed is undergoing dramatic perinatal changes (Rudolph 1970). The neonatal pulmonary circulation is less tolerant of stresses than is the adult circulation (Wagenvoort 1960). Even elevation of flow, as in congenital unilateral atresia of a pulmonary artery, seems less well tolerated (Pool, Vogel, and Blount 1962). Acute neonatal pulmonary hypertension, a result of abnormal hemodynamic stresses, must be identified separately from so-called "persistent fetal circulation" syndrome (Gersony 1984). Experiments in adult animals are perhaps not relevant to the situation in which the stresses are present from birth (Friedli, Kent, and Kidd 1975; Rendas, Branthwaite, and Reid 1978; Rendas, Lennox, and Reid 1979; Rendas and Reid 1983). Within hours of birth, pulmonary vascular resistance falls to 10% of the fetal value and progressively declines thereafter to a pulmonary to systematic resistance ratio of approximately 0.18 by 2 weeks of age (Haworth 1981a). Further maturation occurs until the age of 4 to 8 years. The pulmonary arterial medial thickness relative to external diameter decreases during the first 3 weeks of extrauterine life and continues to regress until 18 months of age (Haworth and Hislop 1981). The rapidity of regression of the medial hypertrophy changes, more rapid changes occurring in the smallest muscular arteries.

While pulmonary vascular impedance has rarely been measured in human neonates, such measurements in pigs demonstrate a correlative fall in the spectra and Z_0 (Figs. 3.10A,B; Greenwald, Johnson, and Haworth 1985) as the piglets mature. These important experiments demonstrate that after the first 24 hours there is a progressive fall in Z_0 and a thinning of muscle mass. This results in a fall in power input and a corresponding decrease in the ratio of pulsatile to steady power that is not a function simply of heart rate but represents structural vessel changes (Table 3.2). Thus the pulmonary circulation has a rapid phase of adaptation followed by a slower remodeling and growth. Quantitative morphometric analysis has shown that stresses applied during the neonatal period associated with the development of pulmonary vascular disease tend to retard the involution of proximal medial hypertrophy and to exacerbate the extension of muscularization to the smaller arterioles (Haworth 1981a, 1983a). In addition, the arteriolar-to-alveolar ratio can be reduced in some conditions (Rabinovitch and Reid 1981; Rabinovitch et al. 1979, 1984). All of this would increase impedance and vascular stiffness and exacerbate the distal regional energy dissipation.

Progression of Pulmonary Vascular Disease Following Repair of Congenital Cardiac Lesions

Perioperative morbidity and mortality increase in cases complicated by pulmonary hypertension. Pulmonary vascular disease has been demonstrated to regress following repairs of ventricular septal defect, but also to progress (Clarkson et al. 1976, Friedli et al. 1974; Rosengart, Fishbein, and Emmanouilides 1975; Sigmann et al. 1977).

Progressive pulmonary vascular disease after atrial-level correction for transposition of the great arteries has been documented in many cases (Berman et al. 1978, Yamaki and Horiuchi 1979). As previously described, once proximal vessels lose the compliance adaptability for flow, a positive

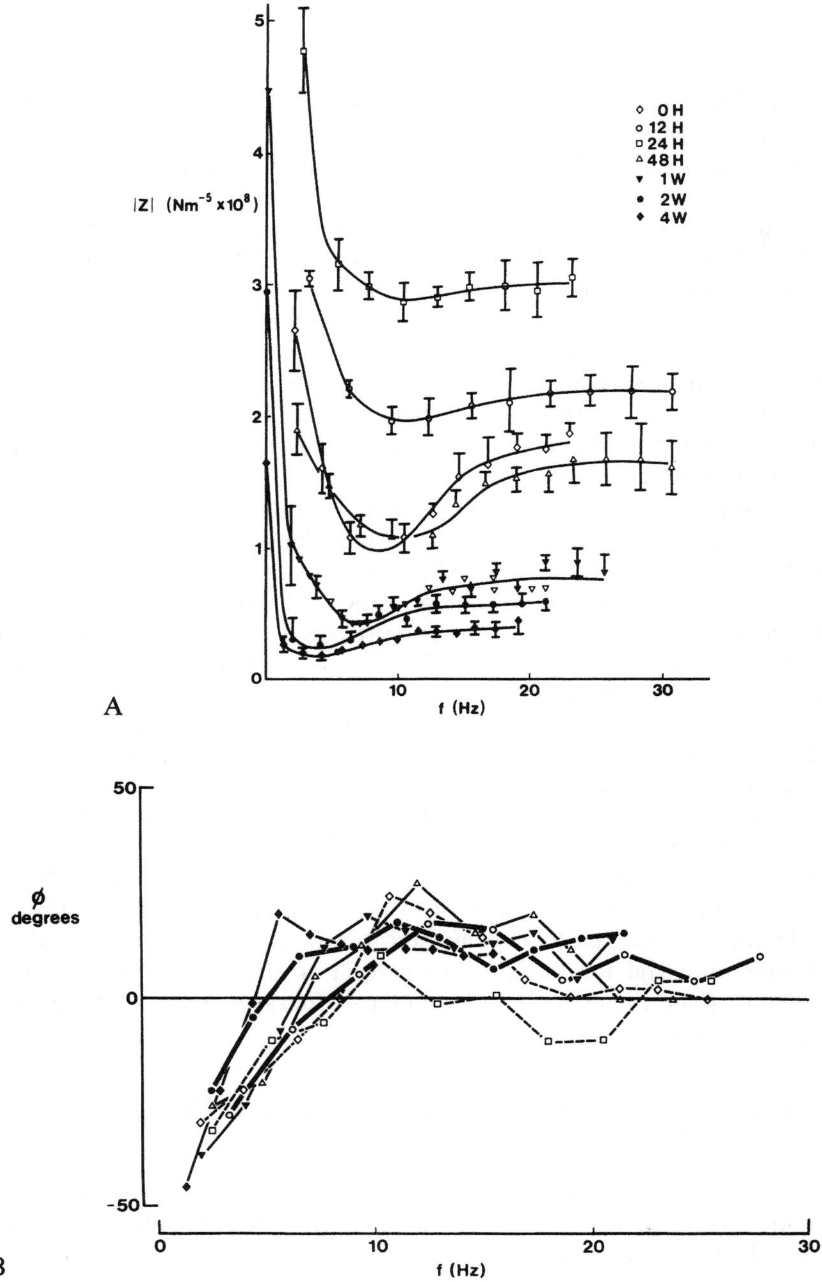

FIGURE 3.10. Data obtained in neonatal pigs showing progressive rapid fall in pulmonary vascular impedance with normal maturation of the pulmonary vessels over the first few weeks of life. This fall demonstrates a decrease not only in pulmonary vascular resistance but also in characteristic impedance. CGS units (dyne-sec cm^{-5} = SI units × 10^{-5}) (N-sec m^{-5}). A: modulus; B: phase. [From Greenwald, Johnson and Haworth 1985; by permission.]

TABLE 3.2. Pulmonary vascular impedance in piglets.

Age	Characteristic impedance (Newton sec m^{-s})	Total input power (mW · Kg^{-1})[a]	Pulsatile/steady input power ratio
1 hr	1.90 ± 0.31	10.4	0.50
12 hr	2.10 ± 0.81	18.9	0.56
24 hr	3.00 ± 0.33	8.7	0.65
48 hr	1.83 ± 0.19	14.6	0.69
1 wk	0.73 ± 0.05	7.1	0.46
2 wk	0.69 ± 0.03	5.4	0.39
4 wk	0.36 ± 0.02	4.8	0.31

[a]mW = milliwatt
Adapted from Greenwald, Johnson, and Hayworth 1985.

feedback process begins. For example, in the case of transposition, although correction reduces the cyanosis, there would not necessarily be any immediate change in the wall thickness of the pulmonary artery, and left atrial pressures tend to be modestly elevated following atrial repair. The stiffer pulmonary arteries could continue to cause increased energy dissipation in the smaller arterioles, invoking progressive vascular disease.

Summary of the Role of Regional Energy Dissipation in the Development of Pulmonary Vascular Disease

The following assumptions regarding energy dissipation form a unifying physiologic concept for the pathogenesis of PVOD.

1. Dissipation of hydraulic energy is the mechanism of injury and is an important stimulus for structural, biochemical, and humoral responses.
2. The higher the arterial impedance, the greater the total energy input into the pulmonary circulation (for a given flow).
3. Pulmonary arterial input impedance or the total opposition to pulsatile flow is affected by proximal vascular geometry and viscoelastic wall properties as well as by distal vasoconstriction, capillary perfusion density, and wave reflections.
4. The stiffer the proximal arteries, the greater the transmission of pulsatile pressure and flow energy to the distal arterioles and the more focal the regional energy dissipation.
5. Modeling the pulmonary circulation only in terms of mean flow/pressure loses information and ignores 30% to 50% of energy input.

This approach postulates two general "sites" of vasomotor activity in the pulmonary vascular bed: (1) a distal arteriolar region, which is assessed by pulmonary vascular resistance (Z_m) and is where the earliest morphologic changes of pulmonary vascular disease occur, and (2) the proximal pulmonary arteries, which can be assessed by characteristic impedance

(Z_0) and which powerfully affect the total input energy, the ratio of mean-to-pulsatile components, and total hydraulic energy transmission distally. The development of PVOD occurs with congenital cardiac defects in proportion to the effects of the hemodynamics on the regional distribution and rate of energy dissipation.

Bibliography

Abbott, M. E., and Dawson, W. T. 1924. The clinical classification of congenital cardiac disease with remarks upon its pathological anatomy, diagnosis and treatment. *Intl. Clin.* 34:156–188.

Abraham, A. S., Kay, J. M., Cole, R. B., and Pincock, A. C. 1971. Haemodynamic and pathological study of the effect of chronic hypoxia and subsequent recovery of the heart and pulmonary vasculature of the rat. *Cardiovasc. Res.* 15:95–102.

Artman, M., Parrish, M. D., Boerth, R. C., Boucek, R. J. Jr., and Graham, T. P. Jr. 1984. Short-term hemodynamic effects of hydralazine in infants with complete atrioventricular canal defects. *Circulation* 5:949–954.

Attinger, E. O. 1963. Pressure transmission in pulmonary arteries related to frequency and geometry. *Circ. Res.* 12:623–641.

Aziz, K. U., Paul, M. H., and Rowe, R. D. 1977. Bronchopulmonary circulation in d-transposition of the great arteries: possible role in genesis of accelerated pulmonary vascular disease. *Am. J. Cardiol.* 39:432–438.

Babb, J. D., McGlynn, T. J., Pierce, W. S., and Kirkman, P. M. 1981. Isolated partial anomalous venous connection: a congenital defect with serious complications. *Ann. Thorac. Surg.* 31:540–543.

Bargainer, J. D. 1967. Pulse wave velocity in the main pulmonary artery of the dog. *Circ. Res.* 20:630–637.

Baumgartner, E. A., and Abbott, M. E. 1929. Interventricular septal defect with dextroposition of aorta and dilatation of the pulmonary artery ("Eisenmenger complex") terminating by cerebral abscess. *Am. J. Med. Sci.* 177:639–647.

Bergel, D. H. 1961. The static elastic properties of the arterial wall. *J. Physiol.* 156:445–452.

Bergel, D. H., and Milnor, W. R. 1965. Pulmonary vascular impedance in the dog. *Circ. Res.* 16:401–415.

Bergofsky, E. H. 1974. Mechanisms underlying vasomotor regulation of regional pulmonary blood flow in normal and disease states. *Am. J. Med.* 57:378–394.

Bergofsky, E. H., Bass, B. G., Ferretti, R., and Fishman, A. P. 1963. Pulmonary vasoconstriction in response to precapillary hypoxemia. *J. Clin. Invest.* 42:1201–1215.

Bergofsky, E. H., Haas, F., and Porcelli, R. 1968. Determination of the sensitive vascular sites from which hypoxia and hypercapnia elicit rises in pulmonary arterial pressure. *Fed. Proc.* 27:1420–1425.

Bergofsky, E. H., and Holtzman, S. 1967. A study of the mechanisms involved in the pulmonary arterial pressor response to hypoxia. *Circ. Res.* 20:506–519.

Berman, W. Jr., Whitman, V., Pierce, W. S., and Waldhausen, J. A. 1978. The development of pulmonary vascular obstructive disease after successful Mustard operation in early infancy. *Circulation* 58:181–185.

Bisse, G. S., and Hirschfield, S. S. 1983. Severe pulmonary hypertension associated with a small ventricular septal defect. *Circulation* 67:470–473.

Blalock, A., and Taussig, H. B. 1945. The surgical treatment of malformations of the heart in which there is pulmonary stenosis or pulmonary atresia. *JAMA* 128:189–202.

Blank, R. H., Muller, W. H. Jr., and Dammann, J. F. Jr. 1961. Experimental pulmonary arterial hypertension. *Am. J. Surg.* 101:143–153.

Blount, S. G. Jr. 1977. Comparison of patients with ventricular septal defect at high altitude and sea level. *Circulation* 56 (Suppl. I):79–82.

Borst, H. G., McGregor, M., Whittenberger, J. L., and Berglund, E. 1956. Influence of pulmonary arterial and left atrial pressures on pulmonary vascular resistance. *Circ. Res.* 9:393–399.

Boughner, D. R. 1971. Pressure-radius relationships (letter). *Circulation* 64:491–492.

Boughner, D. R., and Roach, M. 1971. Etiology of pulmonary artery dilatation and hilar dance in atrial septal defect. *Circ. Res.* 28:415–425.

Brammell, H. L., Vogel, J. H. K., Pryor, R., and Blount, S. G. Jr. 1971. The Eisenmenger syndrome: a clinical and physiologic reappraisal. *Am. J. Cardiol.* 28:679–692.

Brody, J. S., Stemmler, E. J., and DuBois, A. B. 1968. Longitudinal distribution of vascular resistance in the pulmonary arteries, capillaries and veins. *J. Clin. Invest.* 47:783–799.

Bull, C., Deanfield, J., de Leval, M., Stark, J., Taylor, J. F. N., and Macartney, F. J. 1981. Correction of isolated secundum atrial septal defect in infancy. *Arch. Dis. Child.* 56:784–786.

Byar, D., Fiddian, R. V., Querean, M., Hobbs, J. T., and Edwards, E. A. 1965. The fallacy of applying the Poiseuille equation to segmental arterial stenosis. *Am. Heart J.* 70:216–224.

Caro, C. G., and Harrison, G. K. 1962. Observations on pulse wave velocity and pulsatile blood pressure in the human pulmonary circulation. *Clin. Sci.* 23:317–329.

Caro, C. G., and McDonald, D. A. 1961. The relation of pulsatile pressure and flow in the pulmonary vascular bed. *J. Physiol.* 117:426–453.

Caro, C. G., Pedley, T. J., Schroter, R. C., and Seed, W. A. 1978. *The Mechanics of the Circulation.* Oxford, New York, Toronto: Oxford University Press, p. 30.

Chi, T. P., and Krovetz, J. 1975. The pulmonary vascular bed in children with Down's syndrome. *J. Pediatr.* 86:533–538.

Clarkson, P. M., Frye, R. L., DuShane, J. W., Burchell, H. B., Wood, E. H., and Weidman, W. H. 1968. Prognosis for patients with ventricular septal defect and severe pulmonary vascular obstructive disease. *Circulation* 38:129–135.

Clarkson, P., Neutze, J. M., Wardill, J. C., and Barratt-Boyes, B. G. 1976. The pulmonary vascular bed in patients with complete transposition of the great arteries. *Circulation* 53:439–443.

Cole, R. B., Muster, A. J., Fixler, D. E., and Paul, M. H. 1971. Long-term results of aortopulmonary anastomosis for tetralogy of Fallot: morbidity and mortality (1946–1969). *Circulation* 63:263–271.

Collins, G., Calder, L., Rose, V., Kidd, L., and Keith, J. O. 1972. Ventricular septal defect: clinical and hemodynamic changes in the first five years of life. *Am. Heart J.* 84:695–705.

Davies, G., and Reid, L. 1970. Growth of the alveoli and pulmonary arteries in children. *Thorax* 25:669–681.

Dexter, L. 1979. Pulmonary vascular disease in acquired and congenital heart disease. *Arch. Intern. Med.* 139:922–928.

DiSesa, V. J., Cohn, L. H., and Grossman, W. 1983. Management of adults with congenital bidirectional cardiac shunts, cyanosis and pulmonary vascular obstruction: successful operative repair in 3 patients. *Am. J. Cardiol.* 51:1495–1497.

Dobrin, P. B. 1978. Mechanical properties of arteries. *Physiol. Rev.* 58:397–460.

Dugard, A., and Naimark, A. 1967. Effect of hypoxia on distribution of pulmonary blood flow. *J. Appl. Physiol.* 23:663–671.

Ebert, P. A., Robinson, S. J., Stanger, P., and Engle, M. A. 1976. Pulmonary artery conduits in infants younger than six months of age. *J. Thorac. Cardiovasc. Surg.* 72:351–356.

Eisenmenger, V. 1897. Congenital defects of the ventricular septum. *Z. Klin. Med. Suppl.* 32:1–28.

Elkins, R. C., Peyton, M. D., and Greenfield, L. J. 1974. Pulmonary vascular impedance in chronic pulmonary hypertension. *Surgery* 76:57–64.

Esterly, J. A., Glagov, S., and Ferguson, D. J. 1968. Morphogenesis of intimal obliterative hyperplasia of small arteries in experimental pulmonary hypertension. *Am. J. Pathol.* 52:325–337.

Fishman, A. P. 1977. Hypoxia on the pulmonary circulation: how and where it acts. *Am. Heart Assoc. Mono.* 54:17–27.

Folkow, B. 1971. The haemodynamic consequences of adaptive structural changes of the resistance vessels in hypertension. *Clin. Sci.* 41:1–12.

Fowler, N. O. 1969. The normal pulmonary arterial pressure-flow relationships during exercise (editorial). *Am. J. Med.* 47:1–6.

Friedli, B., Kent, G., and Kidd, L. B. S. 1975. The effect of increased pulmonary blood flow on the pulmonary vascular bed in pigs. *Pediatr. Res.* 9:547–553.

Friedli, B., Kidd, L. B. S., Mustard, W. T., and Keith, J. D. 1974. Ventricular septal defect with increased pulmonary vascular resistance: late results of surgical closure. *Am. J. Cardiol.* 33:403–409.

Friedman, M., and Wanner, A. 1977. Volume characteristics of extra- and intraparenchymal segments of the canine pulmonary artery. *J. Appl. Physiol.* 42:519–524.

Fry, D. L. 1968. Acute vascular endothelial changes associated with increased blood velocity gradients. *Circ. Res.* 22:165–197.

Fry, D. L., Patel, D. J., and De Freitas, F. M. 1962. Relation of geometry to certain aspects of hydrodynamics in larger pulmonary arteries. *J. Appl. Physiol.* 17:492–496.

Fung, Y. C., and Sobin, S. S. 1972. Pulmonary alveolar blood flow. *Circ. Res.* 30:470–490.

Gersony, W. M. 1984. Neonatal pulmonary hypertension: pathophysiology, classification, and etiology. *Clin. Perinatol.* 11:517–524.

Gessner, U. 1972. Vascular input impedance. In Bergel, D. H., ed. *Cardiovascular Fluid Dynamics,* Vol. 1. London: Academic, pp. 315–349.

Gow, B. S. 1972. The influence of vascular muscle on the viscoelastic properties of blood vessels. In Bergel, D. H., ed. *Cardiovascular Fluid Dynamics,* Vol. 2. London: Academic, pp. 66–110.

Gow, B. S., and Taylor, M. G. 1968. Measurement of viscoelastic properties of arteries in the living dog. *Circ. Res.* 23:111–122.

Graham, T. P. Jr. 1979. The Eisenmenger reaction and its management. In Roberts, W. C., ed. *Congenital Heart Disease in Adults*. Philadelphia: Davis, pp. 531–542.

Greenfield, J. C. Jr., and Griggs, D. M. Jr. 1963. Relation between pressure and diameter in main pulmonary artery of man. *J. Appl. Physiol.* 18:557–559.

Greenwald, S. E., Johnson, R. J., and Haworth, S. G. 1985. Pulmonary vascular input impedance in the newborn and infant pig. *Cardiovasc. Res.* 19:44–50.

Grover, R. F., Reeves, J. T., and Blount, S. G. 1961. Tolazoline hydrochloride (Priscoline): an effective pulmonary vasodilator. *Am. Heart J.* 61:5–15.

Grover, R. F., Vogel, J. H. K., Averill, K. H., and Blount, S. G. Jr. 1963. Pulmonary hypertension: individual and species variability relative to vascular reactivity. *Am. Heart J.* 66:1–3.

Guntheroth, W. G., Gould, R., Butler, J., and Kinnen, E. 1974. Pulsatile flow in pulmonary artery, capillary, and vein in the dog. *Cardiovasc. Res.* 8:330–337.

Hallman, G. L., Cooley, D. A., Wolfe, R. R., and McNamara, D. G. 1964. Surgical treatment of ventricular septal defect associated with pulmonary hypertension. *J. Thorac. Cardiovasc. Surg.* 48:588–600.

Hammon, J. W. Jr., Smith, P. K., McHale, P. A., Van Benthuysen, K. M., and Anderson, R. W. 1981. Analysis of pulsatile pulmonary artery blood flow in the unanesthetized dog. *Am. Physiol. Soc.* 805–813.

Haworth, S. G. 1981a. Normal pulmonary vascular development and its disturbance in congenital heart disease. In Godman, M. J., ed. *Pediatric Cardiology,* Vol. 4. Edinburgh: Churchill Livingstone, p. 50.

Haworth, S. G. 1981b. Normal structural and functional adaptation to extrauterine life. *J. Pediatr.* 98:911–918.

Haworth, S. G. 1983a. Morphologic changes and progression of pulmonary vascular disease. In Anderson, R. H., Macartney, F. J., Shinebourne, E. A., and Tynan, M., eds. *Pediatric Cardiology,* Vol. 5. Edinburgh: Churchill Livingstone, pp. 51–61.

Haworth, S. G. 1983b. Pulmonary vascular disease in secundum atrial septal defect in childhood. *Am. J. Cardiol.* 51:265–272.

Haworth, S. G., and Hislop, A. A. 1981. Adaptation of the pulmonary circulation to extra-uterine life in the pig and its relevance to the human infant. *Cardiovasc. Res.* 15:108–119.

Haworth, S. G., and Hislop, A. A. 1983. Pulmonary vascular development: normal values of peripheral vascular structure. *Am. J. Cardiol.* 52:579–583.

Haworth, S. G., Rees, P. G., Taylor, J. F. N., Macartney, F. J., de Leval, M., and Stark, J. 1981. Pulmonary atresia with ventricular septal defect and major aortopulmonary collateral arteries: effect of systemic pulmonary anastomosis. *Br. Heart J.* 45:133–141.

Haworth, S. G., and Reid, L. 1977. Structural study of pulmonary circulation and of the heart in total anomalous pulmonary venous return in early infancy. *Br. Heart J.* 39:80–92.

Haworth, S. G., Sauer, U., Bühlmeyer, K., and Reid, L. 1977. Development of the pulmonary circulation in ventricular septal defect: a quantitative structural study. *Am. J. Cardiol.* 40:781–788.

Heath, D., and Edwards, J. E. 1958. The pathology of hypertensive pulmonary vascular disease: a description of six grades of structural changes in the pulmonary arteries with special reference to congenital cardiac septal defect. *Circulation* 18:533–547.

Hislop, A., Haworth, S. G., Shinebourne, E. A., and Reid, L. 1975. Quantitative structural analysis of pulmonary vessels in isolated ventricular septal defect. *Br. Heart J.* 37:1014–1021.

Hislop, A., and Reid, L. 1976. New findings in pulmonary arteries of rats with hypoxia-induced pulmonary hypertension. *Br. J. Exp. Pathol.* 57:542–554.

Hoffman, J. I. E., and Rudolph, A. M. 1965. The natural history of ventricular septal defects in infancy. *Am. J. Cardiol.* 16:634–653.

Hoffman, J. I. E., and Rudolph, A. M. 1966. Increasing pulmonary vascular resistance during infancy in association with ventricular septal defect. *Pediatrics* 38:220–230.

Hoffman, J. I. E., Rudolph, A. M., and Heymann, M. A. 1981. Pulmonary vascular disease with congenital heart lesions: pathologic features and causes. *Circulation* 64:873–877.

Hopkins, R. A., Hammon, J. W. Jr., McHale, P. A., Smith, P. K., and Anderson, R. W. 1979. Adaptation to chronically elevated blood flow in the awake dog. *Circ. Res.* 45:267–274.

Hopkins, R. A., Hammon, J. W. Jr., McHale, P. A., Smith, P. K., and Anderson, R. W. 1980. Analysis of the pulsatile hemodynamic responses of the pulmonary circulation to acute and chronic pulmonary venous hypertension in the awake dog. *Circ. Res.* 47:902–910.

Hutchins, G. M., and Ostrow, P. T. 1976. The pathogenesis of the two forms of hypertensive pulmonary vascular disease. *Am. Heart J.* 92:797–803.

Ingram, R. H., Szidon, J. P., Skalak, R., and Fishman, A. P. 1968. Effects of sympathetic nerve stimulation on the pulmonary arterial tree of the isolated lobe perfused in situ. *Circ. Res.* 22:801–815.

Jarmakani, J. M. M., Graham, T. P. Jr., Benson, W. D., Canent, R. V. Jr., and Greenfield, J. C. Jr. 1971. In vivo pressure-radius relationships of the pulmonary artery in children with congenital heart disease. *Circulation* 43:585–592.

Kaplan, S., Helmsworth, J. A., Ahearn, E. N., Benzing, G. III, Daoud, G., and Schwartz, D. C. 1968. Results of palliative procedures for tetralogy of Fallot in infants and young children. *Ann. Thorac. Surg.* 5:489–497.

Karatzas, N. B., and Lee, G. de J. 1970. Instantaneous capillary blood flow in patients with heart disease. *Cardiovasc. Res.* 4:265–273.

Levy, A. M., Naeye, R. L., Tabakin, B. S., and Hanson, J. S. 1965. Far-advanced intimal proliferation and severe pulmonary hypertension secondary to total anomalous pulmonary venous drainage. *Am. J. Cardiol.* 16:280–285.

Li, J. K.-J., Melbin, J., Riffle, R. A., and Noordergraaf, A. 1981. Pulse wave propagation. *Circ. Res.* 49:442–452.

Linde, L. M., Goldberg, S. J., Momma, K., Awa, S., and Hall, V. E. 1972. Pulmonary vascular pressure response as a function of blood flow. *Cardiovasc. Res.* 6:415–422.

Lindesmith, G. G., Stanton, R. E., Lurie, P. R., Takahashi, M., Tucker, B. L., Stile, Q. R., and Meyer, B. W. 1975. An assessment of Mustard's operation as a palliative procedure for transposition of the great vessels. *Ann. Thorac. Surg.* 19:514–520.

Lloyd, T. C. Jr., and Schneider, A. J. L. 1969. Relation of pulmonary arterial pressure to pressure in the pulmonary venous system. *J. Appl. Physiol.* 27:489–497.

McDonald, D. A. 1974. *Blood Flow in Arteries.* 2nd Ed. Baltimore: Williams & Wilkins.

Macartney, F. J., Leanage, R., Agnetti, A., Graham, G. R., and Taylor, J. F. N. 1981. The fate of 144 patients with complete transposition after balloon septostomy. In Godman, M. J., ed. *Pediatric Cardiology,* Vol. 4. Edinburgh: Churchill Livingstone, pp. 417–429.

Maloney, J. E., Bergel, D. H., Glazier, J. B., Hughes, J. M. B., and West, J. B. 1968. Transmission of pulsatile blood pressure and flow through the isolated lung. *Circ. Res.* 23:11–24.

Marshall, R. J., Wang, Y., Semler, H. J., and Shepherd, J. T. 1961. Flow, pressure, and volume relationships in the pulmonary circulation during exercise in normal dogs and dogs with divided left pulmonary artery. *Circ. Res.* 9:53–59.

Maseri, A., Caldini, P., Harward, P., Joshi, R. C., Permutt, S., and Zierler, K. L. 1972. Determinants of pulmonary vascular volume: recruitment versus distensibility. *Circ. Res.* 31:218–228.

Mills, C. J., Gabe, I. T., Gault, J. H., Mason, D. T., Ross, J. Jr., Braunwald, E., and Shillingford, J. P. 1970. Pressure-flow relationships and vascular impedance in man. *Cardiovasc. Res.* 4:405–417.

Milnor, W. R. 1972a. Pulmonary hemodynamics. In Bergel D. H., ed. *Cardiovascular Fluid Dynamics,* Vol 2. London: Academic.

Milnor, W. R. 1972b. Pulsatile blood flow. *N. Engl. J. Med.* 287:27–34.

Milnor, W. R. 1982. *Hemodynamics*. Baltimore: Williams & Wilkins.

Milnor, W. R., Bergel, D. H., and Bargainer, J. D. 1966. Hydraulic power associated with pulmonary blood flow and its relation to heart rate. *Circ. Res.* 19:467–480.

Milnor, W. R., and Bertram, C. D. 1978. The relation between arterial viscoelasticity and wave propagation in the canine femoral artery in vivo. *Circ. Res.* 43:970–979.

Milnor, W. R., Conti, C. R., Lewis, K. B., and O'Rourke, M. F. 1969. Pulmonary arterial pulse wave velocity and impedance in man. *Circ. Res.* 25:637–649.

Morgan, B. C., Church, S. C., and Guntheroth, W. G. 1968. Hypoxic constriction of pulmonary artery and vein in intact dogs. *J. Appl. Physiol.* 25:356–361.

Moschowitz, E. 1927. Hypertension of the pulmonary circulation: its causes, dynamics and relation to other circulatory states. *Am. J. Med. Sci.* 174:388–406.

Nadas, A. S., and Fyler, D. C., eds. 1972. *Pediatric Cardiology*. 3rd Ed. Philadelphia: Saunders.

Newfeld, E. A., Waldman, J. D., Paul, M. H., Muster, A. J., Cole, R. B., Idriss, F., and Riker, W. 1977. Pulmonary vascular disease after systemic-pulmonary arterial shunt operations. *Am. J. Cardiol.* 39:715–720.

Nichols, W. W., Conti, C. R., Walker, W. E., and Milnor, W. R. 1977. Input impedance of the systemic circulation in man. *Circ. Res.* 40:451–458.

Oakley, C., Glick, G., Luria, M. N., Schreiner, B. F. Jr., and Yu, P. N. 1962. Some regulatory mechanisms of the human pulmonary vascular bed. *Circulation* 26:917–930.

O'Rourke, M. F. 1967. Steady and pulsatile energy losses in the systemic circulation under normal conditions and in simulated arterial disease. *Cardiovasc. Res.* 1:313–326.

Parker, F. Jr., and Weiss, S. 1936. The nature and significance of the structural changes in the lungs in mitral stenosis. *Am. J. Pathol.* 12:573–598.

Patel, D. J., De Freitas, F. M., and Fry, D. L. 1963. Hydraulic input impedance of aorta and pulmonary artery in dogs. *J. Appl. Physiol.* 18:134–140.

Patel, D. J., De Freitas, F. M., and Mallos, A. J. 1962. Mechanical function of the main pulmonary artery. *J. Appl. Physiol.* 17:205–208.

Patel, D. J., Schilder, D. P., and Mallos, A. J. 1960. Mechanical properties and dimensions of the major pulmonary arteries. *J. Appl. Physiol.* 15:92–96.

Pietro, D. A., LaBresh, K. A., Shulman, R. M., Folland, E. D., Parisi, A. F., and Sasahar, A. A. 1984. Sustained improvement in primary pulmonary hypertension during six years of treatment with sublingual isoproterenol. *N. Engl. J. Med.* 310:1032–1034.

Plauth, W. H. Jr., Nadas, A. S., Bernhard, W. F., and Fyler, D. C. 1970. Changing hemodynamics in patients with transposition of the great arteries. *Circulation* 42:131–142.

Pool, P. E., Vogel, J. H. K., and Blount, S. G. Jr. 1962. Congenital unilateral absence of a pulmonary artery. *Am. J. Cardiol.* 11:706–732.

Rabinovitch, M., Gamble, W., Nadas, A. S., Miettinen, O. S., and Reid, L. 1979. Rat pulmonary circulation after chronic hypoxia: hemodynamic and structural features. *Am. J. Physiol.* 236:H818–H827.

Rabinovitch, M., Haworth, S. G., Castaneda, A. R., Naas, A. S., and Reid, L. M. 1978. Lung biopsy in congenital heart disease: a morphometric approach to pulmonary vascular disease. *Circulation* 58:1107–1122.

Rabinovitch, M., Herrera-DeLeon, V., Castaneda, A. R., and Reid, L. 1981. Growth and development of the pulmonary vascular bed in patients with tetralogy of Fallot with or without pulmonary atresia. *Circulation* 64:1234–1249.

Rabinovitch, M., Keane, J. F., Norwood, W. I., Castaneda, A. R., and Reid, L. 1984. Vascular structure in lung tissue obtained at biopsy correlated with pulmonary hemodynamic findings after repair of congenital heart defects. *Circulation* 69:655–667.

Rabinovitch, M., and Reid, L. M. 1981. Quantitative structural analysis of the pulmonary vascular bed in congenital heart defects. In Engle, M. A., ed. *Pediatric Cardiovascular Disease*. Philadelphia: F. A. Davis Co., pp. 149–169.

Reid, J. M., Coleman, E. N., and Stevenson, J. G. 1979. Pulmonary hypertension accompanying ventricular septal defect and patent ductus arteriosus: management in infancy and early childhood. *Thorax* 34:454–458.

Rendas, A., Branthwaite, M., and Reid, L. 1978. Growth of pulmonary circulation in normal pig: structural analysis and cardiopulmonary function. *J. Appl. Phys.* 45:806–817.

Rendas, A., Lennox, S., and Reid, L. 1979. Aorta-pulmonary shunts in growing pigs: functional and structural assessment of the changes in the pulmonary circulation. *J. Thorac. Cardiovasc. Surg.* 77:109–118.

Rendas, A., and Reid, L. 1983. Pulmonary vasculature of piglets after correction of aorta-pulmonary shunts. *J. Thorac. Cardiovasc. Surg.* 85:911–916.

Reuben, S. R. 1971. Compliance of the human pulmonary arterial system in disease. *Circ. Res.* 29:40–50.

Reuben, S. R., Gersh, B. J., Swadling, J. P., and Lee, G. de J. 1970. Measurement of pulmonary arterial distensibility in the dog. *Cardiovasc. Res.* 4:473–481.

Reuben, S. R., Swadling, J. P., Gersh, B. J., and Lee, G. de J. 1971. Impedance and transmission properties of the pulmonary arterial system. *Cardiovasc. Res.* 5:1–9.

Rich, S., Ganz, R., and Levy, P. S. 1983. Comparative actions of hydralazine, nifedipine and amrinone in primary pulmonary hypertension. *Am. J. Cardiol.* 52:1104–1107.

Roach, M. R. 1972. Poststenotic dilatation in arteries. In Bergel, D. H., ed. *Cardiovascular Fluid Dynamics,* Vol. 2. London: Academic. pp. 111–139.
Rosedale, R. S. 1935. Interventricular septal defect, dextroposition of aorta, and dilatation of pulmonary artery: report of a case with structural pathogenesis. *Am. J. Pathol.* 11:333–341.
Rosengart, R., Fishbein, M., and Emmanouilides, G. C. 1975. Progressive pulmonary vascular disease after surgical correction (Mustard procedure) of transposition of great arteries with intact ventricular septum. *Am. J. Cardiol.* 35:107–111.
Ross, R. S., Taussig, H. B., and Evans, M. H. 1958. Late hemodynamic complications of anastomotic surgery for treatment of the tetralogy of Fallot. *Circulation* 18:553–561.
Rudolph, A. M. 1970. The changes in the circulation after birth: their importance in congenital heart disease. *Circulation* 41:343–359.
Saalouke, M. G., Shapiro, S. R., Perry, H. W., and Scott, L. P. III. 1977. Isolated partial anomalous pulmonary venous drainage associated with pulmonary vascular obstructive disease. *Am. J. Cardiol.* 39:439–444.
Samaan, H. A. 1970. Surgery of ventricular septal defect and pulmonary vascular resistance. *Thorax* 25:665–668.
Shephard, R. B., Simpson, D. C., and Sharp, J. F. 1966. Energy equivalent pressure. *Arch. Surg.* 93:730–740.
Sigmann, J. M., Perry, B. L., Behrendt, D. M., Stern, A. M., Kirsh, M. M., and Sloan, H. E. 1977. Ventricular septal defect: results after repair in infancy. *Am. J. Cardiol.* 39:66–71.
Silove, E. D., Tavernor, W. D., and Berry, C. L. 1972. Reactive pulmonary arterial hypertension after pulmonary venous constriction in the calf. *Cardiovasc. Res.* 6:36–44.
Skalak, R., Wiener, F., Morkin, E., and Fishman, A. P. 1966a. The energy distribution in the pulmonary circulation. I. Theory. *Phys. Med. Biol.* 11:287–294.
Skalak, R., Wiener, F., Morkin, E., and Fishman, A. P. 1966b. The energy distribution in the pulmonary circulation. II. Experiments. *Phys. Med. Biol.* 11:437–449.
Somlyo, A. V., and Somlyo, A. P. 1964. Vasomotor function of smooth muscle in the main pulmonary artery. *Am. J. Physiol.* 206:1196–1200.
Streeter, V. L., Keitzer, W. F., and Bohr, D. F. 1963. Pulsatile pressure and flow through distensible vessels. *Circ. Res.* 13:3–20.
Szidon, J.P. 1969. Autonomic control of the pulmonary circulation. In Fishman, A. P., and Hecht, H. H., eds. *The Pulmonary Circulation and Interstitial Space.* Chicago: University of Chicago Press, pp. 239–268.
Taylor, M. G. 1966a. The input impedance of an assembly of randomly branching elastic tubes. *Biophys. J.* 6:29–51.
Taylor, M. G. 1966b. Wave transmission through an assembly of randomly branching elastic tubes. *Biophys. J.* 6:697–716.
Van Benthuysen, K. M., Hammon, J. W. Jr., Mitchener, J. S., and Anderson, R. W. 1977. Electron microscopic alterations in experimental pulmonary hypertension. *J. Surg. Res.* 22:398–418.
Virmani, R., and Roberts, W. C. 1979. Pulmonary arteries in congenital heart disease: a structure-function analysis. In Roberts, W. G., ed. *Congenital Heart Disease in Adults.* Philadelphia: Davis, p. 455.

Vogel, J. H. K., Averill, K. H., Pool, P. E., and Blount, S. G. Jr. 1963. Experimental pulmonary arterial hypertension in the newborn calf. *Circ. Res.* 13:557–571.

Vogel, J. H. K., McNamara, D. G., and Blount, S. G. Jr. 1967. Role of hypoxia in determining pulmonary vascular resistance in infants with ventricular septal defects. *Am. J. Cardiol.* 20:346–349.

Wagenvoort, C. A. 1960. Vasoconstriction and medial hypertrophy in pulmonary hypertension. *Circulation* 22:535–546.

Wagenvoort, C. A., Nauta, J., Van der Schaar, P. J., Weda, H. W. H., and Wagenvoort, N. 1967. Effect of flow and pressure on pulmonary vessels: a semiquantitative study based on lung biopsies. *Circulation* 35:1028–1037.

Westerhof, N., Sipema, P., Van den Bos, G. C., and Elzinga, G. 1972. Forward and backward waves in the arterial system. *Cardiovasc. Res.* 6:648–656.

Womersley, J. R. 1957. Oscillatory flow in arteries: the constrained elastic tube as a model of arterial flow and pulse transmission. *Phys. Med. Biol.* 2:178–187.

Wood, P. 1958. The Eisenmenger syndrome or pulmonary hypertension with reversed central shunt. *Br. Med. J.* 2:701–709, 755–762.

Yamaki, S., and Horiuchi, T. 1979. Quantitative analysis of postoperative changes in the pulmonary vasculature of patients with complete transposition of the great arteries and pulmonary hypertension. *Am. J. Cardiol.* 44:284–289.

Yamaki, S., Horiuchi, T., and Sekino, Y. 1983. Quantitative analysis of pulmonary vascular disease in simple cardiac anomalies with the Down's syndrome. *Am. J. Cardiol.* 51:1502–1506.

4
Age-Related Changes in Left Ventricular/Arterial Coupling

WILMER W. NICHOLS, MICHAEL F. O'ROURKE,
ALBERT P. AVOLIO, TOSHIO YAGINUMA,
JOSEPH P. MURGO, CARL J. PEPINE,
AND C. RICHARD CONTI

Introduction

To understand the influence of the aging process on left ventricular/arterial coupling (or interaction), it is important first to consider the effects of age on ventricular and arterial functions separately. The function of the left ventricle is to pump blood (the cardiac output) through the systemic arterial system to the organs and tissues in an amount sufficient to meet their metabolic requirements. The arterial system has two distinct and separate functions: (1) It serves as a vascular conduit system that delivers blood at high pressure to the various peripheral beds and (2) it serves as a buffering system (Windkessel) that smooths out pulsations resulting from intermittent ventricular ejection so that the capillaries receive a steady or continuous flow of blood.

It is well known that alterations in the arterial system have a profound influence on ventricular function (Elzinga and Westerhof 1973; Imperial, Levy, and Zieski 1961; Ishide et al. 1980; Milnor 1975, 1978; Nichols and Pepine 1982; Nichols et al. 1980b; Noble 1979; Pouleur, Covell, and Ross, 1979; O'Rourke 1967; O'Rourke and Avolio 1981; Salisbury, Cross, and Rieben 1962; Sunagawa, Maughan, and Sagawa 1985; Urschel et al. 1968; Walker 1975; Wilcken et al. 1964). The amount of energy and thus the oxygen used by the left ventricle to produce the cardiac output is dependent not only on the contractile properties of the myocardium but also on the physical properties of the blood and arterial system (Monroe 1964; Sarnoff et al. 1958). Thus the arterial system constitutes the external vascular load placed on the ventricle during fiber shortening and ejection. Since the ventricle ejects a pulsatile blood flow into a distensible arterial buffering system, the load has both static (nonpulsatile) and dynamic (pulsatile) components. The static (or resistive) component is dependent primarily on blood viscosity and arteriolar caliber, while the dynamic (or compliant) component is dependent on the physical properties of the larger arteries and pulse wave reflections (Milnor 1975, 1978; Nichols and Pepine 1982; Nichols et al. 1980b; O'Rourke and Avolio 1981; Walker 1975). Changes in any (or all) of these variables can alter ventricular function. Therefore any change in properties of the arterial system (with its contained blood) as a result of the aging process could have a profound influence on ventricular function.

Effects of aging on cardiovascular function have received the attention of physicians and physiologists since 1809, when Young introduced his elastic modulus to study arterial function, and Roy (1880) and Mahomed (1892) conducted sphygmographic studies of the pressure pulse. Still, remarkably little is known about the precise effects of aging on cardiovascular pulsatile hemodynamics. Left ventricular external arterial load is known to increase with increase in arterial pressure (Quinones, Gaasch, and Alexander 1976; Sagawa 1967), peripheral resistance (Elzinga and Westerhof 1973; Imperial, Levy, and Zieski 1961; Salisbury, Cross, and Rieben 1962; Sunagawa, Maughan, and Sagawa 1985; Walker 1975), and arterial stiffness (Elzinga and Westerhof 1973; O'Rourke 1967; Salisbury, Cross, and Rieben 1962; Urschel et al. 1968; Walker 1975; Wilcken et al. 1964), all of which are alterations associated with aging (Nichols et al. 1985a). Aging is also known to be associated with mild left ventricular hypertrophy (Bafitis and Sargent 1977, Gardin et al. 1977, Gerstenblith et al. 1977, Shock 1961, Yin et al. 1980), increased myocardial cell diameter (Unverferth et al. 1983), and prolonged or delayed left ventricular relaxation (Harrison et al. 1964, Lakatta et al. 1975, Templeton et al. 1979, Yin et al. 1980). Changes in arterial load and cardiac properties are associated with decreased cardiac output and stroke volume (Elzinga and Westerhof 1973; Gerstenblith, Lakatta, and Weisfeldt 1976; Nichols and Pepine 1982; Nichols et al. 1980b; Noble 1979; Walker 1975), decreased capacity for increasing cardiac output with exercise (Gozna et al. 1974; Julius et al. 1967; Yin, Weisfeldt, and Milnor 1981), and increasing incidence of cardiac failure with increasing age, even in the absence of specific causative failure such as myocardial ischemia, hypertension, or valvular disease.

In this chapter, left ventricular arterial load is characterized by the aortic input impedance, since it is a function of systemic vascular resistance, central aortic elastance (the inverse of compliance) or stiffness, and wave reflections and since it selectively describes the arterial system (Finkelstein and Collins 1982; Gessner 1972; McDonald 1974; Milnor 1982; O'Rourke 1982a,b; O'Rourke and Taylor 1967; Westerhof et al. 1979). Each load component is defined in terms of the impedance spectra and is related to geometric and structural variables that are likely to change with advancing age. Also, the effects of increasing individual load components on left ventricular function are briefly discussed.

This chapter also reviews changes with age from maturation to senescence in structure and function of the arterial system, and an attempt is made to relate the age-associated alterations in structure to change in physical properties of the artery. The effects of these changes on pulse wave velocity, elastic modulus, pulse wave contour, and aortic input impedance spectra are discussed. Finally, since the primary focus of this book concerns ventricular/vascular interaction (or coupling), the age-associated changes that occur in the left ventricle are discussed and related to changes in arterial load (or aortic impedance).

Left Ventricular Arterial Load

Peripheral resistance, the ratio of mean pressure drop across a system to mean flow through the system, is often considered to represent the vascular load of the ventricle during ejection (Imperial, Levy, and Zieski 1961). In the systemic circulation, the venous pressure is normally very low compared with aortic pressure and can be disregarded in resistance *(R)* calculations (McDonald 1974).
Thus

$$R = \frac{P_o}{Q_o} = \frac{8\mu L}{\pi r^4} \tag{4.1}$$

where P_o = mean aortic pressure
Q_o = mean ascending aortic blood flow or cardiac output
μ = blood viscosity
L = length of system
r = internal radius

It can be seen that resistance is linearly and directly related to blood viscosity and inversely related to the fourth power of the radius. In the intact circulation, since blood viscosity remains relatively constant, the variable with the strongest influence on resistance is the caliber (or cross-sectional area) of the arteriolar bed. Rearranging Equation 4.1 gives

$$Q_o = \frac{\mathrm{P}_o}{R} = \mathrm{SV} * \text{heart rate} \tag{4.2}$$

so that an inverse relation exists between the resistance of the vascular bed and left ventricular ejection or stroke volume (SV). This relation has been confirmed in experiments designed to test the influence of individual components of arterial load on ventricular performance (Elzinga and Westerhof 1973; Nichols and Pepine 1982; Sunagawa, Maughan, and Sagawa 1985; Walker 1975). In a steady-flow system, the resistance represents the total load placed on the pump by the system. The power that the pump must generate to overcome this load is given by $P_o \times Q_o$. Since the ventricle ejects a pulsatile blood flow into a distensible arterial reservoir, the total opposition to ejection takes on a different meaning. In this type of hydraulic system, the total hydraulic load (external) placed on the ventricle during ejection must include (in addition to peripheral resistance) the viscoelastic properties of the aorta and other major arteries and wave reflections within the system (Milnor 1975, 1978; Nichols and Pepine 1982; Nichols et al. 1980b; O'Rourke and Avolio 1981; Walker 1975).

The aortic input impedance is the best available index to describe all of these properties (Avolio et al. 1976; Finkelstein and Collins 1982; McDonald 1974; Mills et al. 1970; Milnor 1982; O'Rourke 1982a,b; O'Rourke and Taylor 1967; Patel et al. 1965; Westerhof et al. 1979). In general, "impedance" implies a measure of the opposition to flow presented by a system. By conventional use in physics, and in its most familiar form in the terminology of electric current theory, impedance is confined to oscillatory motions or alternating current. Thus impedance is a frequency-dependent function. The term "resistance" conveys a meaning similar to that of impedance but is by conventional definition confined to nonoscillatory or steady motions. Resistance is the direct current component of impedance and may thus be considered as the impedance at zero frequency (McDonald 1974).

Application of this terminology to the vascular system requires certain considerations. In the arterial system, blood pressure and flow exist as oscillatory (or pulsatile) waveforms superimposed on a mean (or nonpulsatile) component. Thus total opposition to arterial blood flow includes both frequency-independent static (nonpulsatile) and frequency-dependent dynamic (pulsatile) components. If the left ventricle pumped a nonpulsatile blood flow through the systemic circulation, then the total external load opposing ventricular ejection would be the static component. On the other hand, if the left ventricle pumped a sinusoidal blood flow into the aorta, then the total external load of the ventricle would be the aortic input impedance *(Z)* calculated as the ratio of oscillatory pressure $(\dot{P})$ and flow $(\dot{Q})$:

$$Z = \frac{\dot{P}}{\dot{Q}} = \frac{|P|\ e^{i(\omega t - \alpha)}}{|Q|\ e^{i(\omega t - \beta)}} = \frac{|P|}{|Q|} e^{i(\beta - \alpha)} \tag{4.3}$$

The real part of the impedance is

$$Z = \frac{|P|}{|Q|} \cos(\beta - \alpha) \tag{4.4}$$

where

$$|Z| = \frac{|P|}{|Q|} \tag{4.5}$$

is the impedance modulus and

$$\theta = (\beta - \alpha) \tag{4.6}$$

is the phase.

Rearranging Equation 4.5 gives

$$|Q| = \frac{|P|}{|Z|} \tag{4.7}$$

so that flow modulus is inversely related to impedance modulus; if impedance increases flow decreases (Elzinga and Westerhof 1973; O'Rourke 1967; Salisbury, Cross, and Rieben 1962; Urschel et al. 1968; Walker 1975; Wilcken et al. 1964). Impedance calculated in this manner depends on the physical properties of large conduit arteries and pulse wave reflections within the arterial system (McDonald 1974; Milnor 1982; O'Rourke 1982a,b). Since pressure and flow waves generated by the left ventricle are actually summations of sinusoidal harmonics superimposed on mean values, moduli and phases of impedance are frequency dependent. Plots of aortic input impedance spectra (modulus and phase) with the individual load components are shown in Figure 4.1.

In a hydraulic conducting system (e.g., a fluid-filled elastic pipe) of some length, the ratio of pulsatile pressure to flow is termed the characteristic impedance if only centrifugal waves are present at the origin, that is, if no reflected waves are present or, if created at the terminations of the system, the system is of such length that they are completely attenuated before returning to the origin (McDonald 1974). Under these circumstances, characteristic impedance (Z_o) is directly related to the blood density (ρ) and pulse wave velocity (c_o) and is inversely related to the lumen radius (r) squared of the pipe:

$$Z_o = \frac{\rho c_o}{\pi r^2} \tag{4.8}$$

For large arteries, pulse wave velocity may be approximately related to the physical properties of the vessel by the Moens–Korteweg or Bramwell–Hill (1922a) equations:

$$c_o = \sqrt{\frac{Eh}{2r\rho}} = \sqrt{\frac{VdP}{\rho dV}} \tag{4.9}$$

where E = Young's modulus of elasticity of wall material
h = wall thickness of vessel
V = arterial volume
dP = pulse pressure
dV = change in arterial volume

In most animals (including humans), since the arterial system is relatively short, aortic input impedance moduli oscillate around the characteristic value because of reflected waves (McDonald 1974; Milnor 1982; O'Rourke 1982a,b; Taylor 1957). Since wave velocity varies with frequency (Nichols and McDonald 1972), wave reflections exert a prominent influence on

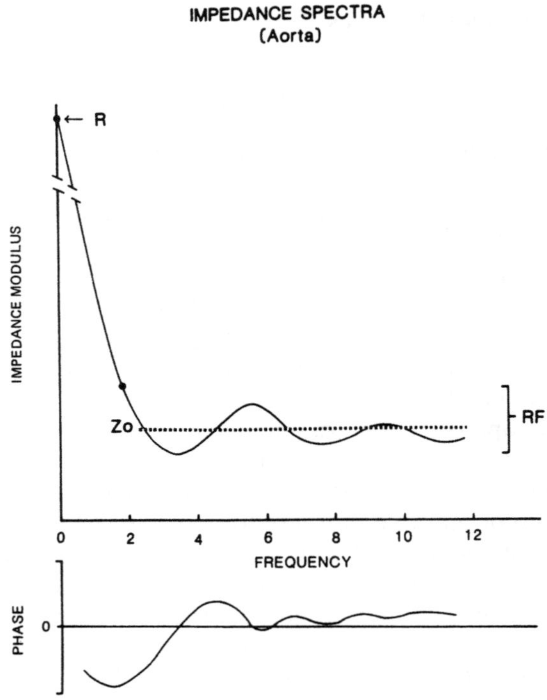

FIGURE 4.1. Hypothetical aortic input impedance spectra. Impedance modulus values decline from a high value at 0 Hz (*R*, peripheral resistance) to a minimum at approximately 3.5 Hz. This is approximately the same frequency at which phase crosses zero. Negative phase values indicate that flow harmonics lead pressure harmonics; positive phase values indicate that flow harmonics lag pressure harmonics. Impedance moduli oscillate around a characteristic value (Z_0 average of moduli > 2 Hz) because of wave reflections. A wave reflection index (RF) is calculated as the difference between maximum and minimum impedance moduli.

aortic input impedance moduli, especially at low frequencies (< 2 Hz). For high frequencies, however, input impedance approaches characteristic impedance (McDonald 1974). Therefore characteristic impedance can be estimated as the arithmetic mean of the input impedance moduli above 2 Hz (Nichols et al. 1977). A wave reflection index can be estimated as the difference between maximum and minimum input impedance moduli (Nichols et al. 1980b) (Fig. 4.1). A shift in the impedance moduli minimum and phase zero crossing indicates an apparent shift in the predominant reflecting site and/or a change in arterial elastance (McDonald 1974). This can be accounted for by an alteration in the vascular bed, such that the major reflecting site alters its position with respect to the site of measurement, a change in pulse wave velocity, or a combination of both effects. The distance from the site of measurement to the major site of wave reflection is obtained from the following relation:

$$\lambda = \frac{c_o}{f} \tag{4.10}$$

where λ is the wavelength of the arterial system in centimeters at a frequency of f hertz. The distance (d) from the site of measurement to the reflecting site is a quarter of the wavelength at the frequency of the first impedance moduli minimum. Thus

$$d = \frac{\lambda}{4} = \frac{c_o}{4f} \tag{4.11}$$

If the average pulse wave velocity along the aorta is 750 cm/sec, (Avolio et al. 1984), for example, and the aortic input impedance minimum occurs at 3.5 Hz (Fig. 4.1), then the distance from the aortic root to the major reflecting site is 53.6 cm.

Since pulsatile ascending aortic pressure and flow fluctuate around mean values, the total (arterial) load the left ventricle must overcome to eject blood must include both frequency-independent static (resistive) and frequency-dependent dynamic (elastic and reflective) components (Fig. 4.2). Changes in these components individually or in any combination will alter ventricular ejection and function (Elzinga and Westerhof 1973; Nichols et al. 1979; Sunagawa, Maughan, and Sagawa 1985; Walker 1975). Elzinga and Westerhof used an isolated cat heart model constructed so that resistive and compliant load components could be changed independently.

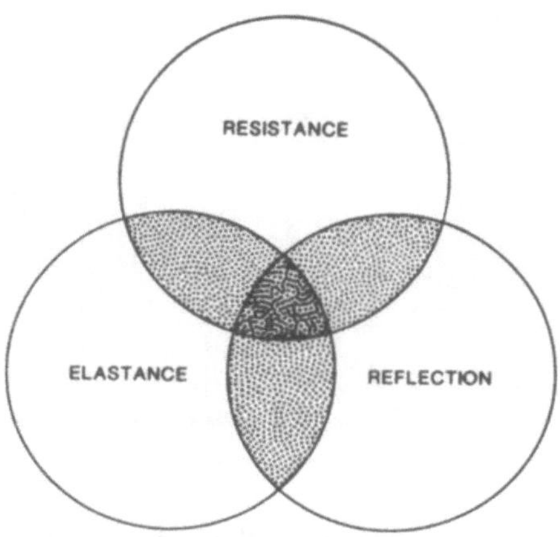

FIGURE 4.2. Venn diagram illustrating the different components of left ventricular vascular load. The nonpulsatile or static component is resistance and the pulsatile or dynamic components are elastance and reflection. Changes in any (or all) of these load components can alter ventricular performance.

They found that increased elastance (characteristic impedance) increased pulse pressure and decreased stroke volume, while resistance was maintained constant. Increasing resistance alone caused an increase in mean aortic pressure and a decrease in stroke volume. Sunagawa, Maughan, and Sagawa (1985) investigated the effect of changing input impedance over three selected frequency ranges (low, transitional, and high) on stroke volume in isolated canine left ventricles. They found that the modulus of impedance in the low-frequency range was, by far, a more important determinant of stroke volume than moduli in the transitional and high-frequency ranges. In addition to studying the influence of changes in the individual arterial load components on ventricle function, Walker (1975) investigated effects of increasing total aortic input in intact dogs. All three load components were increased by this maneuver, resulting in a 21% decrease in stroke volume. Similar results have been reported in humans (Nichols et al. 1979). Aortic input impedance was elevated by inflating a balloon in the descending aorta. This intervention caused an increase in mean, systolic, and pulse pressure and a decrease in cardiac output, peak flow, and stroke volume (Fig. 4.3). In addition to increasing resistance, characteristic impedance, and wave reflection index, balloon inflation caused the impedance moduli minimum and the phase crossover to shift to a higher frequency (Fig. 4.4). These changes in impedance spectra indicate an increase in arterial load and result from increased resistance, elastic modulus, pulse wave velocity, and wave reflections.

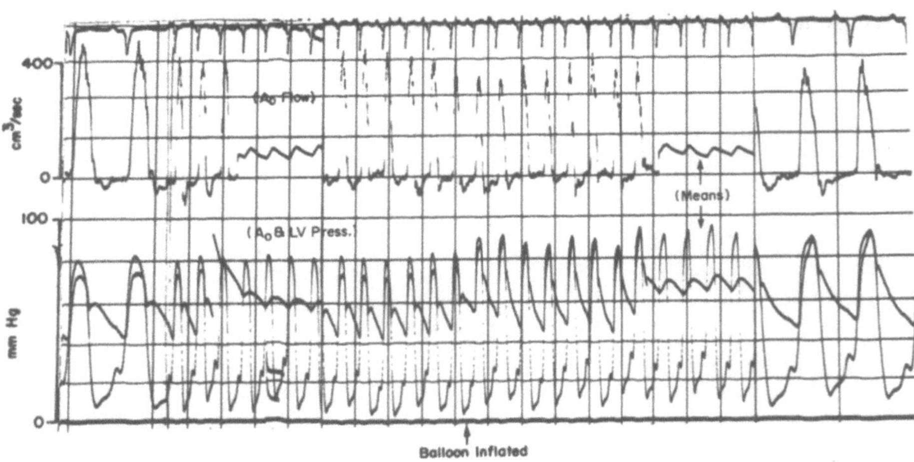

FIGURE 4.3. Changes in ascending aortic flow and pressure and left ventricular pressure (Millar multisensor catheter) during inflation of a balloon in the descending thoracic aorta of a patient. Balloon inflation increased mean, systolic, and pulse pressure and decreased peak flow, stroke volume, and cardiac output. *Top,* aortic flow; *bottom,* aortic and left ventricular pressure.

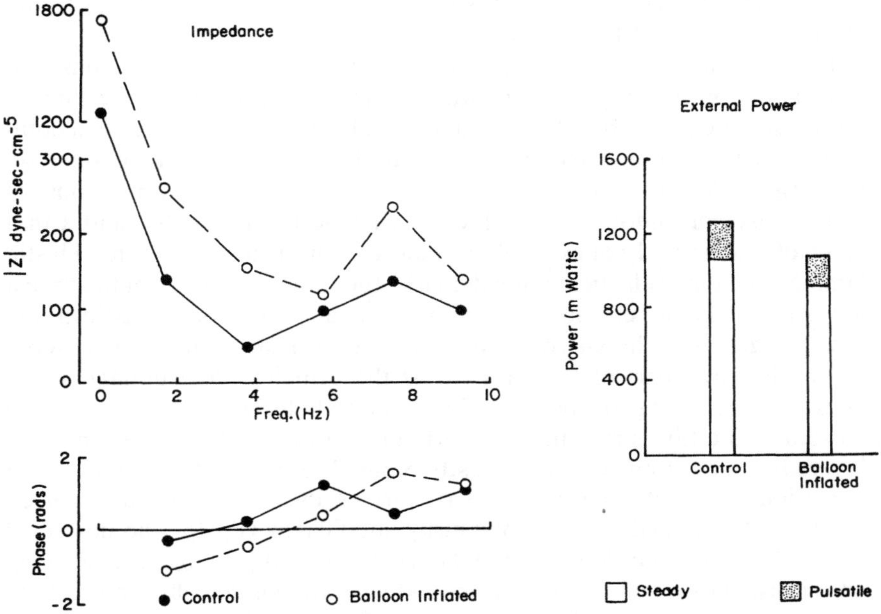

FIGURE 4.4. Aortic input impedance spectra (*left*) and total left ventricular external power (*right*) before and after inflation of a balloon in the descending aorta of a patient. Balloon inflation increased both static (resistance) and dynamic (elastance and reflection) components and shifted the impedance moduli minimum and phase crossover to higher frequencies. These changes in impedance produced decreases in both steady and pulsatile components of left ventricular power.

Aorta and Large Arteries

An extensive review of effects of age on matrix properties of the arterial wall was published by Milch (1965) and was summarized and updated in 1980 by Yin. Age-related changes in the physical properties of the arterial wall are briefly discussed here after the structure of the normal arterial wall is reviewed.

Wall Structure and Geometry

The predominant elastic materials of the arterial wall are collagen and elastin. They are fibrous in nature, although elastin tends to occur in more or less continuous sheets. Another major constituent of the arterial wall is smooth muscle which, while contributing to the tension in the wall, cannot properly be regarded as a true elastic material (McDonald 1974). Collagen and elastin are complexed with an amorphous substance formed of mucoprotein. This substance is not itself elastic but contributes to the elastic properties of the artery.

The arterial wall is divided into three concentric zones or regions; the tunicas intima, media, and adventitia. Demarcation between the tunicas intima and media is provided by the internal elastic lamina. This is a complex structure that consists of a fenestrated membrane of elastin lined on the intimal side by a coarse fibrous network. The tunica intima consists of the vascular endothelium, which is a single layer of cells, and a thin layer of elastin and collagen fibers that anchors it to the internal elastic lamina. The outer elastic lamina demarks the tunica media from the tunica adventitia. The adventitia is a region of collagen and some elastin tissue that merges with the surrounding connective tissue. The tunica media forms the large part of the wall and is the principal determinant of the vessel's mechanical properties. The intermediate layers have a fibrous structure, the fibers running circularly or in a tight helix. Between these layers lie smooth-muscle cells mostly parallel to the elastin, though some are oriented longitudinally. The structure of the tunica media has been subjected to detailed study by Wolinsky and Glagov (1964), who described it as made up of an orderly array of lamellar units. At physiologic distending pressures, elastin and collagen fibers and smooth-muscle cells are precisely oriented and form well-defined layers. Relatively thick elastin bands are seen as concentric plates, while finer elastin fibers form networks between them. Collagen fibers are dispersed in the interstices and are arranged circumferentially. Although the finer elastin fibers connect with the elastin lamellas, no attachments between collagen and elastin fibers can be demonstrated. The smooth-muscle cells extend circumferentially between adjacent lamellas among the fine elastin and collagen fibers.

The relative amounts of the various components of the arterial wall have been measured in a number of dog studies (Apter, Rabinowitz, and Cummings 1966; Fischer and Llaurado 1966; Harkness, Harkness, and McDonald 1957) and human studies (Cleary 1963; Faber and Moller-Hou 1952; Hass 1942; Myers and Lang 1946). Results showed that, as in most living tissue, water comprises most of the wet weight of the wall. Of the dry weight, about 50% is made up of elastin and collagen; the remainder consists of smooth muscle and nonfibrous matrix. The distribution of elastin and collagen differs strikingly between the central and the peripheral arteries. In the proximal aorta elastin is the dominant component, while in the distal aorta the content reverses and in peripheral arteries collagen dominates. The elastic modulus of collagen is much higher than that of elastin, so that as the distance from the heart increases, the arteries become stiffer and the elastic modulus (Learoyd and Taylor 1966) and pulse wave velocity (Nichols and McDonald 1972, Fig. 4.5) increase. This relation is verified by the Moens–Korteweg equation (Eq. 4.9).

With the passage of time, certain histologic changes are seen in the arterial wall. These occur independently of changes that are attributed to

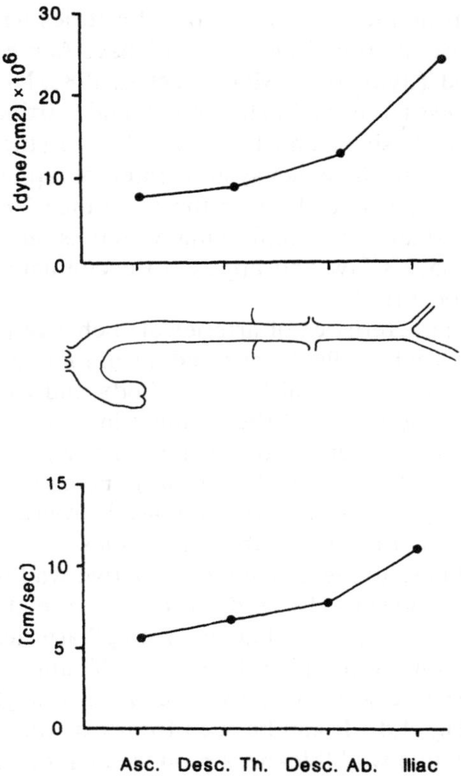

FIGURE 4.5. Elastic modulus (*top*) and pulse wave velocity (*bottom*) measurements in the ascending aorta, descending thoracic aorta, abdominal aorta, and iliac artery of a young normal subject. Both elastic modulus and pulse wave velocity increase as the artery becomes stiffer, owing to decreased elastin and increased collagen content.

atherosclerosis. Several detailed studies of human arterial structural changes with age (Cotton and Wartman 1961; Gray et al. 1953; Guard and Bhende 1953; Milch 1965; Moritz and Oldt 1937; Movat, More, and Haust 1958; Stehbens 1979) have shown that the principal changes occur in the intima and media. Endothelial cells of the intima become more irregular in size and shape and the number of giant multinucleated cells increases. The subendothelial layer becomes thickened with an increase in connective tissue content. In the media, the elastic fibers and laminas lose the orderly arrangement seen in earlier life and display thinning, splitting, fraying, and fragmentation. Degeneration of elastic fibers is associated with an increase in collagenous fibers and in ground substance, and often with calcium deposition in degenerate elastic material. The most striking macroscopic finding is an increase in the size of arteries with age. There is a progressive increase in arterial radius (Ben-Ishay, Abramowitz, and Ungar 1962; Learoyd and Taylor 1966; Merillon et al. 1982; Nichols et al. 1985a) and in wall thickness (Auerbach, Hammond, and Garfinkel 1968; Learoyd and Taylor 1966; Nakashima and Tanikawa 1971). In the study by Nichols and colleagues (1985a), the mean systolic internal radius of the ascending aorta increased 9% per decade over the age range 20 to 60 years (Fig. 4.6). As the artery dilates, circumferential wall tension and stress increase according to Laplace's law and appear to accentuate the degeneration that has already occurred.

Longitudinal characteristics of arteries also change progressively with age. Learoyd and Taylor (1966) measured the amount by which a segment of artery shortened on removal from the body and expressed the value as a percentage of the length of the segment in situ. Arteries taken from "young" subjects (<35 years) showed significantly greater retraction than those taken from "old" subjects (>35 years), indicating increased arterial stiffness. In both groups retraction increased progressively toward the periphery (Fig. 4.7). Instead of absolute values for wall thickness, the ratio of wall thickness to mean diameter (relative wall thickness) was determined in the two groups. In "old" arteries the relative wall thickness was approximately 50% greater than in "young" arteries and the increase was progressive in more peripheral vessels. "Young" arteries showed a fairly constant relative wall thickness except for a slight increase in the femoral artery (Fig. 4.8). It can be seen from Equations 4.8 and 4.9 that an increase in relative wall thickness produces an increase in pulse wave velocity and characteristic impedance.

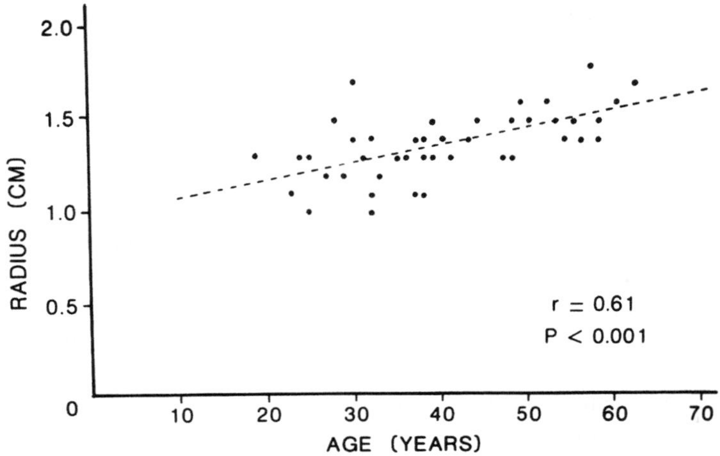

FIGURE 4.6. Change in mean systolic internal radius of ascending aorta with age. The radius was measured 3 to 5 cm above the aortic valve. (From Nichols et al. 1985a.)

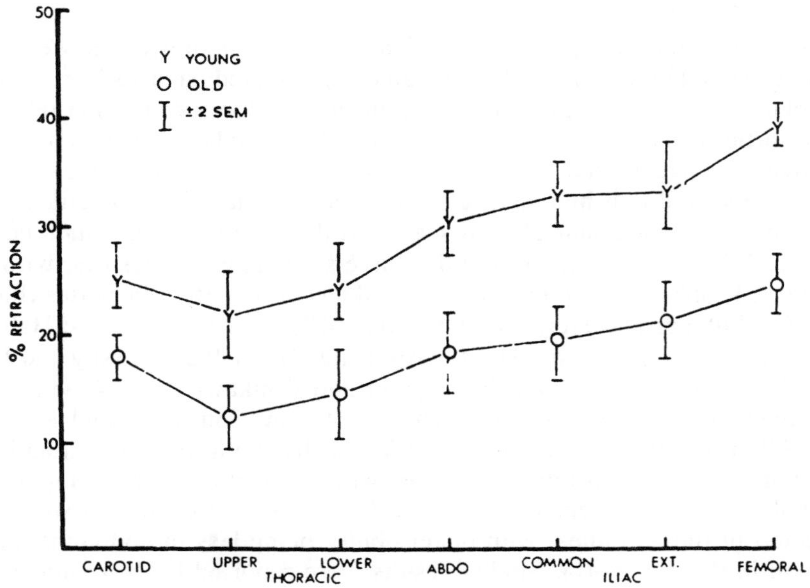

FIGURE 4.7. Retraction of "young" (Y subject < 35 years old) and "old" (O subject > 35 years old) arterial segments expressed as a percentage of the length in situ. Percent retraction is less in old, stiffer arteries. (From Learoyd and Taylor 1966; by permission of the American Heart Association, Inc.)

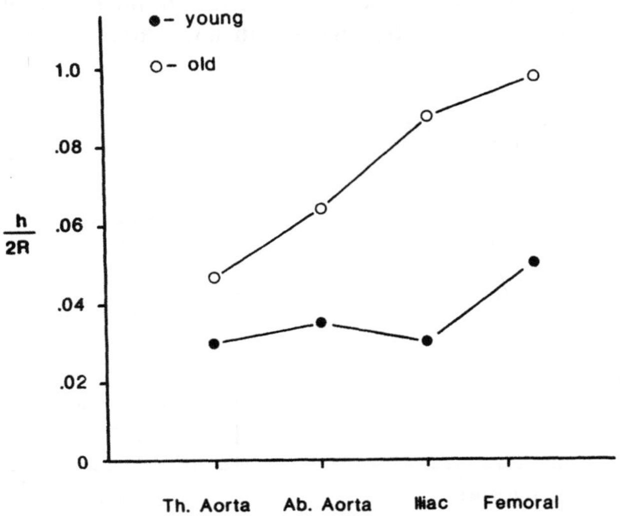

FIGURE 4.8. Changes in relative wall thickness (h = wall thickness; R = vessel radius) measurements in the thoracic aorta, abdominal aorta, iliac artery, and femoral artery of "young" (< 35 years old) and "old" (> 35 years old) subjects. Measurements were made at a distending pressure of 100 mm Hg. (From Learoyd and Taylor 1966; by permission of the American Heart Association, Inc.)

Elastic Properties

Changes in the elastic properties of the arterial system with age were first investigated by Roy in 1880. He studied age-related changes in pressure–volume relations of intact arterial segments and force–elongation relations of aortic strips from humans and rabbits. These studies indicated that with advancing age, elastance (or stiffness) of arterial segments increases. Roy also observed that human arteries were stiffer than those obtained from adult experimental animals and suggested that arterial degeneration begins in childhood. Subsequent studies have shown similar results, with increased stiffness apparent as elastic modulus of the artery increases (Bader 1967; Butcher and Newton 1958; Cox 1977; Gozna et al. 1974; Hallock and Benson 1937; Hass 1942; Learoyd and Taylor 1966; Mirsky and Janz 1976; Mozersky et al. 1973; Nakashima and Tanikawa 1971; Newman and Lallemand 1978; Roach and Burton 1959; Yin, Spurgeon, and Kallman 1983). Butcher and Newton (1958) found that both longitudinal and circumferential distensibility decreased with age and that maximum distensibility occurred at age 8. Regional differences in distensibility were also noted in these studies, with distensibility being less in abdominal aorta than in thoracic aorta. Similar results were reported by Nakashima and Tanikawa (1971) (Fig. 4.9). Gozna and colleagues (1974) measured aortic pressure and diameter in humans and, using the pressure–strain elastic modulus (Peterson, Jensen, and Parnell 1960), found a linear relationship with age (Fig. 4.10). The elastic modulus increased 156% over the age range 20 to 60 years.

In summary, these studies demonstrate that in normal human subjects there are definite viscoelastic changes in large arteries with advancing

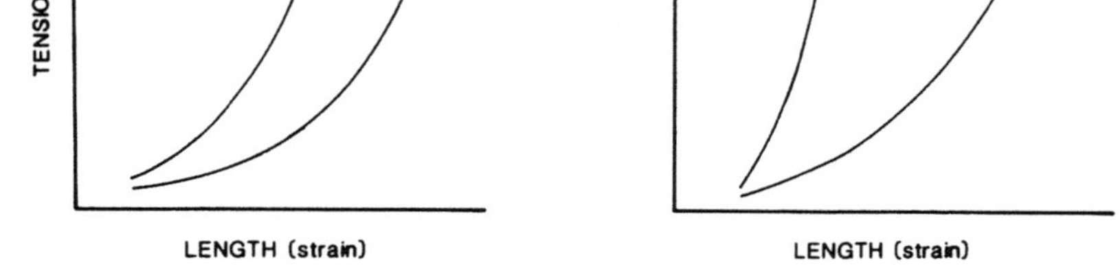

FIGURE 4.9. Effects of location (*left*) and age (*right*) on length–tension relation of the proximal (abdominal and thoracic) aorta. The curve shifts to the left and the slope becomes steeper as the artery becomes stiffer (From Nakashima and Tanikawa 1971.)

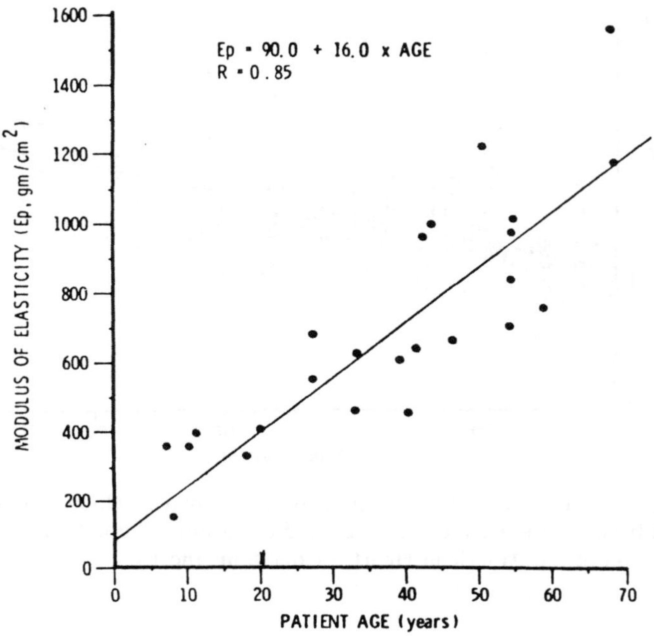

FIGURE 4.10. Age-related changes in the pressure–strain elastic modulus of the ascending aorta. (From Gonza et al. 1974.)

age. The elastic modulus (or elastance) increases or distensibility (and compliance) decreases (i.e., the vessels become stiffer) in old age. An increase in the elastic modulus influences an increase in pulse wave velocity and characteristic impedance (Eq. 4.8 and 4.9).

Pulse Wave Velocity

The viscoelastic properties (elastance or compliance) of arteries can be estimated from measurements of pressure or flow wave propagation along the vessel (McDonald 1974, Nichols and McDonald 1972). Pulse wave velocity between two measuring points is related to the physical properties of the artery by the Moens–Korteweg or Bramwell–Hill (1922a,b) equations (see Eq. 4.9). Since the introduction of the more useful form of the Moens–Korteweg equation by Bramwell-Hill, several studies have been published that show a positive relation between pulse wave velocity and increasing age (Avolio et al. 1983,1984; Bramwell, Hill, and McSwiney 1923; Eliakim, Sapozinikov, and Weinman 1971; Hallock 1934; Landowne 1958; Nielsen, Nielsen, and Fabricius 1968; Schimmler 1966; Simonson and Nakagawa 1960). The majority of these investigators recorded arterial pressure and

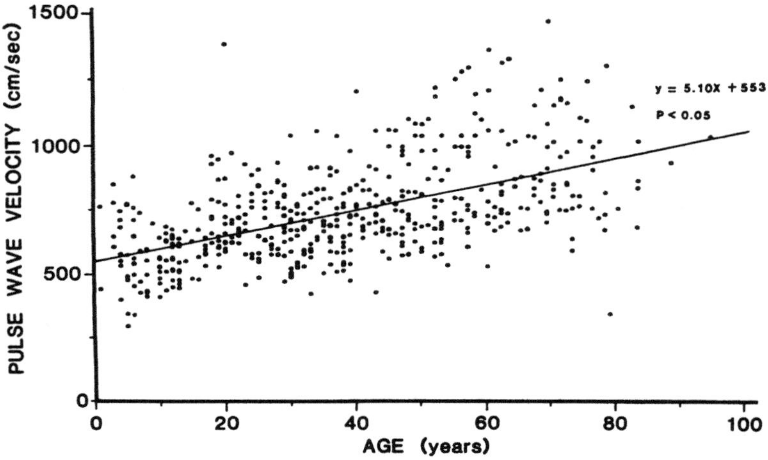

FIGURE 4.11. Age-related changes in aortic pulse wave velocity. Velocity was measured between the base of the neck and the groin. (From Avolio et al. 1984; by permission of the American Heart Association, Inc.)

estimated pulse wave velocity from propagation time of the foot or upstroke of the pressure pulse. Wave velocity calculated in this manner is not influenced by wave reflections and is directly related to characteristic impedance (Eq. 4.8). More recent studies have used the delay time between simultaneously recorded external flow velocity pulses to obtain pulse wave velocity. This method is attractive because large numbers of subjects can be studied (Avolio et al. 1983,1984). In these studies, performed in oriental subjects, pulse wave velocity increased linearly with age, indicating increased arterial stiffness (Fig. 4.11). The fact that similar changes in pulse wave velocity with age are seen in different populations with different prevalences of atherosclerosis argues against atherosclerotic involvement of the wall being a major factor in such change. Medial degeneration, a consequence of aging, appears to be the most important factor (Avolio et al. 1983,1984).

Arterial Pressure Pulse Wave Contour

The effects of arterial stiffening and of increased pulse wave velocity on pressure pulse wave contour are predictable: Reflected (or backward) waves from peripheral reflecting sites return earlier and merge (or sum) with the incident (or forward) wave generated by left ventricular ejection. Such changes of the pulse contour with age were described by O'Rourke and co-workers (1968), who also noted that the changes are associated with disappearance of the diastolic pressure wave. Freis and colleagues

(1967) noted that with increasing age, the early systolic pressure peak of the carotid pulse is replaced by a peak in late systole and that the diastolic pressure wave seen in young subjects becomes inapparent. In another study of the volume pulse in the leg, Puls and Heizer (1967) also showed disappearance of the diastolic wave with age. The most detailed and accurate studies of pressure pulse contour in the human ascending aorta were performed by Murgo and co-workers (1980; 1981a,b) using high-fidelity Millar multisensor cardiac catheters (Murgo, Giolma, and Altobelli 1977; Nichols et al. 1980a,1985b). The measured pressure wave was quite different from that seen in experimental animals, being characterized by a well-defined anacrotic notch (or inflection point) in the mid-to-late part of systole, followed by a secondary wave. The configuration or contour of the pressure wave was classified according to the following criteria (Fig. 4.12):

Type A. Peak systolic pressure (P_{pk}) occurs in late systole after an inflection point (P_i) and ($P_{pk} - P_i$)/pulse pressure is greater than 0.12.
Type B. Peak systolic pressure also occurs in late systole following an inflection point, but ($P_{pk} - P_i$)/pulse pressure is between 0.0 and 0.12.
Type C. Peak systolic pressure precedes an inflection point, and ($P_{pk} - P_i$)/pulse pressure is less than 0.0.

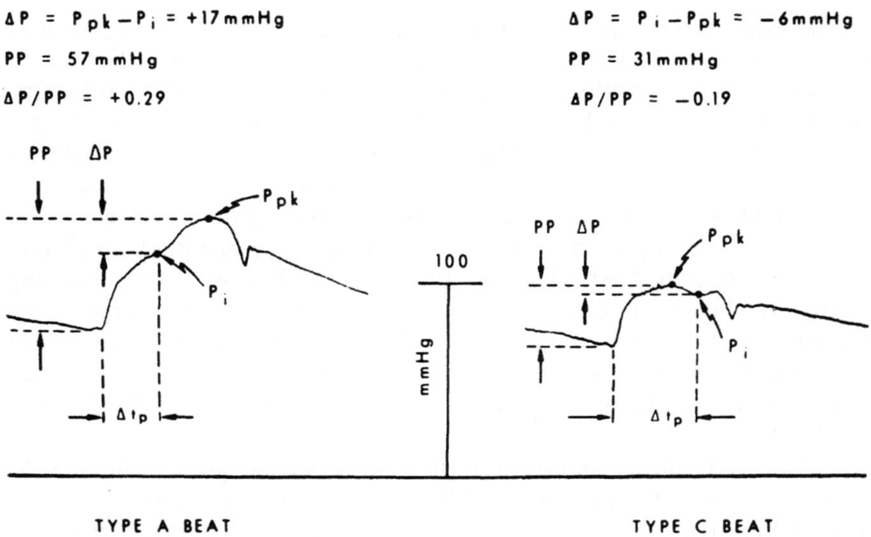

FIGURE 4.12. Criteria used to classify the systolic configuration of the ascending aortic pressure wave form. An example of how the ratio $\Delta P/PP$ is calculated is given for both type A and a type C (see text) beat. (From Murgo et al. 1980; by permission of the American Heart Association, Inc.)

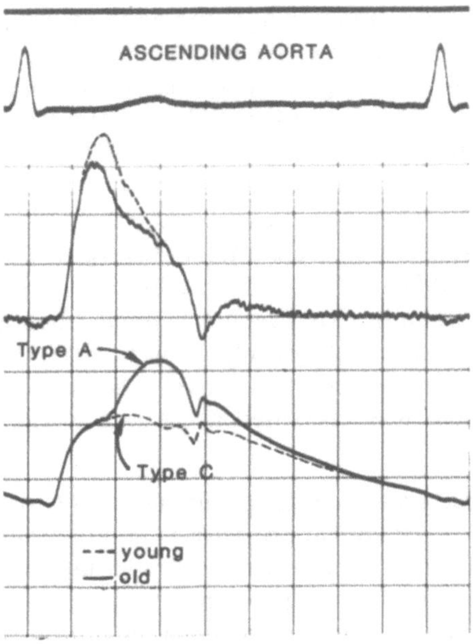

FIGURE 4.13. Ascending aortic blood flow velocity and pressure waveforms recorded in a young (type C pressure wave) and an old (type A pressure wave) subject. (From Nichols et al. 1985a.).

The type A aortic pressure wave contour is almost invariably seen in older subjects (>30 years; Fig. 4.13) and is attributable to earlier return of reflected waves, with the reflected wave from the lower body returning during mid systole so that pressure in late systole is augmented (Murgo et al. 1980, Nichols et al. 1985a). Younger subjects (<30 years) almost invariably exhibit a type C contour, with the foot of the diastolic wave corresponding with or following the incisura caused by aortic valve closure. This is the type of pattern usually seen in large experimental animals and is explicable on the basis of wave reflection from the lower body returning at the conclusion of ventricular ejection (O'Rourke 1982a). Subjects between 30 and 50 years of age usually show type B contour as a result of intermediate timing of wave reflection; however, all three types of pressure waves have been recorded in this age group (Murgo et al. 1980, Nichols et al. 1985a).

Pulse wave reflections from peripheral reflecting sites influence the contour of measured pressure and flow waves. Reflected pressure waves add to the incident wave, while reflected flow waves subtract from the incident wave (McDonald 1974, Milnor 1982, O'Rourke 1982a). The measured pressure and flow waves can be dissected into their forward (or incident) and reflected (or backward) components with use of mathematical techniques (Murgo et al. 1981a, Van den Bos et al. 1976, Westerhof et al. 1972):

$$P_m = P_f + P_r \tag{4.12}$$

$$Q_m = Q_f + Q_r \tag{4.13}$$

$$P_f = Z_o \cdot Q_f \tag{4.14}$$

$$P_b = -Z_o \cdot Q_r \tag{4.15}$$

where P = pressure wave
Q = flow wave
m = measured wave
f = forward wave
r = reflected wave
Z_o = characteristic impedance

Thus the measured pressure and flow waves sensed by transducers are equal to the sum of a forward wave and a reflected wave. The forward pressure and flow waves are related by the characteristic impedance of the aorta. In a reflectionless system, Z_o is constant and measured pressure and flow waves are identical in shape (McDonald 1974; Milnor 1982; O'Rourke 1982a,b). Since measured pressure and flow waves are altered by reflected waves, resulting in a frequency-dependent impedance spectrum, Equations 4.12 through 4.15 may be solved to yield the forward and reflected pressure and flow waves (Fig. 4.14). At a given location in the ascending aorta, a forward-traveling pressure wave first appears, but while systole is still in progress, a backward-traveling (reflected) pressure wave arrives at this location. This reflected wave is added to the forward wave

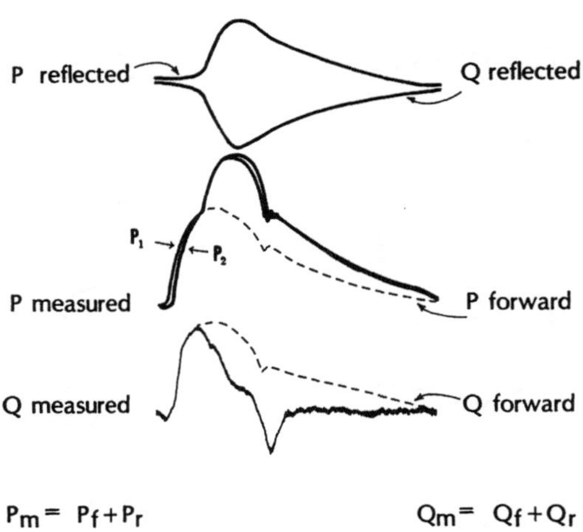

FIGURE 4.14. The influence of pulse wave reflections on ascending aortic pressure (P) and flow (Q) waveforms. Incident or forward (f) and backward or reflected (r) pressure and flow waveforms are summed to yield measured (m) pressure and flow waveforms. Two pressures (P_1 and P_2) were measured so that pulse wave velocity could be calculated.

and yields the typical measured ascending aortic pressure waveform with a late secondary systolic wave. Similarly, the forward flow wave (which has the same shape as the forward pressure wave) is added to the reflected flow wave (also identical in shape to the backward pressure wave, but inverted). Both reflected pressure and flow waves are less in type C beats than in type A beats (Fig. 4.15). The large secondary systolic wave in older individuals can be attributed to a large-amplitude reflected wave and is primarily responsible for the substantial increase in systolic pressure and pulse pressure (Granath, Jonsson, and Strandell 1970; Kannel et al. 1981; Merillon et al. 1982; Murgo et al. 1980; Nichols et al. 1985a) (Fig. 4.16). Change in characteristic impedance alone (resulting from aortic stiffness) can cause only increase in aortic pulse pressure (for a given stroke volume) but no change in wave contour. Change in wave contour is due to change in timing of wave reflection. These changes in arterial pressure with age are a little different from those of epidemiologic studies in which pressure is recorded indirectly in the brachial artery. A greater change occurs with age in central aortic pressure than in peripheral artery pressure because of decreased amplification of the pressure wave during transmission (O'Rourke et al. 1968) (Fig. 4.17). The increase in aortic (and left ventricular) systolic pressure with age increases systolic pressure time index (Schatz, Pasipoularides, and Murgo 1981), which in turn increases myocardial oxygen requirements (Monroe 1964, Sarnoff et al. 1958).

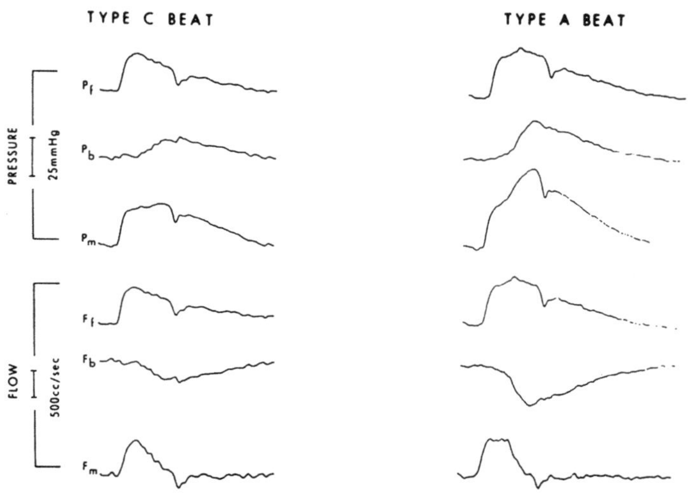

FIGURE 4.15. Forward (*f*), backward (*b*), and measured (*m*) pressure and flow waveforms of type C beat (usually seen in young subjects) and a type A beat (usually seen in old subjects). (From Murgo et al. 1981a; by permission of the American Heart Association, Inc.)

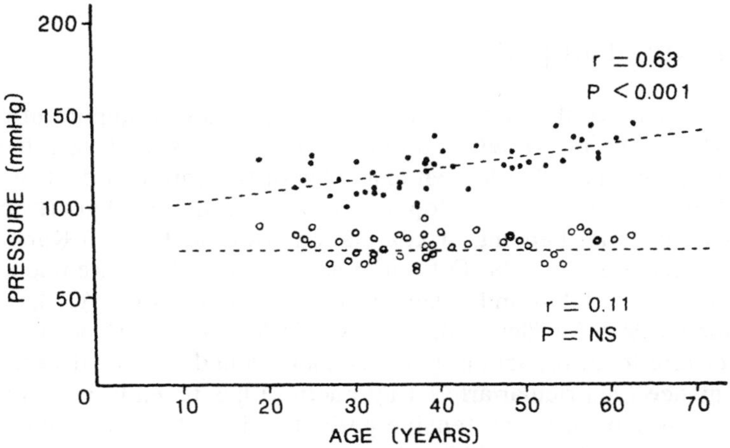

FIGURE 4.16. Change in ascending aortic systolic pressure (closed circles), diastolic pressure (open circles) and pulse pressure (difference between systolic and diastolic pressures) with age. (From Nichols et al. 1985a.)

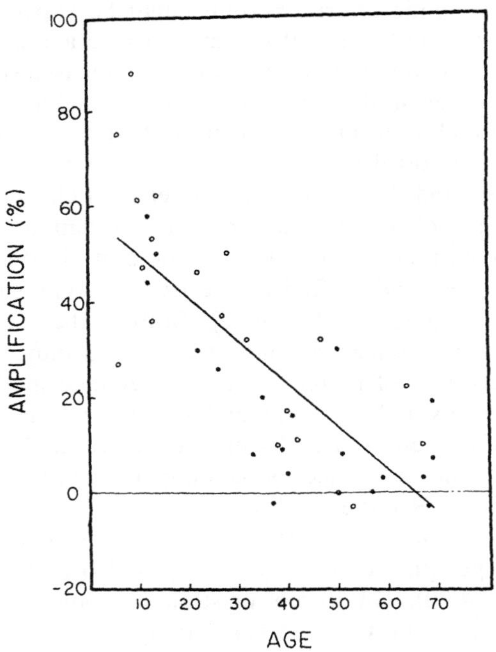

FIGURE 4.17. Age-related changes in amplification of the pressure wave between aortic arch and iliac artery (iliac pulse pressure − aortic arch pulse pressure)/aortic arch pulse pressure. Approximately a 50% amplification of the pulse pressure is seen in young subjects. Amplification of the pulse pressure falls progressively with age. Open circles indicate patients without clinical evidence or any vascular lesion, and closed circles indicate patients with known vascular lesions. (From O'Rourke et al. 1968; by permission of the American Heart Association, Inc.)

Aortic Input Impedance

That properties of the systemic arterial system play an important role in the performance of the left ventricle is well established (see references already given). The best known examples of the chronic effect of an elevated arterial load are development of left ventricular hypertrophy in response to aortic banding or coarctation (Beznak 1956, O'Kane et al. 1973, O'Keefe et al. 1978, O'Rourke and Cartwell 1971, Rembert et al. 1978, Yin et al. 1980) and systemic hypertension (Koyanagi, Eastham, and Marcus 1982; Mueller et al. 1978). Clinically, the importance of arterial load (or afterload) reduction with vasodilators and improved ventricular performance (ventricular/arterial interaction) has recently been realized in the treatment of normotensive patients with left ventricular failure (Chattergee and Parmley 1977, Cohn 1973, Gundel et al. 1981, Miller et al. 1976, Pepine et al. 1979). The concept of altering the aortic pulse mechanically to the heart's advantage with arterial counterpulsation is also based on the principle of ventricular/arterial interaction (O'Rourke, Sammel, and Chang 1979; O'Rourke et al. 1981). With regard to aging, changes that affect the systemic arterial system might be expected to affect left ventricular function independent of any intrinsic age alterations in myocardial function. As pointed out previously, aging is associated with certain structural changes in the systemic arteries, including increase in vessel diameter, wall thickness, and collagen content and loss of integrity of elastin. These structural changes are associated with stiffening of the proximal vessels, and they lead to an increase in elastic modulus, pulse wave velocity, and characteristic impedance. When arterial stiffening is generalized, as with advancing age (Nichols et al. 1985a), impedance spectra are elevated and shifted to the right so that minima of moduli occur at higher frequencies (Fig. 4.18). Both of these effects—increase in characteristic impedance and shifting of the impedance spectrum to the right—have the effect of increasing the impedance modulus and its in-phase component over that frequency band that normally contains most of the energy of the left ventricular ejection wave, and so causing a mismatch between ventricular ejection and impedance presented by the systemic circulation (O'Rourke 1982b).

The cause of the increase in arterial load with age can be inferred from an increase in peripheral resistance (Fig. 4.19) and in characteristic impedance (Fig. 4.20). The former or static component is determined by opposition to steady flow in the small peripheral vessels, whereas the latter or dynamic components are determined by arterial stiffness. Also arterial stiffness increases pulse wave velocity and wave reflection. A slight change occurs in resistance which can be attributed to decreased caliber or to decreased total cross-sectional area of small vessels. A greater change occurs in arterial stiffness, attributable to asymptomatic arterial

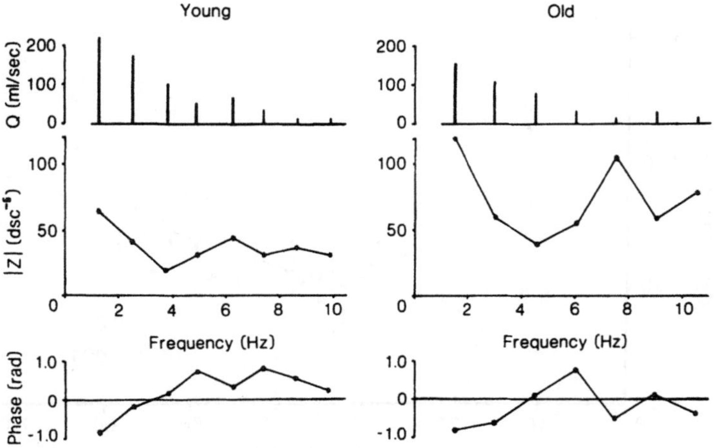

FIGURE 4.18. Aortic input impedance spectra and flow modulus versus frequency in a young subject (*left*) and an old subject (*right*). (From Nichols et al. 1985a.)

degeneration resulting in an increase in the collagen-to-elastin ratio (Cox 1977).

The interaction between the left ventricle and its arterial load is best considered by comparing the modulus of ascending aortic impedance with the frequency components of the left ventricular ejection (flow) wave (O'Rourke 1982a,b). A favorable relationship is apparent from the low

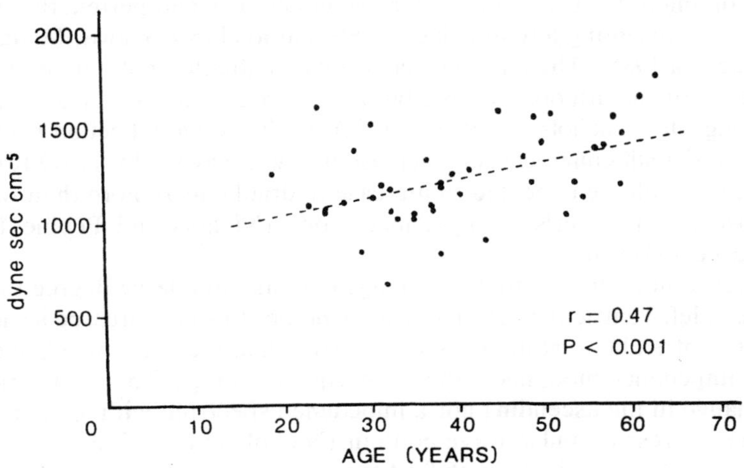

FIGURE 4.19. Change in nonpulsatile or static component of left ventricular arterial load (peripheral resistance) with age. (From Nichols et al. 1985a.)

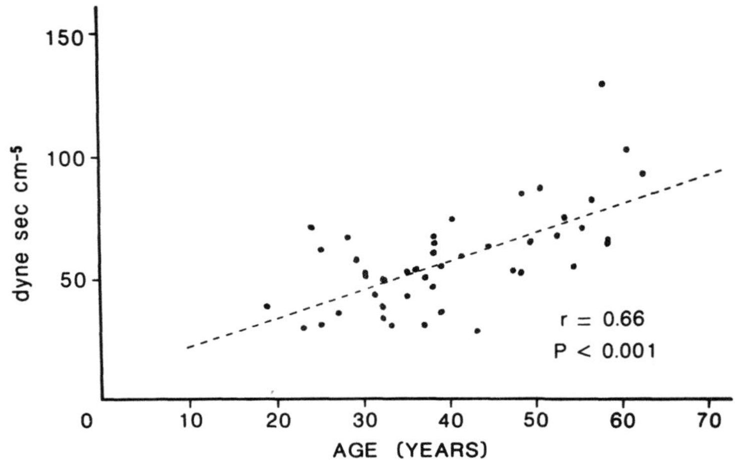

FIGURE 4.20. Change in pulsatile or dynamic component of left ventricular arterial load (characteristic impedance) with age. (From Nichols et al. 1985a.).

impedance values over the frequency range that normally contains most of the energy of the left ventricular ejection wave. Age disturbs this relationship by displacing the impedance spectrum to the right, so that the impedance modulus is even greater at the frequency of the first flow harmonic than would be predicted from consideration of change in characteristic impedance alone (Nichols et al. 1985a). This point illustrates a shortcoming in the use of characteristic impedance and peripheral resistance as representing left ventricular external load (Sunagawa, Maughan, and Sagawa 1985). The left ventricular load is affected not only by aortic distensibility and arteriolar caliber but also by reflected waves from arterial reflecting sites (Nichols and Pepine 1982). Earlier wave reflection resulting from arterial stiffening, causing increase in pulse wave velocity (O'Rourke 1982a,b), would be expected to increase hydraulic load more than would increase in characteristic impedance alone (Nichols and Pepine 1982; O'Rourke 1982a,b).

Increase in arterial stiffness with age explains to a large degree the increase in left ventricular arterial load—whether this is expressed as aortic or left ventricular systolic pressure, aortic characteristic impedance, or aortic impedance modulus at the low-frequency range. This also explains the change in the ascending aorta impedance spectrum with age, and the change in pressure pulse wave contour (Nichols et al. 1985a).

Aging is associated with a shift of the impedance spectrum to the right, with the minimum of impedance modulus and phase crossover occurring at higher frequencies with greater age (see Fig. 4.18). This is explicable on the basis of earlier return of reflected waves as a result of increased arterial pulse wave velocity. Since pulse wave velocity increases and the impedance modulus minimum shifts to a higher frequency, the distance *(d)* to the major reflecting site does not change. For example, using values of wave velocity from Avolio and colleagues (1984) and the impedance minimum from Nichols and co-workers (1985a), we find, using Equation 4.10, that in young subjects d = 53.6 cm and in old subjects d = 54.1 cm. Another feature of the spectra is the increase in fluctuation of the impedance modulus with increasing age (Fig. 4.18). This can be explained by differences in pulse wave velocity in the brachiocephalic and subclavian arterial system on the one hand and the descending aortic and femoral arterial system on the other (O'Rourke and Avolio 1980). In young subjects, reflections return much earlier from the upper body than from the lower body, thus attenuating fluctuations in modulus (O'Rourke 1982a,b; O'Rourke and Avolio 1980). With increasing age, wave velocity increases to a lesser degree in upper body arteries than in the descending aorta and lower body arteries, so that major reflections return at much the same time, with the systemic circulation appearing to present a single (rather than two) discrete reflecting site, and the impedance modulus showing considerable fluctuation, as predicted in such a condition (O'Rourke and Avolio 1980; Pepine, Nichols, and Conti 1978).

The findings described here in the aortic input impedance spectra were from occidental subjects without apparent cardiovascular disease. As shown by coronary angiography, none had significant coronary atherosclerosis or direct evidence of atherosclerosis elsewhere. Although findings might have been influenced by subclinical atherosclerosis, it is much more likely that the changes were only those of aging. In vitro studies indicate that even major atherosclerotic lesions have little influence on aortic stiffness (Langewouters 1982). Similar changes in pulse wave velocity and in arterial pressure with age have been found in occidental subjects, in whom atherosclerosis is prevalent, and in oriental subjects, in whom it is not (Avolio et al. 1983,1984).

Left Ventricular Function

In general, mild hypertrophy of the heart accompanies advancing age (Gardin et al. 1977, Gerstenblith et al. 1977, Linzbach and Akuomoa-Boateng 1973, Yin 1980). Linzbach and Akuomoa-Boateng reviewed heart weight in 7112 patients ranging in age from birth to 109 years. The hearts included those affected by a variety of diseases and a wide range of weights. For the entire group of normal and abnormal hearts there was an overall increase in heart weight with age in both males and females. These results only suggest that there is an age-related increase in left ventricular mass. In another study, by Gerstenblith, Lakatta, and Weisfeldt (1976), echocardiograms were performed on 105 men (24–84 years of age) with no evidence of hypertension or cardiovascular disease. Increasing age was found to correlate with left ventricular wall thickness. Over the age range studied, systolic wall thickness increased 48% and diastolic wall thickness increased 50%. Studies in rats (Yin 1980) suggest that age-related hypertrophy may not necessarily result from elevation in mean arterial pressure, because it occurs in the absence of an age-related increase in blood pressure. Since hypertrophy results from increased loading on the ventricle, one possible explanation for age-related hypertrophy is that the aged arterial system has a higher impedance than the young one (Nichols et al. 1985a). This would place a greater load on the ventricle and result in mild hypertrophy. As mentioned earlier, the left ventricle has been shown to hypertrophy in response to chronic aortic constriction (Beznak 1956, O'Kane et al. 1973, O'Keefe et al. 1978, O'Rourke and Cartwell 1971, Rembert et al. 1978, Yin et al. 1980) or to generalized systemic hypertension (Koyanagi, Eastham, and Marcus 1982; Mueller et al. 1978). Previous studies have shown that these cardiovascular abnormalities influence an increase in aortic input impedance (Nichols et al. 1977; O'Rourke 1970; O'Rourke and Cartwell 1971). The age-related increase in arterial impedance may also explain the increase in myocardial cell diameter observed by Unverferth and colleagues (1983).

Invasive studies in supine subjects have shown that resting cardiac output and stroke volume (determined by Fick or dye dilution techniques) decline with advancing age (Brandfonbrener, Landowne, and Shock 1955; Granath, Jonsson, and Strandell 1970; Julius et al. 1967; Merillon et al. 1982; Nichols et al. 1985a). In a group of 67 men (age range 19–86 years), cardiac output decreased from a mean of 6.49 l/min in the third decade to 3.87 l/min in the ninth. This represents a decline of approximately 7% per decade. Mean stroke volume fell from 85.6 to 60.1 ml in the group (Brandfonbrenner, Landowne, and Shock 1955). In a different study on 45 adults (age range 19–62 years) without apparent cardiovascular disease, average cardiac output fell from 7.1 to 5.7 l/min (Nichols et al. 1985a)—a decline of 5% per decade (Fig. 4.21). Since heart rate did not change with age, the decrease in stroke volume with increasing age was similar to that observed for cardiac output. It is unlikely that age-related decrease in cardiac output (or stroke volume) is a result of depressed myocardial contractile state, because basal contractile function is well maintained in both isolated cardiac muscle (Weisfeldt, Loeven, and Shock 1971) and intact hearts from senescent rats (Abu-Erreish et al. 1977) and humans (Gerstenblith et al. 1977, Port et al. 1980). A more likely explanation for the age-related decline in cardiac output is an increase in left ventricular arterial load or aortic input impedance.

Studies in the rat (Lakatta et al. 1975; Weisfeldt, Loeven, and Shock 1971; Yin et al. 1980), the dog (Templeton et al. 1979), and humans (Harrison et al. 1964) have shown the presence of delayed or prolonged relaxation of cardiac muscle with advancing age. In rats, contraction duration

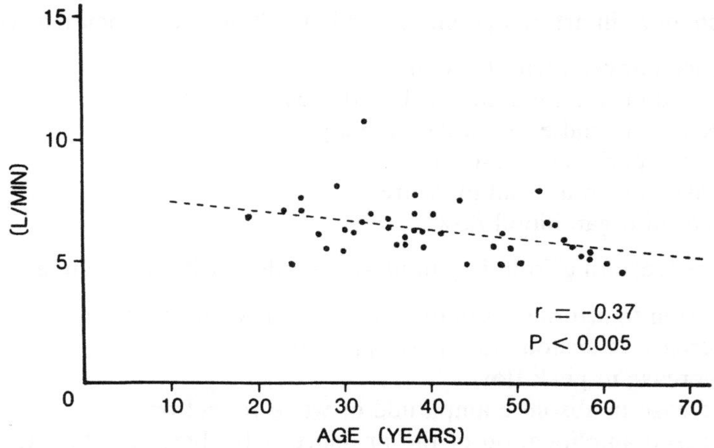

FIGURE 4.21. Change in cardiac output with age. (From Nichols et al. 1985a.).

was approximately 11% greater in aged animals. In dog and human, prolongation of contraction duration with aging was increased 20% to 25%. Although the precise mechanism responsible for these changes are not known, there is great similarity between the observations on aged rat myocardium and on myocardium from rats with left ventricular hypertrophy induced by elevated aortic impedance in response to aortic banding or systemic hypertension (Bing et al. 1971, Sordahl et al. 1973).

Conclusion

This chapter has presented a brief overview of the age-related alterations in structure and functions of the left ventricle and the arterial system in the resting state. The ability of the aged cardiovascular system to respond to stresses such as exercise and drug intervention was not examined.

In summary, the major changes that take place in the arterial system with age are:

1. Dilatation of the aorta and large arteries.
2. Increase in arterial wall thickness.
3. Increase in number of collagenous fibers in the arteries.
4. Degeneration of elastin of the arteries.
5. Decrease in percent longitudinal retraction of the arteries.
6. Increase in arterial stiffness.
7. Decrease in arteriolar caliber.
8. Decrease in vascularity of the organs and tissues.

These changes in arterial geometry and structure are associated with:

1. Increased arterial wall tension.
2. Increased elastic modulus of the arteries.
3. Increased arterial pulse wave velocity.
4. Increased peripheral resistance.
5. Increased mean arterial pressure.
6. Decreased organ blood flow.

Changes resulting from impaired arterial distensibility include:

1. Alteration in arterial pressure and flow pulse wave contour
   a. Increase in systolic and pulse pressure.
   b. Decrease in peak flow.
   c. Increase in absolute amplitude of wave reflections.
2. Decreased amplification of the pressure pulse between the ascending aorta and the peripheral arteries.
3. Alteration in aortic input impedance spectra
   a. Increase in characteristic impedance.
   b. Increase in maximum–minimum impedance moduli.
   c. Shift of impedance modulus minimum and phase crossover to higher frequency.
4. Mismatch between aortic input impedance and energy of left ventricular ejection wave.
5. Increased systolic pressure time index.

Changes that take place in the left ventricle with age include:

1. Mild hypertrophy.
2. Increase in myocardial cell diameter.
3. Prolongation or delay in relaxation.
4. Decrease in stroke volume and cardiac output.
5. Increase in stiffness.
6. Decrease in mitral valve E-F slope.

Those aspects of left ventricular function that appear to show little or no age-related alterations are:

1. Ejection fraction.
2. Fractional shortening of the minor semiaxis.
3. Velocity of circumferential fiber shortening.
4. Maximum rate of change of pressure.
5. Active length–tension relation.
6. Maximum rate of tension development.
7. Heart rate.

From these measurements it is clear that myocardial contractile state is preserved in senescence. Therefore, it is probable that age-related alterations in myocardial function occur because the ventricle is coupled to an abnormally high arterial load or input impedance.

Bibliography

Abu-Erreish, G. M., Neely, J. R., Whitman, J. T., Whitman, V., and Sanadi, D. R. 1977. Fatty acid oxidation by isolated perfused working hearts of aged rats. *Am. J. Physiol.* 232:E258–E262.

Apter, J. T., Rabinowitz, M., and Cummings, D. H. 1966. Correlation of viscoelastic properties of large arteries with microscopic structure. I. Collagen, elastin and muscle determined chemically, histologically, and physiologically. *Circ. Res.* 19:104–121.

Auerbach, O., Hammond, E. C., and Garfinkel, L. 1968. Thickening of walls of arterioles and small arteries in relation to age and smoking habits. *N. Engl. J. Med.* 278:980–984.

Avolio, A. P., Chen, S., Wang, R., Zhang, C., Li, M., and O'Rourke, M. F. 1983. Effects of aging on changing arterial compliance and left ventricular load in a northern Chinese urban community. *Circulation* 68:50–58.

Avolio, A. P., Deng, F. D., Li, W., Lou, Y., Huang, Z., Xing, L., and O'Rourke, M. F. 1984. Effects of aging on arterial distensibility in populations with high and low prevalence of hypertension: comparison between urban and rural communities in China. *Circulation* 71:202–210.

Avolio, A. P., O'Rourke, M. F., Mang, K., Bason, P. T., and Gow, B. S. 1976. A comparative study of pulsatile arterial hemodynamics in rabbits and guinea pigs. *Am. J. Physiol.* 230:868–975.

Bader, H. 1967. Dependence of wall stress in human thoracic aorta on age and pressure. *Circ. Res.* 20:354–361.

Bafitis, H., and Sargent, F. 1977. Human physiological adaptability through the life sequence. *J. Gerontol.* 32:402–410.

Ben-Ishay, Z., Abramowitz, A., and Ungar, H. 1962. Atherosclerosis and aging of the aorta in adult Jewish population of Israel. *Am. J. Cardiol.* 10:407–415.

Beznak, M. 1956. Changes in heart weight and blood pressure following aortic constriction in rats. *Canad. J. Biochem. Physiol.* 33:995–1002.

Bing, O. H. L., Matsushita, S., Fanburg, B. L., and Levine, H. J. 1971. Mechanical properties of rat cardiac muscle during experimental hypertrophy. *Circ. Res.* 28:234–245.

Bramwell, J. C., and Hill, A. V. 1922a. The velocity of the pulse wave in man. *Proc. Roy. Soc. Lond. (Biol.)* 93:298–306.

Bramwell, J. C., and Hill, A. V. 1922b. Velocity of transmission of the pulse wave and elasticity of arteries. *Lancet* 1:891–892.

Bramwell, J. C., Hill, A. V. and McSwiney, B. A. 1923. The velocity of the pulse wave in man in relation to age as studied by the hot-wire sphygmograph. *Heart* 10:233–249.

Brandfonbrener, M., Landowne, M., and Shock, N. W. 1955. Changes in cardiac output with age. *Circulation* 12:557–566.

Butcher, H. B., and Newton, W. T. 1958. The influence of age, arteriosclerosis and homotransplantation upon the elastic properties of major human arteries. *Ann. Surg.* 148:1–20.

Chattergee, K., and Parmley, W. W. 1977. The role of vasodilator therapy in heart failure. *Prog. Cardiovasc. Dis.* 19:301–325.

Cleary, E. 1963. A correlative and comparative study of the nonuniform arterial wall. M. D. Thesis, University of Sydney.

Cohn, J. N. 1973. Vasodilator therapy for heart failure: influence of impedance on left ventricular performance. *Circulation* 48:5–7.

Cotton, R., and Wartman, W. B. 1961. Endothelial patterns in human arteries, their relationship to age, vessel site, and atherosclerosis. *Arch. Pathol.* 2:15–24.

Cox, R. H. 1977. Effects of age on the mechanical properties of rat carotid artery. *Am. J. Physiol.* 233:H256–H263.

Eliakim, M., Sapozinikov, D., and Weinman, J. 1971. Pulse wave velocity in healthy subjects and in patients with various disease states. *Am. Heart J.* 82:448–457.

Elzinga, G., and Westerhof, N. 1973. Pressure and flow generated by the left ventricle against different impedances. *Circ. Res.* 32:178–186.

Faber, M., and Moller-Hou, G. 1952. The human aorta. V. Collagen and elastin in the normal and hypertensive aorta. *Acta Pathol. Microbiol. Scand.* 31:377–382.

Finkelstein, S. M., and Collins, V. R. 1982. Vascular hemodynamic impedance measurement. *Prog. Cardiovasc. Dis.* 24:401–418.

Fischer, G. M., and Llaurado, J. G. 1966. Collagen and elastin content in canine arteries selected from functionally different vascular beds. *Circ. Res.* 19:394–399.

Freis, E. D., Heath, W. C., Luchsinger, P. C., and Snell, R. E. 1967. Changes in the carotid pulse which occur with age and hypertension. *Am. Heart J.* 71:757–765.

Gardin, J. M., Henry, W. L., Savage, D. D., and Epstein, S. E. 1977. Echocardiographic evaluation of an older population without clinically apparent heart disease. *Am. J. Cardiol.* 39:277.

Gerstenblith, G., Fredericksen, J., Yin, F. C. P., Fortuin, N. J., Lakatta, E. G., and Weisfeldt, M. L. 1977. Echocardiographic assessment of a normal adult aging population. *Circulation* 56:273–278.

Gerstenblith, G., Lakatta, E. G., and Weisfeldt, M. L. 1976. Age changes in myocardial function and exercise response. *Prog. Cardiovasc. Dis.* 19:1–21.

Gessner, U. 1972. Vascular input impedance. In Bergel, D. H., ed. *Cardiovascular Fluid Dynamics.* New York: Academic, pp. 315–349.

Gozna, E. R., Marble, A. E., Shaw, A., and Holland, J. G. 1974. Age-related changes in the mechanics of the aorta and pulmonary artery of man. *J. Appl. Physiol.* 36:407–411.

Granath, A., Jonsson, B., and Strandell, T. 1970. Circulation in healthy old men studied by right-heart catheterization at rest and during exercise in supine and sitting position. In Brunner, D., and Jokl, E., eds. *Medicine and Sport: Physical Activity and Aging.* Baltimore: University Park Press, pp. 48–79.

Gray, R. J., Handler, F. P., Blanche, J. O., Zuckner, J., and Blumenthal, H. T. 1953. Aging process of aorta and pulmonary artery in negro and white races. *Arch. Pathol.* 56:238–253.

Guard, H. R., and Bhende, J. M. 1953. Changes due to aging in the abdominal aorta. *Ind. J. Med. Res.* 41:267–276.

Gundel, W., Cherry, G., Rajgopalan, B., Tan, L. P., Lee, G., and Schultz, D. 1981. Aortic input impedance in man: acute response to vasodilator drugs. *Circulation* 63:1305–1314.

Hallock, P. 1934. Arterial elasticity in man in relation to age as evaluated by the pulse wave velocity method. *Arch. Intern. Med.* 54:770–778.

Hallock, P., and Benson, I. C. 1937. Studies on the elastic properties of human isolated aorta. *J. Clin. Invest.* 16:595–602.

Harkness, M. L. R., Harkness, R. D., and McDonald, D. A. 1957. The collagen and elastin of the arterial wall in the dog. *Proc. Roy. Soc. Biol.* 146:541–551.

Harrison, T. R., Dixon, K., Russel, R. O., Bedwai, P. S. and Coleman, H. N. 1964. The relation of age to the duration of contraction, ejection, and relaxation of the normal heart. *Am. Heart J.* 67:189–199.

Hass, G. M. 1942. Elastic tissue. II. A study of the elasticity and tensile strength of elastic tissue isolated from the human aorta. *Arch. Pathol.* 34:971–981.

Imperial, E. S., Levy, M. N., and Zieski, H. 1961. Outflow resistance as an independent determinant of cardiac performance. *Circ. Res.* 9:1148–1155.

Ishide, N., Shimizu, Y., Maruyama, Y., Koiwa, Y., Nunokawa, T., Isoyama, S., Kitaoka, S., Tamaki, K., Inooka, E., and Takishima, T. 1980. Effects of changes in aortic input impedance on systolic pressure-ejected volume relationships in the isolated supported canine left ventricle. *Cardiovasc. Res.* 14:229–243.

Julius, S., Amery, A., Whitlock, L. S., and Conway, J. 1967. Influence of age on the hemodynamic response to exercise. *Circulation* 36:222–230.

Kannel, W. B., Wolfe, P. A., McGee, D. L., Dawber, T. R., McNamara, P., and Castelli, W. P. 1981. Systolic blood pressure, arterial rigidity and risk of stroke: the Framingham study. *JAMA* 245:1225–1229.

Koyanagi, S., Eastham, C., and Marcus, M. L. 1982. Effects of chronic hypertension and left ventricular hypertrophy on the incidence of sudden cardiac death after coronary artery occlusion in conscious dogs. *Circulation* 65:1192–1197.

Lakatta, E. G., Gerstenblith, G., Angell, C. S., Shock, N. W., and Weisfeldt, M. L. 1975. Prolonged contraction duration in aged myocardium. *J. Clin. Invest.* 55:61–68.

Landowne, M. 1958. The relation between inta-arterial pressure and impact pulse wave velocity with regard to age and arteriosclerosis. *J. Gerontol.* 13:153–162.

Langewouters, G. J. 1982. Visco-elasticity of the human aorta in vitro in relation to pressure and age. Ph.D. Thesis, Free University of Amsterdam.

Learoyd, B. M., and Taylor, M. G. 1966. Alterations with age in the viscoelastic properties of human arterial walls. *Circ. Res.* 18:278–292.

Linzbach, A. J., and Akuomoa-Boateng, E. 1973. Die Alternsveranderungen des menschlichen Herzens. I. Das Herzgewicht im Alter. *Klin. Wochenschr.* 51:156–163.

McDonald, D. A. 1974. *Blood Flow in Arteries*. London: Edward Arnold.

Mahomed, F. A. 1892. The physiology and clinical use of the sphygmograph. *Medical Times and Gazette* 1:62–64, 128–130, 220–222.

Merillon, J. P., Motte, G., Masquet, C., Azancot, I., Guiomard, A., and Gourgon, R. 1982. Relationship between physical properties of the arterial system and left ventricular performance in the course of aging and arterial hypertension. *Eur. Heart J.* 3 (Suppl. A):95–101.

Milch, R. A. 1965. Matrix properties of the aging arterial wall. *Mono. Surg. Sci.* 2:261–341.
Miller, R. R., Vismara, L. A., Williams, D. O., Amsterdam, E. A., and Mason, D. T. 1976. Pharmacological mechanisms for left ventricular unloading in clinical congestive heart failure. *Circ. Res.* 39:127–133.
Mills, C. J., Gabe, I. T., Gault, J. H., Mason, D. T., Ross, J. Jr., Braunwald, E., and Shillingford, J. P. 1970. Pressure-flow relationships and vascular impedance in man. *Cardiovasc. Res.* 4:405–417.
Milnor, W. R. 1975. Arterial impedance as ventricular afterload. *Circ. Res.* 36:565–570.
Milnor, W. R. 1978. Influence of arterial impedance on ventricular function. *In* Bauer, R. O., and Busse, R. eds. *The Arterial System.* Berlin: Springer-Verlag, pp. 227–235.
Milnor, W. R. 1982. *Hemodynamics.* Baltimore: Williams & Wilkins.
Mirsky, I., and Janz, R. F. 1976. The effect of age on the wall stiffness of the human thoracic aorta: a large deformation anisotropic elastic analysis. *J. Theoret. Biol.* 59:467–484.
Monroe, R. G. 1964. Myocardial oxygen consumption during ventricular contraction and relaxation. *Circ. Res.* 24:294–300.
Moritz, A. R., and Oldt, M. R. 1937. Arteriolar sclerosis in hypertensives and non-hypertensives. *Am. J. Pathol.* 13:679–728.
Movat, H. Z., More, R. H., and Haust, M. D. 1958. The diffuse intimal thickening of the human aorta with aging. *Am. J. Pathol.* 34:1023–1030.
Mowat, D. H. R., Haites, N. E., and Rawles, J. M. 1983. Aortic blood flow velocity measurement in healthy adults using a simple ultrasound technique. *Cardiovasc. Res.* 17:75–80.
Mozersky, D. J., Sumner, D. S., Hokanson, D. E., and Standress, D. E. 1973. Transcutaneous measurement of arterial wall properties as a potential method of estimating age. *J. Am. Geriatr. Soc.* 21:18–20.
Mueller, T. M., Marcus, M. L., Kerber, R. E., Young, J. A., Barnes, R. W., and Abboud, F. M. 1978. Effect of renal hypertension and left ventricular hypertrophy on the coronary circulation in dogs. *Circ. Res.* 42:543–549.
Murgo, J. P., Giolma, J. P., and Altobelli, S. A. 1977. Signal acquisition and processing for human hemodynamic research. *Proc. IEEE* 65:696–702.
Murgo, J. P., Westerhof, N., Giolma, J. P., and Altobelli, S. A. 1980. Aortic input impedance in normal man: relationship to pressure wave shape. *Circulation* 62:105–116.
Murgo, J. P., Westerhof, N., Giolma, J. P., and Altobelli, S. A. 1981a. Manipulation of ascending aortic pressure and flow wave reflections with the Valsalva maneuver: relationship to input impedance. *Circulation* 63:122–132.
Murgo, J. P., Westerhof, N., Giolma, J. P., and Altobelli, S. A. 1981b. Effects of exercise on aortic input impedance and pressure wave forms in normal humans. *Circ. Res.* 48:334–343.
Myers, V. C., and Lang, W. W. 1946. Some chemical changes in the human thoracic aorta accompanying the aging process. *J. Gerontol.* 1:441–444.
Nakashima, T., and Tanikawa, J. 1971. A study of human aortic distensibility with relation to atherosclerosis and aging. *Angiology* 22:477–490.
Newman, D. L., and Lallemand, R. C. 1978. The effect of age on the distensibility of the abdominal aorta of man. *Surg. Gyn. Obst.* 147:211–214.

Nichols, W. W., Conti, C. R., Walker, W. E., and Milnor, W. R. 1977. Input impedance of the systemic circulation in man. *Circ. Res.* 40:451–458.

Nichols, W. W., and McDonald, D. A. 1972. Wave velocity in the proximal aorta. *Med. Biol. Eng. Comp.* 10:327–335.

Nichols, W. W., O'Rourke, M. F., Avolio, A. P., Yaginuma, T., Murgo, J. P., Pepine, C. J., and Conti, C. R. 1985a. Effects of age on ventricular-vascular coupling. *Am. J. Cardiol.* 55:1179–1184.

Nichols, W. W., Paley, D. M., Thompson, L. V., and Lambert, C. R. 1985b. Experimental evaluation of a multisensor velocity-pressure catheter. *Med. Biol. Eng. Comp.* 23:79–83.

Nichols, W. W., and Pepine, C. J. 1982. Left ventricular afterload and aortic input impedance: implications of pulsatile blood flow. *Prog. Cardiovasc. Dis.* 24:293–306.

Nichols, W. W., Pepine, C. J., Conti, C. R., and Christie, L. G. 1980a. Evaluation of a new catheter mounted electromagnetic velocity sensor during cardiac catheterization. *Cath. Cardiovasc. Diag.* 6:97–113.

Nichols, W. W., Pepine, C. J., Feldman, R. L., Whittle, J., Selby, J. H., Kelly, T., and Conti, C. R. 1979. Influence of changes in pulsatile components of vascular load on left ventricular function: power-load relations in patients without heart failure. *Circulation* 59, 60: 362.

Nichols, W. W., Pepine, C. J., Geiser, E. A., and Conti, C. R. 1980b. Vascular load defined by the aortic input impedance spectrum. *Fed. Proc.* 39:196–201.

Nielsen, B. L., Nielsen, J. S., and Fabricius, J. 1968. Pulse wave velocity in the human aorta. *J. Am. Geriatr. Soc.* 16:647–657.

Noble, M. I. M. 1979. Left ventricular load, arterial impedance and their relationships. *Cardiovasc. Res.* 13:183–148.

O'Kane, H. D., Geha, A. S., Kleiger, R. E., Abe, T., Salaymeh, M. T., and Malik, A. B. 1973. Stable left ventricular hypertrophy in the dog. *J. Thorac. Cardiovasc. Surg.* 65:264–271.

O'Keefe, D. D., Hoffman, J. I. E., Cheitin, R., O'Neill, M. J., Allard, J. R., and Shapkin, E. 1978. Coronary blood flow in experimental canine left ventricular hypertrophy. *Circ. Res.* 43:43–51.

O'Rourke, M. F. 1967. Steady and pulsatile energy losses in the systemic circulation under normal conditions and simulated arterial disease. *Cardiovasc. Res.* 1:313–326.

O'Rourke, M. F. 1970. Arterial hemodynamics in hypertension. *Circ. Res.* 26 (Suppl. 2):123–133.

O'Rourke, M. F. 1982a. *Arterial Function in Health and Disease.* London: Churchill Livingstone.

O'Rourke, M. F. 1982b. Vascular impedance in studies of arterial and cardiac function. *Physiol. Rev.* 62:570–623.

O'Rourke, M. F., and Avolio, A. P. 1980. Pulsatile flow and pressure in human systemic arteries: studies in man and in a multi-branched model of the human systemic arterial tree. *Circ. Res.* 46:363–372.

O'Rourke, M. F., and Avolio, A. P. 1981. Ascending aortic impedance as the load presented to the left ventricle: effects of change in mean pressure, arterial compliance and peripheral resistance. In Baumann, D., ed. *Les Alpha Bloquants.* New York: Masson, pp. 104–117.

O'Rourke, M. F., Blazek, J. V., Morreels, C. L., and Krovetz, L. T. 1968. Pressure wave transmission along the human aorta: changes with age and in arterial degenerative disease. *Circ. Res.* 23:567–579.

O'Rourke, M. F., and Cartwell, T. B. 1971. Influence of aortic coarctation on pulsatile hemodynamics in the proximal aorta. *Circulation* 44:281–292.

O'Rourke, M. F., Norris, R. M., Campbell, T. C., Chang, V. P., and Sammel, W. L. 1981. Randomized controlled trial of intraaortic balloon counterpulsation in early myocardial infarction with acute heart failure. *Am. J. Cardiol.* 47:815–820.

O'Rourke, M. F., Sammel, N. L. and Chang, V. P. 1979. Arterial counterpulsation in severe refractory heart failure complicating acute myocardial infarction. *Br. Heart J.* 41:308–316.

O'Rourke, M. F., and Taylor, M. G. 1967. Input impedance of the systemic circulation. *Circ. Res.* 20:365–380.

Patel, D. J., Greenfield, J. C. Jr., Austen, W. G., Morrow, A. G., and Fry, D. L. 1965. Pressure-flow relationships in the ascending aorta and femoral artery of man. *J. Appl. Physiol.* 20:459–463.

Pepine, C. J., Nichols, W. W., and Conti, C. R. 1978. Aortic input impedance in heart failure. *Circulation* 58:460–465.

Pepine, C. J., Nichols, W. W., Curry, R. C. and Conti, C. R. 1979. Aortic input impedance during nitroprusside infusion: a reconsideration of afterload reduction and beneficial action. *J. Clin. Invest.* 64:643–654.

Peterson, L. H., Jensen, R. E., and Parnell, J. 1960. Mechanical properties of arteries in vivo. *Circ. Res.* 8:622–639.

Port, S., Cobb, F. R., Coleman, E., and Jones, R. H. 1980. Effect of age on the response of the left ventricular ejection fraction to exercise. *N. Engl. J. Med.* 303:1133–1137.

Pouleur, H., Covell, J. W., and Ross, J. Jr. 1979. Effects of alterations in aortic input impedance on force-velocity-length relationship in the intact canine heart. *Circ. Res.* 45:126–135.

Puls, R. J., and Heizer, R. W. 1967. Pulse wave changes with aging. *J. Am. Geriatr. Soc.* 15:153–165.

Quinones, M. A., Gaasch, W. H., and Alexander, T. K. 1976. Influence of acute changes in preload, afterload, contractile state and heart rate on ejection and isovolumic indices of myocardial contractility in man. *Circulation* 53:293–302.

Rembert, J. C., Kleinman, L. H., Fedor, J. M., Wechsler, A. S. and Greenfield, J. C. 1978. Myocardial blood flow distribution in concentric left ventricular hypertrophy. *J. Clin. Invest.* 62:379–386.

Roach, M., and Burton, A. C. 1959. The effect of age on the elasticity of the human iliac arteries. *Canad. J. Biochem. Physiol.* 37:557–570.

Roy, C. S. 1880. The elastic properties of the arterial wall. *J. Physiol. (Lond.)* 3:125–159.

Sagawa, K. 1967. Analysis of the ventricular pumping capacity as a function of input and output pressure loads. In Reeve, E. B., and Guyton, A. C., eds. *Physiological Bases of Circulatory Transport*. Philadelphia: Saunders, pp. 141–149.

Salisbury, P. F., Cross, C. E., and Rieben, P. A. 1962. Ventricular performance modified by elastic properties of outflow system. *Circ. Res.* 11:319–328.

Sarnoff, S. J., Braunwald, E., Welch, G. H. Jr., Case, R. B., Stainsby, W. N., and Macruz, R. 1958. Hemodynamic determinants of oxygen consumption of the heart with special reference to the time tension index. *Am. J. Physiol.* 192:148–156.

Schatz, R. A., Pasipoularides, A., and Murgo, J. P. 1981. The effect of arterial pressure reflections on myocardial supply-demand dynamics (abstr.). *Circulation* 64 (Suppl IV):324.

Schimmler, W. 1966. Correlation between the pulse wave velocity in the aortic iliac vessel and age, sex and blood pressure. *Angiology* 17:314–322.

Shock, N. W. 1961. Physiological aspects of aging in man. *Ann. Rev. Physiol.* 23:97–122.

Simonson, E., and Nakagawa, K. 1960. Effect of age on pulse wave velocity and aortic ejection time in healthy men and in men with coronary artery disease. *Circulation* 22:126–129.

Sordahl, L. A., McCollum, W. B., Wood, W. G., and Schwartz, A. 1973. Mitochondric and sarcoplasmic reticulum function in cardiac hypertrophy and failure. *Am. J. Physiol.* 224:497–502.

Stehbens, W. E. 1979. *Hemodynamics and the Blood Vessel Wall.* Springfield, Ill.: Thomas.

Sunagawa, K., Maughan, W. L., and Sagawa, K. 1985. Stroke volume effect of changing arterial input impedance over selected frequency ranges. *Am. J. Physiol.* 248:H477–H484.

Taylor, M. G. 1957. An approach to the analysis of the arterial pulse wave. II. Fluid oscillations in an elastic tube. *Phys. Med. Biol.* 1:321–329.

Templeton, G. H., Platt, M. R., Willerson, J. T. and Weisfeldt, M. L. 1979. Influence of aging on left ventricular hemodynamics and stiffness in beagles. *Circ. Res.* 44:189–194.

Unverferth, D. V., Felters, J. K., Unverferth, B. J., Leier, C. V., Magorien, R. D., Arn, A. R. and Baker, P. B. 1983. Human myocardial histologic characteristics in congestive heart failure. *Circulation* 68:1194–1200.

Urschel, C. W., Covell, J. W., Sonnenblick, E. H., Ross, J. Jr., and Braunwald, E. 1968. Effects of decreased aortic compliance on performance of the left ventricle. *Am. J. Physiol.* 214:298–304.

Van den Bos, G. C., Westerhof, N., Elzinga, G., and Sipkema, P. 1976. Reflection in the systemic arterial system: effects of aortic and carotid occlusion. *Cardiovasc. Res.* 10:565–576.

Walker, W. E. 1975. The influence of changes in aortic input impedance on the dynamics of left ventricular performance. Ph.D. Thesis, Johns Hopkins University, Baltimore.

Weisfeldt, M. L., Loeven, W. A., and Shock, N. W. 1971. Resting and active mechanical properties of trabeculae carneae from aged male rats. *Am. J. Physiol.* 220:1921–1927.

Westerhof, W., Sipkema, ,P., Elzinga, G., Murgo, J. P., and Giolma, J. P. 1979. Arterial impedance. In Hwang, N. H. C., Gross, D. R., and Patel, D. J. eds. *Quantitative Cardiovascular Studies: Clinical and Research Applications of Engineering Principles*. Baltimore: University Park Press, pp. 111–150.

Westerhof, N., Sipkema, P., Van den Bos, G. C., and Elzinga, G. 1972. Forward and backward waves in the arterial system. *Cardiovasc. Res.* 6:648–658.

Wilcken, D. E. L., Charlier, A. A., Hoffman, J. I. E., and Guz, A. 1964. Effects of alterations in aortic impedance on the performance of the ventricle. *Circ. Res.* 14:283–293.

Wolinsky, H., and Glagov, S. 1964. Structural basis for the static mechanical properties of the aortic media. *Circ. Res.* 14:400–413.

Yin, F. C. P. 1980. The aging vasculature and its effects on the heart. In Weisfeldt, M. L., ed. *The Aging Heart.* New York: Raven Press, pp. 137–213.

Yin, F. C. P., Spurgeon, H. A., and Kallman, C. H. 1983. Age-associated alterations in viscoelastic properties of canine aortic strips. *Circ. Res.* 53:464–472.

Yin, F. C. P., Spurgeon, H. A., Weisfeldt, M. L., and Lakatta, E. G. 1980. Mechanical properties of myocardium from hypertrophied rat hearts: a comparison between hypertrophy induced by senescence and by aortic banding. *Circ. Res.* 46:292–300.

Yin, F. C. P., Weisfeldt, M. L., and Milnor, W. R. 1981. Role of aortic input impedance in the decreased cardiovascular response to exercise with aging in dogs. *J. Clin. Invest.* 68:28–38.

Young, T. 1809. On the functions of the heart and arteries: the Croonian Lecture. *Phil. Trans. Roy. Soc.* 99:1–31.

5
Aging and Vascular Impedance

FRANK C. P. YIN

Introduction

The assessment of age-associated changes in the cardiovascular system has long interested clinicians and physiologists (see reviews by Gerstenblith, Lakatta, and Weisfeldt 1976; Lakatta 1979; Lakatta and Yin 1982; Yin 1980). This interest exists not only because aging is a physiologic intervention that affects every individual but also because aging can provide insight into the mechanisms underlying normal or pathologic function.

Despite the recognition that alterations in vascular function can dramatically affect ventricular function, only a few experimental studies address the effects of aging on ventricular/vascular interaction (Brandfonbrenner, Landowne, and Shock 1955; Cox 1977; Gundel et al. 1981; Merillon et al. 1978; Nichols et al. 1985; Yin 1980; Yin, Weisfeldt, and Milnor 1981; Yin et al. 1980). Most of these studies are descriptive rather than quantitative assessments of this interaction, owing in large part to lack of a satisfactory means of quantitatively expressing ventricular/vascular interaction. As the various chapters in this book and several recent publications (Sunagawa, Sagawa, and Maughan 1984; Van den Horn, Westerhof, and Elzinga 1984) illustrate, however, there are some new and exciting approaches to the quantitative description of ventricular/vascular interaction. Application of these approaches should provide considerable new insight in terms of aging as well as other interventions.

Quantitative description of ventricular/vascular interaction is complicated because such a description requires not only information about both ventricular and vascular function but also assumptions about the dynamics of the valve that connects the two systems. Moreover, as some of the succeeding chapters illustrate, quantitative approaches require that ventricular and vascular function be described very specifically in order for the subtleties of the interaction to be appreciated. For ventricular function we need the pressure–volume or –elastance relationships (Chapters 3, 8, and 10) and perhaps the internal resistance as well (Chapter 11). For vascular function we need the stroke volume–mean pressure relationship

(Chapter 9), compliance (Chapter 16), resistance (Chapters 1, 10, and 11), reflections (Chapter 6), input impedance (Chapters 1, 3, 4, and 8), or the impulse response (Chapter 15).

Many recent studies look at the effect of aging on myocardial muscle and some aspects of intact ventricular function (Froehlich et al. 1978; Gerstenblith, Lakatta, and Weisfeldt 1976; Gerstenblith et al. 1979; Guarnieri et al. 1979; Janz et al. 1976; Kane et al. 1976; Lakatta, Gerstenblith, and Angell 1975; Lee, Karpeles, and Downing 1972; Rumberger and Timmermann 1976; Shreiner, Weisfeldt, and Shock 1969; Spurgeon et al. 1977; Templeton et al. 1979; Weisfeldt, Loeven, and Shock 1971). To my knowledge, however, there are no data on how aging affects ventricular pressure–volume or internal resistance.

The effects of aging on many different aspects of vascular function have been examined and recently reviewed (Yin 1980, Ch. 4). The arterial system increases in size (Bader and Kapal 1957), becomes less distensible (Avolio et al. 1983, 1985; Bader 1967; Band, Goedhard, and Knoop 1972; Bramwell, Hill, and McSwiney 1923; Butcher and Newton 1958; Cox 1977; Gonza et al. 1974; Guard and Bhende 1953; Hallock and Benson 1937; Krafka 1940; Landowne 1958; Learoyd and Taylor 1966; Mirsky and Janz 1976; Nakashima and Tanikawa 1971; Roy 1880; Saxton 1942; Schimmler 1966; Simonson and Nakagawa 1960; Wilens 1937; Yin, Spurgeon, and Kallman 1983), and becomes less responsive to adrenergic vasoactive agents (Cohen and Berkowitz 1974; Fleisch and Hooker 1976; Fleisch, Maling, and Brodie 1970; Tuttle 1966). While each of these descriptions of the effect of aging on vascular function has provided some insight into the basic mechanisms of aging, almost none is in a format suitable for specifically assessing the effects of aging on ventricular/vascular interaction.

Because vascular impedance is a concise yet complete means of quantifying the pressure/flow relations and the hydraulic load that the vasculature presents to the ejecting ventricle, as several chapters in this book illustrate, arterial impedance occupies a central role in any discussion of ventricular/vascular interaction (McDonald 1974; Milnor 1965,1982). However, since aortic size, stiffness, reflections, and other characteristics each influence impedance differently and can be altered to different degrees by aging, it is difficult to predict the influence of aging on impedance using the existing vascular aging data.

Only a few studies deal specifically with the effects of aging on arterial impedance (Cox 1977; Gundel 1981; Merillon et al. 1978; Nichols et al. 1985; Yin, Weisfeldt, and Milnor 1981; Yin et al. 1980). Before reviewing these studies I will first discuss several clinical and physiologic aspects of aging and ventricular/vascular interaction. Since few data are available, of necessity most of this discussion is speculative and is intended to indicate fruitful areas for further work. Some other aspects of aging and ventricular/vascular coupling are discussed in the preceding chapter.

Aging and Ventricular/Vascular Coupling

One can envision several circumstances in which aging could be an important modifier of ventricular/vascular interaction. For example, many studies have shown that there is a mild degree of left ventricular hypertrophy with aging (Linzbach and Boateng 1976; Yin et al. 1980,1982). In the young individual, the load presented by the vasculature is nearly optimal—the proportion of pulsatile to steady components of load is very low (O'Rourke 1982). If the steady-state and/or the pulsatile components of the vascular load increased with aging, ventricular hypertrophy could result. My colleagues and I have examined this hypothesis by comparing aortic impedances and the function of hypertrophied left ventricular muscles from aged rats with those in which left ventricular hypertrophy was induced by aortic banding (Yin et al. 1980). We found comparable degrees of hypertrophy in the two groups. The left ventricular hypertrophy in the aged rats, however, could not be attributed to an elevation in either mean blood pressure or aortic impedance. The function of the aged, hypertrophied myocardium, while qualitatively similar to that of the pressure-overload hypertrophied muscle, differed quantitatively. This study was performed in anesthetized animals, and thus we cannot state whether the vascular state in the conscious animal would demonstrate similar results. Clearly more studies are needed before one can ascertain, with any degree of assurance, how much if any of the hypertrophy can be attributed to an increased vascular load produced by aging.

In addition to the direct effects of interaction between the ventricle and its own vasculature, which are perhaps exemplified by myocardial hypertrophy, it has become evident in recent years that interdependence between the two ventricles may have a profound effect on the functioning of each one, particularly in the presence of an intact pericardium (Bemis et al. 1974; Elzinga et al. 1974; Glantz et al. 1978; Santamore et al. 1976; Weber et al. 1981). For example, the pulmonary vasculature, by virtue of its physical location, directly affects right ventricular function via its input impedance and indirectly affects left ventricular function via interdependence. Clinically, this interdependence is the postulated mechanism for the normal decrease in arterial pressure during deep inspiration (pulsus paradoxus). This effect is exaggerated in conditions in which the ventricular volumes are constrained more than normally by the pericardium, as in tamponade or constrictive pericarditis (Shabetai, Fowler, and Guntheroth 1970). Recent studies (Brinker et al. 1980, Weiss et al. 1978, Weyman et al. 1977) have demonstrated that the interventricular septum shifts toward the left ventricle during conditions that cause an overload on the right ventricle. Whether this septal shift influences left ventricular function is not yet clear. With aging, if there are age-associated alterations in the properties of the pericardium or the septum, interdependence may be exaggerated or diminished. The role of aging in modulating ventricular interdependence has not been studied. From what has already been discussed, one might speculate that, during exercise, the pulmonary vascular bed in the aged might not be able to decrease its total power as much as in the young (perhaps owing to less efficient oxygenation or diminished pulmonary vascular response to catecholamines). This would cause a relatively higher load on the right ventricle in the old, and thus might impair left ventricular function, leading to diminished exercise capacity.

The efficacy of vasodilators in treating heart failure has been attributed primarily to direct unloading of the ventricles owing to lowered peripheral resistance (Chatterjee et al. 1973, Cohn and Franciosa 1977, Cohn et al. 1974, Miller et al. 1977). While this mechanism is undoubtedly operative, another possible mechanism involving interdependence has recently been suggested (Ludbrook, Byrne, and McKnight 1979; Pepine et al. 1979; Sivak et al. 1977; Yin et al. 1983). Sivak and colleagues (1977) demonstrated that during nitroprusside infusion in dogs, left ventricular cardiac output increased in conjunction with an increase in pulmonary blood volume and a decrease in pulmonary vascular resistance. The rise in output occurred prior to any change in mean arterial pressure. Presumably, nitroprusside dilated the pulmonary bed and thereby decreased the load presented to the right ventricle. Via the interdependence mechanism, this unloading of the right side allowed the left ventricular end-diastolic pressure to decrease and output to increase even before a direct drug effect on aortic resistance or impedance occurred. This study was only inferential, however, since no measurements of impedance were made. Data supportive of a similar mechanism were obtained recently in humans with severe left heart failure, and impedance measurements were made (Pepine et al. 1979; Yin et al. 1983). In Pepine's study a sustained improvement in cardiac output was seen during nitroprusside infusion, even after resistance had been returned to control levels with simultaneous infusion of phenylephrine. In Yin's study both pulmonary arterial and aortic input impedances were measured during infusion of nitroprusside. Left ventricular forward output increased and filling pressure decreased at the same time that pulmonary resistance decreased relatively more than aortic resistance. Pulmonary but not aortic characteristic impedance decreased at all doses. Although these data do not prove cause–effect relationships, they do suggest that right ventricular unloading has some role in this ventricular response. If aging alters the pulmonary vasculature response to these drugs such that input impedance is less affected than in the young, the aged right ventricle would be less unloaded and the beneficial effect on the aged left ventricle might be diminished. On the other hand, if pulmonary vasodilatation actually has a deleterious effect on right ventricular function and the effect is accentuated with age, this therapeutic modality might not be useful in older patients.

As a final example of how age may affect ventricular/vascular interaction, let's consider the effect of the left ventricle on the peripheral pulmonary system. It is known that ischemia causes both an increase in stiffness of myocardial tissue and a decrease in distensibility of the ventricular chamber. How age affects the response to ischemia is not known, but as was discussed earlier, age causes changes similar to those caused by ischemia in mechanical properties in the normal state. Thus during ischemia (due to coronary artery disease, for example), the aged left ventricle (and atrium) might become much less distensible than a comparably diseased young left ventricle. This would be reflected in elevated end-diastolic pressure and pulmonary capillary pressure, making the aged person more prone to development of pulmonary edema. Thus the aged group might be more likely to present with pulmonary edema in addition to (or instead of) angina pectoris during ischemia. The relationship of this so-called "angina equivalence" to aging has not been examined. Similarly, any other factor that causes an overload of the left ventricle would be prone to manifest itself clinically with pulmonary symptoms in the aged because of the decreased distensibility of the ventricle (and presumably of the atrium) with increasing age.

Aging and Arterial Impedance

The few studies that have specifically assessed the effects of age on arterial impedance have yielded conflicting results. Cox (1977) measured the pressure–diameter relationships of excised segments of carotid arteries from Charles River rats aged 2, 12, and 24 months. He calculated the incremental elastic modulus and the characteristic impedance derived from this parameter during both maximal smooth-muscle activation with norepinephrine and complete inactivation with cyanide, iodoacetate, and dinitrophenol. He found that the passive characteristic impedance (1) was pressure dependent in all three groups, (2) had the highest magnitude in the 2-month-old group at all pressures, and (3) did not differ between the 12- and 24-month-old groups (Fig. 5.1). Whether these results from carotid arteries can be extrapolated to the aorta is questionable. Nevertheless, Cox interpreted his findings as indicating that the decreased efficiency of an aging heart was compensated to some extent by a decreased hydraulic load.

Merillon and co-workers (1978) measured pressure–diameter relationships in humans at the time of cardiac catheterization. From the systolic and diastolic diameters obtained during aortic angiography, they calculated a pressure–strain elastic modulus from which they estimated characteristic impedance. They found a significant positive correlation between age and characteristic impedance in one group of patients but no correlation in a group with hypertension (Fig. 5.2). It is difficult to interpret their results. First, the method they used to estimate characteristic impedance is of questionable accuracy. Second, their first group of patients included sev-

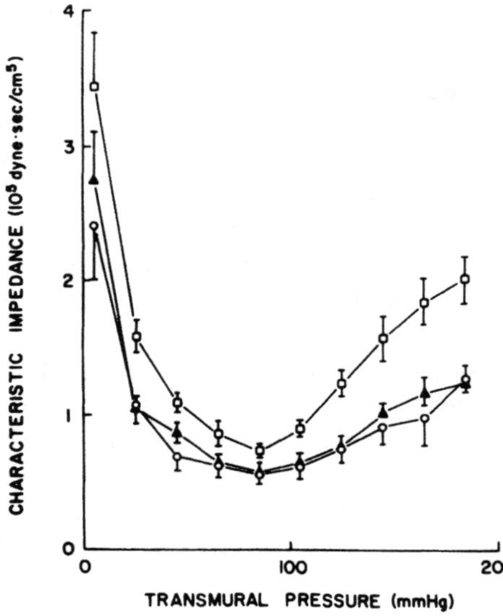

FIGURE 5.1. Average values of characteristic impedance of rat carotid arteries as a function of transmural pressure for rats aged 2 (open square), 12 (solid triangle), and 24 (open circle) months. Impedance was calculated from the incremental elastic modulus obtained from pressure–diameter measurements. (From Cox 1977).

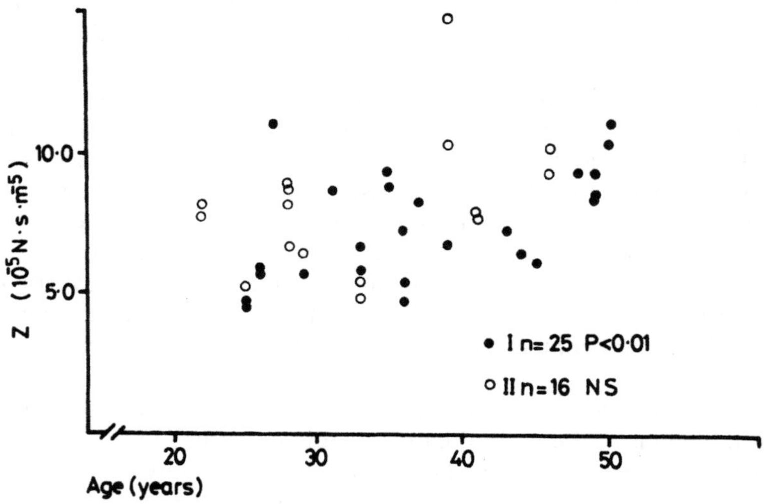

FIGURE 5.2. Relationship between characteristic impedance and age in patients with hypertension (group II) and those with either normal function or congestive heart failure (group I). Impedance was calculated from a pressure–strain modulus obtained from pressure–diameter measurements at cardiac catheterization. (From Merillon et al. 1978).

eral with congestive heart failure secondary to myocardial infarction. The effect of aging in the absence of disease is thus difficult to delineate.

Gundel and colleagues (1981) measured flow velocity and pressure using fluid-filled catheters in the ascending aorta of patients with angina pectoris undergoing cardiac catheterization. They calculated the input and characteristic impedances using standard Fourier analysis methods (Milnor 1982) and found no correlation between characteristic impedance and age over an age range of 36 to 62 years. Their results are also difficult to interpret because only three normal patients were in their study, because the age range examined was small, and because the frequency response of their pressure measurement system was limited. This latter limitation is of some concern because the characteristic impedance is an estimated quantity derived from the high-frequency components of the input impedance. The estimate is quite sensitive to noise in the data and to the exact harmonics used in the estimation.

Nichols and co-workers (1985) measured aortic flow velocity and pressure using micromanometers in a group of patients without discernible vascular disease. Using basically the same data analysis techniques as those Gundel used, they found that the systolic aortic diameter, the resistance, and the characteristic impedance all had a positive correlation with age (see Ch. 4, Figs. 4.18–4.20). In addition, the pressure waveforms in the older patients tended to have a mid-systolic rise (type A, see Ch. 4), whereas those in the younger patients tended to be rounder (type C). These investigators concluded that both the steady and pulsatile components of the hydraulic load increased and that arterial reflections returned earlier with increasing age. As they pointed out, however, one cannot exclude the possibility that, despite the apparently normal populations examined, the effects of both age and clinically inapparent arteriosclerosis were manifested. As discussed in Chapter 1, the changes ascribable to arteriosclerosis may overshadow those due to aging per se. Indeed, Gundel and colleagues (1981) in another study in normal humans found no relationship between age and characteristic impedance.

We felt that it was essential to measure the aortic impedance in aging dogs in whom arteriosclerosis is not normally present. In an earlier study we chronically instrumented beagle dogs with an electromagnetic flow probe and an indwelling micromanometer in the ascending aorta (Yin, Weisfeldt, and Milnor 1981). After recovery from surgery each dog was subjected to a graded treadmill exercise protocol while pressure and flow were recorded. Aortic input impedance was calculated with standard Fourier analysis techniques. We found no difference in resistance or characteristic impedance between the young and old dogs at rest (Figs. 5.3–5.5), but at all levels of exercise, including a slow walk, the old but not the young dogs had an increase in characteristic impedance. Both groups displayed a large decrease in peripheral resistance at all levels of exercise. We postulated that the rise in characteristic impedance produced enough

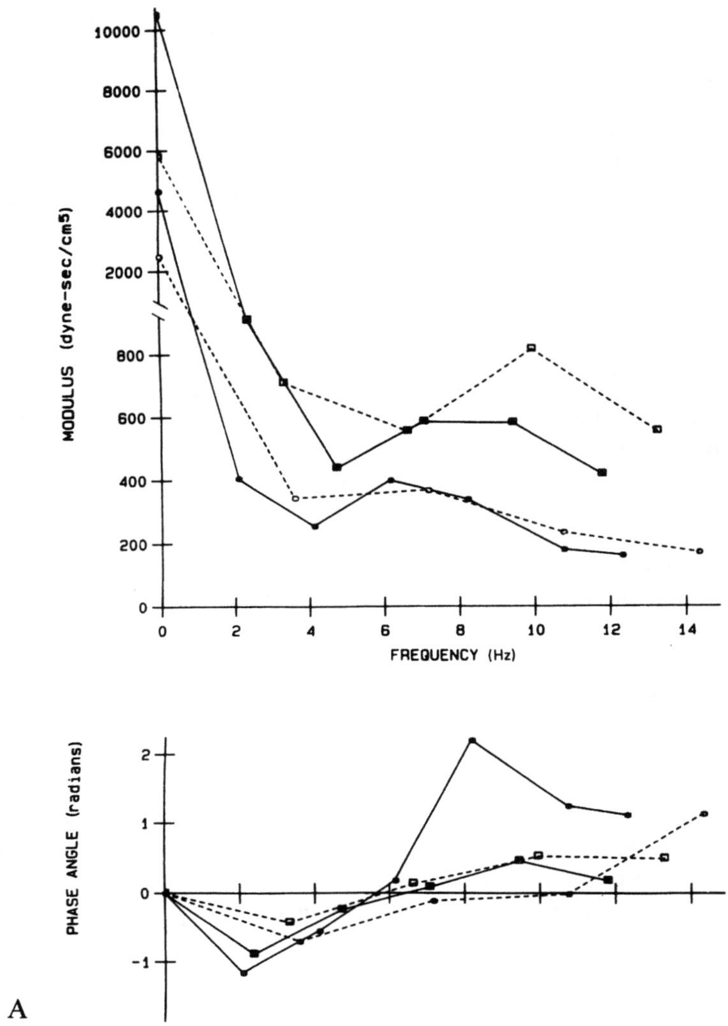

FIGURE 5.3. Typical aortic impedance responses to exercise in a senescent (open squares) and a young (open circles) dog and respective responses at rest (solid squares and circles) before and after propranolol administration. A. Before propranolol administration.

of an increase in vascular load to limit the stroke volume in the older dogs and thus limit their exercise capacity. Without direct measurement of ventricular responses, however, this speculation could not be verified. The aortic characteristic impedance response appeared to be due, in part, to a decreased beta-adrenergically mediated aortic relaxation, since beta blockade in the young dogs resulted in a response that resembled that in the old dogs. This would agree with the previously described decreased

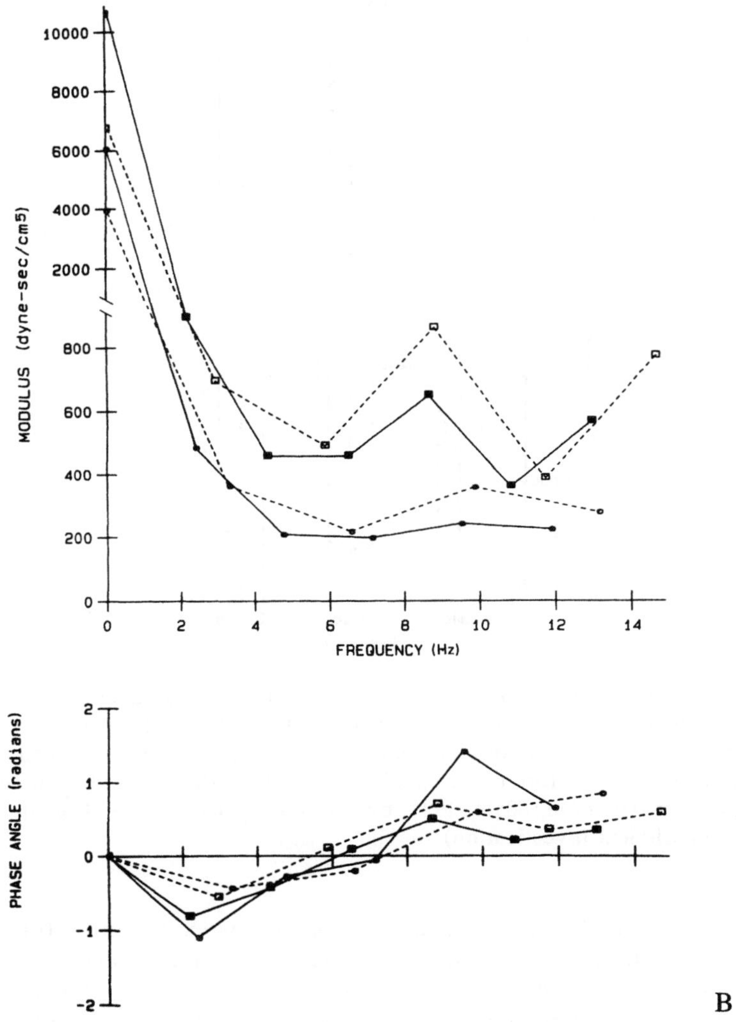

FIGURE 5.3. B. After propranol administration.

beta-adrenergically mediated vasodilatory response in strips of aged rabbit aortas (Fleisch and Hooker 1976; Fleisch, Maling, and Brodie 1970; Tuttle 1966). Since vasodilatation is also due in part to metabolic factors in the exercising muscles, however, the interpretations are not unequivocal.

All of these examples point out the clear need for more data on the effect of aging on aortic impedance, particularly with respect to the role of adrenergically active agents. We therefore compared the direct effect of several different vasoactive agents on the aortic impedance in these same senescent and young dogs. Specifically, we examined the aortic impedance response to beta-adrenergically mediated vasodilatation (iso-

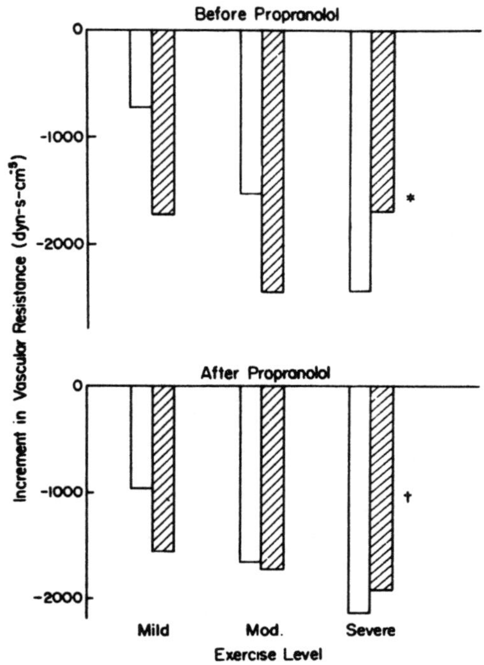

FIGURE 5.4. Averaged response of vascular resistance to graded levels of exercise in young (open bars) and senescent (striped bars) dogs before (*top*) and after (*bottom*) intravenous propranolol administration. The age difference in response is significant ($P = 0.05$) before propranolol but not after. (Reproduced from *Journal of Clinical Investigation,* 68:34, 1981; by copyright permission of The American Society for Clinical Investigation).

proterenol), to alpha-adrenergically mediated vasoconstriction (phenylephrine), and to intrinsic aortic tone (alpha- or beta-adrenergic blockade and combined blockade). To ascertain whether the in vivo vasodilatory response to a nonadrenergic vasodilator was independent of the adrenergic system, we also examined the response to nitroprusside. The remainder of this chapter details our findings.

Effects of Vasoactive Agents on Aortic Impedance in Dogs

Methods

To minimize sex and strain differences, we studied only purebred, exbreeder, female beagle dogs. To examine aging and not developmental effects, we limited the study to mature (1–3-year-old) and senescent (10–

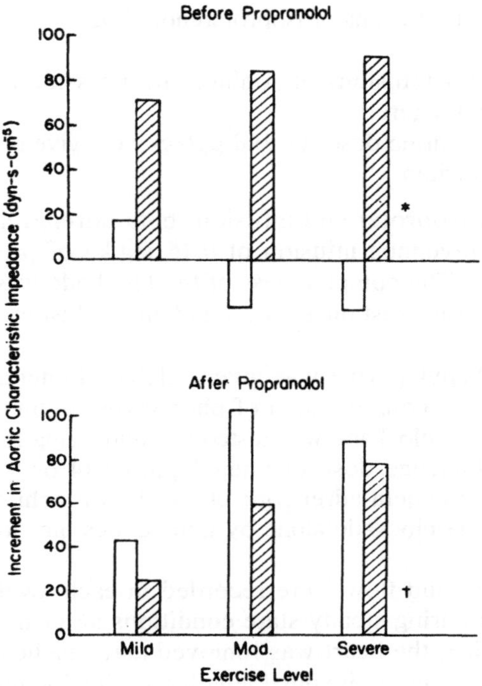

FIGURE 5.5. Averaged response of characteristic impedance to graded levels of exercise in senescent (striped bars) and young dogs (open bars) before (*top*) and after (*bottom*) intravenous propranolol administration. The age difference in response was significant ($P = 0.03$) before propranolol but was obliterated after. (Reproduced from *Journal of Clinical Investigation* 68:35, 1981, by copyright permission of The American Society for Clinical Investigation).

12-year-old) animals. All dogs were free of apparent cardiovascular disease at the time of the study and had no history of cardiovascular problems. With the animals under pentobarbital anesthesia (30 mg/kg i.v.), a left thoracotomy was performed and the ascending aorta was exposed. An electromagnetic cuff flow probe (Biotronex Labs, Silver Spring, Md.) that fit loosely around the ascending aorta was positioned just above the coronary ostia. A miniature pressure transducer (Konigsberg Instruments, Inc; Pasadena, Calif.) was placed in the aortic arch via an aortotomy just beyond the junction of the left subclavian artery. An indwelling right atrial line was also placed for later drug infusion. The chest was then closed and the animal allowed to recover for at least one week before any data were obtained.

The following protocols were carried out in each dog, the sequence varying and at least 48 and usually 72 hours elapsing between any studies involving adrenergic-blocking drugs:

1. Graded constant infusions of isoproterenol (0.02 to 0.64 μg/kg/min) were given.
2. Graded constant infusions of sodium nitroprusside (0.2 to 4.8 μg/kg/min) were administered.
3. Graded phenylephrine doses (1 to 8 μg/kg) were given by rapid injection into the right atrium.

On the day of isoproterenol infusion, beta-adrenergic blockade was produced by intravenous infusion of 0.15 mg/kg of propranolol hydrochloride (Ayerst). The completeness of the blockade was tested by monitoring the heart rate response to a challenge infusion of isoproterenol (0.64 μg/kg/min).

On the day phenylephrine was given, alpha-adrenergic blockade was produced by intravenous infusion of phenoxybenzamine (1 mg/kg). The completeness of the blockade was tested by monitoring the blood pressure response to a challenge dose of phenylephrine (4 or 8 μg/kg). Finally, combined alpha- and beta-adrenergic blockade was achieved shortly after production of alpha blockade alone by intravenous infusion of propranolol (0.15 mg/kg).

Aortic pressure and flow were recorded in each awake, unrestrained dog prior to and during steady-state conditions at each dose level of the drugs. On sacrifice, the heart was removed and weighed. Body size was assessed by measuring both spinal length and tibial length. Spinal length was estimated by measuring the distance from the first cervical spinal process to the last lumbar process. Tibial length was measured directly after dissection.

Data Analysis

All flow probes used in this study were calibrated by infusing saline through an excised length of aorta around which the probe was placed. Earlier measurements in our laboratory showed the flow probes and flowmeter amplifier system to have amplitude response that was flat up to 20 Hz and down 10% at 50 Hz. The phase lag was almost linear with frequency (1.3 degrees/Hz), and the damping ratio was 0.75. The frequency response of the Konigsberg pressure transducer was flat to at least 1000 Hz. Appropriate corrections for the frequency response of the flow system were applied to the data. The pressure transducers were calibrated at 37°C in vitro with a mercury column both before surgery and immediately after removal from the dog. Since only minimal baseline drift occurred (1–3 mm Hg/mo) and these studies were completed within 1 month, the post-study atmospheric pressure was assumed to have been the zero pressure level throughout the course of the recordings in vivo.

All analog signals were recorded on FM tape and subsequently digitized at 4-msec intervals. The calculations were made on PDP-11/10 (Digital Equipment Corp., Maynard, Mass.). As described previously (Yin, Weisfeldt, and Milnor 1981), at least three nonconsecutive beats for each baseline and intervention period were analyzed and averaged. Zero flow was assumed to be the value of flow in mid-to-late diastole. Standard hemodynamic parameters consisting of aortic pressure and flow, heart rate, stroke volume, and cardiac output were recorded and/or calculated directly. Aortic input impedance was calculated as described previously by decomposing the pressure and flow signals into their Fourier harmonics. The characteristic impedance (Z_o) was derived by averaging only the significant impedance moduli of frequencies greater than 5 Hz. External left ventricular hydraulic power was calculated from the pressure and flow harmonics by previously reported methods (Milnor, Bergel, and Bargainer 1966). The efficiency with which the external power of the left ventricle was converted into useful work was assessed by calculating the percentage of total power that was pulsatile.

Two measures of the reflection characteristics of the aorta and vasculature were examined: (1) The first zero crossing of the impedance angle was estimated by linear interpolation and (2) a reflection index was derived by dividing the maximum minus the minimum value of impedance by the characteristic impedance for frequencies greater than 5 Hz: $(Z_{max}-Z_{min})/Z_o$. This latter index estimates the fraction of the traveling wave that is reflected to the proximal aorta from peripheral sites.

Preliminary calculations indicated that there could be a variation of up to 10% in some variables (characteristic impedance, stroke volume, and the power terms) between peak inspiration and peak expiration either at rest or during exercise. These variations were usually 5% in aortic pressure, flow, cardiac output, and resistance. Therefore the beats selected for analysis were those near the middle of the respiratory cycle, and three such nonconsecutive beats were selected as representative of the resting or intervention conditions.

The data for the baseline condition consisted of the mean of the averages of three nonconsecutive beats recorded during stable hemodynamic periods before each of the drug infusions. The variables for these 9 to 15 beats were tested for statistically significant age differences by using an unpaired t test. During steady-state conditions at each drug level, the average values for three nonconsecutive beats were compared with the averaged baseline values. The effects of age and drug were analyzed by repeated measures analysis of variance. When interactions between effects were present, the interacting effects were analyzed separately to allow delineation of the influence of the individual underlying effects. All data are presented as mean plus or minus standard error.

Results and Discussion

The young and senescent dogs did not differ significantly in body weight (10.4 ± 0.6 vs. 11.1 ± 0.6 kg), tibial length (12.3 ± 0.3 vs. 12.6 ± 0.3 cm), spine length (44.6 ± 1.2 vs. 47.5 ± 1.0 cm), left ventricular wet weight

(49.5 ± 2.8 vs. 44.9 ± 3.1 g), aortic wall thickness (1.33 ± 0.11 vs. 1.33 ± 0.15 mm), or aortic diameter (15.5 ± 0.33 vs. 15.7 ± 0.29 mm).

The hemodynamic and impedance responses to isoproterenol are summarized in Table 5.1. There were no age-associated differences in the baseline values of heart rate, aortic pressure, external power, stroke volume, or reflection index. The flow was higher and both resistance and characteristic impedance were lower in the young compared with the senescent group, but these age differences were of borderline statistical significance. Isoproterenol produced highly significant dose-related increases in heart rate, aortic flow, and external power and a decrease in resistance and pressure in both groups. Characteristic impedance and the reflection index were not changed in either group. The only demonstrated age-dependent differences were a larger fall in the absolute values of resistance and a smaller increase in flow in the senescent compared with the young group. When these changes were expressed as a proportion of the baseline values, however, there was no longer an age difference. The finding that, across a range of doses, isoproterenol decreased resistance; increased total power, oscillatory power, and the fraction of oscillatory power; and did not alter characteristic impedance or the reflection index is in agreement with prior reports in both humans and dogs (Cox 1974, Gundel et al. 1981, O'Rourke and Taylor 1967). An earlier study, however, found an increase in characteristic impedance with isoproterenol (Patel, De Freitas, and Fry 1963), but this study did not contain enough data to allow statistical evaluation of the significance of the changes in the higher frequency components of the aortic impedance modulus.

The lack of age difference in response was somewhat surprising in view of the age difference in beta-adrenergically mediated relaxation response in isolated rabbit aortas (Fleisch and Hooker 1976; Fleisch, Maling, and Brodie 1970) and our previous demonstration of an age difference in characteristic impedance during exercise (Yin, Weisfeldt, and Milnor 1981). However, since the exact site of the beta-adrenergically mediated effect in either the aorta or the periphery is not known, it is difficult to compare the isolated tissue response with that of characteristic impedance. It is also possible that dog and rabbit aortas differ in their relaxation response to beta-adrenergic stimulation. Since the blood pressure responses during exercise and isoproterenol infusion may differ, it is also difficult to compare the responses of characteristic impedance seen here with our previously described exercise response.

The responses to the nonadrenergically mediated vasodilator nitroprusside are summarized in Table 5.2. There were no age differences in the baseline values or in the response to the drug in any of the parameters. This lack of effect of nitroprusside on characteristic impedance at rest is in agreement with results of previous studies in humans (Gundel et al. 1981). To my knowledge there are no other published data in dogs with which to compare our results.

TABLE 5.1. Hemodynamic response to isoproterenol (IS) in seven young and seven senescent dogs.

Group (dosage, μg/kg/min)	Heart rate (beats/min)	Mean aortic flow (ml/sec)	Mean aortic pressure (mm Hg)	Vascular resistance (dyne sec cm^{-5})	Impedance (dyne sec cm^{-5})		ϕ_o	Power (m Watt)			
					Z_o	Z_1	(Hz)	W_t	W_o	$\%W_o$	$\Delta Z/Z_o$
Young											
Control	130±6.3	27.4±2.8[a]	90.8±4.7	4824±615[b]	267.8±28.2[b]	453.4±40	6.7±0.6	422±46	99±16	21.3±2.9	0.48±0.04
IS (0.08)	165±6.7	37.6±2.9	68.6±5.2	2595±334	283.4±35.2	322.7±27	8.3±1.0	533±44	196±39	35.2±4.4	0.38±0.06
IS (0.16)	192±5.8	41.9±2.1	63.0±6.6	2066±267	260.9±36.9	297.4±29	8.6±0.5	577±39	229±39	39.0±5.2	0.42±0.06
IS (0.32)	209±5.1	47.5±2.2	57.9±7.8	1666±238	245.5±30.6	314.0±42	8.9±0.6	644±49	278±36	44.2±5.9	0.40±0.07
IS (0.64)	222±6.4	52.1±3.5	65.3±8.1	1706±204	245.6±30.0	312.6±43	9.3±0.6	798±53	339±43	43.9±5.9	0.40±0.08
P, Drug	0.0001	0.0001	0.0001	0.0001	NS	0.0005	0.02	0.0001	0.0001	0.0001	NS
Senescent											
Control	139±17	20.0±1.5	103.7±10.8	7328±945	416.9±50.4	663.6±90	6.6±0.4	375±45	100±12	27.1±1.8	0.53±0.08
IS (0.08)	179±12	28.0±3.0	89.4±10.7	4598±722	453.9±64.1	503.3±67	7.6±1.1	487±58	156±17	33.0±2.7	0.55±0.11
IS (0.16)	185±10	30.7±3.5	87.0±11.3	4116±685	418.3±60.1	486.0±71	7.5±1.0	536±70	184±23	35.1±2.6	0.44±0.07
IS (0.32)	196±10	36.0±3.4	88.2±9.9	3493±500	385.7±60.1	428.8±56	7.9±1.5	666±70	243±35	36.4±2.5	0.35±0.09
IS (0.64)	208±11	38.3±3.8	80.7±8.9	3011±499	399.1±65.5	399.7±57	8.3±1.2	688±74	286±31	42.1±2.5	0.33±0.10
P, Drug	0.0001	0.0001	0.0001	0.0001	NS	0.0005	NS	0.0001	0.0001	0.0001	NS
P, Age	NS	0.02	NS	0.02	NS	NS	NS	NS	NS	NS	NS

[a] $P = 0.038$ for young vs. senescent control values.
[b] $P = 0.05$ for young vs. senescent control values.
Z_o, characteristic impedance; Z_1, first harmonic of impedance modulus; ϕ_o, first zero crossing of impedance phase angle; W_t, total external power; W_o, oscillatory power; $\%W_o$, percentage of total power that is oscillatory; $\Delta Z/Z_o$, reflection index.

TABLE 5.2. Hemodynamic response to nitroprusside (NP) in seven young and six senescent dogs.

Group (dosage, μg/kg/min)	Heart rate (beats/min)	Mean aortic flow (ml/sec)	Mean aortic pressure (mm Hg)	Vascular resistance (dyne sec cm^{-5})	Impedance (dyne sec cm^{-5}) Z_o	Z_1	ϕ_o (Hz)	Power (m Watt) W_t	W_o	$\%W_o$	$\Delta Z/Z_o$
Young											
Control	114±3	22.9±1.8	92.3±5.0	5793±633	288.8±36.4	533.9±78	6.8±0.4	357±30	74±11	21.0±1.8	0.57±0.08[a]
NP (0.4)	123±10	23.9±2.4	90.6±5.6	5491±585	284.0±42.8	483.0±89	7.0±0.7	364±38	80±11	21.6±1.9	0.53±0.10
NP (0.8)	127±8	22.9±1.9	83.4±5.8	5280±690	272.9±27.3	474.7±89	7.6±0.5	329±29	80±14	24.2±3.0	0.54±0.11
NP (1.6)	135±6	22.8±1.4	81.7±6.5	4869±450	249.8±27.1	409.7±67	7.1±0.8	345±37	93±16	26.8±3.4	0.53±0.07
NP (2.4)	131±7	23.8±2.2	76.0±7.3	4536±604	274.1±30.4	396.6±75	7.2±0.8	327±58	83±14	25.8±3.3	0.47±0.08
NP (4.8)	139±13	25.2±3.0	72.5±9.6	4189±645	264.5±30.3	397.3±89	6.5±0.7	327±58	87±14	27.4±3.6	0.44±0.03
P, Drug	NS	NS	NS	0.0009	NS	0.003	NS	NS	NS	0.02	NS
Senescent											
Control	127±16	20.6±1.6	89.9±9.7	6248±1115	353.9±61.5	701.8±78	6.7±0.4	327±26	85±9	26.2±2.4	0.40±0.04
NP (0.4)	141±17	19.7±1.6	86.9±11.7	6117±1054	358.9±55.6	639.4±106	6.5±0.5	309±36	84±9	27.1±2.4	0.51±0.08
NP (0.8)	156±20	19.9±1.9	78.8±9.3	5611±1056	351.6±70.0	492.0±74	6.1±0.5	286±29	80±12	27.9±3.4	0.40±0.07
NP (1.6)	165±14	20.2±1.3	78.8±11.3	5658±1314	335.2±54.2	496.3±84	5.8±0.5	281±36	73±11	26.4±3.0	0.49±0.12
NP (2.4)	161±16	19.8±1.3	74.5±10.6	5140±952	320.9±61.2	425.4±99	6.4±0.8	255±27	71±7	27.3±2.4	0.45±0.06
NP (4.8)	158±14	21.6±1.5	65.8±9.4	4165±935	286.9±50.5	352.5±53	6.2±1.1	271±22	73±10	28.2±3.0	0.43±0.06
P, Drug	0.017	NS	0.0001	0.0001	0.0006	0.003	NS	0.08	NS	NS	NS
P, Age	NS	NS	NS	NS	NS	NS	NS	NS	NS	NS	NS

[a]$P = 0.06$ for young vs. senescent control values.

Z_o, characteristic impedance; Z_1, first harmonic of impedance modulus; ϕ_o, first zero crossing of impedance phase angle; W_t, total external power; W_o, oscillatory power; $\%W_o$, percentage of total power that is oscillatory; $\Delta Z/Z_o$, reflection index.

The responses to phenylephrine are summarized in Table 5.3. As with the nitroprusside, there were no differences between the groups in any of the parameters under baseline conditions. In both groups the drug induced significant vasoconstriction, manifested by rise in aortic pressure and resistance. This was accompanied by a reflex bradycardia, a fall in aortic flow, an increase in characteristic impedance, a decrease in pulsatile power and fraction of pulsatile power, and no change in reflection index. None of the responses differed between the two age groups.

To my knowledge only one other study has examined the effect of phenylephrine on aortic impedance in dogs (Gersh et al. 1972). In contrast to our findings, these investigators found no change in characteristic impedance or fraction of pulsatile power, although there was a significant increase in resistance. The discrepancy in findings might be attributed to the fact that Gersh's study was performed under open-chest conditions after sympathectomy and bilateral vagotomy. Other studies have examined the effect of norepinephrine on aortic impedance, and results can be compared with our results using phenylephrine. Two of the studies (Cox 1974; Patel, De Freitas, and Fry 1963), however, employed a dose that was insufficient to produce vasoconstriction. Another study (O'Rourke and Taylor 1967), like ours, found an increase in resistance and a fall in fraction of pulsatile power, but unlike ours, found no change in characteristic impedance. This discrepancy might be attributable to the rather low dose of norepinephrine used, since our findings indicate that the increase in characteristic impedance is dose related.

Since phenylephrine elevates aortic distending pressure by vasoconstriction, and since the senescent aorta is stiffer than the young aorta, we expected to see a larger increase in aortic characteristic impedance during administration of this drug in old compared with young aortas. However, we found no age differences. Even when the changes in characteristic impedance for the two age groups were compared for a similar increase in mean aortic pressure (33.3 ± 7.2 mm Hg for the 4-μg dose in the young compared with 37.1 ± 2.7 mm Hg for the 8-μg dose in the senescent) there was no age difference in the increment in characteristic impedance (77.8 ± 35.4 vs. 100.6 ± 24.3 dyne sec cm^{-5}, P not significant). We have no clear explanation for this unexpected finding.

The responses to the adrenergic blocking drugs are summarized in Table 5.4. There were no age differences in any of the parameters in the control state. Alpha-adrenergic blockade produced only a borderline significant fall in aortic pressure and total power in the young group. In the old group there was, in addition, a small decrease in flow and oscillatory power but there were no significant age-related differences in any of these responses. Beta blockade alone decreased heart rate and aortic flow and increased resistance in the young animals. The percentage of oscillatory power was decreased in both age groups. There were, however, no statistically significant age differences in the response to beta-adrenergic blockade. When

TABLE 5.3. Hemodynamic response to phenylephrine in six young and six senescent dogs.

Group (dosage, μg/kg)	Heart rate (beats/min)	Mean aortic flow (ml/sec)	Mean aortic pressure (mm Hg)	Vascular resistance (dyne sec cm^{-5})	Impedance (dyne sec cm^{-5})		ϕ_o (Hz)	Power (m Watt)		$\%W_o$	$\Delta Z/Z_o$
					Z_o	Z_1		W_t	W_o		
Young											
Control	128±9	23.7±4.4	106.1±7.0	7004±1315	333.9±42.7	641.8±125	5.9±0.5	431±94	89±25	20.3±3.3	0.46±0.10
PE (1)	124±13	23.4±5.1	117.9±9.9	8066±1571	351.9±50.6	756.2±145	6.7±0.7	486±138	97±29	19.7±3.8	0.60±0.15
PE (4)	110±8	19.2±3.6	139.4±10.1	11487±2429	411.6±52.4	1046.8±172	6.2±0.5	460±111	88±26	18.4±2.9	0.65±0.15
PE (8)	102±3	15.8±2.8	157.9±12.2	16700±4673	494.5±81.6	1364.8±273	6.6±0.3	412±96	71±22	16.3±2.3	0.65±0.18
P, Drug	0.013	0.0007	0.0001	0.003	0.002	0.0003	NS	NS	0.04	0.07	NS
Senescent											
Control	125±11	17.8±1.4	98.9±8.0	7894±996	385.9±71.4	718.0±110	7.0±0.2	307±24	73±6	24.6±2.0	0.43±0.04
PE (1)	115±9	17.2±1.2	108.3±9.5	9027±1479	430.8±87.5	969.7±396	6.9±0.3	330±15	88±10	26.7±2.0	0.58±0.09
PE (4)	104±8	13.9±1.5	124.5±8.6	13477±2745	450.6±89.5	1283.1±225	6.4±0.3	288±19	63±7	22.1±1.6	0.62±0.15
PE (8)	106±4	11.7±2.0	136.0±7.6	19956±4876	486.5±58.1	1637.6±379	6.3±0.2	261±41	54±10	20.4±1.1	0.64±0.10
P, Drug	0.045	0.0001	0.0001	0.0018	0.046	0.0003	NS	NS	0.04	0.04	NS
P, Age	NS	NS	NS	NS	NS	NS	NS	NS	NS	NS	NS

Z_o, characteristic impedance; Z_1, first harmonic of impedance modulus; ϕ_o, first zero crossing of impedance phase angle; W_t, total external power; W_o, oscillatory power; $\%W_o$, percentage of total power that is oscillatory; $\Delta Z/Z_o$, reflection index.

TABLE 5.4. Hemodynamic response to adrenergic blockade in five young and six senescent dogs.

Group	Heart rate (beats/min)	Mean aortic flow (ml/sec)	Mean aortic pressure (mm Hg)	Vascular resistance (dyne sec cm^{-5})	Impedance (dyne sec cm^{-5}) Z_o	Z_1	ϕ_o (Hz)	Pressure (m Watt) W_t	W_o	$\%W_o$	$\Delta Z/Z_o$
Alpha-adrenergic blockade											
Young											
Control	134±13	24.9±5.3	109.3±7.1	7144±1677	364.0±72.4	671.6±136	5.4±0.9	463±108	96±29	20.6±3.6	0.49±0.12
α-Blockade	132±16	21.1±3.5	92.8±7.6	6615±1300	337.6±66.3	575.6±124	5.2±0.9	329±58	70±16	21.2±3.0	0.41±0.07
P, Drug	NS	NS	0.058	NS	NS	NS	NS	0.076	NS	NS	NS
Senescent											
Control	125±11	17.8±1.4	98.9±8.0	7894±996	385.9±71.4	700.9±102	7.0±0.2	307±24	73±6	24.6±2.0	0.43±0.04
α-Blockade	124±6	13.0±1.9	73.2±5.1	9014±1722	417.4±75.5	686.5±185	7.5±0.8	174±18	50±8	29.6±4.3	0.69±0.16
P, Drug	NS	0.06	0.014	NS	NS	NS	NS	0.012	0.038	NS	NS
P, Age	NS	NS	NS	NS	NS	NS	NS	0.068	NS	NS	NS
Beta-adrenergic blockade											
Young											
Control	130±6	27.4±2.8[a]	90.8±4.7	4824±615[a]	267.7±28.2[b]	501.7±78	6.7±0.6	422±46	99±16	21.3±2.9	0.48±0.04
β-Blockade	119±6	24.4±2.3	95.4±5.9	5644±687	279.4±29.0	524.6±73	6.6±0.7	381±41	70±11	18.9±2.4	0.54±0.05
P, Drug	0.017	0.02	NS	0.012	NS	NS	NS	NS	NS	0.048	NS
Senescent											
Control	134±15	21.3±0.9	108.1±11.7	7161±1100	395.7±54.1	702.4±94	6.4±0.4	408±29	106±12	26.2±1.8	0.55±0.09
β-Blockade	121±6	19.5±2.1	112.5±7.1	8140±1889	362.5±64.5	742.3±151	6.1±0.2	340±35	77±10	22.7±2.1	0.51±0.07
P, Drug	NS	NS	NS	NS	NS	NS	NS	NS	0.049	0.018	NS
P, Age	NS	NS	NS	NS	NS	NS	NS	NS	NS	NS	
Combined alpha- and beta-adrenergic blockade											
Young											
Control	134±13	24.8±5.3	109.3±7.1	7144±1677	364.0±72.4	671.6±136	5.4±0.9	463±108	96±29	20.6±3.6	0.49±0.12
α,β-Blockade	105±13	18.6±3.1	83.7±9.8	7121±1988	328.6±59.5	710.8±163	4.8±0.8	253±45	48±10	19.9±3.4	0.48±0.08
P, Drug	0.031	0.071	0.028	NS	NS	NS	NS	0.047	NS	NS	NS
Senescent											
Control	125±11	17.8±1.4	98.9±8.0	7894±998	385.9±71.4	700.9±102	7.0±0.2	307±24	73±6	24.6±2.0	0.43±0.04
α,β-blockade	96±5	12.1±1.3	77.9±6.7	9752±1829	412.0±66.6	745.9±147	7.4±0.9	168±16	46±8	26.9±3.6	0.60±0.08
P, Drug	0.063	0.025	0.027	NS	NS	NS	NS	0.007	0.03	NS	0.06
P, Age	NS	NS	NS	NS	NS	NS	NS	NS	NS	NS	NS

[a] $P = 0.08$ for young vs. senescent control values.
[b] $P = 0.05$ for young vs. senescent control values.
Z_o, characteristic impedance; Z_1, first harmonic of impedance modulus; ϕ_o, first zero crossing of impedance phase angle; W_t, total external power; W_o, oscillatory power; $\%W_o$, percentage of total power that is oscillatory; $\Delta Z/Z_o$, reflection index.

alpha- and beta-adrenergic blockade were combined, both groups demonstrated a decrease in heart rate, flow, pressure, total power, and fraction of oscillatory power. There were, again, no age differences in this response. These findings suggest that the lack of age difference in aortic impedance at rest is not due to an intrinsic difference in underlying adrenergic tone, that is, the young animal does not have an increased beta-adrenergic tone at rest that masks the effect of an increased alpha tone. The lack of age differences during each adrenergic blockade is further evidence for this contention. There are few reports in the literature with which our results can be compared.

One study demonstrated examples of a decrease in resistance, increase in characteristic impedance, and increases in both oscillatory power and fraction of oscillatory power after ganglionic blockade with trimethaphan (Gersh et al. 1972). However, not enough data were presented to allow statistical evaluation of the results. In addition, the study was performed in open-chest dogs after sympathectomy and vagotomy, so that direct comparisons with our study are difficult to make. Regardless, one possible reason for the discrepant results may be the direct vasodilatory action of trimethaphan, which would be expected to increase the fraction of pulsatile power. Another possibility is that our level of alpha-adrenergic blockade was not hemodynamically equivalent to the alpha blockade achieved with the dose of ganglionic blocking agent used.

Conclusion

The new pharmacologic data presented in this chapter generally confirm our previous observation of no age difference in aortic impedance during resting conditions in dogs. On only 1 of 4 days were resistance and characteristic impedance somewhat higher in the aged group, and even these differences were of borderline significance. Apparently there are enough counterbalancing effects under resting conditions to produce no age difference in impedance despite the increased intrinsic stiffness of the aged aorta (Yin, Spurgeon, and Kallman 1983). Thus to date, the only significant age difference in impedance that we have been able to elicit in dogs is the response of aortic characteristic impedance during exercise (Yin, Weisfeldt, and Milnor 1981). The differences between our results in dogs and those previously described in humans and other animals point out both the need for further studies and the potential for marked species variations.

Bibliography

Avolio, A. P., Chen, S., Wang, R., Zhang, C., Li, M., and O'Rourke, M. F. 1983. Effects of aging on changing arterial compliance and left ventricular load in a northern Chinese urban community. *Circulation* 68:50–58.

Avolio, A. P., Deng, F. Q., Li, W. Q., Luo, Y. F., Huang, Z. D., Xing, L. F., and O'Rourke, M. F. 1985. Effects of aging and arterial distensibility in populations with high and low prevalence of hypertension: comparison between urban and rural communities in China. *Circulation* 71:202–210.

Bader, H. 1967. Dependence of wall stress in human thoracic aorta on age and pressure. *Circ. Res.* 20:354–361.

Bader, H., and Kapal, E. 1957. Relationship between radius and wall thickness of human thoracic aorta at different ages and different pressures. *Physiologist* 2:4.

Band, W., Goedhard, W. J. A., and Knoop, A. A. 1972. Effects of aging on dynamic viscoelastic properties of the rat's thoracic aorta. *Pflueg. Arch.* 331:357–364.

Bemis, C. E., Serur, J. R., Borkenhagen, D., Sonnenblick, E. H., and Urschel, C. E. 1974. Influence of right ventricular filling pressure on left ventricular pressure and dimension. *Circ. Res.* 34:498–504.

Bramwell, J. C., Hill, A. V., and McSwiney, B. A. 1923. The velocity of the pulse wave in man in relation to age as studied by the hot-wire sphygmograph. *Heart* 10:233–249.

Brandfonbrenner, M., Landowne, M., and Shock, N. W. 1955. Changes in cardiac output with age. *Circulation* 12:557–556.

Brinker, J. A., Weiss, J. L., Lappe, D. L., Rabson, J. R., Summer, W. R., Permutt, S., and Weisfeldt, M. L. 1980. Leftward septal displacement during right ventricular loading in man. *Circulation* 61:626–633.

Butcher, H. B., and Newton, W. T. 1958. The influence of age, arteriosclerosis, and homotransplantation upon the elastic properties of major human arteries. *Ann. Surg.* 148:1–20.

Chatterjee, J., Parmley, W. W., Ganz, W., Forrester, J., Walinsky, P., Crexells, C., and Swan, H. J. C. 1973. Hemodynamic and metabolic response to vasodilator therapy in acute myocardial infarction. *Circulation* 48:1183–1193.

Cohen, M. L., and Berkowitz, B. A. 1974. Age-related changes in vascular responsiveness to cyclic nucleotides and contractile agonists. *J. Pharmacol. Exp. Ther.* 191:147–155.

Cohn, J., and Franciosa, J. A. 1977. Vasodilator therapy of cardiac failure. *N. Engl. J. Med.* 297:27–31, 254–258.

Cohn, J., Mathew, K. J., Franciosa, J. A., and Snow, J. A. 1974. Chronic vasodilator therapy in the management of cardiogenic shock and intractible left ventricular failure. *Ann. Intern. Med.* 81:777–780.

Cox, R. H. 1974. Determinants of systemic hydraulic power in unanesthetized dogs. *Am. J. Physiol.* 226:579–587.

Cox, R. H. 1977. Effects of age on the mechanical properties of rat carotid artery. *Am. J. Physiol.* 233:H256–H263.

Elzinga, G., von Grondelle, R., Westerhof, N., and van den Bos, G. C. 1974. Ventricular interference. *Am. J. Physiol.* 226:941–947.

Fleisch, J. H., and Hooker, C. S. 1976. The relationship between age and relaxation of vascular smooth muscle in the rabbit and rat. *Circ. Res.* 38:243–250.

Fleisch, J. H., Maling, H. M., and Brodie, B. B. 1970. Beta-receptor activity in aorta: variations with age and species. *Circ. Res.* 26:151–162.

Froehlich, J. P., Lakatta, E. G., Beard, E., Spurgeon, H. A., Weisfeldt, M. L., and Gerstenblith, G. 1978. Studies of sarcoplasmic reticulum function and contraction duration in young adult and aged rat myocardium. *J. Molec. Cell. Cardiol.* 10:427–438.

Gersh, J., Prys-Roberts, C., Reuben, S. R., and Schultz, D. L. 1972. The effects of halothane in the interactions between myocardial contractility, aortic impedance and left ventricular performance. II. Aortic input impedance, and the distribution of energy during ventricular ejection. *Br. J. Anesth.* 44:767–774.

Gerstenblith, G., Lakatta, E. G., and Weisfeldt, M. L. 1976. Age changes in myocardial function and exercise response. *Prog. Cardiovasc. Dis.* 19:1–21.

Gerstenblith, G., Spurgeon, H. A., Froehlich, J. P., Weisfeldt, M. L., and Lakatta, E. G. 1979. Diminished inotropic responsiveness to ouabain in aged rat myocardium. *Circ. Res.* 44:517–523.

Glantz, S. A., Misbach, G., Moores, W., Mathey, D., Lekven, J., Stowe, D., Parmley, W., and Tyberg, J. 1978. The pericardium substantially affects the diastolic pressure–volume relations in the canine left ventricle. *Circ. Res.* 42:433–441.

Gonza, E. R., Marble, A. E., Shaw, A., and Holland, J. G. 1974. Age-related changes in the mechanics of the aorta and pulmonary artery of man. *J. Appl. Physiol.* 36:407–411.

Guard, H. R., and Bhende, Y. M. 1953. Changes due to aging in the abdominal aorta. *Ind. J. Med. Res.* 41:267–276.

Guarnieri, T., Spurgeon, H., Froehlich, J. P., Weisfeldt, M. L., and Lakatta, E. G. 1979. Diminished inotropic response but unaltered toxicity to acetylstrophanthidin in the senescent beagle. *Circulation* 60:1548–1554.

Gundel, W., Cherry, G., Rajagopalan, B., Tan, L. B., Lee, G., and Schultz, D. 1981. Aortic input impedance in man: acute response to vasodilator drugs. *Circulation* 63:1305–1314.

Hallock, P., and Benson, I. C. 1937. Studies on the elastic properties of human isolated aorta. *J. Clin. Invest.* 16:595–602.

Janz, R. F., Kubert, R. B., Mirsky, I., Korecky, B., and Taichman, G. C. 1976. Effect of age on passive elastic stiffness of rat heart muscle. *Biophys. J.* 16:281–290.

Kane, R. L., McMahon, T. A., Wagner, R. L., and Abelmann, W. H. 1976. Ventricular elastic modulus as a function of age in the Syrian golden hamster. *Circ. Res.* 38:74–80.

Krafka, J. 1940. Changes in the elasticity of the aorta with age. *Arch. Pathol.* 29:303–309.

Lakatta, E. G. 1979. Alterations in the cardiovascular system that occur in advanced age. *Fed. Proc.* 38:163–167.

Lakatta, E. G., Gerstenblith, G., and Angell, C. S. 1975. Prolonged contraction duration in aged myocardium. *J. Clin. Invest.* 55:61–68.

Lakatta, E. G., Gerstenblith, G., Angell, C. S., Shock, N. W., and Weisfeldt, M. L. 1976. Diminished inotropic response of aged myocardium to catecholamines. *Circ. Res.* 36:262–269.

Lakatta, E. G., and Yin, F. C. P. 1982. Myocardial aging: functional alterations and related cellular mechanisms. (Invited review) *Am. J. Physiol.* 242:H927–941.

Landowne, M. 1958. The relation between intra-arterial pressure and impact pulse wave velocity with regard to age and arteriosclerosis. *J. Gerontol.* 13:153–162.

Learoyd, B. M., and Taylor, M. G. 1966. Alterations with age in the viscoelastic properties of human arterial walls. *Circ. Res.* 18:278–292.

Lee, J. C., Karpeles, L. M., and Downing, S. E. 1972. Age-related changes of cardiac performance in male rats. *Am. J. Physiol.* 222:432–438.

Linzbach, A. J., and Boateng, E. A. 1976. The ageing human heart. (abstr.) Presented at European Congress of Cardiology, Amsterdam, p. 171.

Ludbrook, P. A., Byrne, J. D., and McKnight, R. C. 1977. Influence of right ventricular hemodynamics on left ventricular diastolic pressure–volume relations. *Circulation* 59:21–31.

McDonald, D. A. 1974. *Blood Flow in Arteries*. Baltimore: Williams & Wilkins, pp. 351–388.

Merillon, J. P., Motte, G., Fruchard, J., Masquet, C., and Gourgon, R. 1978. Evaluation of the elasticity and characteristic impedance of the ascending aorta in man. *Cardiovasc. Res.* 12:401–406.

Miller, R. R., Awan, N. A., Maxwell, K. S., and Mason, D. T. 1977. Sustained reduction of cardiac impedance and preload in congestive heart failure with the antihypertensive vasodilator prazosin. *N. Engl. J. Med.* 297:303–307.

Milnor, W. R. 1965. Arterial impedance as ventricular afterload. *Circ. Res.* 36:565–570.

Milnor, W. R. 1982. *Hemodynamics*. Baltimore: Williams & Wilkins, pp. 157–191.

Milnor, W. R., Bergel, D. H., and Bargainer, J. D. 1966. Hydraulic power associated with pulmonary blood flow and its relation to heart rate. *Circ. Res.* 19:467–471.

Mirsky, I., and Janz, R. F. 1976. The effect of age on the wall stiffness on the human thoracic aorta: a large deformation anisotropic elastic analysis. *J. Theoret. Biol.* 59:467–484.

Nakashima, T., and Tanikawa, J. 1971. A study of human aortic distensibility with relation to atherosclerosis and aging. *Angiology* 22:477–490.

Nichols, W. W., O'Rourke, M. G., Avolio, A. P., Yaginuma, T., Murgo, J. P., Pepine, C. J., and Conti, C. R. 1985. Effects of age on ventricular-vascular coupling. *Am. J. Cardiol.* 55:1179–1184.

O'Rourke, M. F. 1982. *Arterial Function in Health and Disease*. New York: Churchill Livingstone, pp. 67–184.

O'Rourke, M. F., and Taylor, M. G. 1967. Input impedance of the systemic circulation. *Circ. Res.* 20:365–380.

Patel, D. J., DeFreitas, F. M., and Fry, D. L. 1963. Hydraulic input impedance to aorta and pulmonary artery in dogs. *J. Appl. Physiol.* 18:134–140.

Pepine, C. J., Nichols, W. W., Curry, R. C., and Conti, C. R. 1979. Aortic input impedance during nitroprusside infusion. *J. Clin. Invest.* 64:643–654.

Roy, C. S. 1880. The elastic properties of the arterial wall. *J. Physiol.* 3:125–159.

Rumberger, E., and Timmermann, J. 1976. Age-changes of the force-frequency relationship and the duration of action potential in isolated papillary muscles of guinea pig. *Eur. J. Appl. Physiol. Occup. Physiol.* 34:277–284.

Santamore, W. P., Lynch, P. R., Meier, G., Heckman, J., and Bove, A. A. 1976. Myocardial interaction between the ventricles. *J. Appl. Physiol.* 41:362–368.

Saxton, J. A. 1942. Elastic properties of the rabbit aorta in relation to age. *Arch. Pathol.* 34:262–274.

Schimmler, W. 1966. Correlation between the pulse wave velocity in the aortic iliac vessel and age, sex, and blood pressure. *Angiology* 17:314–322.

Shabetai, R., Fowler, N. O., and Guntheroth, W. G. 1970. The hemodynamics of cardiac tamponade and constrictive pericarditis. *Am. J. Cardiol.* 26:480–489.

Shreiner, D. P., Weisfeldt, M. L., and Shock, N. W. 1969. Effects of age, sex, and breeding status on the rat heart. *Am. J. Physiol.* 217:176–180.

Simonson, E., and Nakagawa, K. 1960. Effect of age on pulse wave velocity and aortic ejection time in healthy men and in men with coronary artery disease. *Circulation* 22:126–129.

Sivak, E. D., Gray, B. A., McCurdy, H. T., and Phillips, A. K. 1977. Pulmonary vascular response to nitroprusside. *Circulation* 55:III–163.

Spurgeon, H. A., Thorne, P. R., Yin, F. C. P., Shock, N. W., and Weisfeldt, M. L. 1977. Increased dynamic stiffness of trabeculae carneae from senescent rats. *Am. J. Physiol.* 232 (*Heart Circ. Physiol.* 1):H373–H380.

Sunagawa, K., Sagawa, K., and Maughan, W. L. 1984. Ventricular interaction with the loading system. *Am. Biomed. Eng.* 12:163–190.

Templeton, G. H., Platt, M. R., Willerson, J. T., and Weisfeldt, M. L. 1979. Influence of aging on left ventricular hemodynamics and stiffness in beagles. *Circ. Res.* 44:189–194.

Tuttle, R. S. 1966. Age-related changes in the sensitivity of rat aortic strips to norepinephrine and associated chemical and structural alterations. *J. Gerontol.* 21:510–516.

Van den Horn, C. J., Westerhof, N., and Elzinga, G. 1984. Interaction of heart and arterial system. *Am. Biomed. Eng.* 12:151–162.

Weber, K. T., Janicki, J. S., Shroff, S., and Fishman, A. P. 1981. Contractile mechanics and interaction of the right and left ventricles. *Am. J. Cardiol.* 47:686–695.

Weisfeldt, M. L., Loeven, W. A., and Shock, N. W. 1971. Resting and active mechanical properties of trabeculae carneae from aged male rats. *Am. J. Physiol.* 220:1921–1927.

Weiss, J. L., Brinker, J. A., Lappe, D. A., Rabson, J. L., Summer, W. L., Permutt, S., and Weisfeldt, M. L. 1978. Leftward septal displacement during right ventricular loading in man: demonstration by two-dimensional echo. *Am. J. Cardiol.* 41:362.

Weyman, A. E., Heeger, J. J., Kronik, G., Wann, L. S., Dillon, J. C., and Feigenbaum, H. 1977. Mechanism of paradoxical early diastolic septal motion in patients with mitral stenosis: a cross-sectional echocardiographic study. *Am. J. Cardiol.* 40:691–699.

Wilens, S. L. 1937. The postmortem elasticity of the adult human aorta: its relation to age and to the distribution of intimal atheroma. *Am. J. Pathol.* 13:811–833.

Yin, F. C. P. 1980. The aging vasculature and its effects on the heart. In Weisfeldt, M. L., ed. *The Heart in Old Age: Its Function and Response to Stress.* New York: Raven pp. 137–217.

Yin, F. C. P., Guzman, P. A., Brin, K. P., Maughan, W. L., Brinker, J. A., Traill, T. A., Weiss, J. L., and Weisfeldt, M. L. 1983. Effect of nitroprusside on hydraulic vascular loads on the right and left ventricle of patients with heart failure. *Circulation* 67:1330–1359.

Yin, F. C. P., Spurgeon, H. A., and Kallman, C. H. 1983. Age-associated alterations in viscoelastic properties of canine aortic strips. *Circ. Res.* 53:464–472.

Yin, F. C. P., Spurgeon, H. A., Rakusan, K., Weisfeldt, M. L., and Lakatta, E. G. 1982. Use of tibial length to quantify cardiac hypertrophy: application in the aging rat. *Am. J. Physiol.* 243:H941–947.

Yin, F. C. P., Spurgeon, H. A., Weisfeldt, M. L., and Lakatta, E. G. 1980. Mechanical properties of myocardium from hypertrophied rat hearts: a comparison between hypertrophy induced by senescence and by aortic banding. *Circ. Res.* 46:293–300.
Yin, F. C. P., Weisfeldt, M. L., and Milnor, W. R. 1981. Role of aortic impedance in the decreased cardiovascular response to exercise with aging in dogs. *J. Clin. Invest.* 68:28–38.

6
Arterial Reflections and Pressure Waveforms in Humans

JOSEPH P. MURGO AND NICOLAAS WESTERHOF

Introduction

The character of the arterial pulse as palpated by the physician at the bedside, and laboratory counterpart measurements, such as pulse or pressure recordings, historically have been a critical part of the clinical evaluation of patients with cardiovascular disease. A fine review of the history of the nomenclature and interpretation of the arterial pulse, based on extensive studies of hemodynamics, has recently been provided by O'Rourke (1982).

Over the last few years we have studied the human arterial pressure pulse (Murgo et al. 1980, 1981a,b; Murgo and Westerhof 1984), keeping in mind the knowledge available from animal studies (McDonald 1974; Milnor 1982; Westerhof, Elzinga, and Van den Bas 1973; Westerhof et al. 1972). The magnitude and shape of arterial pressure waveforms result from interaction between the heart and the arterial system, that is, the cardiovascular coupling. Thus the "personality" of the arterial pulse is affected by the peripheral circulation, usually expressed in terms of input impedance (McDonald 1974, Milnor 1982, O'Rourke 1982, Westerhof et al. 1979), and by cardiac function, as expressed in terms of pump function (Elzinga and Westerhof 1973). Modern developments in the technology of pressure measurement, especially in cardiac catheterization, have led to the recognition that major differences in the configuration of the aortic pressure wave exist among patients even in the absence of cardiovascular disease (Murgo, Giolma, and Altobelli 1977; Murgo and Westerhof 1984; Murgo et al. 1975, 1980, 1981a,b). The advent of pressure micromanometry and catheter-mounted miniature flow velocity probes now makes feasible the study of arterial dynamics in humans in more detail.

This chapter describes the systemic and pulmonary circulation qualitatively in terms of pressure and flow waveforms, and quantitatively in terms of input impedance. We focus on the relationship between pressure waveforms and impedance spectra to indicate the importance of the architecture of the arterial bed and the role of reflections in the arterial tree.

The systemic circulation is discussed first, including the results of several clinical interventions. The pulmonary circulation is then described, and finally, the two systems are compared.

All pressure waveforms discussed here were obtained using special multisensor catheters (Millar Mikrotip), the details of which have been presented in previous publications (Murgo, Giolma, and Altobelli 1977; Murgo et al. 1975, 1980). Flow velocity waveforms were obtained from miniaturized catheter-mounted electromagnetic flow velocity probes, so that in some figures velocities rather than volume flows are reported. The technical details of the instrumentation utilized and the details of the patient population have also been presented elsewhere (Murgo et al. 1980,1981a,b).

Systemic Circulation

Typical Waveforms in Humans

The archetypical ascending aortic pressure waveform in the normal adult is shown in Figure 6.1 in conjunction with a left ventricular pressure signal and an ascending aortic flow velocity waveform. A striking characteristic of this pressure waveform is the presence of a mid-systolic inflection followed by a late-systolic rise in pressure, in the absence of any such rise in the flow waveform. These characteristic waveforms stimulated our interest, since they were not seen in the experimental animal in the control condition but were found during occlusion of the thoracic or abdominal aorta (Westerhof, Elzinga, and Van den Bos 1973). In addition, this late-systolic pressure rise in humans can be augmented or diminished by a variety of clinical states or interventions. Figure 6.2 illustrates the effects

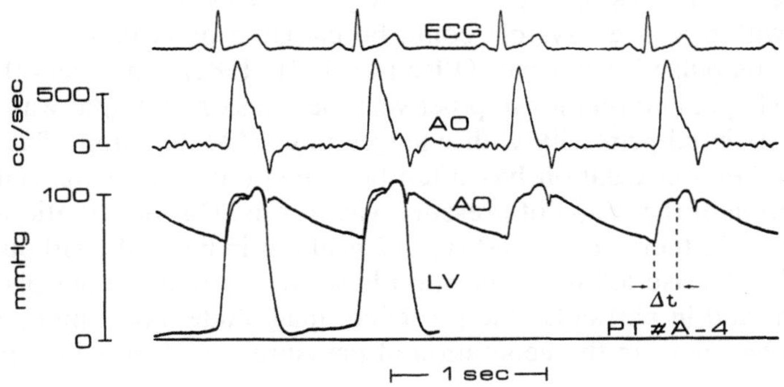

FIGURE 6.1. The electrocardiogram (ECG)), ascending aortic flow, and pressures in the ascending aorta (AO) and left ventricle (LV) of a normal person. The pressure records show a secondary rise in systole. Δt = interval between onset of pressure waveform and the systolic inflection point. (From Murgo and Westerhof 1980; by permission of the American Heart Association, Inc.)

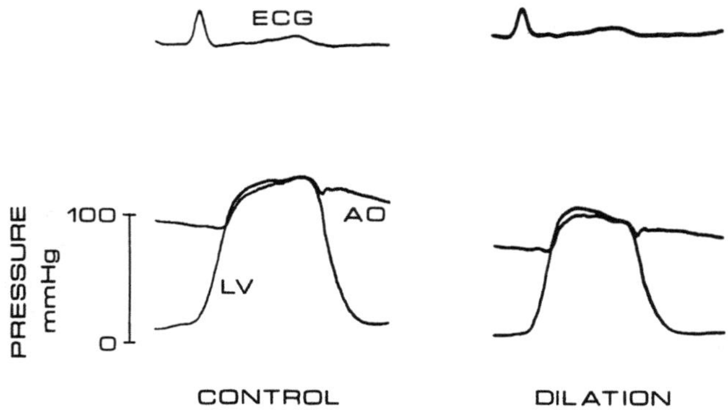

FIGURE 6.2. Electrocardiogram (ECG) and ascending aortic (AO) and left ventricular (LV) pressures during control and sublingual nitroglycerin. Note the change from a type A to a type C beat.

of a vasodilator: The late-systolic pressure rise present in the control state disappears after the administration of sublingual nitroglycerin.

We previously have described in some detail the differences between these pressure waveforms, and have arbitrarily labeled them as "type A" beats (when peak systolic pressure occurs late in systole), and "type C" beats (when peak systolic pressure occurs in early systole) (Murgo et al. 1980). In this same publication we also reported that type C beats normally appear to occur in a young population, whereas type A beats tend to occur in adults, even in the absence of cardiovascular disease.

Earlier clinical literature often used the term "percussion wave" to describe the initial portion of the pressure waveform and "tidal wave" to describe the late-systolic portion of the pressure pulse. When two distinct pulses within a given systole could be palpated or visualized, the term "bisferiens pulse" was used (O'Rourke 1971, 1982). For years the magnitude, shape, and relationship between percussion and tidal waves have been associated primarily with changes in cardiac function. The role of the peripheral circulation has often been ignored or downplayed in the explanation of the shape of pressure waveforms. Our observations of the presence of either type A and type C beats in individuals without cardiovascular disease led us to question how large a role the peripheral circulation, and in particular the presence, magnitude, and timing of wave reflections, plays in the generation of pressure waveforms in man.

Systemic Impedance Counterparts

Methods that selectively describe the arterial system can be used to determine whether the differences in wave shape are the result of cardiac function or are due to the peripheral system. One method is to calculate the input impedance of the arterial system in question. This requires measurement of pulsatile pressure and flow waves and derivation of their respective sinusoidal components by Fourier analysis. Detailed descriptions of the techniques and methodology utilized have been presented (McDonald 1974, Milnor 1982, Westerhof et al. 1979). We derived impedance spectra in the absence of cardiovascular disease for both type A and type C beats. Figure 6.3 depicts the average normalized impedance spectra (with respect to characteristic impedance, Murgo et al. 1980) for seven patients with type A beats and four patients with type C beats. At 0 Hz the peripheral resistance is found. For high frequencies the modulus becomes rather independent of frequency. The average input impedance for these high frequencies equals characteristic impedance (Z_o), which is the characterization of the proximal aorta only. It holds that for a system without reflections, input impedance equals characteristic impedance. Or inversely, when input impedance deviates from characteristic impedance, reflections are present. The input impedance of patients with type A beats demonstrated a well-defined minimum, followed by a maximum and relatively large oscillations about the characteristic impedance, while those

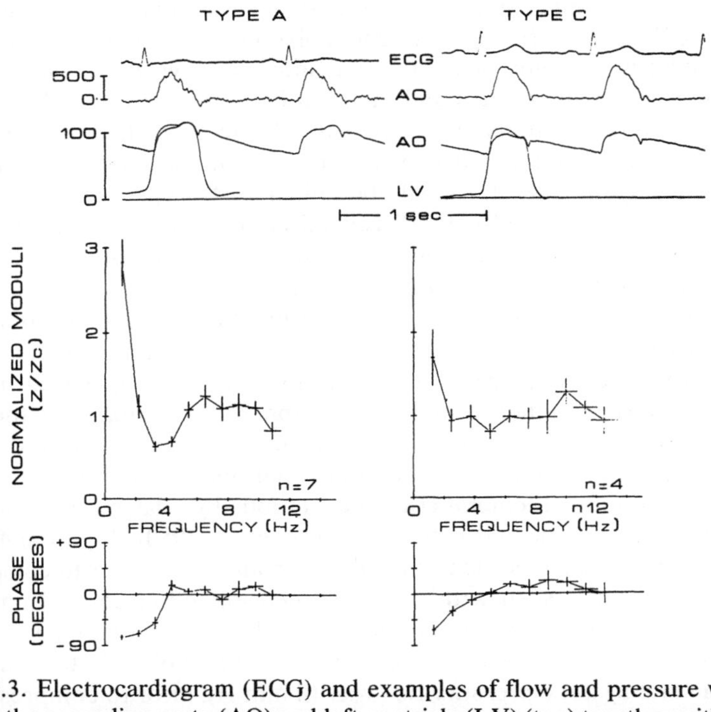

FIGURE 6.3. Electrocardiogram (ECG) and examples of flow and pressure wave shapes in the ascending aorta (AO) and left ventricle (LV) (top) together with the averaged normalized systemic input impedance (bottom) for type A waves *(left)* and type C waves *(right)*. *n*, number of patients.

with type C beats revealed a flat or less well-organized amplitude spectrum. In patients with type A beats, the phase angles of the first three harmonics were also more negative than in patients with type C beats.

Implications of Waveforms and Impedance Spectral Patterns

The simplest analog model that displays many features of the impedance data just described is an elastic tube connected to a pump (the heart) and terminated in a tube of different properties or a (peripheral) resistance. When the second tube has a characteristic impedance different from that of the first tube, or when the resistance differs from the characteristic impedance of the first tube, wave reflections arise at the junction. Reflections result in a pattern of minima and maxima in the modulus of the input impedance associated with zero crossings of the phase angle. Thus the differences between patients with type A and type C beats would suggest that the secondary rise in systolic pressure may be due to more prominent reflections in the arterial systems of the patients with type A beats.

In this study an attempt was also made to derive an "effective" reflection site, using the quarter wavelength principle (McDonald 1974), utilizing measured foot-to-foot pulse wave velocity and the first minimum of the impedance modulus and first zero crossing of the impedance phase angle. The "effective" reflection site could also be calculated from the foot-to-foot pulse wave velocity and the time interval between *the onset* ("foot") of the pressure waveform and the systolic *inflection point* (Fig. 6.1, Δt). These two calculations led to the same "effective" length. The foot-to-foot pulse wave velocity used was derived from the ascending aortic root, and the region of the terminal abdominal aorta in the adult appeared to be the dominating "effective" reflection site. The differences between type A and type C beats could therefore be explained by differences in the magnitude and timing of reflections.

While it is recognized that a single tube model is much too simplistic to describe the complex arterial system, we were convinced in our early work that the waveform of the pressure pulse that appeared in the ascending aorta resulted from diffuse peripheral reflections together with a major "effective" reflection site, causing the inflection in the pressure wave near the terminus of the abdominal aorta. To further investigate this, we performed femoral arterial occlusions by bilateral external manual compression in selected patients (Murgo et al. 1980). Figure 6.4 demonstrates one example of the results of such maneuvers. Prior to compression, a type A beat is seen with a well-defined inflection point. Immediately after the onset of external compression, the late secondary systolic rise in pressure is augmented by approximately 100% of its initial value. That is, the late-systolic pressure rise was 10 mm Hg prior to compression and rose to a value of 20 mm Hg following compression. These results appeared to confirm our initial impressions that the effective reflection site was in the region of the terminal abdominal aorta and its major bifurcation.

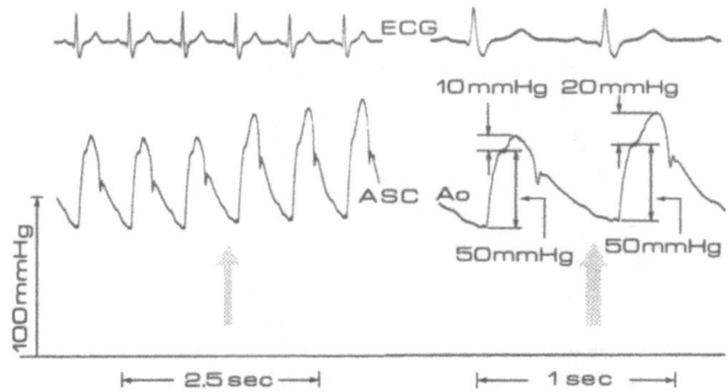

FIGURE 6.4. Electrocardiogram (ECG) and ascending aortic pressure (ASC) before and after bilateral occlusion of the femoral arteries (see text) displayed at two paper speeds. The start of compression is shown by the arrows. Note that the initial wave is not altered in amplitude, while the secondary systolic wave is doubled. (From Murgo and Westerhof 1980; by permission of the American Heart Association, Inc.)

However, later findings on pressure wave shapes during the Valsalva maneuver (Murgo et al. 1981a) and the complex branching of the abdominal aorta in the proximity of the origin of several major arterial trees (celiac, mesenteric, renal, etc.) led us to design a special catheter that would allow us to study pressure waves in the entire thoracic and abdominal aorta simultaneously (Latham et al. 1986). A catheter with six micromanometers equally spaced at 10-cm intervals was constructed and used to investigate regional wave travel and reflection phenomena along the entire human aorta. Figure 6.5 gives an example of six pressure waveforms obtained with this catheter (plus ascending aortic pressure waveform obtained separately) from the aortic valve to the femoral artery of a normal patient, with the location of each pressure sensor shown in a diagrammatic representation of the aorta. Note that the foot of the pressure pulse appears progressively later as the sensor is located more distally. In contrast, the inflection point at which the late systolic rise begins occurs progressively earlier, so that a distinct inflection point becomes less obvious as the incident and reflected waves merge near the level of the origin of the renal arteries.

These data, combined with more sophisticated analyses of regional pulse wave velocities (Latham et al. 1986) yielded important new information. First, as suspected, pulse wave velocity varied significantly as a function of location along the aorta. Second, considerable change in aortic diameter existed at the site of the origin of the renal arteries: The diameter of the aorta above and below the renal arteries was 20 and 13 mm, respectively ($n = 9$). Pulse wave velocity was significantly higher distal to the renal arteries, as compared with velocity at sites proximal to the renal arteries. Accordingly, our original interpretation that the major effective reflecting

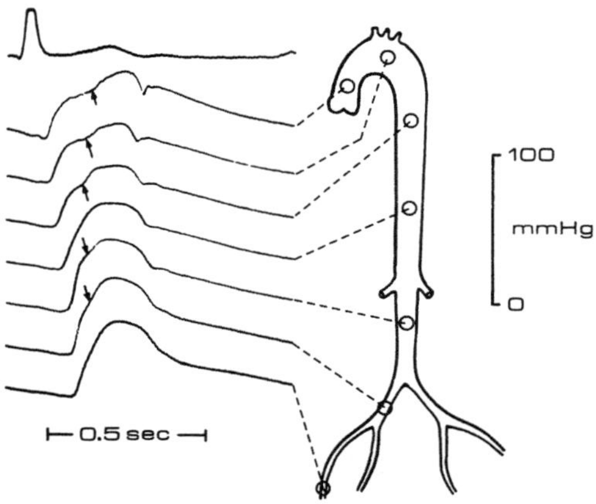

FIGURE 6.5. Example of pressure waveform along the human aorta. Simultaneous pressures were obtained with a six-sensor micromanometer. The pressure in the ascending aorta is determined separately, but also with a micromanometer. Except for ascending aortic pressure, all distances between sensors are 10 cm.

sites in humans appeared to originate in the termination of the abdominal aorta has been modified. From the new data we conclude that the archetypical type A pressure waveform is generated by the diffuse reflection, together with the reflected waves originating primarily from the site of origin of the renal arteries.

Forward and Backward Waves

The role of reflected waves in determining the waveform of arterial pressures and flows may be more easily conceptualized by separating the measured pressure and flow waveforms into their forward (incident) and backward (reflected) components (Murgo et al. 1981a, Westerhof et al. 1972). This separation may be accomplished using mathematical techniques described in detail earlier; the pertinent equations are summarized as follows (Murgo et al. 1981a):

$$P_m = P_f + P_b \tag{6.1}$$
$$F_m = F_f + F_b \tag{6.2}$$
$$P_f = Z_o \cdot F_f \tag{6.3}$$
$$P_b = -Z_o \cdot F_b \tag{6.4}$$

where P = pressure
F = flow
m = measured wave
f = forward wave
b = backward wave
and Z_o = characteristic impedance

The measured pressure wave (P_m) is equal to the sum of forward (P_f) and backward (P_b) pressure waves. The measured flow wave (F_m) is also equal to the sum of the forward (F_f) and backward (F_b) flow waves. The forward pressure and flow waves are related by the characteristic impedance of the aorta, as given by Equation 6.3. The backward pressure and flow waves are also related to the characteristic impedance of the aorta, as given by Equation 6.4. In larger arteries such as the aorta, Z_o is primarily determined by inertial and compliance effects, the combination of which lead to Z_o being a real number. If Z_o is in fact a real number, the relationship given by Equation 6.3 predicts that the forward pressure and flow waves are identical in shape. Similarly, Equation 6.4 would predict that the same relationship holds for the backward waves, but flow is inverse with respect to pressure, as indicated by the minus sign. Since P_m and F_m are measured quantities, and the characteristic impedance (Z_o) may be derived from the impedance spectra, Equations 6.1 through 6.4 may be solved to yield the forward and backward flow and pressure waves:

$$P_f = Z_oF_f = (P_m + Z_oF_m)/2 \quad (6.5)$$

$$P_b = -Z_oF_b = (P_m - Z_oF_m)/2 \quad (6.6)$$

Using a digital computer, the forward pressure and flow waves may be found from the measured waves. Thus measured waves can be dissected into their forward and backward components. An example of a type A beat is given in Figure 6.6. In the ascending aorta, a forward pressure wave first appears, but the wave speed is of such a magnitude that a backward (reflected) pressure wave returns to this location while systole is still in progress. The reflected wave and the forward wave are added to yield the archetypical ascending aortic pressure waveform with a late systolic wave. Similarly, the forward flow wave is added to the backward flow wave yielding the archetypical flow wave. Mainly because the reflected waves return in systole, the measured pressure wave is markedly different from the measured flow wave.

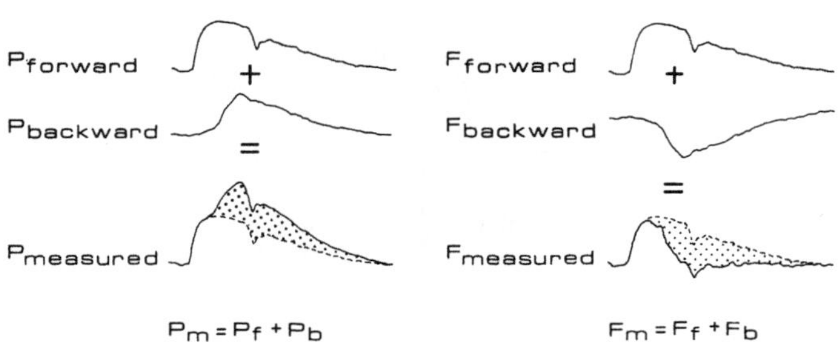

FIGURE 6.6 Pressure (*left*) and flow (*right*) both consist of forward (initial) (P_f, F_f) and backward (reflected) (P_b, F_b) waves. The forward waves of pressure and flow have the same shape; the backward pressure and flow waves also have the same shape but the flow wave is inverted. P_m = measured pressure; F_m = measured flow.

This explanation of forward and backward pressure and flow waves differs considerably from the concept of reflections given by others. While reflections in pressure phenomena have been discussed in considerable detail in the literature, the contribution of reflected *flow* waves has been ignored or its importance downplayed (Kouchoukos, Sheppard, and McDonald 1970; McDonald 1974; McDonald and Taylor 1956; O'Rourke 1982; Wetterer 1954). Thus, while we explain differences between measured pressure and flow waveforms as a result of reflection phenomena in both parameters, others have focused on the difference as primarily a result of reflections in pressure phenomena alone.

Manipulation of Reflections

To further understand the role of reflected waves in the generation of arterial pressure and flow waveforms, it is of interest to examine interventions that cause, or result in, major changes in arterial waveforms. One such intervention is the Valsalva maneuver. In a study of eight normal subjects, we have demonstrated that manipulation of ascending aortic pressure and flow wave reflections by means of the Valsalva maneuver also correlates with significant changes in input impedance (Murgo et al. 1981a).

An example of a typical Valsalva maneuver is given in Figure 6.7. The three phases include control, a strain phase, and the postrelease phase. Note that marked changes in systolic pressure, pulse pressure, diastolic pressures, heart rate, and aortic flow velocity all occur during this maneuver.

Figure 6.8 illustrates in more detail pressure and flow waveforms in the three phases shown in Figure 6.7 combined with the average normalized

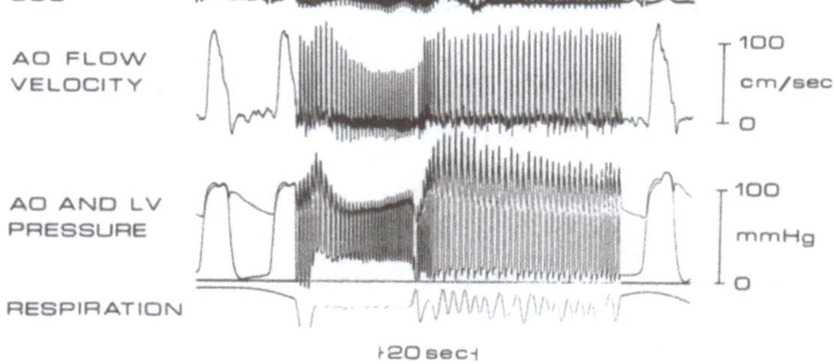

FIGURE 6.7. Pressure in the left ventricle (LV) and the ascending aorta (AO), and ascending aortic flow, together with electrocardiogram (ECG) and respiration record before, during, and immediatlely after the Valsalva maneuver. Note the increased late-systolic rise in pressure during the postrelease phase.

input impedance spectra for the eight patients studied. During the strain phase (middle) the pressure waveform has been markedly altered from a type A beat (seen during control) to a type C beat. In addition, during the postrelease phase, when hypertension and reflex brachycardia are present, an accentuated type A beat appears. Associated with these changes in wave shapes are marked differences in the impedance spectra. During the control and postrelease phases the modulus spectra demonstrate considerable oscillations about the characteristic impedance. In contrast, during the strain phase the modulus spectrum loses its oscillatory character and becomes extremely flat, that is, equal to characteristic impedance. Similarly, during strain the phase plots reveal a much less negative angle in the first harmonic, with zero phase angles already appearing by the second and third harmonics. These findings imply that the reflections existing in the control period are reduced during the strain phase and augmented in the postrelease period. The pressure and flow waveforms shown in the strain phase in Figure 6.8 are much more similar in appearance than those shown in either the control or the postrelease phase, also implying reduced reflection in the strain phase (see the following).

To pursue this issue further, it is of interest to dissect the measured pressure and flow waveforms into their respective forward and backward

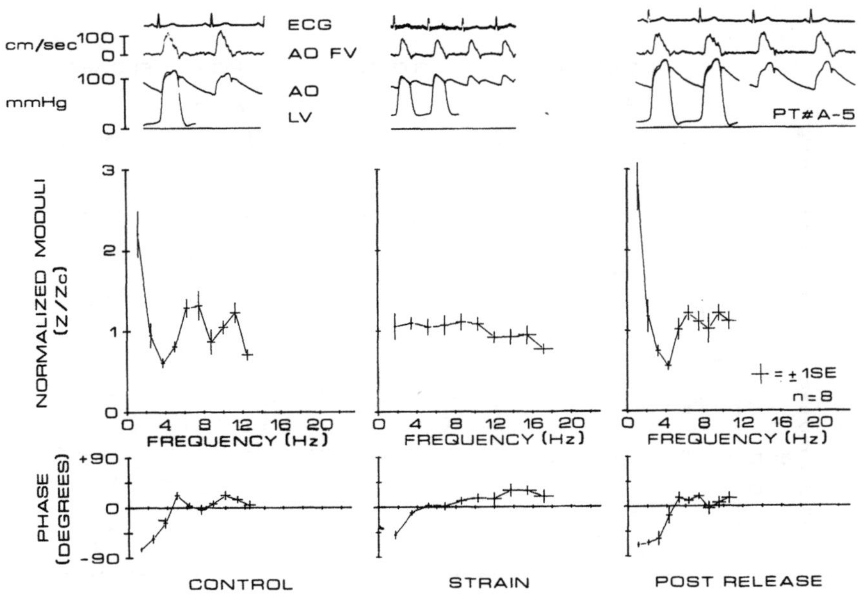

FIGURE 6.8. Examples of pressures and flow wave shapes in the normalized ascending aorta (AO) and left ventricle (LV) (top) together with the average systemic input impedance *(bottom)* for control, Valsalva strain, and postrelease phase in eight patients *(n)*.

components during each of the phases of the Valsalva maneuver (Figure 6.9). Focusing on the backward pressure and flow components, it can be seen tht during the strain phase almost no reflections exist. This finding is consistent with the flat impedance spectrum shown during the strain phase in Figure 6.8. Absence of reflections results in a measured pressure wave P_m) that is almost identical to the forward pressure wave (P_f). Similarly, the absence of flow reflections results in a measured flow wave (F_m) that is very similar to the forward flow wave (F_f). Since the forward pressure and flow waves are identical in shape, the measured pressure and flow waves also are similar in shape in a reflectionless system. In contrast, significant reflections occur during the control and postrelease phases, and the measured pressure and flow waveforms are considerably different from each other. The backward waves occurring during the postrelease phase are also larger than those during control, so that the differences in pressure and flow waveforms are accentuated.

While the Valsalva maneuver appears to decrease the amount of reflection in the system considerably, studies using a six-sensor catheter placed in the descending aorta have revealed that some reflection still exists, but that it is much diminished in amplitude and delayed in timing (Figure 6.10). This delayed return of the reflected wave is no doubt secondary to the marked reduction in distending pressure that exists within the aorta due to the rise in intrapleural and abdominal pressures and the

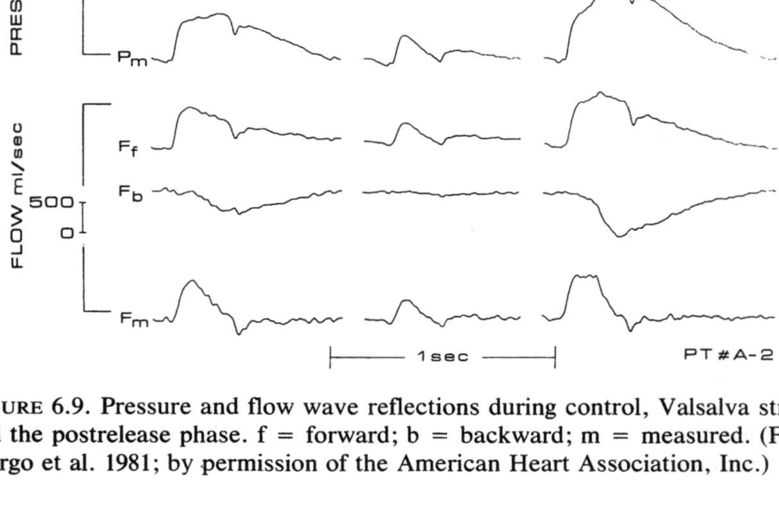

FIGURE 6.9. Pressure and flow wave reflections during control, Valsalva strain, and the postrelease phase. f = forward; b = backward; m = measured. (From Murgo et al. 1981; by permission of the American Heart Association, Inc.)

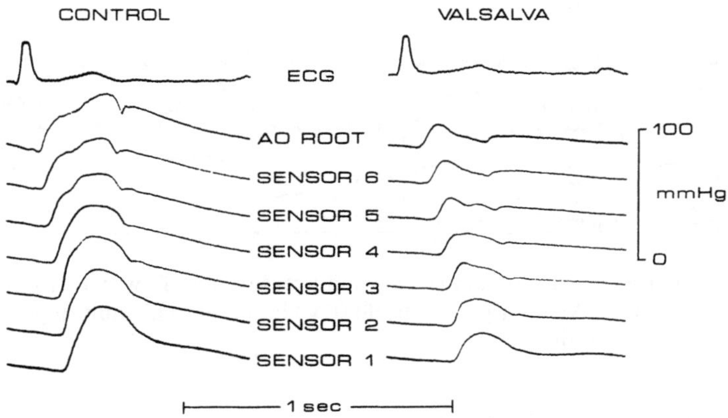

FIGURE 6.10. Electrocardiogram (ECG) and pressure wave shapes along the human aorta (AO) before and during Valsalva strain. Measurement locations are the same as those shown in Figure 6.5.

decrease in intraluminal pressures during the strain phase. As a result, pulse wave velocity decreases considerably, as shown in the right hand panel of Figure 6.10.

Use of the Valsalva maneuver has added to our understanding of the distribution of pulse wave velocities and of the major "effective" reflection site in man. While the origin of the renal vessels appears to be the site responsible for the archetypical type A waveforms seen in normal adults, additional reflection sites may be found with progression distally into the iliac and femoral bifurcations.

Pulmonary Circulation

Typical Waveforms in Humans

The archetypical main pulmonary artery pressure waveform in normal adults is shown in Figure 6.11 in conjunction with a right ventricular pressure signal and a main pulmonary artery flow velocity waveform (Murgo and Westerhof 1984). In contrast to the type A beat shown in the ascending aorta (Fig. 6.1), the pressure waveform in the main pulmonary artery of normal human subjects is a type C. That is, peak pressure occurs in early systole and, if secondary reflection is seen, it ordinarily occurs after valve closure. Thus the normal configuration of the pulmonary artery pressure waveform in humans at rest appears more like the beat from the vasodilated systemic circulation shown in Figure 6.2.

In contrast, the presence of pulmonary hypertension causes a dramatic change from a type C beat in the pulmonary artery to a typical type A

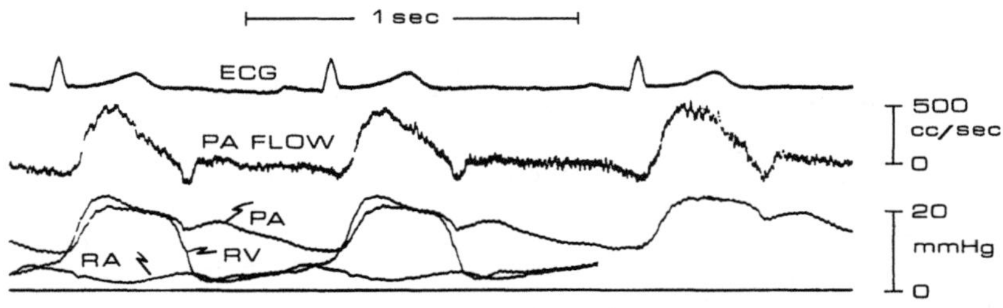

FIGURE 6.11. Electrocardiogram (ECG) and pressure and flow waveforms in the pulmonary artery (PA) and right ventricle (RV) in the normal adult, together with right atrial (RA) pressure.

beat, as illustrated in Figure 6.12. The presence of pulmonary hypertension causes the pulmonary arterial waveform of humans to look much like the type A beat seen in the systemic circulation under control conditions.

Pulmonary Impedance Counterparts

We studied 10 patients without evidence of cardiovascular or pulmonary disease to evaluate the input impedance of the pulmonary circulation (Murgo and Westerhof 1984). Figure 6.13 shows the average input impedance spectrum of these 10 patients. Noteworthy are the low peripheral vascular resistance and rather flat nonoscillatory nature of the modulus spectrum. Similarly, the phase angle plot shows phase angles close to zero degrees already in the second and third harmonics.

While we have not yet completed a systematic study of a significant group of patients with pulmonary hypertension, we have studied individuals. Figure 6.14 illustrates the impedance spectrum of one patient with

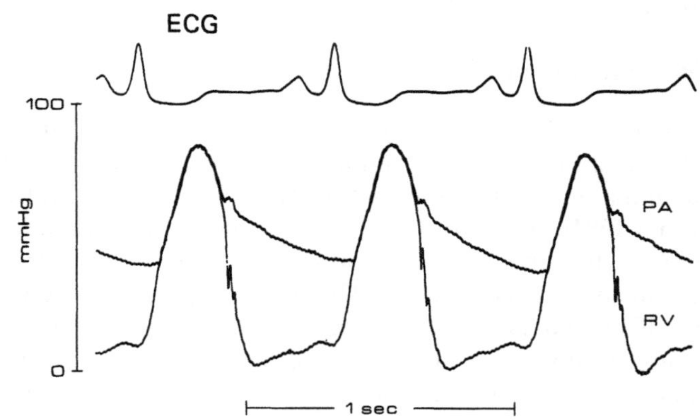

FIGURE 6.12. Electrocardiogram (ECG) and pressure waveforms in the pulmonary artery (PA) and right ventricle (RV) in pulmonary hypertension.

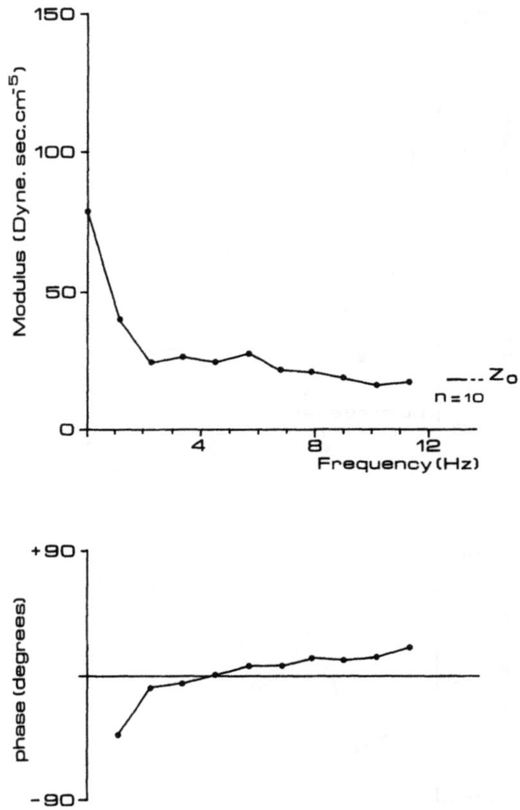

FIGURE 6.13. Input impedance spectrum of the pulmonary circulation during control in 10 normal humans.

pulmonary hypertension. Several obvious differences exist between this spectrum and spectra derived from normal patients. First, the peripheral vascular resistance is considerably higher. Second, more oscillations occur around the characteristic impedance. Finally, the phase angles of the first few harmonics are considerably more negative than those for patients with a normotensive pulmonary circulation. The differences in the impedance spectrum of the patient with pulmonary hypertension imply a greater degree of reflection and are consistent with the same explanations invoked for the systemic circulation and the cause of the late-systolic pressure rise in the typical type A beat.

Effects of Respiration

Since the pulmonary system in normal individuals is a low-pressure system that is entirely exposed to changes in intrathoracic pressure during res-

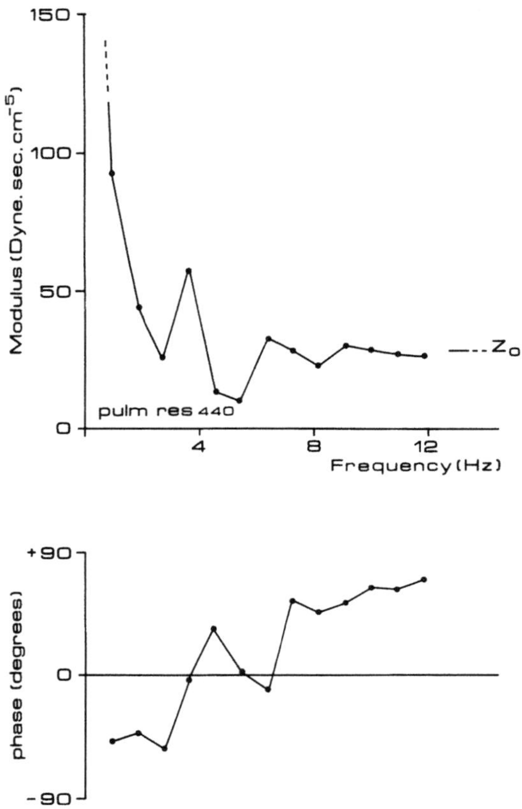

FIGURE 6.14. Example of the input impedance of the human pulmonary circulation in pulmonary hypertension. Pulmonary resistance is 440 dyne sec cm^{-5}.

piration, we also investigated the effects of respiration on the input impedance of the human pulmonary circulation (Murgo and Westerhof 1984). This was of particular interest because of alterations that occur in right ventricular output as a consequence of respiration. Are these changes secondary to changes in right ventricular function, or due to alterations of pulmonary input impedance? We found that no differences in pulmonary vascular resistance, characteristic impedance, or the overall appearance of the impedance modulus and phase spectra occur during quiet respiration. It should be mentioned that we calculated pulmonary vascular resistance utilizing a *difference* between mean pulmonary artery pressure and mean left atrial (or pulmonary capillary wedge) pressure (Westerhof et al. 1979), and the flow through the vascular bed. While this subtraction in usually not performed in the systemic circulation because of the insignificant magnitude of right atrial pressure compared with aortic pressure, the levels of pressure obtained in the left atrium as compared with the pulmonary artery pressure may be quite significant in both normotensive and disease states. Our results with this subtraction showed no differences in pulmonary vascular resistance. We concluded that alterations in right ventricular function are due to changes in filling (preload), rather than to changes in pulmonary input impedance.

Comparison of the Systemic and Pulmonary Circulations

Differences in Pressure Waveforms

We have shown that in adults the ascending aortic pressure waveform is ordinarily a type A beat, which may be manipulated into a type C beat by interventions (see Figs. 6.1, 6.2, and 6.8). In contrast, the typical waveform in the pulmonary artery of the human is a type C beat, which may be changed to a type A beat by the occurrence of pulmonary hypertension (compare Figs. 6.11 and 6.12). The differences in these two circulations may be further understood by examining the differences in the input impedance spectra of the pulmonary and systemic trees.

Comparison of Pulmonary and Systemic Input Impedances

In the same study of pulmonary input impedance (Murgo and Westerhof 1984), 5 of the 10 patients also had systemic input impedance derived from simultaneous measurements performed in the ascending aorta. Figure 6.15 illustrates the averaged data from the 10 patients for the pulmonary vascular tree (thin lines), and from five patients for the systemic vascular tree (thick lines). Since the impedance moduli are plotted on the same scale to better show the differences in the two circulations, systemic resistance could not be plotted and is given as a number. There are significant differences between the systemic and pulmonary circulations. For example, the ratio of peripheral resistance and characteristic impedance is smaller in the pulmonary bed than in the systemic bed; the moduli of the impedance of the systemic circulation oscillate more than the impedance moduli of the pulmonary circulation; and the phase angles in the low-frequency range of the pulmonary impedance spectrum are less negative than those in the systemic tree spectrum. In the dog, separation of the forward and backward waves in the systemic and pulmonary circulations has shown that the backward waves in the pulmonary circulation are much smaller than those in the systemic tree (Van den Bos, Westerhof, and Randall 1982).

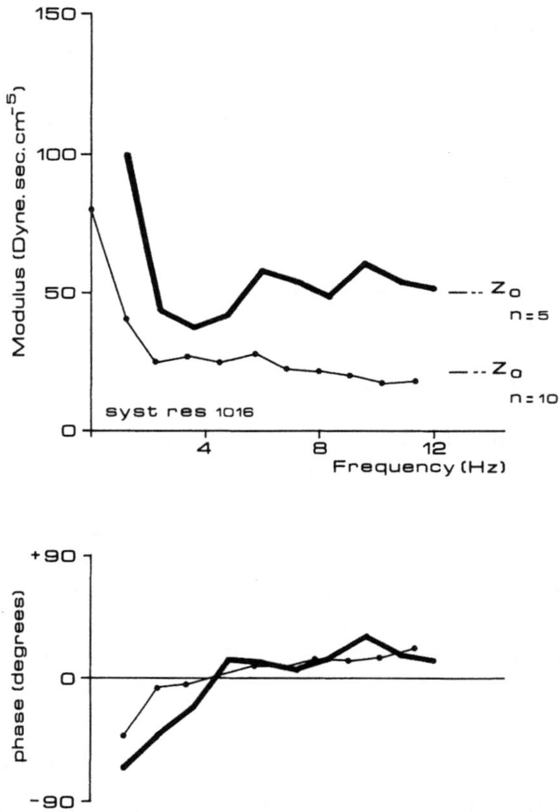

FIGURE 6.15. Averaged input impedance of the systemic ($n = 5$) and pulmonary ($n = 10$) circulations in normal patients. Systemic resistance is 1016 dyne sec cm^{-5}.

Conclusions

In this chapter we have attempted to emphasize one particular aspect of the interaction between the heart and the peripheral system by focusing on the role of reflections in generating the shapes of pressure and flow waveforms in both the systemic and pulmonary circulations of humans. While diffuse reflections no doubt occur from multiple sites within the systemic and pulmonary vascular circulations, it appears that in adults, the typical ascending aortic waveform has a distinct inflection point followed by a rise in late-systolic pressure. We believe that this phenomenon is a result of reflection from the region of the origin of the renal arteries. More recent and ongoing studies using a catheter with multiple sensors distributed through the entire length of the thoracic and abdominal aorta have greatly assisted in this interpretation.

In contrast to the situation in the systemic circulation, the interaction between the right heart and the normally low-pressure pulmonary circulation is of such a nature that pulmonary artery pressure waveforms are ordinarily type C in appearance, implying more diffuse and delayed reflections. However, in the presence of pulmonary hypertension, the right heart and pulmonary circulation behave more like the systemic circulation, generating archetypical type A beats. The presence of increased reflections is also illustrated in the impedance counterparts to these various examples.

We have also illustrated how reflections may be manipulated utilizing certain maneuvers. Such manipulation contributes to an increased understanding of the role of reflection phenomena in the generation of pressure waveforms in humans.

Bibliography

Elzinga, G., and Westerhof, N. 1973. Pressure and flow generated by the left ventricle against different impedances. *Circ. Res.* 32:178–186.

Kouchoukos, N. T., Sheppard, L. C., and McDonald, D. A. 1970. Estimation of stroke volume in the dog by a pulse contour method. *Circ. Res.* 26:611–623.

Latham, R. D., Westerhof, N., Sipkema, P., Rubal, B. J., Reuderink, P., and Murgo, J. P. 1985 1986. Regional wave travel and reflections along the human aorta: a study with simultaneous micromanometry. *Circulation,* 72:1257–1269.

McDonald, D. A. 1974. *Blood Flow in Arteries*. 2nd Ed. London: Arnold.

McDonald, D. A., and Taylor, M. G. 1956. An investigation of the arterial system using a hydraulic oscillator. *J. Physiol. (Lond.)* 133:74–75.

Milnor, W. R. 1982. *Hemodynamics*. Baltimore: Williams & Wilkins.

Murgo, J. P., Altobelli, S. A., Dorethy, J. F., Logsdon, J. R., and McGranahan, G. M. 1975. Normal ventricular ejection dynamics in man during rest and exercise. In Leon, D. F., and Shaver, J. A., eds. *Physiologic Principles of Heart Sounds and Murmurs*. Dallas: American Heart Association, Monograph 46, pp. 92–101.

Murgo, J. P., Giolma, J. P., and Altobelli, S. A. 1977. Signal acquisition and processing for human hemodynamic research. *Proc. IEEE* 65:696–702.

Murgo, J. P., and Westerhof, N. 1984. Input impedance of the pulmonary arterial system in normal man. *Circ. Res.* 54:666–673.

Murgo, J. P., Westerhof, N., Giolma, J. P., and Altobelli, S. A. 1980. Aortic input impedance in normal man: relationship to pressure wave forms. *Circulation* 62:105–116.

Murgo, J. P., Westerhof, N., Giolma, J. P., and Altobelli, S. A. 1981a. Manipulation of ascending aortic pressure and flow wave reflections with the Valsalva maneuver: relationship to input impedance. *Circulation* 63:122–132.

Murgo, J. P., Westerhof, N., Giolma, J. P., and Altobelli, S. A. 1981b. Effects of exercise on aortic input impedance and pressure waveforms in normal humans. *Circ. Res.* 48:334–343.

O'Rourke, M. F. 1971. The arterial pulse in health and disease. *Am. Heart J.* 82:687–702.

O'Rourke, M. F. 1982. *Arterial Function in Health and Disease*. Edinburgh/London: Churchill Livingstone.

Van den Bos, G. C., Westerhof, N., and Randall, O. S. 1982. Pulse wave reflection: can it explain the differences between systemic and pulmonary pressure and flow waves? *Circ. Res.* 51:479–485.

Westerhof, N., Elzinga, G., and Van den Bos, G. C. 1973. Influence of central and peripheral changes on the hydraulic input impedance of the systemic arterial tree. *Med. Biol. Eng.* 11:710–723.

Westerhof, N., Murgo, J. P., Sipkema, P., Giolma, J. P., and Elzinga, G. 1979. Arterial Impedance. In Hwang, N. H. C., Gross, D. R., and Patel, D. J., eds. *Quantitative Cardiovascular Studies*. Baltimore: University Park Press, pp. 110–131.

Westerhof, N., Sipkema, P., Van den Bos, G. C., and Elzinga, G. 1972. Forward and backward waves in the arterial system. *Cardiovasc. Res.* 6:648–656.

Wetterer, E. 1954. Flow and pressure in the arterial system, their hemodynamic relationship and the principles of their measurement. *Minn. Med.* 37:77–86.

7
The Control of Cardiac Output Through Coupling of Heart and Blood Vessels

SOLBERT PERMUTT AND ROBERT A. WISE

At constant blood volume, the flow of blood from the blood vessels to the heart, the so-called venous return, is a function of the atrial pressures. The atrial pressures act as back pressures to flow from the blood vessels to the heart: The greater the atrial pressures, the less the flow. At the same time, the atrial pressures act in an opposite way on ejection of blood from the heart: The greater the atrial pressures, the greater the ventricular filling and stroke volume. The opposite effect of the atrial pressures on venous return and ventricular ejection is the focal point of the mechanical coupling of the heart and circuit as determinants of the steady-state cardiac output.

In this chapter we derive analytic expressions for the relation between cardiac output and atrial pressures acting either as back pressures to venous return or as ventricular filling pressures, and then combine these expressions in a way that eliminates the atrial pressures and defines cardiac output exclusively in terms of circuit and heart parameters.

Atrial Pressures as Back Pressures to Venous Return

Let us consider the systemic circulation as a whole and isolated from the rest of the circulation. The inflow is from the aorta and the outflow is to the right atrium. The relation between change in blood volume, blood flow, and right atrial pressure is

$$dV_s = \left(\frac{\partial V_s}{\partial \dot{Q}_s}\right)_{P_{ra}} d\dot{Q} + \left(\frac{\partial V_s}{\partial P_{ra}}\right)_{\dot{Q}_s} dP_{ra} \qquad (7.1)$$

where V_s = blood volume of systemic circulation
$\dot{Q}_s$ = systemic blood flow
P_{ra} = right atrial pressure

At constant blood volume ($dV_s = 0$)

$$(\partial P_{ra}/\partial \dot{Q}_s)_{V_s} = -(\partial V_s/\partial \dot{Q}_s)_{P_{ra}}/(\partial V_s/\partial P_{ra})_{\dot{Q}_s} \qquad (7.2)$$

Thus any two ratios determine the third. The three ratios $(\partial V_s/\partial \dot{Q}_s)_{P_{ra}}$, $(\partial V_s/\partial P_{ra})_{\dot{Q}_s}$, and $(\partial P_{ra}/\partial \dot{Q}_s)_{V_s}$ are in units of time, compliance, and resistance, respectively. Let $(\partial V_s/\partial \dot{Q}_s)_{P_{ra}} = \tau_s$, the *time constant* of the systemic circulation; $(\partial V_s)/\partial P_{ra})_{\dot{Q}_s} = C_s$, the *compliance* of the systemic circulation, and $-(\partial P_{ra}/\partial \dot{Q}_s)_{V_s} = R_s$, the *resistance to venous return of the systemic circulation*. Let us consider these ratios as constants over a range small enough that linearity is approximated. It is apparent that

$$\tau_s = C_s R_s \tag{7.3}$$

Consider the relation between right atrial pressure and blood flow at constant volume. From integration of Equation 7.1 with $dV_s = 0$,

$$\dot{Q}_s = \frac{P_s - P_{ra}}{R_s} \tag{7.4}$$

where $P_s = P_{ra}$ at $\dot{Q}_s = 0$, the *systemic static pressure*. When $\dot{Q}_s > 0$, the location in the systemic circulation where the pressure equals P_s is the *systemic static locus*, and R_s is the resistance to flow between this locus and the right atrium (Permutt and Caldini 1978). Equation 7.4 is similar to the relation between venous return and right atrial pressure obtained from a right heart bypass preparation such as that used by Guyton (Guyton, Jones, and Coleman 1973) to define *mean systemic pressure* and *resistance to venous return*, but the P_s and R_s of Equation 7.4 differ slightly from Guyton's parameters in that they are strictly defined at constant systemic blood volume, whereas the values obtained from a right heart bypass preparation allow shifts of blood between the pulmonary and systemic circulations.

Integration of Equation 7.1 at $\dot{Q}_s = 0$ and thus $P_{ra} = P_s$ yields

$$V_s = C_s P_s + V_{s_o} \tag{7.5}$$

where V_{s_o}, the constant of integration, is the volume of the systemic circulation at $P_s = 0$, the *unstressed volume of the systemic circulation*. This is the static pressure–volume curve of the systemic circulation. If both the numerator and denominator of Equation 7.4 are multiplied by C_s, we obtain

$$\dot{Q}_s = \frac{C_s P_s - C_s P_{ra}}{C_s R_s} \tag{7.6}$$

Solving Equation 7.5 for $C_s P_s$ and substituting in Equation 7.6 yields

$$\dot{Q}_s = \frac{V_s - V_{so} - P_{ra} C_s}{C_s R_s} \tag{7.7}$$

This is the basic equation of venous return of the systemic circulation.

Exactly the same considerations apply to the pulmonary circulation, except that the pulmonary vessels are assumed to be exposed to pleural pressure rather than to atmospheric pressure. Thus

$$\left[\frac{\partial V_p}{\partial(P_{\text{la}} - P_{\text{pl}})}\right]_{\dot{Q}_p} = C_p \tag{7.8}$$

where P_{la} = left atrial pressure
P_{pl} = pleural pressure
$\dot{Q}_p$ = steady-state pulmonary blood flow
C_p = pulmonary vascular compliance

This modifies the basic equation of venous return of the pulmonary circulation to

$$\dot{Q}_p = \frac{V_p - V_{p_o} - (P_{\text{la}} - P_{\text{pl}})C_p}{C_pR_p} \tag{7.9}$$

where $C_pR_p = \tau_p$, the *time constant of the pulmonary circulation*. $V_{p_o} = V_p$ at $\dot{Q}_p$ and $(P_{\text{la}} - P_{\text{pl}}) = 0$ and is the *unstressed volume of the pulmonary circulation*.

Solving Equation 7.7 for V_s and Equation 7.9 for V_p and adding yields, after rearranging,

$$\dot{Q} = \frac{V - V_o - P_{\text{pl}}C_s - (P_{\text{ra}} - P_{\text{pl}})C_s - (P_{\text{la}} - P_{\text{pl}})C_p}{C_sR_s + C_pR_p} \tag{7.10}$$

where $\dot{Q} = \dot{Q}_s = \dot{Q}_p$
$V = V_s + V_p$
$V_o = V_{s_o} + V_{p_o}$.

This equation is an analytic solution of the relation between steady-state cardiac output and atrial pressures acting as back pressures to venous return in terms of circuit parameters: V, V_o, P_{pl}, C_s, C_p, R_s, and R_p. V_o is affected by gravity, body position, and external pressure. It is also affected by effective back pressures other than the atrial pressures (see chapter appendix). C_s and C_p depend significantly on the elastic properties of the blood vessels but also on the magnitude and distribution of effective back pressures affecting the transmission of changes in atrial pressures to compliant regions of the circulation (see chapter appendix). R_s and R_p are determined by venous resistance and also the distribution of resistance between parallel vessels (see chapter appendix). The effect of the distribution of resistance and effective back pressures is considered later.

Atrial Pressures as Determinants of Ventricular Ejection

Because of the close relationship between mean atrial pressure and ventricular filling pressure, stroke volume is correlated with atrial pressure. The ratio of change in atrial pressure to change in cardiac output has the units of resistance, and we shall show that this ratio is additive to the resistance to venous return as a determinant of cardiac output. The relationship between cardiac output and right atrial pressure was used by Guyton (Guyton, Jones, and Coleman 1973), along with the venous return curve, to provide a graphic solution of steady-state cardiac output, but the relationship was empirical. If the concepts of cardiac function of Sagawa et al. (Suga and Sagawa 1974) are applied to the relationship between cardiac output and atrial pressure, an analytic expression can be derived.

Suga and Sagawa analyze cardiac contractility in terms of a change in the compliance of the ventricle. During diastole the ventricular compliance is large; at end systole the compliance is at its lowest level (elastance at its maximum level, E_{max}). At each time interval during systole there is a ventricular pressure–volume curve that is intermediate in slope between the curves at end diastole and end systole. The volume at the end of ejection is very close to the ventricular pressure–volume curve of minimal compliance or maximum elastance; thus for stroke volume we need be concerned only with two pressure–volume curves. The stroke volume is determined by the point on the end-diastolic pressure–volume curve where contraction begins and the point near the end-systolic pressure–volume curve where contraction ends.

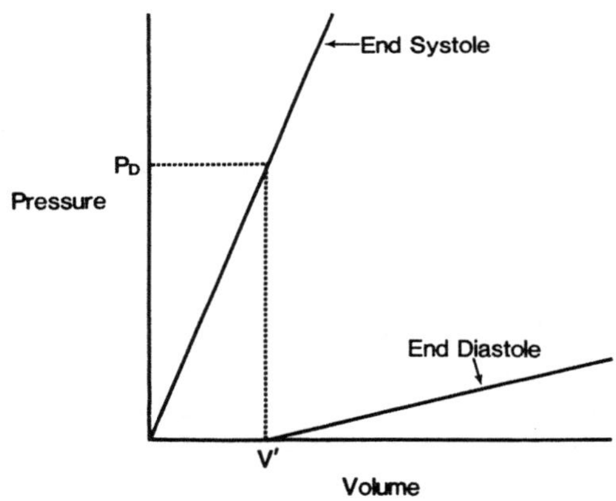

FIGURE 7.1. Schematic diagram of the end-diastolic and end-systolic pressure–volume relationships of the ventricle. The ordinate is the transmural pressure of the ventricle relative to surrounding pleural pressure. The abscissa is the volume of the ventricle above the unstressed volume at end systole. The ventricle becomes progressively stiffer (less compliant) between end diastole and end systole. The end-systolic curve is approximated by a straight line. The end-diastolic curve can be approximated by a straight line over a small range, so that the compliance may be characterized by the slope of a tangent to the curve and an unstressed volume intercept. V' is the difference between the end-systolic and end-diastolic unstressed volumes. P_D is the end-systolic pressure of the ventricle at the end-diastolic unstressed volume (V').

We shall assume that the end-systolic pressure–volume curve is linear and begins at the origin (no unstressed volume). The end-diastolic pressure–volume curve is linear and is characterized by a slope and a volume intercept, V', the unstressed diastolic ventricular volume (Fig. 7.1). The end-diastolic pressure–volume curve may be curvilinear, but we shall limit the range to a portion where linearity is approximated. We shall assume that the ventricle is surrounded by pleural pressure. While there might be an unstressed volume at end systole, this volume does not influence heart function per se, but the difference between the unstressed diastolic and systolic curves (V') has a significant influence.

The stroke volume is the difference between the heart volume at end diastole and at end systole:

$$\mathrm{SV} = V' + C_{\mathrm{ED}}(P_{\mathrm{ED}} - P_{\mathrm{pl}}) - C_{\mathrm{ES}}(P_{\mathrm{ES}} - P_{\mathrm{pl}}) \tag{7.11}$$

where SV = stroke volume
V' = end-diastolic volume intercept
C_{ED} = ventricular compliance at end diastole
P_{ED} = ventricular pressure at end diastole
C_{ES} = compliance at end systole
P_{ES} = pressure at end systole
C_{ES} = $1/E_{\max}$

Left Heart

We shall assume that the end-diastolic pressure is close to the mean left atrial pressure and that the end-systolic pressure is close to the mean aortic pressure. Thus

$$\dot{Q} = f\mathrm{SV} = fV'_l + fC_l(P_{\mathrm{la}} - P_{\mathrm{pl}}) - \frac{P_a - P_{\mathrm{pl}}}{E_{\max_l}} \tag{7.12}$$

where f = heart rate
V'_l = diastolic left ventricular volume intercept
C_l = diastolic left ventricular compliance
P_a = mean aortic pressure
$E_{\max_l}$ = reciprocal of left ventricular compliance at end systole

We shall assume that the relationship between mean aortic pressure and steady-state cardiac output can be characterized by a straight line where $dP_a/d\dot{Q} = R_a$, and the pressure intercept = P_{Ba}, the effective back pressure of the systemic circulation. While the actual relation between $\dot{Q}$ and P_a may be curvilinear, we shall consider a small enough portion that linearity is approximated. Thus

$$P_a = \dot{Q}R_a + P_{\mathrm{Ba}} \tag{7.13}$$

Substitution of P_a of Equation 7.13 into Equation 7.12 and solving for $\dot{Q}$ yields

$$\dot{Q} = \frac{P_{\mathrm{la}} - P_{\mathrm{pl}} - [(P_{\mathrm{Ba}} - P_{\mathrm{pl}})/\lambda_l] + P_{D_l}/\lambda_l}{(T/C_l) + R_a/\lambda_l} \tag{7.14}$$

where $T = 1/f$ = time between heart beats

λ_l = ratio of left ventricular diastolic to systolic compliance ($C_l E_{\mathrm{max}_l}$)

P_{D_l} = systolic left ventricular distending pressure at V'_l

Right Heart

We assume that the relationship between pulmonary arterial pressure (P_p) and $\dot{Q}$ is

$$P_p = \dot{Q}R'_p + P_{\mathrm{Bp}} \tag{7.15}$$

where P_{Bp} is effective back pressure of pulmonary circulation and R_{p}' is resistance of pulmonary circulation (the prime symbol distinguishes this resistance to flow from R_p, the resistance to venous return). Following the same steps as for the left heart

$$\dot{Q} = \frac{P_{\mathrm{ra}} - P_{\mathrm{pl}} - [(P_{\mathrm{Bp}} - P_{\mathrm{pl}})/\lambda_r] + P_{D_r}/\lambda_r}{(T/C_r) + R'_p/\lambda_r} \tag{7.16}$$

where λ_r is the ratio of right ventricular diastolic to systolic compliance ($C_r E_{\mathrm{max}_r}$) and P_{D_r} is systolic right ventricular distending pressure at diastolic volume intercept V_r'.

Atrial Pressure of Opposite Heart as Back Pressure to Flow

The effective back pressures to flow (P_{Ba} and P_{Bp}) may or may not be the atrial pressure of the opposite heart. For example, in the pulmonary circulation alveolar pressure and not left atrial pressure is the effective back pressure when alveolar pressure is higher than left atrial pressure due to a Starling resistor or waterfall effect (Permutt, Bromberger-Barnea, and Bane 1962). There is considerable evidence that vascular waterfalls and Starling resistors may be operative in the systemic circulation (Bellamy 1978, Ehrlich et al. 1980, Jachman and Green 1977, Permutt and Riley 1963, Sylvester et al. 1980). The ratio of change in the effective back pressure to change in the downstream atrial pressure determines the extent to which one side of the heart may influence the ejection of the other side. The degree of coupling can be expressed by the following differential equations:

$$dP_{\mathrm{Ba}} = F_{\mathrm{Ca}}dP_{\mathrm{Ca}} + F_{\mathrm{ra}}dP_{\mathrm{ra}} \tag{7.17}$$

$$dP_{\mathrm{Bp}} = F_{\mathrm{Cp}}dP_{\mathrm{Cp}} + F_{\mathrm{la}}dP_{\mathrm{la}} \tag{7.18}$$

where dP_{Ca} and dP_{Cp} are the changes in critical back pressures due to Starling resistors or vascular waterfalls in the systemic and pulmonary circulations, respectively. F_{Ca} and F_{Cp} are the fraction of the blood vessels in the systemic and pulmonary circulations, respectively, whose flow is not influenced by changes in atrial pressure. Thus $F_{Ca} + F_{ra} = 1$ and $F_{Cp} + F_{la} = 1$. For example, if the pulmonary circulation were completely in zone 2 (West, Dollery, and Naimark 1964), $F_{Cp} = 1$ and $F_{la} = 0$; if it were completely in zone 3, $F_{Cp} = 0$ and $F_{la} = 1$ (Brower et al. 1985). Integration of Equations 7.17 and 7.18 yields

$$P_{Ba} = F_{Ca}P_{Ca} + F_{ra}(P_{ra} + P'_{Ba}) \tag{7.19}$$

$$P_{Bp} = F_{Cp}P_{Cp} + F_{la}(P_{la} + P'_{Bp}) \tag{7.20}$$

P'_{Ba} and P'_{Bp} are constants of integration that are added to account for the effect of alinearity of the arterial pressure–flow relations. The effective back pressure may be greater than atrial pressure owing either to a vascular waterfall or alinearity. In the former case a change in atrial pressure does not have an effect, but in the latter case it does.

Resistance and Back Pressure to Ejection

If we substitute P_{Ba} of Equation 7.19 into Equation 7.14 and P_{Bp} of Equation 7.20 into Equation 7.16, we have two simultaneous equations in terms of P_{la} and P_{ra} whose solutions are

$$\dot{Q} = \frac{P_{la} - P_{pl} - P_{H_l}}{R_{H_l}} \tag{7.21}$$

$$\dot{Q} = \frac{P_{ra} - P_{pl} - P_{H_r}}{R_{H_r}} \tag{7.22}$$

where

$$P_{H_l} = \frac{F_{Ca}(P_{Ca} - P_{pl}) + F_{ra}P'_{Ca} - P_{D_l}}{\lambda_l - (F_{la}F_{ra}/\lambda_r)} + \frac{F_{ra}[F_{Cp}(P_{Cp} - P_{pl}) + F_{la}P'_{Cp} - P_{D_r}]}{\lambda_l\lambda_r - F_{la}F_{ra}}$$

$$P_{H_r} = \frac{F_{Cp}(P_{cp} - P_{pl}) + F_{la}P'_{Cp} - P_{D_r}}{\lambda_r - (F_{la}F_{ra}/\lambda_l)} + \frac{F_{la}[F_{Ca}(P_{Ca} - P_{pl}) + F_{ra}P'_{Ca} - P_{D_l}]}{\lambda_l\lambda_r - F_{la}F_{ra}}$$

$$R_{H_l} = \frac{(T/C_l) + R_a/\lambda_l}{1 - F_{la}F_{ra}/\lambda_l\lambda_r}$$

$$+ \frac{(F_{ra}/\lambda_l)\,[(T/C_r) + R'p/\lambda_r]}{1 - F_{la}F_{ra}/\lambda_l\lambda_r}$$

$$R_{H_r} = \frac{(T/C_r) + R'_p/\lambda_r}{1 - F_{la}F_{ra}/\lambda_l\lambda_r}$$

$$+ \frac{(F_{la}/\lambda_r)\,[(T/C_l) + R_a/\lambda_l]}{1 - F_{la}F_{ra}/\lambda_l\lambda_r}$$

Equations 7.21 and 7.22 define the relations between the transmural atrial pressures and steady-state cardiac output in terms of *back pressure to ejection, P_H,* and *resistance to ejection, R_H*. Both P_H and R_H are determined by heart and circuit parameters and the degree of coupling between the heart-sides. Within the conventional context, the circuit parameters of P_H and R_H are the components of *afterload,* and the heart parameters are the components of *cardiac function* and *contractility*. The parameters of preload, afterload, cardiac function, and coupling are itemized in Table 7.1.

The afterload to ejection is determined by the position and slope of the arterial pressure–flow curves. The position is a determinant of back pressure to ejection; the slope, the resistance to ejection. Pleural pressure affects afterload through its effect on P_{Ca} relative to the pressure around the heart (Permutt, Wise, and Sylvester 1985). Similarly, cardiac function is determined by the relative position of the diastolic-to-systolic pressure–volume curves and the slopes of the curves. The position affects back pressure; the slope, resistance. The heart period affects the resistance to ejection through its effect on the rate of ventricular filling during diastole.

The degree of coupling between the ventricles determines the extent to which one side of the heart affects the other. If the effective back pressure to flow in both the pulmonary and systemic circulations are from Starling resistors, one side of the heart has no influence on the ejection of the other side. Such conditions exist when F_{ra} and $F_{la} = 0$ and the arterial pressure–flow relations are those of vascular waterfalls. If, on the other hand, F_{ra} and $F_{la} = 1$, no Starling resistors are affecting flow and P_H and R_H of one side of the heart are functions of the other side. The

TABLE 7.1. Parameters of the heart and blood vessel circuitry

Preload	Afterload	Cardiac function	Coupling
$P_{la} - P_{pl}$	P_{pl}	E_{max_l}, E_{max_r}	F_{ra}
$P_{ra} - P_{pl}$	P_{Ca}, P_{Cp}	C_l, C_r	F_{la}
	P'_{Ca}, P'_{Cp}	V'_l, V'_r	$(F_{Ca} = 1 - F_{ra})$
	R_a, R'_p	T	$(F_{Cp} = 1 - F_{la})$
		$(\lambda = CE_{max})$	
		$(P_D = V'E_{max})$	

See text for definitions.

TABLE 7.2. Determinants of back pressure to ejection, P_H, and resistance to ejection, R_H, in right and left sides of the heart

	Uncoupled (F_{ra} and $F_{\mathrm{la}} = 0$)	Coupled (F_{ra} and $F_{\mathrm{la}} = 1$)
$P_{\mathrm{H_l}}$	$P_{\mathrm{Ca}} - P$p4:5lP l $_{\mathrm{Dl}}$	$\frac{(P'_{\mathrm{Ca}} - P_{\mathrm{Dl}})/\lambda_l + (P'_{\mathrm{CP}} - P_{\mathrm{DR}})/\lambda_l\lambda_r}{1 - 1/\lambda_l\lambda_r}$ s
$P_{\mathrm{H_r}}$	$\frac{P_{\mathrm{Cp}} - P_{\mathrm{pl}} - P_{\mathrm{Dr}}}{\lambda_r}$	$\frac{(P'_{\mathrm{Cp}} - P_{\mathrm{Dr}})/\lambda_r + (P'_{\mathrm{Ca}} - P_{\mathrm{DL}})/\lambda_l\lambda_r}{1 - 1/\lambda_l\lambda_r}$
$R_{\mathrm{H_l}}$	$(T/C_l) + R_a/\lambda_l$	$\frac{(T/C_l) + (R_a/\lambda_l) + (T/C_r + R'_p/\lambda_r)/\lambda_l}{1 - 1/\lambda_l\lambda_r}$
$R_{\mathrm{H_r}}$	$(T/C_r) + R'_p/\lambda_r$	$\frac{(T/C_r) + (R'_p/\lambda_r) + (T/C_l + R_a/\lambda_l)/\lambda_r}{1 - 1/\lambda_l\lambda_r}$

See text for definitions.

determinants of P_{H} and R_{H} for each side are shown in Table 7.2 for the extreme conditions of F_{la} and $F_{\mathrm{ra}} = 0$ and $= 1$.

There are two ways that the sides of the heart can become uncoupled: through development of negative transmural pressures in collapsible blood vessels or through increases in smooth-muscle tone that cause critical closing pressures. Negative transmural pressure uncouples the left heart from the pulmonary circulation when alveolar pressure rises above left atrial pressure. The right heart becomes uncoupled from the systemic circulation when right atrial pressure falls below the pressure in the extrathoracic veins (Permutt, Wise, and Sylvester 1985). While evidence is considerable that smooth-muscle tone may lead to critical pressures acting as Starling resistors, there is still doubt as to the importance of the phenomenon in hemodynamics (Brunner et al. 1983).

Combined Effect of Circuit and Heart Parameters on Steady-State Cardiac Output

Solving Equation 7.21 for $P_{\mathrm{la}} - P_{\mathrm{pl}}$ and Equation 7.22 for $P_{\mathrm{ra}} - P_{\mathrm{pl}}$ and then substituting these values in Equation 7.10 eliminates atrial pressure and provides an analytic solution of cardiac output exclusively in terms of heart and circuit parameters:

$$\dot{Q} = \frac{V - V_o - P_{\mathrm{pl}}C_s - P_{\mathrm{H_l}}C_p - P_{\mathrm{H_r}}C_s}{C_s(R_s + R_{\mathrm{H_r}}) + C_p(R_p + R_{\mathrm{H_l}})} \tag{7.23}$$

There is a striking analogy between Equation 7.23 and the equation for the drainage of liquid from an elastic container through a resistor. The volume of liquid that can drain from an elastic container through a resistor against a fixed back pressure is the stressed volume of the container minus the product of the back pressure and the compliance of the container. This latter product is the stressed volume of the container that remains after all the liquid that is capable of draining has left the container. We define the volume that is capable of draining as the *mobile volume*. The back pressure to drainage of the pulmonary circulation is the back pressure to ejection of the left heart, P_{H_l}. The back pressure to ejection of the systemic circulation is $P_{H_r} + P_{pl}$ because of the pressure difference between the surrounding pressure of the systemic blood vessels and heart. The numerator of Equation 7.23 is thus the total mobile volume of both circulations.

The time taken for the liquid of an elastic container to drain against a fixed back pressure is independent of both the stressed volume and the back pressure. Therefore the magnitude of the mobile volume does not affect the time of drainage. The time of drainage is dependent on the product of the compliance of the container and the resistance through which the container drains, the *time constant of drainage*. (The time constant for a single container is equal to the time taken for the drainage of 63% of the mobile volume.) The denominator of Equation 7.23 is the sum of the time constants of the systemic and pulmonary circulations. Interestingly, the resistance to drainage of each system is the sum of the resistance to venous return and to ejection.

Within the context of the analogy of drainage of elastic containers, the steady-state cardiac output is simply the ratio of the mobile volume to the time constant of drainage. Steady-state cardiac output is mathematically equivalent to the continuous drainage of blood from the elastic elements of the systemic blood vessels to the elastic elements of the pulmonary blood vessels and then back to the systemic blood vessels to complete the circuit. The function of the heart is to provide a conduit of drainage and a pressure gradient for drainage to occur, and to return the blood that is drained so that the process is continuous.

Effect of Changes in Heart and Circuit Parameters on Cardiac Output

The solution of Equation 7.23 allows a comprehensive analysis of the control of steady-state cardiac output. The factors affecting P_{H_l}, P_{H_r}, R_{H_l}, and R_{H_r} are itemized in Table 7.2, and those affecting V_o, C_s, C_p, and R_p are discussed following Equation 7.10. A consideration of how all of the factors interact in the control of cardiac output is beyond the scope of this chapter. Here we will consider two specific relations.

Starling's Law of the Heart

The relationship between cardiac output and atrial pressure is one form of the expression of Starling's law. This relationship was called the *cardiac response curve* by Guyton in his analysis of the control of cardiac output (Guyton, Jones, and Coleman 1972). Because large changes in cardiac output may occur with essentially no change in atrial pressure, as in muscular exercise, the relevance of the Starling relationship has been questioned. The controversies surrounding the role of Starling's law in the control of cardiac output were reviewed by Sarnoff and Stanley (1955). Here we shall show the characteristics of the law with all heart and circuit parameters constant except for a change in blood volume. In the following section we show how this relationship, although operative, may be concealed during exercise because of changes in heart rate and arterial resistance.

The diastolic and systolic pressure–volume relations of the normal dog heart are shown in idealized form in Figure 7.2. Also shown is the effect of decreasing E_{max} moderately and severely. For estimated circuit parameters of a normal dog, the relations between $\dot{Q}$, blood volume, and atrial pressures are shown in Figure 7.3 (coupling present).

We were surprised that there is a descending limb to the Starling curve. Although neither side of the heart alone has a descending limb, their coupling with each other creates a point where there is a maximum $\dot{Q}$. In retrospect the explanation is straightforward. The static pressure of the circuit is not only the upstream end of the venous return gradient but is also the back pressure to flow between the heart and the static locus. Beyond a given level, an increase in static pressure impedes ejection through its effect on afterload more than it aids ejection through its effect on preload. That the vascular volume affects afterload may have implications in the therapy of congestive heart failure.

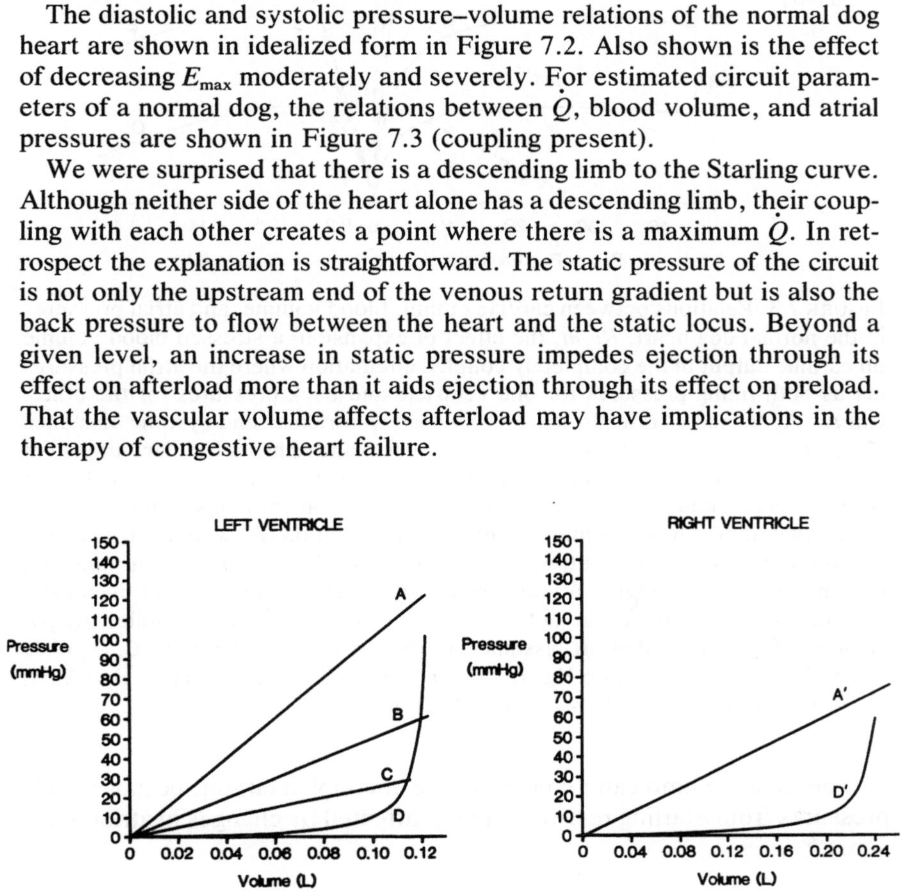

FIGURE 7.2. The end-systolic and end-diastolic pressure–volume relationships of the left and right ventricles of the dog heart (used in the analysis presented in Fig. 7.3). The end-systolic elastances (E_{max}) of the left ventricle are A, 1000; B, 500; and C, 250 mm Hg/l. The end-systolic elastance of the right ventricle is 300 mm Hg/l. The diastolic curves (D and D′) are modeled as hyperbolas which pass through the origin ($V' = 0$): $P = \frac{\alpha V}{B - V}$, where B is the limiting volume of the heart, α is the transmural pressure at half the limiting volume, and V is the volume of the heart. In this figure α is 2.5mm Hg for both ventricles and B is 0.125 and 0.25 L for the left and right ventricles, respectively. In the present analysis, λ, the index of cardiac function, with $V' = 0$, varies with P: $\lambda = BE_{max} / (\alpha + P)$.

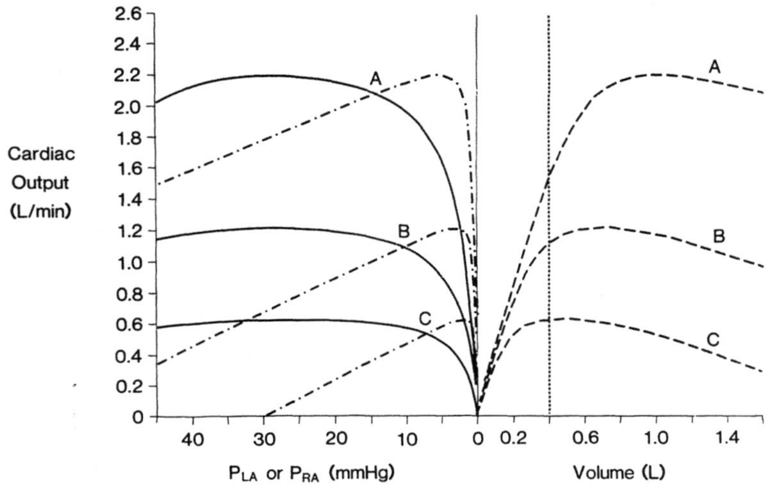

FIGURE 7.3. Relations between cardiac output, blood volume, and atrial pressures in the normal dog heart. *Right,* the effect of expansion of stressed blood volume on cardiac output in the completely coupled circulation where the atrial pressures act as both filling pressures for one ventricle and back pressures for the other. Curves A, B, and C refer to the levels of left ventricular end-systolic elastance (E_{max}) shown in Figure 7.2, with the diastolic curves and right ventricular parameters as shown. Left, the cardiac function curves for the left (———) and right (— • — •)hearts under the three conditions. The vertical line at 0.4 L blood volume shows that adding blood volume in the normal condition (A) would cause a large initial increase in cardiac output, but that the effect would be less in moderate (B) and severe (C) failure. Administration of enough volume would eventually lead to a decrease in cardiac output at high atrial pressures. The following parameters for a 20-kg dog are assumed in this figure: f = 100/min; C_s = 0/.05 L/mm Hg; C_p = 0/.01 L/mm Hg; R_p' = 7 mm Hg/L/min; R_a = 40 mm Hg/L/min; R_s = 3.5 mm Hg/L/min; R_p = 3.5 mm Hg/L/min.

A descending limb cannot occur in the uncoupled circuit, because back pressures from Starling resistors are not affected by changes in atrial pressures.*

Effect of Change in Arterial Resistance and Heart Rate During Exercise

Analysis of Equations 7.21 and 7.22 reveals that a proportional decrease in T, R_a, and R'_p results in an increase in $\dot{Q}$ that is exactly proportional to the increase in heart rate and that the systemic and pulmonary arterial

*It can be shown that a descending limb can result from ventricular interdependence, another form of coupling not considered here.

blood pressures are constant so long as the stressed volume and time constant of drainage of the circuit (Equation 7.10) are sufficient to maintain constant atrial pressures. This is true regardless of the state of cardiac function. Thus the cardiac output during exercise at constant arterial blood pressure and atrial pressures is determined by three factors: the maximum heart rate (minimum T), the minimum R_a and R'_p, and the ability of the circuit to maintain preload through changes in stressed volume and time constant. Let us consider how these three factors interact as determinants of maximum $\dot{Q}$ during exercise.

The primary circulatory event of muscular exercise we shall consider to be vasodilation in the exercising muscles, causing a decrease in R_a. If the heart rate increases appropriately, P_a, $P_{la} - P_{pl}$, and stroke volume can be maintained constant. These relations are shown in Figure 7.4 for a dog with the diastolic pressure–volume curves of Figure 7.2 and normal E_{max}. The resting $\dot{Q}$ of 1.76 L/min (point A) results from estimated heart and circuit parameters of a normal dog with the blood flow distributed predominantly to the nonmuscle circulation.

If the maximum heart rate were 300 beats/min, $\dot{Q}$ would have risen to the value at point B. This would have occurred with no change in cardiac function, but could not have occurred without a change either in stressed volume or the time constant of systemic circulation (C_sR_s). How these changes could occur will be discussed later. While $P_{la} - P_{pl}$ would have remained constant, $P_{ra} - P_{pl}$ would have risen, because we have assumed that the only change in resistance is in the systemic circulation. At constant R'_p there would have been a rise in pulmonary arterial pressure, an increase in pulmonary blood volume $= \Delta\dot{Q}\tau_p$, and an increase in the static pulmonary pressure $= \Delta\dot{Q}\tau_p/C_p$. Thus pulmonary congestion would have occurred, even at constant left atrial pressure, unless there were a decrease in pulmonary vascular resistance. This might limit maximum cardiac output through the sensation of dyspnea.

If the stressed volume and systemic time constant did not change, there would be no significant increase in $\dot{Q}$ regardless of the magnitude of the fall in R_a and increase in heart rate. For example, if R_a had decreased to 10% of its initial value (from 40 to 4 mm Hg/L/min) and heart rate had increased to 300 beats/min, $\dot{Q}$ would have risen only to point C (from 1.76 to 2.23 L/min), and the P_a would have fallen to less than 10 mm Hg. Without a change in those circuit parameters affecting venous return (stressed volume and time constants), the net effect of the fall in R_a and increase in heart rate would be a marked reduction in stroke volume from a decrease in preload. The low $\dot{Q}$ at point C resulted from right and left atrial pressures below 1 mm Hg.

Let us consider the possible ways in which the stressed volume or systemic time constant may change to provide adequate venous return and preload. From Equation 7.10 it can be seen that at constant blood volume and pleural pressure and with the changes limited to the systemic circu-

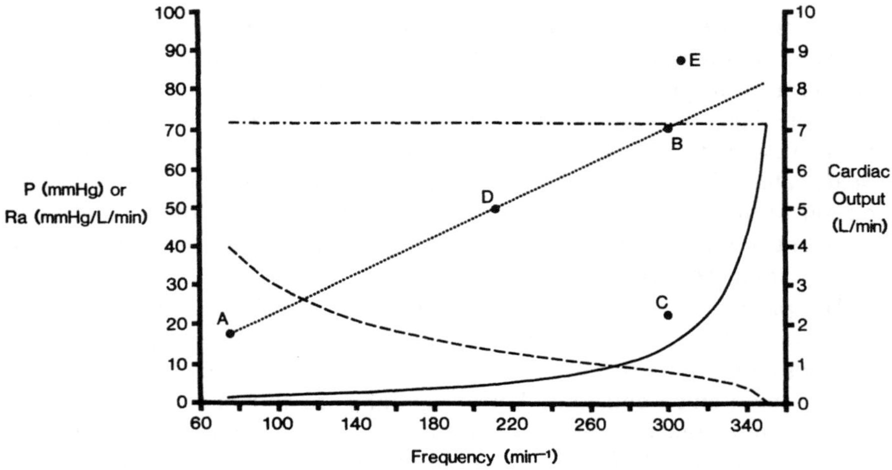

FIGURE 7.4. The effect of changing heart rate with constant arterial pressure (• — • —) and end-diastolic volume (P_{la}-P_{pl} = 8 mm Hg) on cardiac output (·····) in the completely coupled circulation of the dog. (Ra, – – – –); (P_{ra} − P_{pl}, ———) The initial basal condition (point A) assumes the normal cardiac parameters of Figure 7.2, with a heart rate of 75 beats/min, the pulmonary vascular parameters from Figure 7.3, and a peripheral circulation in which all of the blood flow is through the nonmuscle bed (F_n = 1), with τ_n = 0.287 min and a stressed blood volume of 0.72 L. Point B is the cardiac output at a maximum heart rate of 300 beats/min (if the periphery could accommodate the flow by increasing stressed volume or by decreasing the systemic time constant, but this could not be achieved with the systemic parameters considered here). If all the blood flow were directed to the muscle bed (F_m = 1), with τ_m = 0.0436 min, the maximum cardiac output would be achieved at point D. Point C is the effect of reducing the peripheral resistance to 10% of basal value associated with a compensatory increase in heart rate to 300 beats/min, but without a redistribution of peripheral blood flow (no change in τ). Point E is the cardiac output that would be achieved if the cardiac contractility (E_{max}) were increased so that end-systolic volume approached 0 L and all of the blood flow were directed to the muscle bed (F_m = 1, τ_n = 0.0436 min).

lation, there must be either a decrease in unstressed volume, V_o, a decrease in systemic compliance, C_s, or a decrease in the resistance downstream from the static locus, R_s. If the systemic circulation were functioning as a single compartment, there would have to be some change in the state of the systemic veins or venules where the major compliance of the circuit is located. An increase in their external pressure from contracting muscles would effectively increase the stressed volume. An increase in venous smooth-muscle tone would have a similar effect if there were a parallel shift in the pressure–volume characteristics of the veins (Brunner, Shoukas, and MacAnespie 1981). An increase in venous tone could also cause a decrease in C_s. All of these changes would have the net effect of increasing the static pressure of the systemic circulation, P_s, and thus the gradient of venous return (see Equation 7.4). Theoretically it would be possible for the circuit to maintain preload through a decrease in R_s, but the more likely changes from external pressure and tone would be incompatible with a decrease in R_s (Caldini et al. 1974).

The increase in P_s through changes in the state of the systemic veins is the mechanism that has been emphasized by Guyton (Guyton, Jones, and Coleman, 1973). Changes in the distribution of the cardiac output between the exercising muscles and other parts of the systemic circulation might play an even greater role (Caldini et al. 1974).

Consider the systemic circulation divided into two compartments: the circulation to the muscles and the rest of the circulation. Equation 7.7 now applies to each of the compartments:

$$\dot{Q}_m = \frac{V_m - V_{m_o} - P_{ra}C_m}{C_mR_m} \tag{7.24}$$

$$\dot{Q}_n = \frac{V_n - V_{n_o} - P_{ra}C_n}{C_nR_n} \tag{7.25}$$

where the subscripts m and n refer to the muscle and nonmuscle components of the systemic circulation, respectively. Let $C_mR_m = \tau_m$ and $C_nR_n = \tau_n$. It is apparent that τ_m and τ_n are the ratios of the change in blood volume to change in blood flow at constant P_{ra} for the muscle and nonmuscle components, respectively. There is evidence that τ_n is large in relation to τ_m and that the magnitude of τ_n is largely dependent on the splanchnic circulation (Caldini et al. 1974). Thus large increases in muscle blood flow require relatively small increases in muscle blood volume, and a small decrease in splanchnic blood flow causes a relatively large decrease in splanchnic blood volume. The large difference between τ_m and τ_n allows the distribution of flow in the systemic circulation to play a major role in the adjustments of the circuit in the control of cardiac output. The distribution of flow is largely dependent on the state of the arterioles. Thus changes in arteriolar tone may change stressed volume and the systemic time constant; but such changes could be produced only by venous changes if the systemic circulation acted as a single vascular compartment.

The distribution of blood flow is affected both by resistance and effective back pressure. From Equation 7.13, the blood flow to the two compartments is

$$\dot{Q}_m = \frac{P_a - P_{Ba_m}}{R_{a_m}} \tag{7.26}$$

$$\dot{Q}_n = \frac{(P_a - P_{Ba_n})}{R_{a_n}} \tag{7.27}$$

where R_{a_m} and R_{a_n} are the resistances and P_{Ba_m} and P_{Ba_n} are the effective back pressures to flow of the muscle and nonmuscle compartments of the systemic circulation, respectively. A change either in back pressure or resistance that favors a distribution of blood flow to the muscles enhances venous return. These relations can be seen analytically through a modification of Equation 7.10 (see chapter appendix):

$$\dot{Q} = \frac{V - V_o + V_B - P_{\mathrm{pl}}C_s - (P_{\mathrm{ra}} - P_{\mathrm{pl}})C_s - (P_{\mathrm{la}} - P_{\mathrm{pl}})C_p}{F_m\tau_m + F_n\tau_n + \tau_p} \tag{7.28}$$

where

$$V_B = \frac{(P_{\mathrm{Ba}_n} - P_{\mathrm{Ba}_m})(\tau_n - \tau_m)}{R_{a_n} + R_{a_m}}$$

$$F_m = \frac{1/R_{a_m}}{(1/R_{a_m}) + 1/R_{a_n}} \tag{7.29}$$

$$F_n = \frac{1/R_{a_n}}{(1/R_{a_m}) + 1/R_{a_n}}$$

Equation 7.28 is the same as Equation 7.10 except for the addition of a volume term, V_B, to the stressed volume and the partition of C_sR_s to $F_m\tau_m + F_n\tau_n$. It is apparent that a change in the distribution of back pressures affects the stressed volume and a change in the distribution of resistance affects the systemic time constant.

On the basis of the estimated τ_m and τ_n of the dog (Caldini et al. 1974), shifting the distribution of blood flow from the nonmuscle to the muscle circulation would allow $\dot{Q}$ to increase threefold (point D, Fig. 7.4). Any further increase would require changes in the state of the systemic veins in the example under consideration, where the effective back pressures are the atrial pressures. If, however, the effective back pressures were greater than the atrial pressures, even if coupling were present, a shift of the muscle pressure–flow curve to a lower P'_{Ba_m} or an increase in P'_{Ba_n} would increase the stressed volume through an increase in V_B.

An increase in cardiac contractility through an increase in $E_{\max}$ would have essentially no effect on $\dot{Q}$ at constant stressed volume and time constant of drainage. For example, the atrial pressures at point C in Figure 7.4 are below 1 mm Hg, and it is apparent from Equation 7.10 that there is nothing the heart can do to increase output above the level where the atrial transmural pressures are zero. An increase in $E_{\max}$ accompanying changes in circuit parameters, however, can have a very great effect. If there were an increase in $E_{\max_l}$ and $E_{\max_r}$ sufficient to make the end-systolic volumes of both ventricles small (ejection fraction approaching unity), the output would shift from point D to point E, and there now would be a fivefold increase in $\dot{Q}$ and an increase in P_a.

Point E is shifted above the line because stroke volume is increased. In this case the increase is due to a decrease in end-systolic volume. Stroke volume could also increase if the changes in stressed volume and systemic time constant changed out of proportion to the increase in heart rate. The increase in stroke volume would require an increase in end-diastolic ventricular volumes and transmural atrial pressures. There is little evidence that heart volume is increased during exercise (Slutsky 1981). Whether or not this is true, there appears to be little need for such a change. With only a threefold increase in heart rate from rest to maximum exercise, the cardiac output could be expected to increase from 5 to 15 L/min with no change in contractility or left ventricular volume. With some improvement in contractility, which would be expected through sympathetic stimulation and circulating epinephrine, the output would increase still further, owing to a decrease in end-systolic volume.

What the present analysis shows is that the increase in $\dot{Q}$ during exercise requires little change in cardiac function other than an increase in rate. A more complex physiologic adjustment involves the mechanisms through which the circuit maintains adequate preload. The adjustment of the circuit must consist of an increase in stressed volume or a decrease in the time constant of drainage. It appears that the required changes in volume and time constant of drainage can occur in large part through changes in arteriolar tone alone. If this is so, the coupling of the circuit and heart through the systemic resistance and back pressure allows essentially automatic control of blood pressure and blood flow to the exercising muscles so long as there are mechanisms that can increase heart rate in proportion to the fall in arterial resistance.

While the present analysis is based on the fundamental role of the atrial pressures in the coupling of heart and circuit, it is interesting that the coupling maintains virtual constancy of the atrial pressures even with marked changes in cardiac output during normal function. The absence of significant changes in end-diastolic pressures or volumes does not imply that Starling's law is nonoperative—on the contrary, the operation of the law conceals the presence of the law.

Appendix

Let the systemic circulation be composed of a number of blood vessels in parallel, each having a constant inflow pressure of P_a and outflow pressure of P_{ra}. Divide each vessel into a large number of segments in series. Consider the ith segment of the jth vessel. Let us assume that the pressure in this segment, P_{ij}, changes linearly with flow at constant P_{ra} such that

$$\dot{Q}_j = \frac{P_{ij} - P_{B_{ij}}}{R_{ij}} \qquad (7.1a)$$

where $1/R_{ij}$ is slope and $P_{B_{ij}}$ is pressure intercept or effective back pressure.

Let the stressed volume of the segment $V_{ij} = P_{ij}C_{ij}$ where C_{ij} = compliance of segment. Substituting V_{ij}/C_{ij} for P_{ij} in Equation 7.1a and rearranging yields

$$V_{ij} = \dot{Q}_j R_{ij} C_{ij} + C_{ij} P_{B_{ij}} \tag{7.2a}$$

The stressed volume of the jth vessel, V_j, is the sum of the volume in all of the segments and is thus

$$V_j = \dot{Q}_j \sum_i R_{ij} C_{ij} + \sum_i C_{ij} P_{B_{ij}} \tag{7.3a}$$

$$V_j = \dot{Q}_j \tau_j + C_j \overline{P}^c_{B_j} \tag{7.4a}$$

where $\tau_j = \sum_i R_{ij} C_{ij} = (\partial V_j / \partial \dot{Q}_j) =$ time constant of jth vessel

$C_j = \sum_i C_{ij} =$ compliance of jth vessel

$\overline{P}^c_{B_j} = \sum_i (C_{ij}/C_j) P_{B_{ij}} =$ compliance-weighted mean back pressure of jth vessel

The total stressed volume of the systemic blood vessels, V, is the sum of the stressed volumes of all parallel vessels:

$$V = \Sigma V_j = \Sigma \dot{Q}_j \tau_j + C \overline{P}^c_B \tag{7.5a}$$

where $C = \Sigma C_j$ and $\overline{P}^c_B =$ compliance-weighted mean back pressure of all parallel vessels $= \Sigma\, (C_j/C) \overline{P}^c_{B_j}$.

Let the relationship between $\dot{Q}_j$ and P_a be linear with a pressure intercept of $P^{\dot{Q}}_{B_j}$. The superscript $\dot{Q}$ is used to distinguish the effective back pressure to flow from the back pressure to drainage from the compliant regions, $\overline{P}^C_{B_j}$. Thus

$$\dot{Q}_j = (P_a - P^{\dot{Q}}_{B_j}) G_j \tag{7.6a}$$

where $G_j =$ conductance of the jth vessel $= 1/R_{a_j}$. The total systemic flow, $\dot{Q}$, is the sum of the flow in all parallel vessels. From Equation 7.6a,

$$\dot{Q} = \Sigma \dot{Q}_j = P_a \Sigma G_j - \Sigma G_j P^{\dot{Q}}_{B_j}$$

$$\dot{Q} = P_a G - G \Sigma (G_j/G) P^{\dot{Q}}_{B_j}$$

$$\dot{Q} = (P_a - \overline{P}^{\dot{Q}}_B) G = \frac{P_a - P_{Ba}}{R_a} \tag{7.7a}$$

where $G = \Sigma G_j = 1/R_a$ and $\overline{P}^{\dot{Q}}_B = \Sigma (G_j/G) P^{\dot{Q}}_{B_j} = P_{Ba} =$ conductance-weighted mean effective back pressure to flow of the systemic vessels.

Solving Equation 7.7a for P_a, substituting for P_a in Equation 7.6a, and substituting $\dot{Q}_j$ of Equation 7.6a for $\dot{Q}_j$ in Equation 7.5a yields, after rearranging:

$$V = \dot{Q} \Sigma \frac{G_j}{G} \tau_j + G \overline{P}^{\dot{Q}}_B \Sigma \frac{G_j}{G} \tau_j - \Sigma G_j \tau_j P^{\dot{Q}}_{B_j} + C \overline{P}^c_B \tag{7.8a}$$

where $\Sigma(G_j/G)\tau_j$ = the conductance-weighted mean time constant of all the parallel vessels = $\partial V/\partial\dot{Q}$. Let $\bar{\tau} = \Sigma(G_j/G)\tau_j$. Substituting in Equation 7.8a yields

$$V = \dot{Q}\,\bar{\tau} + G\bar{P}_B^{\dot{Q}}\tau - \Sigma G_j\tau_j P_{B_j}^{\dot{Q}} + C\bar{P}_B^C \tag{7.9a}$$

The second and third terms on the right of Equation 7.9a can be combined so that

$$V = \dot{Q}\bar{\tau} + \Sigma G_j P_{B_j}^{\dot{Q}}(\bar{\tau} - \tau_j) + C\bar{P}_B^c \tag{7.10a}$$

Some compliant segments have P_{ra} as the back pressure to drainage, while other segments may have critical back pressures from functioning Starling resistors such that changes in P_{ra} do not affect the drainage. Thus

$$\bar{P}_B^c = \left(\frac{C_{ra}}{C}\right)P_{ra} + \left(\frac{C - C_{ra}}{C}\right)\bar{P}_o \tag{7.11a}$$

where C_{ra}/C = fraction of the total compliance C that drains against P_{ra} and $(C - C_{ra})/C$ = fraction of the total compliance that has critical pressures in excess of P_{ra}. The compliance-weighted mean of these critical pressures is $\bar{P}_o$. Substituting P_B^c of Equation 7.11a into Equation 7.10a and solving for $\dot{Q}$ yields

$$\dot{Q} = \frac{V - \Sigma G_j P_{B_j}^{\dot{Q}}(\bar{\tau} - \tau_j) - (C - C_{ra})\bar{P}_o - C_{ra}P_{ra}}{\bar{\tau}} \tag{7.12a}$$

In comparing Equation 7.12a with Equation 7.7, it is apparent that

$$V_s - V_{s_o} = V - \Sigma G_j P_{B_j}^{\dot{Q}}(\bar{\tau} - \bar{\tau}_j) - (C - C_{ra})\bar{P}_o \tag{7.13a}$$

Thus the effective unstressed volume, V_{s_o}, has a component that depends on the distribution of the effective back pressure to flow between parallel beds of different time constants and another component that depends on the magnitude and extent of critical pressures in excess of P_{ra} that affects drainage from compliant regions. It is also apparent that $C_s = C_{ra}$ and $C_sR_s = \bar{\tau}$.

If we consider the systemic circulation equivalent to only two parallel vessels, m and n,

$$\bar{\tau} = (G_m/G)\tau_m + (G_n/G)\tau_n \tag{7.14a}$$

Equation 7.12a reduces to

$$\dot{Q} = \frac{V + \left(\dfrac{G_mG_n}{G_m + G_n}\right)(P_{B_n}^{\dot{Q}} - P_{B_m}^{\dot{Q}})(\tau_n - \tau_m) - (C - C_{ra})\bar{P}_o - C_{ra}P_{ra}}{\left(\dfrac{G_m}{G_m + G_n}\right)\tau_m + \left(\dfrac{G_n}{G_m + G_n}\right)\tau_n} \tag{7.15a}$$

where

$$\frac{G_m}{G_m + G_n} = \frac{R_a}{R_{a_m}} = F_m$$

$$\frac{G_n}{G_m + G_n} = \frac{R_a}{R_{a_n}} = F_n$$

$$\frac{G_m G_n}{G_m + G_n} = \frac{1}{R_{a_m} + R_{a_n}}$$

$$P^{\dot{Q}}_{B_n} = P_{\mathrm{B_m}}$$

$$P^{\dot{Q}}_{B_m} = P_{\mathrm{B_m}}$$

Bibliography

Bellamy, R. F. 1978. Diastolic coronary artery pressure–flow relations in the dog. *Circ. Res.* 43:92–101.

Brower, R., Wise, R. A., Hassapoyannes, C., Bromberger-Barnea, B., and Permutt, S. 1985. Effect of lung inflation on lung blood volume and pulmonary venous flow. *J. Appl. Physiol.* 58:954–963.

Brunner, M. J., Greene, A. S., Sagawa, K., and Shoukas, A. A. 1983. Determinants of systemic zero-flow arterial pressure. *Am. J. Physiol.* 245:H453–460.

Brunner, M. J., Shoukas, A. A., and MacAnespie, C. C. 1981. The effect of the carotid baroreceptor reflex on blood flow and volume redistribution in the total systemic vascular bed of the dog. *Circ. Res.* 48:274–285.

Caldini, P., Permutt, A., Waddell, J. A., and Riley, R. L. 1974. Effect of epinephrine on pressure, flow and volume relationships in the systemic circulation of dogs. *Circ. Res.* 34:606–623.

Ehrlich, W. E., Baer, R. W., Bellamy, R. E., and Randazzo, R. 1980. Instantaneous femoral artery pressure flow relations in supine anesthetized dogs and the effect of unilateral elevation of femoral venous pressure. *Circ. Res.* 47:88–98.

Guyton, A. C., Jones, C. E., and Coleman, T. G. 1973. *Circulatory Physiology: Cardiac Output and its Regulation.* 2nd Ed. Philadelphia: Saunders.

Jachman, A. P., and Green, J. F. 1977. Arterial pressure–flow relationships in the anesthetized dog. *Ann. Biomed. Eng.* 54:384–394.

Permutt, S., Bromberger-Barnea, B., and Bane, H. N. 1962. Alveolar pressure, pulmonary venous pressure and the vascular waterfall. *Med. Thorac.* 19:239–260.

Permutt, S., and Caldini, P. 1978. Regulation of cardiac output by the circuit: venous return. In Baan, J., Noordergraff, A., and Raines, J., eds. *Cardiovascular System Dynamics.* Cambridge, MA: MIT Press, pp. 465–479.

Permutt, S., and Riley, R. L. 1963. Hemodynamics of collapsible vessels with tone: the vascular waterfall. *J. Appl. Physiol.* 18:924–932.

Permutt, S., Wise, R., and Sylvester, J. T. 1985. Interaction between the ventilatory and circulatory pumps. In Roussos, C., and Macklem, P., eds. *The Thorax.* New York: Marcel Dekker, pp. 701–735.

Sarnoff, S. J. 1955. Myocardial contractility as described by ventricular function curves: observations on Starling's law of the heart. *Physiol. Rev.* 35:107–122.
Slutsky, R. 1981. Response of the left ventricle to stress: effects of exercise, atrial pacing, afterload stress and drugs. *Am. J. Cardiol.* 47:357–364.
Suga, H., and Sagawa, K. 1974. Instantaneous pressure–volume relationships and their ratio in the exercised, supported canine left ventricle. *Circ. Res.* 35:117–126.
Sylvester, J. T., Gilbert, R. D., Traystman, R. J., and Permutt, S. 1981. Effects of hypoxia on the closing pressure of the canine systemic arterial circulation. *Circ. Res.* 49:980.
West, J. B., Dollery, C. T., and Naimark, A. 1964. Distribution of blood flow in isolated lung: relation to vascular and alveolar pressures. *J. Appl. Physiol.* 19:713–724.

8
Matching between Right Ventricle and Pulmonary Bed

HROAR PIENE

Introduction

The concept of ventricle–load matching may create new insight into ventricular vascular coupling. The term "matching" is borrowed from the engineering sciences, where it means the load conditions under which a power generator yields maximum output power to the load. For example, an electrical battery connected to a load resistor achieves maximum output power when the load resistance equals the internal battery resistance. The *efficiency* of the source-load system will be 50%, that is, half the energy drawn from the battery will be developed as heat in the battery itself by energy loss in the source resistance.

Elzinga and Westerhof (1973,1974,1979) have provided experimental evidence that supports the characterization of the ventricle as a pump similar in fundamental respects to a general power generator. In other words, ventricular function can be characterized by descriptive parameters developed by the sciences. Specifically, the terms "ventricular hydromotive power" and "source impedance" can be employed as basic descriptors of ventricular properties. These terms are borrowed from the Thévenin analog of a general electrical power source. This chapter discusses source impedance as related to load impedance to yield predictions of the load at which power output of a ventricle will be maximum with a given inotropy and diastolic filling.

Another description of ventricular function that has received wide attention during the last decade involves the instantaneous relationship of ventricular pressure and volume. Suga, Sagawa, and collaborators (Suga and Sagawa 1974; Suga, Sagawa, and Shoukas 1973) have developed this description for the left ventricle, and studies of the right ventricle have been performed along similar lines (Maughan et al. 1979, Piene and Sund 1979). This chapter discusses how such a description of ventricular function may be linked to arterial load characteristics in an analysis of the conditions for ventricle–load matching.

Pulmonary Impedance

The first measurements of pulmonary input impedance were performed on isolated rabbit lungs perfused by a sinusoidal pulsatile pump of variable frequency (Caro and MacDonald 1961). Through measurement of mean and pulsatile pressure and flow at the inlet of the pulmonary bed, the pulmonary impedance *modulus* was found by dividing pressure and flow amplitudes. According to these measurements, impedance falls rapidly from its value at low oscillation frequency to a minimum value around 5 Hz, and it displays secondary maxima and minima at higher frequencies. (Here the impedance at 0 Hz oscillation frequency is termed "DC impedance," which is equal to mean pressure divided by mean flow. The impedance at frequencies above 0 Hz is termed "AC impedance.") At frequencies higher than that corresponding to the first impedance minimum, AC impedance oscillates around a value commonly denoted as "characteristic impedance." Caro and MacDonald (1961) attributed the impedance oscillations around the characteristic impedance level to pulse wave reflections in the pulmonary bed. If the reflected pressure waves appear at the pulmonary arterial inlet in counterphase to the forward pulse waves, the pressure pulse will be partially cancelled and hence give rise to a low impedance modulus. The frequency at which this partial cancellation is most effective is usually the frequency of the first impedance minimum, which therefore depends on pressure and flow pulse wave velocity and the location in the pulmonary bed where pulse wave reflections predominantly occur.

Bergel and Milnor (1965) performed impedance measurements in dog lungs in a more intact preparation. They examined the applicability of splitting the natural pressure and flow waveforms into series of harmonic (sinusoidal) oscillations by Fourier analysis. A basic requirement for this kind of impedance calculation is that the pulmonary bed behaves as a linear system. That is, the resulting impedance spectrum should not be influenced to any considerable extent by heart rate. After these investigators demonstrated that the requirement of linearity was closely met in the pulmonary circulation system (Bergel and Milnor 1965), the procedure of splitting pressure and flow into their respective Fourier series has become standard in most investigations. The basic results of the Bergel and Milnor study (see Fig. 8.1) show the rapid drop of the impedance level for the DC value to the minimum at around 5 Hz. These authors also demonstrated that the phase shift between pressure and flow oscillations (i.e., *impedance phase*) is negative below the first impedance minimum and positive above it. Compared with similar measurements in the systemic bed (see the review of O'Rourke, 1982), the characteristic impedance level of the pulmonary bed is relatively high, approximately 25% to 30% of the DC impedance (or vascular resistance). In the systemic bed it is only around 10% of systemic vascular resistance.

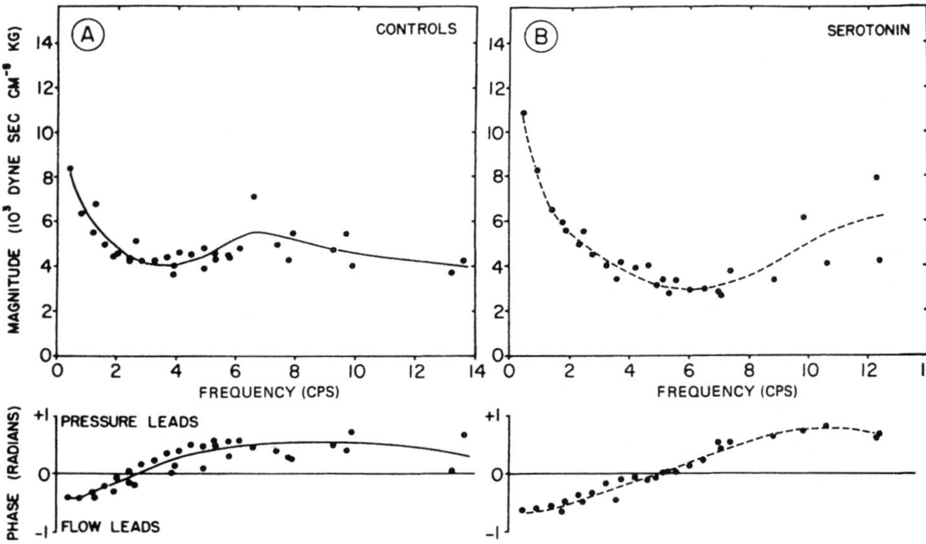

FIGURE 8.1. Input impedance of dog lungs, obtained by Fourier analysis of pressure and flow at the pulmonary arterial inlet, during control conditions *(left)* and during vasoconstriction (serotonin infusion) *(right)*. Note the shift of the impedance minimum. (From Bergel and Milnor 1965; by permission of the American Heart Association, Inc.)

Bergel and Milnor (1965) also examined the influence of vasoconstriction (serotonin infusion) on the impedance spectrum and observed a marked rightward shift of the frequency of the first impedance minimum (Fig. 8.1). This may be interpreted as a shift to a more proximal position of the main reflecting point in the pulmonary bed (Bergel and Milnor 1965) and/ or a higher pulse wave velocity (Bargainer 1967, Reuben et al. 1971).

Pace (1971) examined how pulmonary impedance is influenced by stimulation of sympathetic nerves to the lung and observed that such stimulation significantly raised the characteristic impedance level even at low stimulation rates (1–2 cycles per second), while change in pulmonary resistance was first detectable at 5 to 10 cps. This observation is consistent with other findings that sympathetic influence on the pulmonary bed mainly is on the stiffness of the vessel walls and only to a lesser extent on their caliber (Ingram et al. 1968). Pace (1971) concluded that even a low grade of sympathetic influence on the pulmonary bed would significantly raise the opposition to pulsatile flow through the pulmonary circuit and thereby increase the mechanical load on the right ventricle.

The effects on pulmonary impedance of nonadrenergic vasoconstriction, alveolar and venous pressure, adrenergic influence, and right ventricular flow rate were examined by Piene and Hauge (1976) and Piene (1976a). According to these investigations, the passive effects on impedance of vessel distension during elevated intraarterial pressure has a major influence on the characteristic impedance level. Vessel distension tends to lower impedance by reducing blood flow resistance and inertance in central pulmonary arterial vessels, and also tends to increase impedance by increasing wall stiffness. Hence opposing factors influence impedance during vessel distension. Piene and Hauge (1976) found indications that these factors optimally balance each other, to yield the lowest value of characteristic impedance when the intraarterial pressure corresponds to the normal pulmonary pressure. Figure 8.2 displays the impedance spectra of a completely relaxed, a moderately constricted, and a highly constricted pulmonary bed (nonadrenergic vasoconstriction), and shows that minimum characteristic impedance is attained at a slight degree of constriction yielding an arterial pressure corresponding to normal (Piene and Hauge 1976). According to the same study, effects of alveolar or venous pressure changes could be attributed to the distension of central, or extraalveolar, arterial vessels.

In view of the known adrenergic influence on the pulmonary bed, the question arises whether impedance is controlled by the sympathetic system. Clear indications of this have not been found. The rise of characteristic impedance during exercise (Elkins and Milnor 1971) or during hemorrhage (Dujardin et al. 1983) indicates rather that the sympathetic

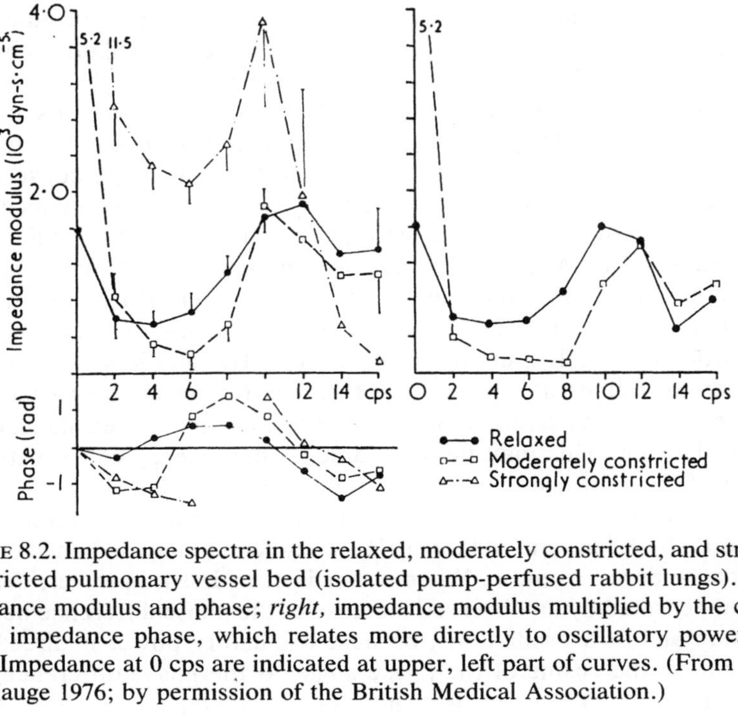

FIGURE 8.2. Impedance spectra in the relaxed, moderately constricted, and strongly constricted pulmonary vessel bed (isolated pump-perfused rabbit lungs). *Left,* impedance modulus and phase; *right,* impedance modulus multiplied by the cosine of the impedance phase, which relates more directly to oscillatory power (see text). Impedance at 0 cps are indicated at upper, left part of curves. (From Piene and Hauge 1976; by permission of the British Medical Association.)

system controls volume of the pulmonary blood depot (Aarseth 1972). Such control should involve reduction of pulmonary compliance to reduce the amount of blood stored in the pulmonary vessels, which again would be reflected by increased characteristic impedance.

One interesting aspect of interaction of the right ventricular flow and pulmonary impedance was observed by Piene (1976a) during sympathetic stimulation of the pulmonary arterial bed. Increase of flow through the sympathetically stimulated bed caused a reduction of the characteristic impedance toward the control value without, however, affecting the vascular compliance reduction that occurred during stimulation. As flow and sympathetic stimulation are expected to increase simultaneously, for example, during the "fight or flight" response, these data point to an interplay between the pulmonary bed and the right heart in defining an impedance level that is favorable in terms of both right ventricular afterload and blood volume control.

The hydraulic power yield of the right ventricle propels blood through the pulmonary vessels. Part of this power, however, is consumed as heat created by oscillatory blood motion in the pulmonary bed and does not serve the net forward motion though the lung vasculature. The oscillatory power is directly related to the AC impedance level: The higher the AC impedance, the greater the amount of oscillatory power that has to be produced by the right ventricle to propel a certain amount of blood through the pulmonary bed. As oscillatory power generally is considered an unavoidable waste linked to the nature of the ventricle as a pulsatile blood pump, it would seem that conditions serving to keep oscillatory power at a low level would be beneficial from an energy viewpoint. The major part of the oscillatory power is carried by the pressure and flow oscillations at the fundamental frequency (i.e., the frequency corresponding to heart rate). Therefore it would seem beneficial in terms of energy conservation if heart rate corresponded to the frequency of the first impedance minimum. The relation between heart rate and pulmonary hydraulic power was examined by Milnor, Bergel, and Bargainer (1966), who observed that minimum oscillatory power is achieved when heart rate and first impedance minimum coincide (Fig. 8.3). In humans the first minimum falls around 2 to 3 Hz, in dogs around 3 to 5 Hz, and in smaller animals like cats or rabbits at about 5 to 8 Hz. The frequencies correspond, in these species, to moderate to high heart rates, as seen, for example, during physical strain. This may be a condition when energy conservation on behalf of the ventricle is of most importance.

Characteristic impedance and oscillatory power during exercise were measured by Elkins and Milnor (1971). They found increased characteristic impedance during exercise. When power is considered, however, this may be compensated by an observed increase of the phase difference between pressure and flow oscillations, because oscillatory power is directly dependent on the cosine of the phase angle. If the phase difference is increased toward 90 degrees, oscillatory power will fall toward zero. There-

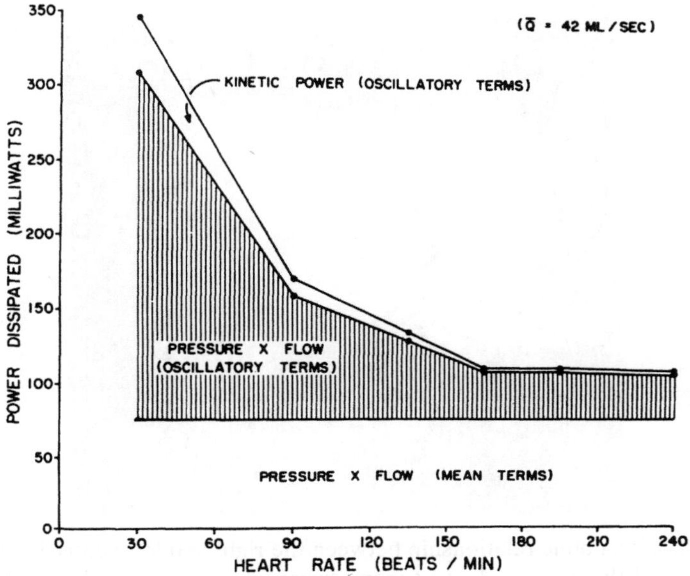

FIGURE 8.3. Oscillatory power according to heart rate in the dog. Note the reduction of oscillatory power as the heart rate approaches the frequency of the impedance minimum (3 Hz or 180 beats/min). $\overline{Q}$ = mean flow. (From Milnor, Bergel, and Bargainer 1966; by permission of the American Heart Association, Inc.)

fore, increased characteristic impedance may be seen at the same time that the ratio of oscillatory to total power is reduced in the pulmonary bed (Elkins and Milnor 1971).

Observations of pulmonary impedance in pathologic lung conditions have shown that characteristic impedance generally is elevated, more or less in parallel with the rise in pulmonary vascular resistance (Hopkins et al. 1979, 1980; Milnor et al. 1969; Reuben and Kitchin 1975; Wilcox and Lucas 1980). This imposes an increased burden on the right heart, both by increasing the mean arterial pressure load and by enhancing the pulsatile pressure swings which in turn increase the peak arterial pressure (Elzinga, Piene, and DeJong 1980). The energy cost for each milliliter of blood passing through the pulmonary bed may thus be considerably increased.

Right Ventricular Function

The right ventricle is like a pocket wrapped around approximately a third of the left ventricle (including the septum) (see Fig. 8.4). Both superficial and deeper myocardial fibers of the right ventricle are continuous with fibers of the left ventricle. Superficial fibers of the right ventricle run un-

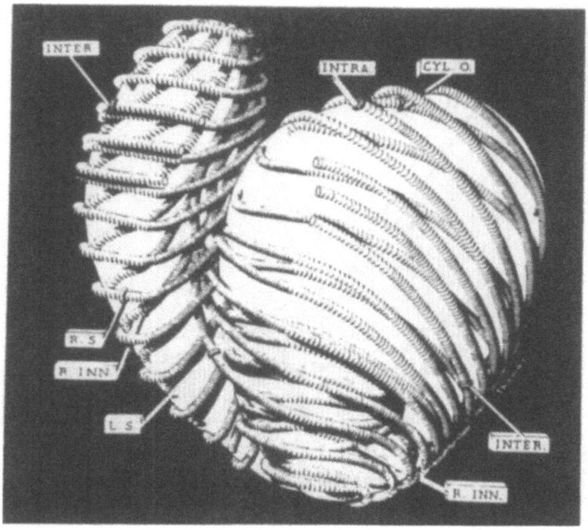

FIGURE 8.4. Anatomic relationship between the right and left ventricle. The fiber orientation of the epicardial and subepicardial fiber layers are shown. Inter, Interventricular band of inner layer; intra, Intraventricular band of middle layer; Cyl. o., Outer portion of cylindrical layer; R. inn., Right inner layers; L.S., Left septal band of inner layer; R.S. Right septal band of middle layer. (From Thomas 1957).

interruptedly onto the left side, and deeper fibers are continuous with septal fibers that further encompass the left ventricular free wall in a figure-eight fashion (Thomas 1957). The right ventricular free wall is divided into the sinus (inflow) and conus (outflow) regions. The regions have different embryologic origins (March, Ross, and Lower 1962) and may have separate functions. In dog hearts, it has been demonstrated that sympathetic stimulation of the right ventricular free wall leads to enhanced contractions of both sinus and conus regions to the extent that a functional stricture between the two exists (Armour, Pace, and Randall 1970). A significant pressure drop has been measured over this stricture (Pace et al. 1969), and it has therefore been speculated that the stricture serves to protect the pulmonary circuit from high pressures developed in the right ventricular cavity (Armour, Pace, and Randall 1970; Pace et al. 1969; Raines, LeWinter, and Covell 1976).

Reduction of right ventricular volume during systole is assumed to occur predominantly because of a translational movement of the right ventricular free wall toward the septum, with little change in local and overall geometry (Rushmer 1976). According to the law of Laplace, this configuration should result in a nearly constant relationship between local right ventricular free wall force and right ventricular cavity pressure. This has been confirmed by direct measurements (Piene and Covell 1983).

The right ventricle has been characterized as a "flow pump," in contrast to the left ventricle which has been considered a "pressure pump," although the exact meaning of these terms is uncertain (see later discussion). The usual interpretation is that the right ventricle is a flow pump because wide variations in afterload have little effect on its output (Abel and Waldhausen 1967). However, this may reflect a compensatory mechanism that regulates diastolic filling of the ventricle, rather than a property of the ventricle itself. As will be shown here, the left and right ventricles respond to afterload alterations in very much the same way.

Sarnoff and Berglund (1954) found that ventricular function curves (i.e., stroke work vs. end-diastolic pressure) appear to have a more linear course in the right than in the left ventricle and that in the right ventricle they often do not show the flattening at high end-diastolic pressure characteristic of the left ventricle (see Fig. 8.5). Another way of describing ventricular performance, by relating time-averaged values of pressure and flow, is the "pump function graph" of Elzinga and Westerhof (1979), which relates mean ventricular pressure to flow at constant end-diastolic filling and inotropy. According to this description of ventricular function, left and right ventricular properties are quite similar when the difference in pressure scales is accounted for (Elzinga, Piene, and DeJong 1980). As demonstrated in Figure 8.6, mean ventricular pressure (at constant diastolic filling and inotropy) is inversely related to flow in both ventricles, although the relationship may have a more linear course in the right than in the left ventricle. Figure 8.6 also shows mean power output versus mean flow of both ventricles. Both ventricles appear to work around the maxima of these curves when the load conditions are around physiologically normal values (aortic and pulmonary artery pressures 80 and 16 mm Hg, and end-diastolic pressures of left and right ventricles 7 and 2 mm Hg, respectively).

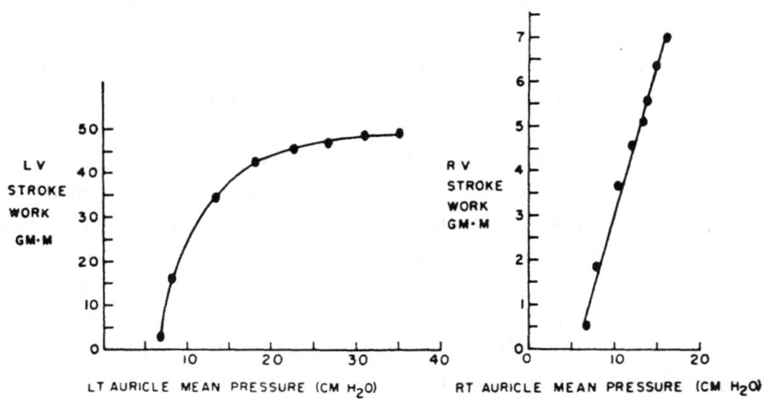

FIGURE 8.5. Ventricular function graphs (stroke work vs. atrial pressure) of left (L.V.) and right (R.V.) ventricle. (From Sarnoff and Berglund 1954; by permission of the American Heart Association, Inc.)

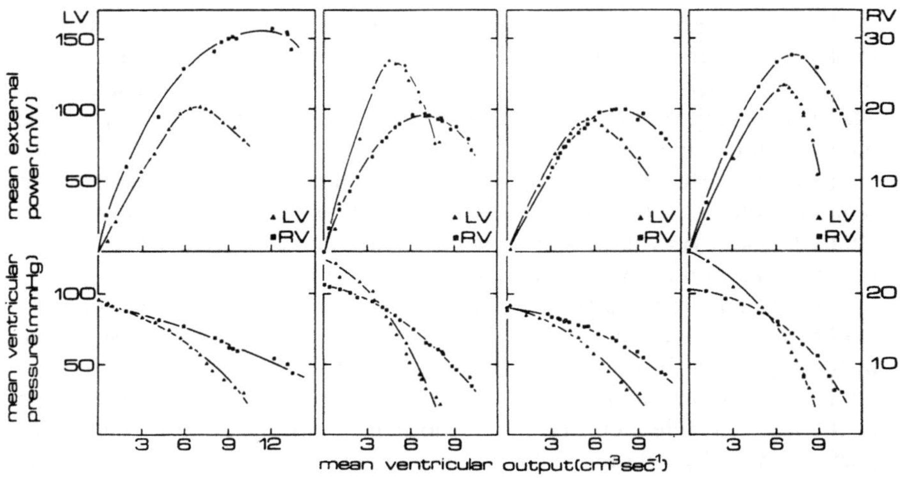

FIGURE 8.6. Left ventricular (LV) and right ventricular (RV) external power and mean ventricular pressure versus ventricular output in four isolated cat hearts pumping into hydraulic models of their respective arterial systems. Output variation was accomplished by beat-to-beat alteration in arterial pressure while end-diastolic pressure and inotropy were maintained constant. Note the similar characteristics of the two ventricles. (From Elzinga, Piene, and DeJong 1980; by permission of the American Heart Association, Inc.)

Another close similarity between left and right ventricular function is the way the ventricles respond to alterations in resistance or compliance of the arterial load. According to experiments of Elzinga, Piene, and DeJong (1980) and Piene and Sund (1979) on isolated cat hearts, a reduction in load compliance leads to higher right ventricular systolic pressure, increased pulse pressure in the pulmonary arterial bed, and flattening and protraction of the flow pulse, as shown in Figure 8.7. The observed changes in pressure and flow contours in the right ventricle produced by load compliance and resistance alterations are completely analogous to those seen in the left ventricle and aorta under similar experimental conditions (Elzinga and Westerhof 1973). In the right ventricle, both reduced arterial compliance and increased arterial resistance lead to reduced cardiac output, but flow is influenced proportionally more by an increase in resistance than by reduction in compliance (Piene and Sund 1979).

These descriptions of ventricular performance—the ventricular function curves of Sarnoff and Berglund (1954) and the pump function graphs of Elzinga and Westerhof (1979)—give no information about the ventricles as pulsatile pumps. To describe the time-varying behavior of the ventricles, it is necessary to relate pulsatile ventricular pressure, flow, or volume in the time or frequency domains. According to Elzinga and Westerhof (1973,1974) a ventricle may be described as a hydromotive pressure source

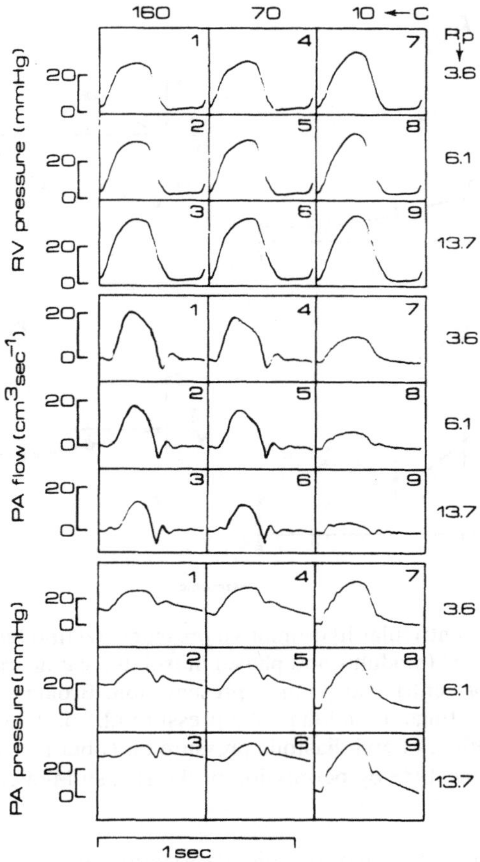

FIGURE 8.7. Right ventricular (RV) pressure, pulmonary artery (PA) pressure, and PA flow in isolated cat hearts pumping into loads of variable resistance R_p and compliance C. Note alterations in flow contour as compliance is reduced—it becomes flatter but protracted. (From Elzinga, Piene, and DeJong 1980; by permission of the American Heart Association, Inc.)

with an internal impedance, in direct analogy with the Thévenin representation of a general power source (Fig. 8.8). Elzinga and Westerhof (1973,1974) devised a method of calculating hydromotive pressure and source impedance from ventricular pressure and flow at two different levels of load impedance. Sund and Piene (1983) extended this method to include regression analysis of a larger set of right ventricular pressure and flow data at different load conditions. In both methods the ventricle is described by a generator function, $G(f)$ (f, frequency), and a source impedance, $Z_s(f)$, for both of which modulus and phase must be specified. $G(f)$ describes the pressure in the right ventricular cavity at infinitely high load impedance;

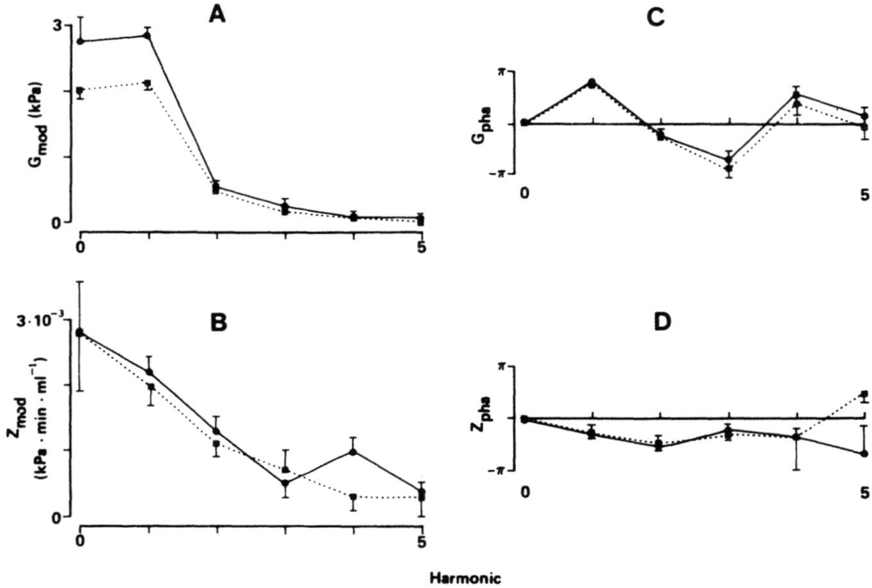

FIGURE 8.8. Right ventricular hydromotive pressure *(G)* (modulus and phase) and source impedance *(Z)* (modulus and phase) in isolated cat hearts according to the Thévenin analog of ventricular pump representation. Broken and full lines show *G* and *Z* for two settings of end-diastolic pressure (2.5 and 4.0 cm H_2O, respectively). Note the effect of end-diastolic pressure on *G* but not on *Z*. (Reproduced from Sund and Piene 1983 by permission of the British Medical Association).

hence it is analogous to isovolumic ventricular pressure and should be sensitive to end-diastolic pressure and volume. $G(f)$ and $Z_s(f)$ for the right ventricle of isolated cat hearts are shown in Figure 8.8 (Sund and Piene 1983). The graphs demonstrate the dependence of $G(f)$ on end-diastolic pressure and show as well that $Z_s(f)$ is little influenced by end-diastolic pressure and volume variations.

Another description of the ventricle as a pulsatile pump relates instantaneous ventricular pressure, $P(t)$, to ventricular volume, $V(t)$. Maughan and colleagues (1979) followed the method devised by Suga, Sagawa, and co-workers (Suga and Sagawa 1974; Suga, Sagawa, and Shoukas 1973) and described the P versus V and t (time) relationship by a set of linear pressure–volume relations at specific time points, t, during systole. Results of this study are shown in Figure 8.9. The linear, time-dependent pressure–volume relations intersect with the volume axis at a time-dependent value, $V_o(t)$. The time dependent pressure–volume relationship is little influenced by the actual load, hence the equation

$$P(t) = E(t)\,[V(t) - V_o(t)] \tag{8.1}$$

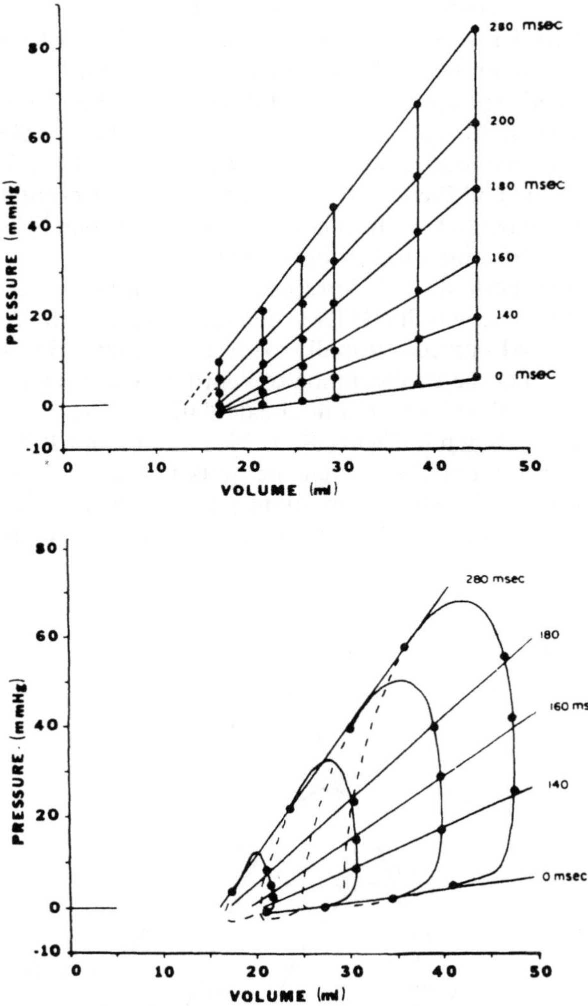

FIGURE 8.9. Relationships of right ventricular pressure, volume, and time in isolated dog hearts. Top, data from isovolumic contractions; bottom, data during physiologic loading conditions. (From Maughan et al. 1979; by permission of the American Heart Association, Inc.)

where $E(t)$ is the slope of the pressure–volume line at time t, gives a close description of ventricular function.

Piene and Sund (1980, Sund and Piene 1983) analyzed the properties of the right ventricle within the framework of instantaneous pressure–volume relations, but preferred to display the relationship in the V versus t coordinate system with P as a parameter. In this way, nonlinear relations among P, V, and t could be better disclosed. Their data indicated that $E(t)$ of Equation 8.1 should rather be a function $E(t, V)$ to account for nonlinear relations among P, V, and t (Sund and Piene 1983). The P-V-t relationship could be expressed as a surface in the three dimensions P, V, and t (see Fig. 8.10). Parametric equations could be established to describe the surface without reliance on a linear equation like, for example, Equation 8.1. Sund and Piene (1983) found that this representation of ventricular function gave better fit to experimental data than linear $P(t)$-$V(t)$ relations or the Thévenin analog representation.

Employing the method of P-V-t representation of ventricular function, Piene (1980) and Piene and Sund (1981) demonstrated that the P-V-t surface could be combined computationally with a representation of the load impedance to predict the configuration of ventricular and pulmonary artery pressure and flow. Results from initial calculations of expected flow and observed flow are shown in Figure 8.11 (Piene and Sund 1979). One conclusion from these studies was that methods to analyze ventricle–load interactions and conditions for optimum pumping were at hand, as the ventricular responses to afterload alterations were closely predictable.

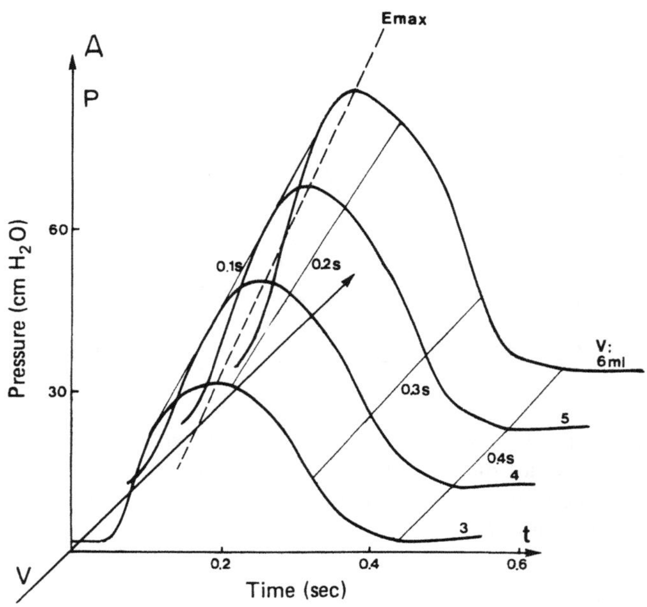

FIGURE 8.10. Systolic right ventricular pressure–volume–time relationship in isolated cat hearts. The relationship is displayed as a three-dimensional surface in the pressure–volume–time *(P-V-t)* coordinate system. The surface, end-diastolic volume, and load impedance define how $V(t)$ and $P(t)$ will vary during systole. E_{max} = maximal slope of pressure-volume relationship. (From Piene and Sund 1982).

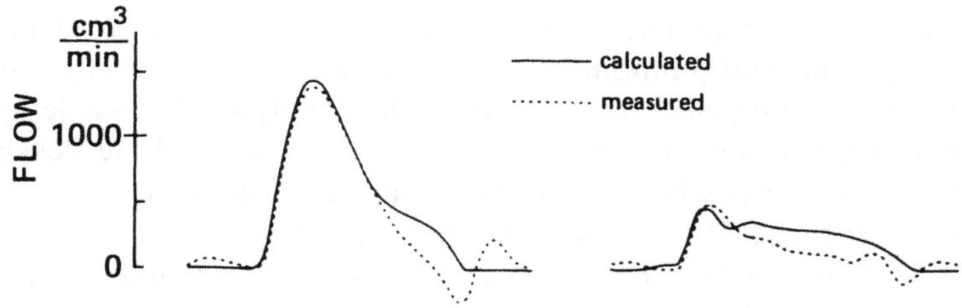

FIGURE 8.11. Calculated and measured flow in cat pulmonary artery. By the calculation, an experimentally obtained pressure–volume–time representation of right ventricular function was combined with a load with discrete components (resistance, inertance, compliance) that were estimated from actual pulmonary artery impedance. (Adapted from Piene and Sund 1980; by permission of the British Medical Association).

It has been questioned whether the instantaneous pressure–volume relationship is independent of flow from the ventricle (Shroff, Janicki, and Weber 1980; Campbell et al. 1982). From measurements of right ventricular flow and pressure, Campbell and colleagues (1982) had to assume an internal resistance of considerable magnitude, in addition to a time-varying compliance, in order to predict right ventricular flow from pressure. These data are in contrast to the results of Sund and Piene (1983), who found negligible direct influence of flow on the pressure–volume relationship.

Right Ventricle and Pulmonary Bed as a Functional Unit

This section discusses characteristics of the right ventricle and the pulmonary bed that suggest that the two may be considered parts of one functional whole. Already from the early measurements of vascular impedance, investigators were struck by the appropriateness of the input impedance spectrum as a measure of the mechanical load in the ventricle (Taylor 1964). In particular, the arterial system as a compliant and nonlinear pulse transmission system was recognized as an uncoupling device between the ventricle and the main vascular resistance offered by the small arteries.

As already mentioned, minimum impedance falls within the physiologic range of heart rates, and the total hydraulic power per milliliter of blood propelled through the pulmonary bed drops to a minimum when heart rate corresponds to the frequency of the first impedance minimum (Milnor, Bergel, and Bargainer 1966). This seems to suggest that either through evolution or development of the individual organism, the pulmonary bed adapted to the right ventricle in such a way as to reduce the mechanical load on the ventricle and conserve energy. The observation of Piene and

Hauge (1976) that characteristic impedance (and oscillatory power) attains a minimum at normal pulmonary artery pressure fits into this picture.

Recently, Lefevre (1982) made an interesting analysis of possible optimal pulmonary impedance by combining the abilities of the lung vessel bed to adapt to two somewhat conflicting demands: the need to provide an effective blood-to-air barrier within the constraints of the thoracic cage, and the need to minimize the mechanical load on the right ventricle over the range of normal cardiac outputs. By simulating the dog lung with a dichotomously branching ("fractal") vascular tree and establishing a cost function that combined the demands of the appropriate vascular and cardiac functions, Lefevre calculated an optimal impedance for the canine pulmonary bed that was almost identical to that measured by Bergel and Milnor (1965).

According to the concept of ventricle–load matching referred to in the introduction, the arterial load should enable the ventricle to work at maximum power. Right ventricular power–load relations have been examined in isolated, supported cat hearts (Elzinga, Piene, and DeJong 1980; Piene and Sund 1979), and these studies demonstrated that the right ventricle during normal load conditions indeed works at the maximum of the power–flow or power–afterload curve (see Fig. 8.6 and 8.12). These observations are analogous to observations of the left ventricle (Elzinga, Piene and DeJong 1980; Wilcken et al. 1964) and have been taken as direct indications that a ventricle and its normal arterial load constitute an optimally matched system (Piene and Sund 1979,1981; Piene 1984).

Piene and Sund (1979) examined power output in relation to resistance and compliance of arterial loading (isolated cat hearts pumping into a load with independently variable resistance and compliance). They observed that arterial compliance significantly influenced the power–load curve (Fig. 8.12). When load compliance was reduced, the power–load curve optimum was depressed, and the maximum shifted to lower values of arterial resistance. At high compliance the power maximum fell within the physiologic range of pulmonary vascular resistance values for the species. This dependence on load compliance is in harmony with predictions based on the Thévenin analog (Piene and Sund 1979). The (apparent) source resistance, R_s, is given by the (negative) slope of the pump function graph (mean ventricular pressure vs. mean flow). According to Piene and Sund (1979) and to Piene (1984), the relation between R_s and load resistance, R_l, for optimum external power yield should be $R_l = R_s T/t_s$, where T is the heart period and t_s is systolic time. Direct observations of R_l at maximum power output yielded a ratio of measured R_l to measured R_s of 2.7, a value that corresponded well with T/t_s. At low load compliance the ratio R_l/R_s was assumed to approach 1.0, which also was confirmed experimentally. Piene and Sund (1979) concluded that the normal high compliance and, consequently, low characteristic impedance of the pulmonary bed serve to enhance power transmission from the right ventricle to the lung vessel system.

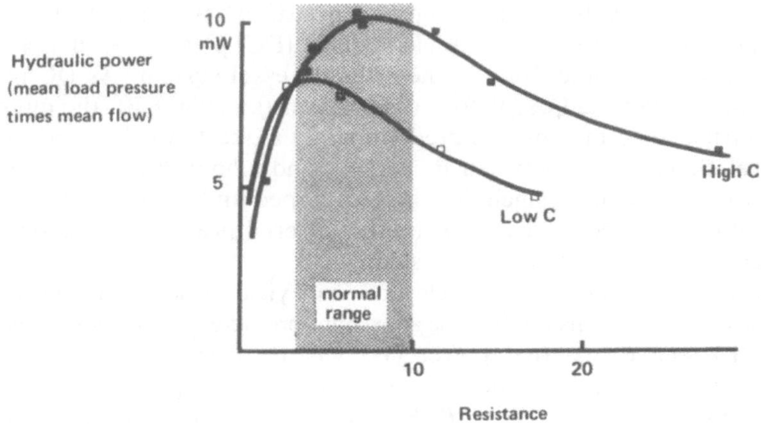

FIGURE 8.12. Right ventricular hydraulic power output versus load resistance at high and low (10 x decrease) load compliance *(C)* in cat lungs. The normal range of pulmonary resistance is indicated. Note the optimum of power output within the normal resistance range, and the shift and depression of the power optimum as *C* is lowered. (Adapted from Piene and Sund 1979.)

An alternative approach to the question of matching between right ventricle and pulmonary bed is given by the pressure–volume–time relationship. External energy output may be calculated from the pressure–volume loop area, which is in turn defined by the end-diastolic filling, the impedance of the load, and the end-systolic pressure–volume relation. Assuming low characteristic impedance and thus small arterial pressure pulsations, a pressure–volume loop will have approximately the shape of a rectangle, with edges defined by arterial pressure, the end-systolic pressure–volume relation (with slope E_{max}), and the diastolic pressure–volume filling curve. Making the approximation that opening and closing pressures at start and end of ejection are close to the mean arterial pressure (as would be the case if arterial pressure oscillations were negligible), the following relation among E_{max}, R_s, and heart rate (HR) should assure optimum external power from the ventricle (Piene 1984)

$$\frac{E_{max}}{\mathrm{HR}} = R_1 \tag{8.2}$$

In cat hearts, E_{max} of the right ventricle was $1.0 - 1.25 \times 10^4$ dyne/sec cm^{-5} (Sund and Piene 1983). Assuming a normal value of $R_{1s} = 5 \times 10^3$ dyne-sec cm^{-5} (Piene 1976a,c) the preceding relationship is balanced for HR $= 2$ to 3 Hz, which is in the physiologic range of heart rates in this species. Thus, according to these analyses, the data indicate that observable parameters characterizing the right heart (R_s, E_{max}, HR) and the pulmonary vascular system (R_1) match each other in the predicted way.

In discussing matching between ventricle and load, it should be realized that the arterial valves and the compliant arterial system act to convert oscillatory power from the ventricle to static (DC) power (i.e., the product of mean pressure and flow) in the arterial vessel system. As DC power is usually higher than oscillatory (AC) power (see Fig. 8.3), the question of ventricle–load matching is predominantly a question of DC matching. The relations shown between R_l and R_s, and among E_{max}, HR, and R_l, refer to DC matching, since they are developed under the constraint that arterial compliance is high, that is, that arterial pressure pulsations and associated oscillatory power are small.

DC matching between ventricle and load yields insight into whether a ventricle acts as a constant flow or a constant pressure source. For a constant flow (F) source, external power (W) will be given as $W = F^2 \cdot R_l$ that is, it is directly proportional to R_l. For a constant pressure (P) source, W would be defined as $W = P^2/R_l$, that is, as inversely proportional to R_l. Therefore, a ventricle may be considered a flow source if R_l is lower than the matching R_l, and a pressure source if R_l is higher than the matching R_l (see Figure 8.13). If R_l is around the value of the matching R_l, P and F respond to R_l alterations by inverse but proportionally equal changes as the product of P and F remains constant. Therefore, the question of whether a ventricle is a pressure or a flow source is largely a question of where on the power–load curve the ventricle is actually working. According to the present data it seems that the right ventricle is neither a flow nor a pressure source but a source matched to the load.

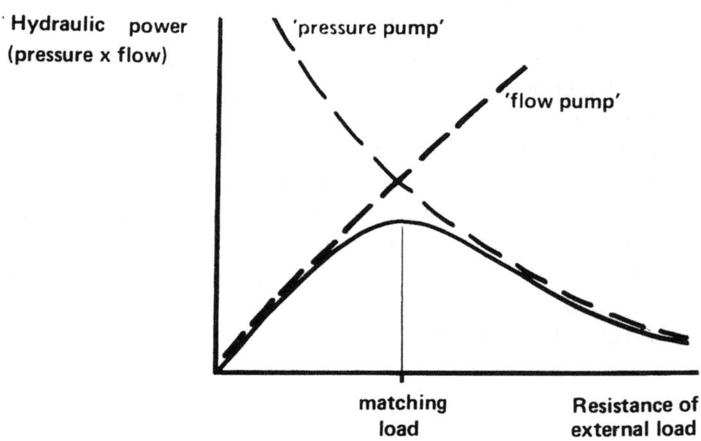

FIGURE 8.13. Analysis of right ventricular characteristics based on the matching concept. The broken curves show the theoretical power output versus load resistance of "flow pump" and "pressure pump," respectively. The fully drawn curve shows the actual power–load curve of the right ventricle (compare Fig. 8.12). Depending on the position of the working point on the power–load curve, the ventricle may act either as "flow pump" or "pressure pump," or may be matched to the load (working point at curve optimum).

It may also be asked whether the AC impedance of the load finds any optimal counterpart in the ventricle. Optimum AC matching, if it exists, may again simply reflect energy-conserving properties of ventricle–load coupling, but as oscillatory (AC) power will be a very small fraction of myocardial oxygen consumption, such optimization will be less likely to occur. AC matching if it did occur would have little effect on total power yield from the ventricle, as DC power is far greater.

Analysis of matching based on the AC components of source and load has not been performed owing to obstacles presented by the nonlinear element, the pulmonary valves. It may be noteworthy, however, that as R_s optimally should be lower than R_1 by a factor of 2 to 3 (see preceding), Z_s for the first harmonic (of value about 2.5×10^{-3} dyne-sec cm^{-5}; see Fig. 8.8) is two to three times *higher* than the load impedance at the same frequency (Piene 1976). Thus if there actually is a mismatch between source and load impedance at frequencies around the heart rate, this could mean that the right ventricle is able to withstand alterations in load impedance at these frequencies, with little change in the pulsatile components of flow. This would be in accordance with the experimental data of Pouleur and co-workers (1978), but on the other hand seems to contradict other observations of the sensitivity of the flow pulse to alterations in AC impedance (Piene and Sund 1979; Elzinga, Piene, and DeJong 1980).

Piene and Sund (1981) have discussed the existence of AC matching between right ventricle and pulmonary bed from another point of view. Assuming fixed right ventricular stroke volume and pulmonary artery pressure (and hence fixed DC power output), they calculated, on the basis of experimentally determined right ventricular pressure–volume–time relations (see Fig. 8.10), how central arterial resistance, R_c (simulating characteristic impedance), arterial blood inertance, and vessel compliance C influenced Q, the ratio of external power from the ventricle to total power developed in the ventricle. Q is a measure of the efficiency of power transfer from the ventricle to the load, but is not identical to the commonly employed ventricular efficiency, that is, external power/myocardial oxygen consumption rate. Optimization of R_c, L, and C with respect to Q is to be regarded as an additional additive feature of ventricle–load coupling already at or around optimal DC matching.

It was found that an increase in C, as expected, reduced total external power (the constant DC plus the variable AC power) and increased Q. At a C above 150 to 200 $\times$ 10^{-3} ml/cm H_2O, Q was little influenced by C, and this value of C accords well with normal compliance in the cat lung (Piene 1976c). An increase in L reduced Q, and according to these calculations a low value for L is favorable. By variation of R_c, which defines the characteristic impedance level, an optimum in Q was observed for R_c around 1.0 cm H_2O sec/ml. This is in accordance with the normal characteristic impedance in cat lungs (Piene 1976c), as shown in Figure 8.14.

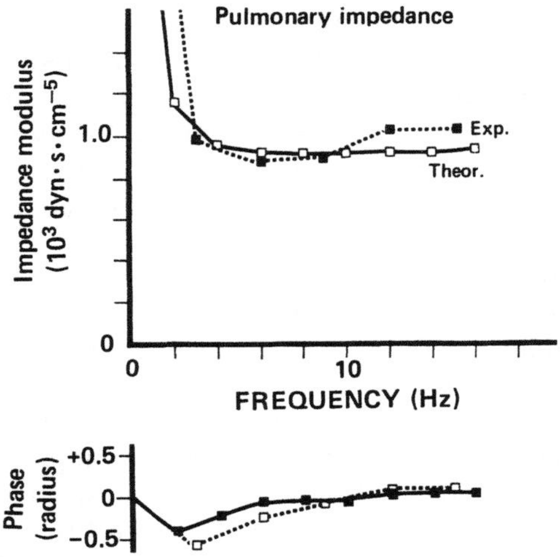

FIGURE 8.14. Calculated pulmonary impedance spectrum and phase for optimum "efficiency" Q (Q = external power/potential power) (see text) of the right ventricle. The theoretically optimal impedance (fully drawn curve) is compared with experimentally obtained impedance (broken curve). (Adapted form Piene and Sund 1982.)

These results, although highly hypothetical and theoretical, could indicate that R_c optimally should have a value different from zero. This could mean that the right ventricle normally is adapted to yield a relatively large amount of oscillatory power to the pulmonary circulation. Oscillatory power is an unavoidable consequence of the heart as a pulsatile pump; it could be reduced to zero only if the central pulmonary arteries had zero resistance and were infinitely compliant, not a favorable situation with respect to blood volume distribution in the body. Also, pulsatile blood motion (and power) may even have beneficial effects on the organism. Experiments indicate that pressure pulsations in the pulmonary circuit have a positive effect on fluid filtration in the lungs (Hauge and Nicolaysen 1979), and may also positively influence the filling of the left atrium (Piene 1976d).

Summary and Conclusion

This presentation has emphasized the characteristics of pulmonary impedance and right ventricular function that indicate that these two interdependent parts of the circulation are mutually adapted to each other. The form of the pulmonary impedance spectrum is such as to reduce the mechanical load on the right ventricle and to uncouple the peripheral resistance from the arterial input. Minimum impedance falls in a frequency range that accords with the physiologic range of the heart rate, and this serves to keep low the ratio of oscillatory ("wasted") power to static ("useful") power in the pulmonary arterial system. Theoretical analysis aiming to optimize the conflicting demands of adequate air-to-blood barrier space within the thoracic cage and low load to the right ventricle leads to an input impedance that is very close to the actual value.

The ventricle may be described by methods that make it possible to correlate ventricular descriptive parameters with pulmonary vascular impedance. It has been found that the apparent source resistance of the right ventricle matches the resistance of the pulmonary bed, if the two are related as systolic time to total heart cycle time. Direct measurements in ventricles and lungs indicate that this is the case in the normal pulmonary circulation. These results also agree with studies in the left heart (Van den Horn, Westerhof, and Elzinga 1984).

Also, the end-systolic pressure–volume relation may be paralleled by vascular parameters, as indicated by the equation

$$\frac{E_{\text{max}}}{\text{heart rate}} = \text{vascular resistance}$$

for optimum DC power yield from the ventricle. This relationship seems to be fulfilled in the feline pulmonary circulation. Further optimization analysis, taking the compliance and characteristic impedance of the lung vessel bed into account, indicates that vascular compliance and characteristic impedance have optimum values that accord with the normal.

The data summarized in this chapter suggest that the organism, during evolution or individual development, has adapted to attain optimal matching between ventricle and load. The fundamental aim of this adaptation seems to be to provide the hydraulic energy necessary to propel blood through the vessels and to carry out other possible functions, in an optimal way. How this adaptation takes place in the organism, however, is beyond current knowledge. Most difficult to understand is how known control processes ruling the function of the cardiovascular system may influence or participate in this adaptation. Further analyses may elucidate whether the body's pressure and flow control systems work in combination to act as a power control system.

Bibliography

Aarseth, P., 1972. Studies on the depot function of the pulmonary vascular bed. Thesis. Universitetsforlaget, Oslo, Norway.

Abel, F. L., and Waldhausen, J. A. 1967. Effects of alterations in pulmonary vascular resistance on right ventricular function. *J. Thorac. Cardiovasc. Surg.* 54:885–894.

Armour, J. A., Pace, J. B., and Randall, W. C. 1970. Interrelationship of architecture and function of right ventricle. *Am. J. Physiol.* 218:174–179.
Bargainer, J. D. 1967. Pulse wave velocity in the main pulmonary artery of the dog. *Circ. Res.* 20:630–637.
Bergel, D. H., and Milnor, W. R. 1965. Pulmonary vascular impedance in the dog. *Circ. Res.* 16:401–415.
Campbell, K. B., Ringo, J. A., Wakav, Y., Klavano, P. A., and Alexander, J. E. 1982. Internal capacitance and resistance allow prediction of right ventricle outflow. *Am. J. Physiol.* 243:H99–H112.
Caro, G. C., and McDonald, D. A. 1961. The relation of pulsatile pressure and flow in the pulmonary vascular bed. *J. Physiol.* 157:426–453.
Dujardin, J. P., Stone, D. N., Forcino, C. D., Paul, L. T., and Pieper, H. P. 1983. Effects of blood volume changes on characteristic impedance of the pulmonary artery. *Am. J. Physiol.* 242:H197–H202.
Elkins, R. C., and Milnor, W. R. 1971. Pulmonary vascular response to exercise in the dog. *Circ. Res.* 29:591–599.
Elkins, R. C., Peyton, M. D., and Greenfield, L. J. 1974. Pulmonary vascular impedance in chronic pulmonary hypertension. *Surgery* 76:57–64.
Elzinga, G., Piene, H., and DeJong, J. P. 1980. Left and right ventricular pump function and consequences of having two pumps in one heart: a study on the isolated cat heart. *Circ. Res.* 46:564–579.
Elzinga, G., and Westerhof, N. 1973. Pressures and flow generated by the left ventricle against different impedances. *Circ. Res.* 32:178–186.
Elzinga, G., and Westerhof, N. 1974. End-diastolic volume and source impedance of the heart. In Guz, A., ed. *The Physiological Basis of Starling's Law of the Heart.* CIBA Foundation Symposium 24. Amsterdam: Elsevier, pp. 241–255.
Elzinga, G., and Westerhof, N. 1979. How to quantify pump function of the heart. *Circ. Res.* 44:303–308.
Hauge, A., and Nicolaysen, G. 1979. The importance of flow pulsatility for the rate of transvascular fluid filtration in lungs. *Physiology* 290:569–576.
Hopkins, R. A., Hammon, J. W., McHale, P. A., Smith, P. K., and Anderson, R. W. 1979. Pulmonary vascular impedance analysis of adaptation to chronically elevated blood flow in the awake dog. *Circ. Res.* 45:267–274.
Hopkins, R. A., Hammon, J. W. Jr., Mettale, P. A., Smith, P. K., and Anderson, R. W. 1980. An analysis of the pulsatile hemodynamic responses of the pulmonary circulation to acute and chronic pulmonary venous hypertension in the awake dog. *Circ. Res.* 47:902–910.
Ingram, R. H., Szidon, J. P., Skalak, R., and Fishman, A. P. 1968. Effects of sympathetic nerve stimulation on the pulmonary arterial tree of the isolated lobe perfused in situ. *Circ. Res.* 22:801–815.
Lefevre, J. 1982. Teleonomical representation of the pulmonary arterial bed of the dog by a fractal tree. In Kenner, T., Busse, R., and Hinghofer-Szalkay, H., eds. *Cardiovascular System Dynamics: Methods and Measurements.* New York: Plenum, pp. 137–146.
Lucas, C. L., Wilcox, B. R., and Coulter, N. A. Jr. 1975. Pulmonary vascular response to arterial septal defect closure in children. *Surg. Res.* 18:571–586.
March, H. W., Ross, J. K., and Lower, R. R. 1962. Observations on the behavior of the right ventricular outflow tract, with reference to its developmental origins. *Am. J. Med.* 32:835–845.

Maughan, W. L., Shoukas, A. A., Sagawa, K., and Weisfeldt, M. L. 1979. Instantaneous pressure–volume relationship of the canine right ventricle. *Circ. Res.* 44:309–315.

Milnor, W. R. 1975. Arterial impedance as ventricular afterload. *Circ. Res.* 36:565–570.

Milnor, W. R., Bergel, D. H., and Bargainer, J. D. 1966. Hydraulic power associated with pulmonary blood flow and its relation to heart rate. *Circ. Res.* 19:467–480.

Milnor, W. R., Conti, C. R., Lewis, K. B., and O'Rourke, M. F. 1969. Pulmonary arterial pulse wave velocity and impedance in man. *Circ. Res.* 25:637–649.

O'Rourke, M. F. 1982. Vascular impedance in studies of arterial and cardiac function. *Physiol. Rev.* 62:570–623.

Pace, J. B. 1971. Sympathetic control of pulmonary vascular impedance in anesthetized dogs. *Circ. Res.* 29:555–568.

Pace, J. B., Cox, R. H., Alvarez-Vera, F., and Karreman, G. 1972. Influence of sympathetic nerve stimulation on pulmonary hydraulic input power. *Am. J. Physiol.* 222:196–201.

Pace, J. B., Keefe, W. F., Armour, J. A., and Randall, W. C. 1969. Influence of sympathetic nerve stimulation on right ventricular outflow tract pressures in anesthetized dogs. *Circ. Res.* 24:397–407.

Piene, H. 1976a. The influence of pulmonary blood flow rate on vascular impedance and hydraulic power in the sympathetically and nonadrenaline stimulated cat lung. *Acta Physiol. Scand.* 98:44–53.

Piene, H. 1976b. Influence of vessel distension and myogenic tone on pulmonary arterial input impedance: a study using a computer model of rabbit lung. *Acta Physiol. Scand.* 98:55–66.

Piene, H. 1976c. Some physical properties of the pulmonary arterial bed deduced from pulsatile arterial flow and pressure. *Acta Physiol. Scand.* 98:295–306.

Piene, H. 1976d. Improved left ventricular performance by the transmission of pulse waves through the pulmonary vascular bed. *Acta Physiol. Scand.* 98:450–456.

Piene, H. 1980. Interaction between the right heart ventricle and its arterial load: a quantitative solution. *Am. J. Physiol.* 238:H932–H937.

Piene, H. 1984. Impedance matching between ventricle and load. *Ann. Biomed. Eng.* 12:191–207.

Piene, H., and Covell, J. W. 1983. Local auxotonic systolic force and work in canine right ventricular free wall. *Am. J. Physiol.* 244:H186–193.

Piene, H., and Hauge, A. 1976. Reduction of pulsatile hydraulic power in the pulmonary circulation caused by moderate vasoconstriction. *Cardiovasc. Res.* 10:503–513.

Piene, H., and Sund, T. 1979. Flow and power output of right ventricle facing load with variable input impedance. *Am. J. Physiol.* 237:H125–H130.

Piene, H., and Sund, T. 1980. Performance of the right ventricle: a pressure plane analysis. *Cardiovasc. Res.* 14:217–222.

Piene, H., and Sund, T. 1981. Does normal pulmonary impedance constitute the optimum load for the right ventricle? *Am. J. Physiol.* 242:H154–H160.

Piene, H., and Sund, T. 1982. Calculation of flow and pressure curves from the ventricular pressure–volume–time relationship and load impedance. In Kenner, T., Busse, R., and Hinghofer-Szalkay H., eds. *Cardiovascular System Dynamics: Models and Measurements*. New York: Plenum, pp. 47–55.

Pouleur, H., Lefevre, J., van Eyll, C., Jaumin, P. M., and Charlier, A. A. 1978. Significance of pulmonary input impedance in right ventricular performance. *Cardiovasc. Res.* 12:617–629.

Raines, R. H., LeWinter, M. M., and Covell, J. W. 1976. Regional shortening patterns in canine right ventricle. *Am. J. Physiol.* 231:1395–1400.

Reuben, S. R., and Kitchin, A. H. 1975. Pulmonary artery input impedance in pulmonary hypertension. *Prog. Resp. Res.* 9:261–266.

Reuben, S. R., Swadling, J. P., Gersh, B. J., and Lee, G. de J. 1971. Impedance and transmission properties of the pulmonary arterial system. *Cardiovasc. Res.* 5:1–9.

Rushmer, R. F. 1976. *Cardiovascular Dynamics*. Philadelphia: Saunders, pp. 91–96.

Sarnoff, S. J., and Berglund, E. 1954. Ventricular function. I. Starlings law of the heart studied by means of simultaneous right and left ventricular function curves in the dog. *Circulation* 9:706–718.

Shroff, S., Janicki, J. S., and Weber, K. T. 1983. Left ventricle systolic dynamics in terms of its chamber mechanical properties. *Am. J. Physiol.* 245:H110–H124.

Suga, H., and Sagawa, K. 1974. Instantaneous pressure–volume relationships and their ratio in the excised supported canine left ventricle. *Circ. Res.* 35:117–126.

Suga, H., Sagawa, K., and Shoukas, A. A. 1973. Load independence of the instantaneous pressure–volume relationship of the canine left ventricle and effects of norepinephrine on the ratio. *Circ. Res.* 37:320–330.

Sund, T., and Piene, H. 1983. Right ventricular mechanics: a comparison of models. *Cardiovasc. Res.* 17:320–330.

Taylor, M. G. 1964. Wave travel in arteries and the design of the cardiovascular system. In: "Pulsatile Blood Flow", (Ed: E. O. Attinger). New York: McGraw Hill.

Thomas, C. E. 1957. The muscular architecture of the ventricles of hog and dog hearts. *Am. J. Anat.* 101:17–58.

Van den Horn, G. J., Westerhof, N., and Elzinga, G. 1984. Interaction of heart and arterial system. *Ann. Biomed. Eng.* 12:151–162.

Wilcken, D. E. L., Charlier, A. A., Hoffman, J. I. E., and Guz, A. 1964. Effects of alterations in aortic impedance on the performance of the ventricles. *Circ. Res.* 14:283–293.

Wilcox, B. R., and Lucas, C. L. 1980. Pulmonary input impedance in children with left-right shunt. *Surg. Res.* 29:40–49.

9
Heart-Arterial Coupling: How to Bypass the Aortic Valves

GERARDUS J. VAN DEN HORN

Introduction

Coupling of the heart and the arterial system has been a subject of investigation for more than half a century. Both systems are usually described in terms of pressure and flow. In the analysis of heart arterial coupling, however, the aortic valves constitute an obstacle. Owing to the nonlinear properties of the aortic valves and the compliance of the arterial system, two distinct pressures exist: left ventricular pressure and aortic pressure. These nonlinear properties prohibit a direct comparison of the quantitative descriptions of the two systems.

We proposed to study heart arterial coupling by characterizing the heart and the arterial system in terms of mean pressure and mean flow (Van den Horn, Westerhof, and Elzinga 1984, 1985). The left heart can be described quantitatively by the relationship between mean left ventricular pressure and mean flow: the pump function graph (Elzinga and Westerhof 1974). From this relationship the apparent source resistance (Westerhof and Elzinga 1978) can be computed. During the steady state the arterial system can be characterized by the relationship between mean aortic pressure and mean flow, from which the peripheral resistance can be calculated.

A link between the characterizations of the heart and the arterial system can be found in the relationship between the mean aortic pressure and the mean left ventricular pressure. We have shown that these pressures in the experimental animal can be related over a wide range and under a variety of experimental conditions (Van den Horn, Westerhof, and Elzinga 1984). This allows for the calculation of peripheral resistance at any point of the pump function graph. The pump function graph can be determined reliably with the heart in situ in the anesthetized open-thorax cat (Van den Horn 1985; Van den Horn, Westerhof, and Elzinga 1985). This implies that this characterization can be used to study the heart arterial coupling in vivo.

This chapter deals with the relationship between mean aortic and mean left ventricular pressures, since this relationship constitutes the key to our approach. We attempt to gain insight into this relationship by comparing the results obtained from electrical analog studies with those obtained from animal experiments.

The Electrical Analog

The analog model shown in Figure 9.1 was used to guide our study of the relationship between mean left ventricular and mean aortic pressure. The model is presented in electrical form—resistances are comparable to hydraulic resistances and capacitances to compliance. The source mimicking the heart is modeled by ideal generator G and finite source resistance R_s. This heart model was proposed by Elzinga and Westerhof (1973). The potential of the generator changes stepwise from zero to V_G and vice versa. The cycle time is T; T_e is the duration of the block pulse delivered by the generator (or the ejection time). The source is connected to the load via the diode D, representing the aortic valves. The load consists of a three-element network: a parallel network of a load resistance R_p and a capacitor (compliance) C in series with a resistive element R_c. This model of the arterial system was proposed by Westerhof (1968). In the network, R_c represents the characteristic impedance of the aorta, C the total arterial compliance, and R_p the peripheral resistance. V_a and V_b represent left ventricular and aortic pressure, respectively. Finally, the flow is represented by I.

We assumed a mean left ventricular pressure of 9 kPa, an output of 5 ml/sec, a frequency of 3 Hz, and an ejection period of 40% of the cardiac cycle to be the physiologic values for the feline left heart; for the systemic arterial bed a mean aortic pressure of 15 kPa and an R_PC time of 1 second

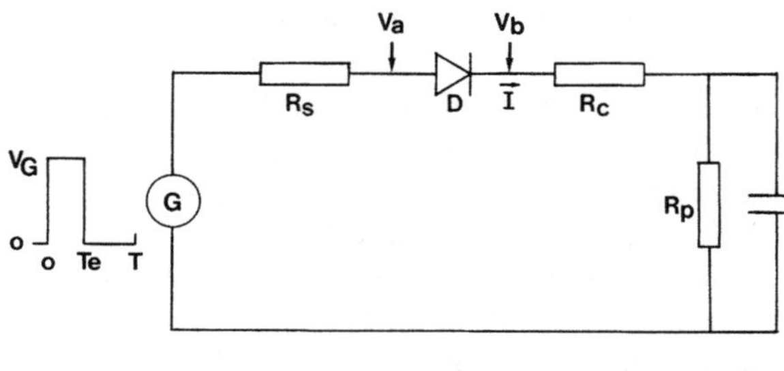

FIGURE 9.1. Analog model representing the heart and the arterial system. G, generator; D, diode; R_s, source resistance; R_c, characteristic resistance; C, compliance; R_p, peripheral resistance; T, cycle time: T_e, ejection time; V_a, left ventricular pressure; V_b, aortic pressure; I, flow; V_G, maximal generator pressure (reached for infinite load, that is, isovolumic contraction). See text for further explanation.

were considered normal. Equating kPa to 1 V, and 1 ml/sec to 1 mA, these assumptions lead to the following "physiologic" values of the components:

$$R_p = 3000 \text{ ohm}$$

$$C = 350 \ \mu\text{farad}$$

The value of the characteristic resistance R_c was taken as 10% of the peripheral resistance (Elzinga and Westerhof 1973).

Experimental Animal

Four male adult cats were anesthetized with sodium pentobarbital (35 mg/kg) and artificially ventilated with a mixture of room air and oxygen (3:1). Following sternotomy and heparinization, the left ventricular cavity and the proximal aorta were cannulated with stiff polyethylene catheters. Pressures were measured with Statham P23Db transducers. A tightly fitting perivascular flow probe (Skalar Transflow 601 system) was put around the root of the ascending aorta and connected with a Skalar flowmeter. Different experimental conditions were induced by administration of halothane, norepinephrine, atrial pacing, volume loading, and combinations of these. In this way a wide range of pressures and flows was obtained in each animal. All signals were converted from analog to digital form at a sampling rate of 500 Hz. Mean pressures were obtained by integrating the respective signals over the heart period, and dividing by this period. Calculations were performed on a DEC PDP 11/34 computer. Integrations were carried out numerically, using Simpson's rule.

Results and Discussion

Figure 9.2 shows mean aortic pressure versus mean left ventricular pressure as measured in the four animals. Mean pressures were computed during a steady state only; each data point thus represents one steady state. No efforts were undertaken in these experiments to keep heart rate or left ventricular filling constant. The data were fitted with straight lines through the origin. This fit can be seen to constitute a good description of the data. Table 9.1 gives regression coefficients, the number of data points, and coefficients of correlation.

Figure 9.3 shows the "pressure" ratio V_b/V_a in the analog model as a function of the peripheral resistance R_p for different values of the compliance C. On the left is the special case when the characteristic resistive element R_c is absent. One can see that the pressure ratio approaches a finite value k, which on theoretical grounds is found to equal T/T_e, that is, the ratio of cycle time to ejection time (see also Piene and Sund 1979). If the value of the compliance is chosen sufficiently large, the ratio V_b/V_a may be approx-

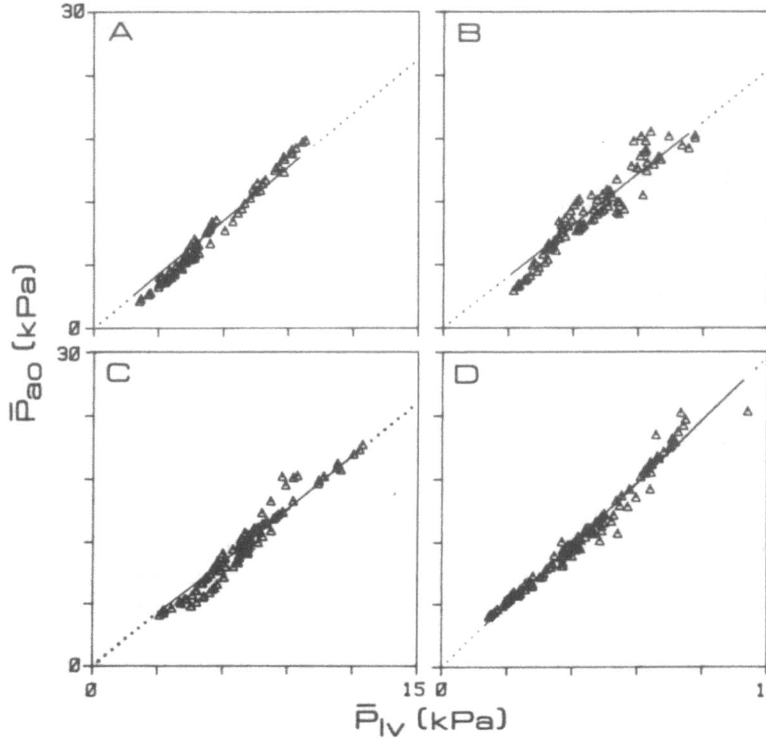

FIGURE 9.2. Relationship between mean aortic pressure and mean left ventricular pressure in four cats. The data were fitted with straight lines through the origin. See Table 9.1 and the text for details.

imated by this constant k for a wide range of values of peripheral resistance. In Figure 9.4, compliances and resistances that yield a V_b/V_a ratio deviating by 10% from k are related. The top right area of this figure thus consists of R_pC values for which V_b/V_a is nearly constant, and amounts to more than 90% of k. The analog model predicts a proportionality between V_b and V_a and for R_pC values of more than 300 msec.

TABLE 9.1. High correlation between mean aortic pressure (P_{ao}) and mean left ventricular pressure (P_{lv}) in cats.

Cat	N	$a(=P_{ao}/P_{lv})$	r	r'
A	74	1.70	0.98	0.98
B	73	1.63	0.95	0.99
C	96	1.68	0.97	0.99
D	123	1.96	0.99	0.99

N, number of data points; a, regression coefficient; r, coefficient of correlation; r', same coefficient after correction for T/T_e (see text).

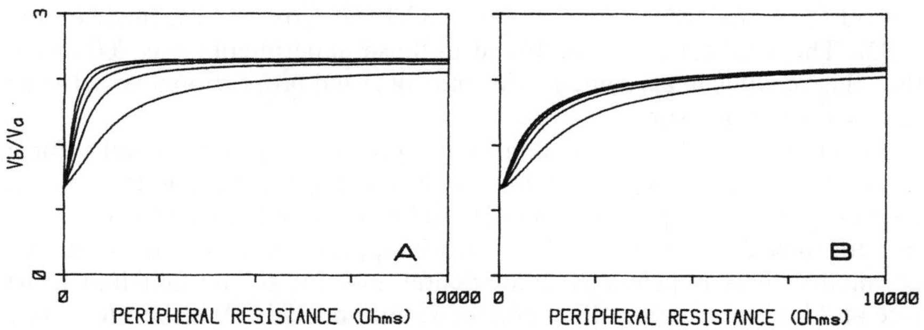

FIGURE 9.3. Ratio of mean potentials on both sides of the diode as a function of load resistance for several compliance values. *Left* (A), the characteristic resistive element is absent; *right* (B), it is present. C = 100, 200, 300, 400, and 500 μfarad; cycle frequency = 500 Hz.

To verify this result in the experimental animal, we evaluated the R_pC time of the cat's systemic arterial bed. For this purpose, 10 beats were selected from each experiment with pressures throughout the entire pressure range. For these beats, the R_pC time was estimated from the diastolic decay of the aortic pressure signal. The arterial compliance was then calculated from the R_pC time and the peripheral resistance. The R_pC time changed with peripheral resistance in the same direction, but in a less than proportional ratio. This is due to the fact that the compliance changed in the opposite direction with peripheral resistance. Such behavior is ob-

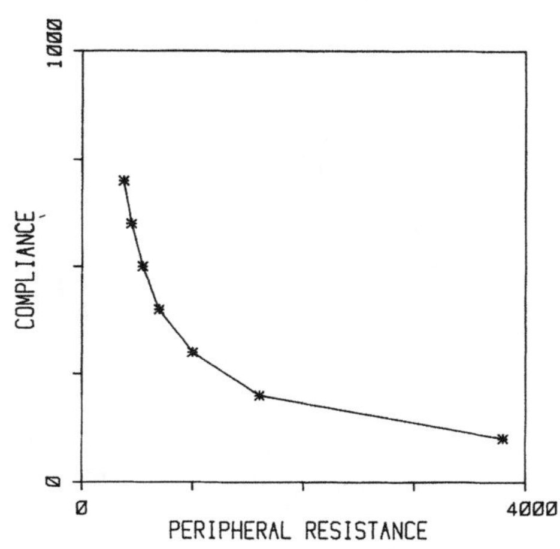

FIGURE 9.4. Load compliances for which the ratio V_b/V_a in Figure 9.3A amounts to 90% of its maximal value k, plotted against their respective load resistances. To the right of the lines interconnecting the data, the R_pC time of the network is sufficiently large to imply a proportionality between V_a and V_b. R_p = peripheral resistance; C = compliance.

served for many types of vessels (Bergel 1972, Cox 1975, Langewouters 1982). The smallest R_pC time found in these experiments was 700 msec; thus this result is in agreement with the observed proportionality between the two mean pressures.

We emphasize that the analog model predicts a proportionality for a constant ratio T/T_e only; this ratio may be expected to vary in the animal. However, since the pressure data could be adequately fitted by a straight line through the origin, the T/T_e ratio is apparently constant in our experiments. This is probably coincidental and due to the fact that heart rate and left ventricular filling change simultaneously. We therefore may expect that part of the scatter in Figure 9.2 is the result of variations in the T/T_e ratio. This scatter should be reduced by plotting mean aortic pressure against mean left ventricular pressure times T/T_e. Such a plot should also eliminate the interindividual differences in the slopes of the

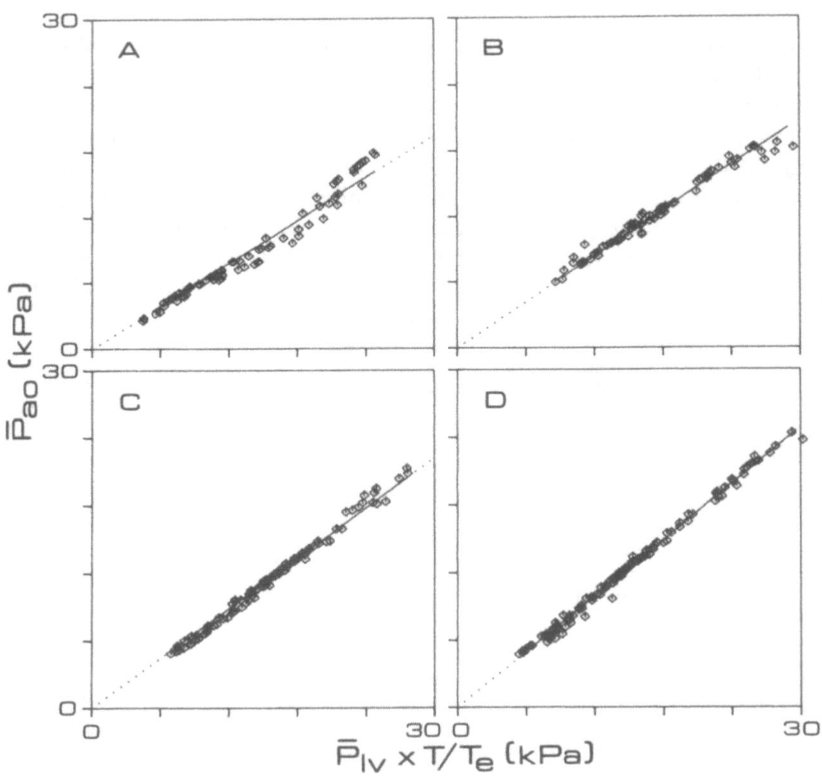

FIGURE 9.5. Mean aortic pressure ($\bar{P}_{ao}$) plotted against mean left ventricular pressure ($\bar{P}_{lv}$) times T/T_e (cycle time/ejection time). The data are from the same experiments as in Figure 9.2. Better linear fits were obtained by this way of plotting (see Table 9.1 for coefficients of correlation, r'). The slopes of the lines are quite uniform, indicating that in the four animals the ratio $\bar{P}_{ao}/\bar{P}_{lv}$ approximates a constant fraction of the ratio T/T_e.

relationship. As one can see from Figure 9.5, a better linear fit is obtained, as measured from the coefficients of correlation r. The slopes, which we would expect to be unity, amount to 0.75. However, with the introduction of a characteristic resistive element in the analog, only 80% of the value of k is reached for peripheral resistances in the physiologic range (see Fig. 9.3B).

Conclusion

We have shown that for infinite peripheral resistance, the ratio of mean aortic pressure to mean left ventricular pressure approaches a constant value k, which is given by the ratio of the cycle time T and the ejection time T_e. For physiologic values of the peripheral resistance, the pressure ratio varies little, and may be approximated by a value of 75% of k. In addition, the ratio T/T_e is nearly constant, probably owing to simultaneous changes in heart rate and left ventricular filling. This implies that, in the experimental animal, mean aortic and mean left ventricular pressure are proportional, even under non-steady-state conditions.

Bibliography

Bergel, D. H. 1972. The properties of blood vessels. In *Y.C.B. Fung*, ed. *Biomechanics: Its Foundations and Objectives*. Englewood Cliffs, NJ: Prentice Hall.

Cox, R. H. 1975. Arterial wall mechanics and composition and the effect of smooth muscle activation. *Am. J. Physiol.* 229:807–812.

Elzinga, G., and Westerhof, N. 1973. Pressure and flow generated by the left ventricle against different impedances. *Circ. Res.* 32:178–186.

Elzinga, G. and Westerhof, N. 1974. End-diastolic volume and source impedance of the heart. In *A.* Guz, ed. *The Physiological Basis of Starling's Law of the Heart*. Amsterdam/New York:Elsevier.

Langewouters, G. J. 1982. Visco-elasticity of the human aorta in vitro in relation to pressure and age. Ph.D. Thesis, Free University of Amsterdam.

Piene, H., and Sund, T. 1979. Flow and power output of right ventricle facing load with variable imput impedance. *Am. J. Physiol.* 237:H125–H130.

Van den Horn, G. J. 1985. Interaction of heart and arterial system. Ph.D. Thesis, Free University of Amsterdam.

Van den Horn, G. J., Westerhof, N., and Elzinga, G. 1984. Interaction of heart and arterial system. *Ann. Biomed. Eng.* 12:151–162.

Van den Horn, G. J., Westerhof, N., and Elzinga, G. 1985. Optimal power generation by the left ventricle: a study in the anesthetized open-thorax cat. *Circ. Res.* 56:252–261.

Westerhof, N. 1968. Analog studies of human systemic hemodynamics. Ph.D. Thesis, University of Pennsylvania, Philadelphia.

Westerhof, N., and Elzinga, G. 1978. The apparent source resistance of heart and muscle. *Ann. Biomed. Eng.* 6:16–32.

10
Ventricular Interaction with the Vascular System in Terms of Pressure–Volume Relationships

KENJI SUNAGAWA, KIICHI SAGAWA, AND
W. LOWELL MAUGHAN

Introduction

Guyton and his associates (Guyton 1963; Guyton, Jones, and Coleman 1973) analyzed the regulation of cardiac output in the closed circulatory loop by experimentally opening the loop at the junction of the vena cava and the right atrium. Reasoning that the cardiac output curve depicts the performance of the ventricle as a pump and the venous return curve represents vascular system properties and circulatory blood volume, they superimposed these two relationship curves and showed that cardiac output in the natural closed-loop condition is found at their intersection. Although this concept was useful in analyzing the global mechanisms that govern cardiac output in the closed circulatory system, it did not allow separation of the contributions of the individual cardiac chambers and the pulmonary vascular components of the circulatory system.

Subsequently, Sagawa (1967) and Herndon and Sagawa (1969) described the flow-generating capability of the heart–lung section as a function of mean aortic pressure and mean right atrial pressure by decomposing the circulatory loop into the heart–lung section and the systemic vascular section. Although this advanced Guyton's original analysis somewhat, defining the individual contributions of the cardiac chambers and vascular components to the control of cardiac output in the closed circulatory system was still difficult.

Recently Piene and associates (Piene 1980, Piene and Sund 1982) showed that instantaneous pulmonary arterial pressure and flow can be determined numerically by expressing the pump properties of the right ventricle by the instantaneous pressure–volume relationship and the properties of the arterial system by the input hydraulic impedance. Although this numerical method does describe the behavior of the ventricular arterial coupled system precisely, it does not reveal the mechanism underlying the interactive effects.

This chapter presents an explicit statement in terms of analytically derived algebraic equations of the effects of ventricular and vascular properties on cardiac output, with the goal of identifying the contributions to cardiac output of the various components of the circulatory system (Sunagawa, Sagawa, and Maughan 1984).

Basic Strategy

The first step was to identify the functional units of the circulatory loop and to decide how extensively the system should be decomposed. Since there are two ventricles and two vascular beds (systemic and pulmonary vascular systems) in the circulatory loop, we decomposed the system into four major functional blocks (Fig. 10.1). Each block was characterized as a bilaterally coupled system. The entire circulatory system could then be reconstructed by recoupling the blocks successively. Since the analytical recoupling between the ventricular and arterial systems appeared simpler than that between the venous system and the ventricles, we first recoupled the left heart with the systemic vascular bed and the right heart with the pulmonary vascular bed. With these recouplings done, the circulatory loop appears as a system comprising two functional blocks. The junctions between these blocks are the systemic venoatrial junction and the pulmonary venoatrial junction. Cardiac output through each block can be represented by a cardiac output surface that is a function of the pressure variables of the inflow and outflow ports. The cardiac output surface of the left heart–systemic vascular bed system, CO_s (systemic cardiac output surface), can be described (on the basis of the mechanical properties of the left heart

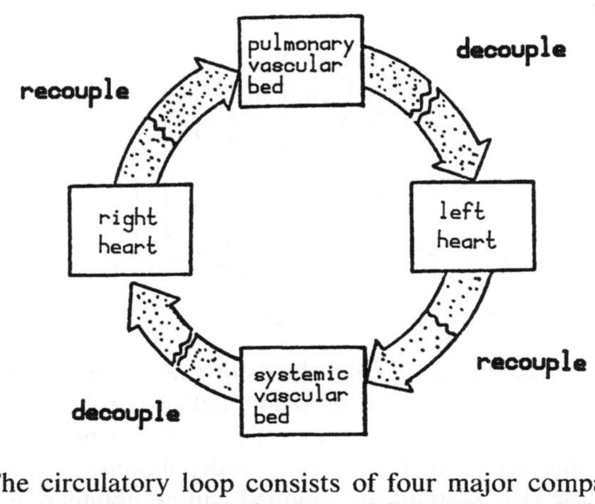

FIGURE 10.1. The circulatory loop consists of four major compartments. Each compartment is separately characterized as a bilaterally coupled system. (From Sunagawa, Sagawa, and Maughan 1984; with permission from Pergamon Press, Ltd.)

and the systemic vascular bed) in general form as

$$CO_s = F_s(P_{ra}, P_{la}) \tag{10.1}$$

where P_{ra} = mean right atrial pressure
P_{la} = mean left atrial pressure
F_s = functional symbol of systemic flow.

Similarly, the cardiac output surface of the right heart–pulmonary vascular bed system, CO_p (pulmonary cardiac output surface), can be described as

$$CO_p = F_p(P_{ra}, P_{la}) \tag{10.2}$$

These systemic and pulmonary cardiac output surfaces are shown in Figure 10.2. Since steady-state cardiac output through both systems must be

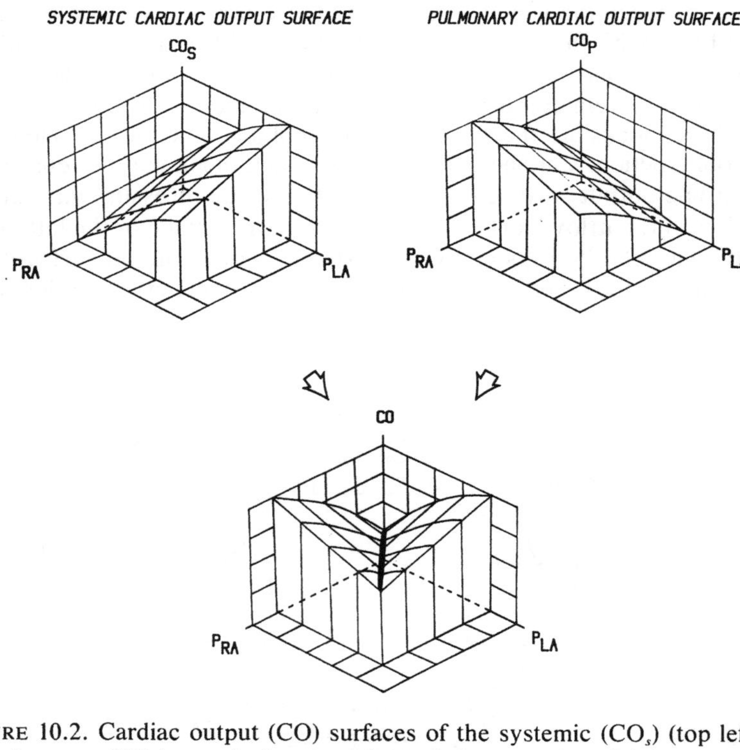

FIGURE 10.2. Cardiac output (CO) surfaces of the systemic (CO_s) (top left) and the pulmonary (CO_p) ventricular-arterial coupled system (top right), expressed as a function of right and left atrial pressure (P_{ra}, and P_{la}, respectively). Since cardiac output is the same for both the systemic and pulmonary systems, the cardiac output when these two systems are coupled can be obtained as the intersection curve between the two surfaces (bottom). (From Sunagawa, Sagawa, and Maughan 1984; with permission from Pergamon Press, Ltd.)

identical, the steady-state cardiac output in the closed circulatory loop can be determined as an equilibrium flow through these two subsystems. This equilibrium condition is identified as the intersection curve (the solid curvilinear line in the bottom part of Fig. 10.2) between the two surfaces. This operating cardiac output curve corresponds to Guyton's cardiac output curve.

One more restraining condition needs to be specified to arrive at a unique operating point on the cardiac output curve just determined: The operating point must also exist on the surface defined by the relationship between vascular properties and total blood volume. This surface is the three-dimensional equivalent of Guyton's venous return curve.

$$CO_v = F_v(P_{ra}, P_{la}) \tag{10.3}$$

where F_v is a functional symbol of venous flow.

By analogy with Guyton's venous return curve, we anticipate that the venous return surface (CO_v) will have negative slopes along both left and right atrial pressures, as shown on the left of Figure 10.3. Cardiac output in a closed circulatory loop can then be obtained as an intersection between the cardiac output curve (bottom of Fig. 10.2) and the venous return surface, as shown on the right of Figure 10.3.

In the following sections we characterize these three major surfaces, which determine cardiac output, on the basis of properties of the ventricular and vascular systems.

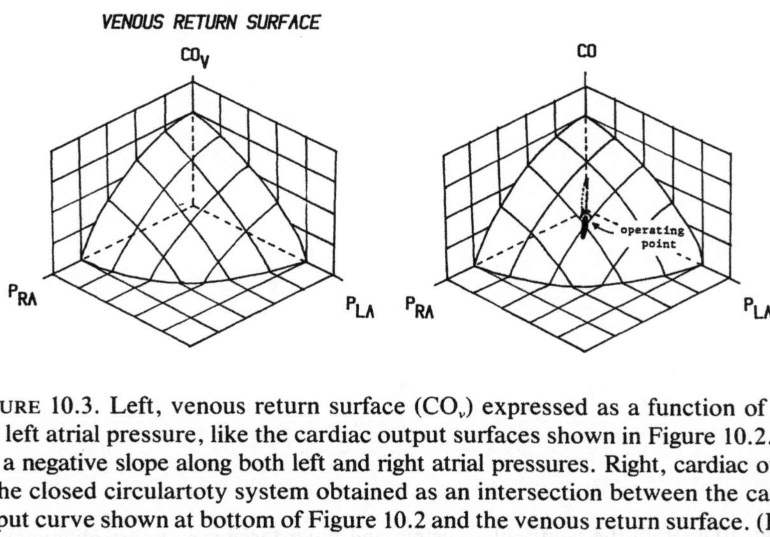

FIGURE 10.3. Left, venous return surface (CO_v) expressed as a function of right and left atrial pressure, like the cardiac output surfaces shown in Figure 10.2. CO_v has a negative slope along both left and right atrial pressures. Right, cardiac output of the closed circulartoty system obtained as an intersection between the cardiac output curve shown at bottom of Figure 10.2 and the venous return surface. (From Sunagawa, Sagawa, and Maughan 1984; with permission from Pergamon Press, Ltd.)

The Pressure–Volume Relationship

Characterization of the Ventricular System

To determine the characteristics of the ventricular output surfaces, we need to quantify the couplings between the ventricles and the arterial systems (ventricular/arterial coupling). We have chosen to use the pressure–volume relationship of the ventricles as the basis of the coupling because of its mathematical simplicity.

The end-systolic pressure–volume relationship (ESPVR) has been shown to be approximately linear over a physiologic range, and while its slope (E_{es}) varies greatly in response to changes in contractility, the ESPVR is relatively insensitive to changes in loading conditions (Maughan and Sunagawa 1984; Maughan, Sunagawa, and Sagawa 1984; Sagawa 1981; Suga and Sagawa 1974; Suga, Sagawa, and Shoukas 1973). Thus the relation of the left ventricular end-systolic pressure (P_{es}) to stroke volume (SV) can be defined as

$$P_{es} = E_{es} (V_{ed} - SV - V_0) \tag{10.4}$$

where V_{ed} is end-diastolic volume and V_o is the volume axis intercept of

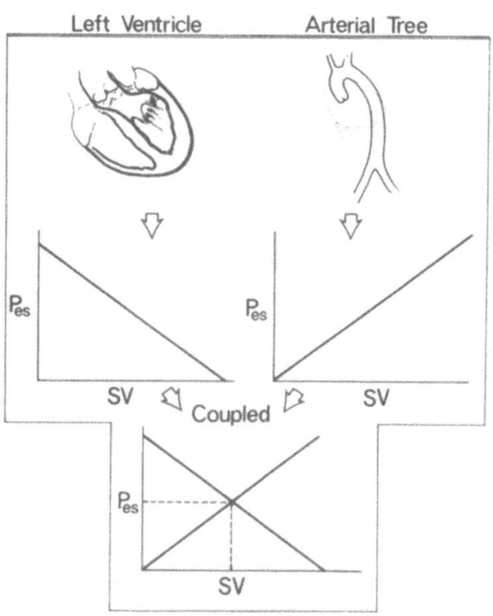

FIGURE 10.4. Both the ventricular system and the arterial system are characterized by the relationship of end-systolic pressure (P_{es}) to stroke volume (SV). The intersection between these two relationship lines gives the resultant stroke volume. (From Sunagawa et al. 1983).

the ESPVR. Therefore, for a given V_{ed}, P_{es} varies inversely and linearly with stroke volume (left of Fig. 10.4).

Characterization of the Arterial System

If, as we did for the ventricle, we characterize the properties of the arterial system in terms of the relationship of end-systolic pressure to stroke volume (P_{es}-SV relationship) (right of Fig. 10.4), an equilibrium stroke volume when both systems are coupled can be determined graphically as the point of intersection between the ventricular and arterial P_{es}-SV relationships (bottom of Fig. 10.4) (Sunagawa, Sagawa, and Maughan 1984; Sunagawa et al. 1983b).

The mean arterial pressure–flow relationship of the systemic vascular bed has been shown to be reasonably linear with a small zero flow pressure over its physiologic range (Sagawa and Eisner 1975). Therefore, if we ignore the zero flow pressure, the mean arterial pressure–flow relation can be expressed as

$$P_m = R_T F \tag{10.5}$$

where P_m = mean arterial pressure
R_T = total arterial resistance
F = mean arterial flow

If we approximate the end-systolic pressure by mean arterial pressure and substitute the flow as SV/T, where T is the cycle length of the cardiac contraction, the end-systolic pressure is expressed as

$$P_{es} = \frac{R_T}{T}\,\mathrm{SV} \tag{10.6}$$

Therefore the slope of the P_{es}-SV relationship, that is, the effective arterial volume elastance (E_a), is approximated by

$$E_a = \frac{R_T}{T} \tag{10.7}$$

Figure 10.5 shows the relationship between experimentally determined effective arterial elastances P_{es}/SV (or E_a) and their approximations by R_T/T under a wide variety of resistance values and heart rates studied in a single isolated ventricle (Sunagawa, Sagawa, and Maughan 1984). The ventricle ejected into a simulated arterial impedance (Sunagawa et al. 1982). The relationship was highly linear (r = .99, n=33, P<0.0001) and the slope was close to unity with a very small y-axis intercept (R_T/T = $0.934E_a$ + 0.48). Therefore, one can estimate the effective arterial elastance reasonably accurately by dividing the total arterial resistance by the duration of one cardiac cycle. This method provides a quick, practical way to associate conventional arterial properties with the arterial P_{es}-SV relationship.

In the actual vascular system, the presence of effective downstream pressure, which varies with venous pressure (Brunner et al. 1983), is

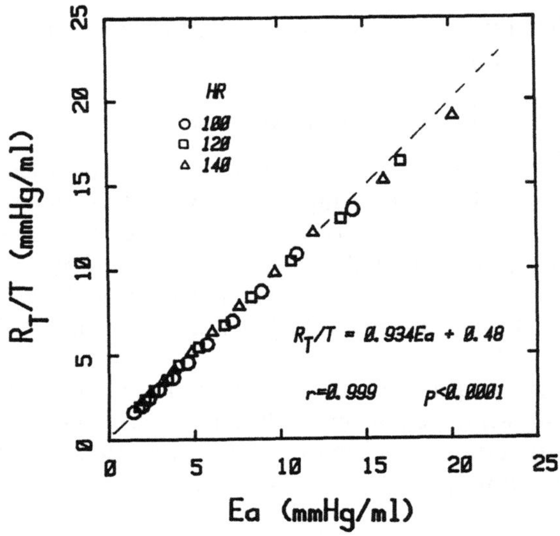

FIGURE 10.5. Correlation between the slope of the P_{es}-SV (end-systolic pressure to stroke volume) relationship (E_a) versus R_T/T (the ratio of total arterial resistance to beat to beat time interval) for a variety of experimental conditions. The total resistance was varied from 1.0 to 8.2 mm Hg sec/ml (From Sunagawa, Sagawa, and Maughan 1984; with permission from Pergamon Press, Ltd.)

known. If we denote the effective downstream pressure by P_d, the P_{es}-SV relationship will be approximated by

$$P_{es} = \frac{R_T}{T}\,\mathrm{SV} + P_d \tag{10.8}$$

Ventricular/Arterial Coupling

Once we have characterized both the ventricular and arterial systems in terms of the P_{es}-SV relationship (Fig. 10.4), we can determine the stroke volume that should result when the ventricle is coupled with the arterial system as the intersection between the two P_{es}-SV relationships (bottom of Fig. 10.4). The equilibrium stroke volume can be obtained analytically from Equations 10.4 and 10.6 as

$$\mathrm{SV} = \frac{E_{es}(V_{ed} - V_o)}{E_{es} + E_a} \tag{10.9}$$

If we include the influence of the effective downstream pressure P_d, the equilibrium stroke volume will be

$$\mathrm{SV} = \frac{E_{es}(V_{ed} - V_o) - P_d}{E_{es} + E_a} \tag{10.10}$$

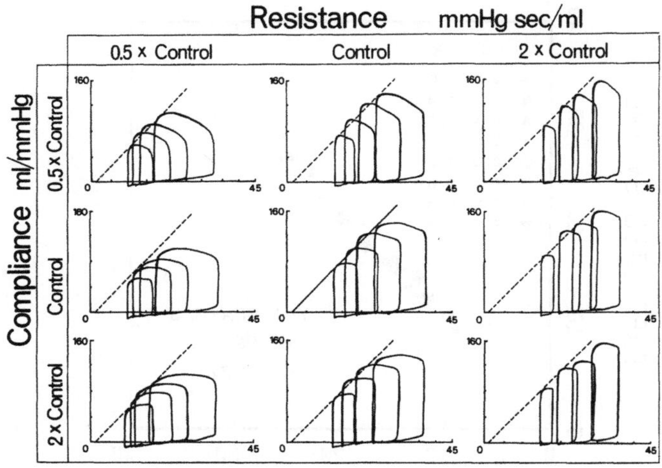

FIGURE 10.6. Multiple pressure–volume loops obtained by varying resistance and compliance from 50% to 200% of their control values at four different end-diastolic volumes under each combination of the resistancce and compliance values. The solid line in the center indicates the end-systolic pressure–volume relationship (ESPVR) under a control condition. The dashed lines elsewhere indicate the transcribed control ESPVR. (From Sunagawa et al. 1983).

Test of Ventricular/Arterial Coupling Framework

To validate the proposed coupling framework, we first obtained pressure–volume loops under the control afterload condition at four different end-diastolic volumes, using an isolated canine heart preparation (center of Fig. 10.6). We then changed the values of the afterload resistance and/or compliance from 50% to 200% of their control values in a random sequence and obtained the left ventricular pressure–volume loops at four different end-diastolic volumes under each of eight different afterload conditions (Fig. 10.6). We compared the stroke volumes predicted by the model with those obtained experimentally (Sunagawa et al. 1983).

Figure 10.7 shows an example of the relationship between the predicted and measured stroke volumes under various values of arterial resistance, compliance, and end-diastolic volume. All the data in this plot were obtained from a single ventricle. The different symbols represent data with different combinations of arterial resistance and compliance. The correlation coefficient in this example was .993 ($P<0.0001$) and the linear regression equation was $y = 1.06x + 0.3$. The slope of the regression line was close to unity with a very small y-axis intercept, indicating the accuracy of the prediction. The average correlation coefficient in the eight ventricles studied was .985. Thus our model was able to predict the stroke volume reasonably well in all the left ventricles studied.

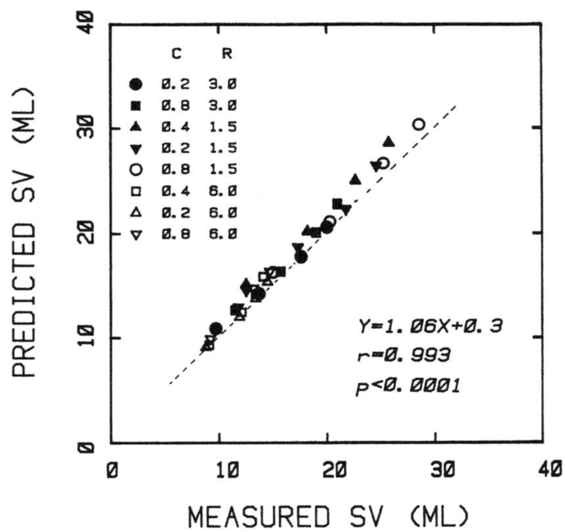

FIGURE 10.7. Example of the relationship between predicted and measured stroke volume (SV). C = compliance (ml/mm Hg); R = resistance (mm Hg sec/ml). Despite a wide range of loading conditions, data were tightly distributed along the line of identity (dashed line) (From Sunagawa et al. 1983).

Physical Implications of the Coupling Model

Since the slopes of the P_{es}-SV relationships (Figure 10.4 bottom) have the units of volume elastance (mm Hg/ml), our model treats the arterial system as an elastic chamber with volume elastance E_a and the ventricle as an elastic chamber with end-systolic elastance E_{es}. Thus our prediction of the equilibrium stroke volume when the ventricle is coupled to the arterial system presupposes the question: How much volume will be transferred if one couples an elastic chamber (left ventricle) with a known volume elastance, dead volume, and initial volume, to a second elastic chamber (arterial system) with a known volume elastance (Fig. 10.8)? Obviously, the distribution of the volume between the two chambers will be determined by the ratio of their volume elastance values. The slope of the arterial P_{es}-SV relationship represents not the physical elastance or compliance of the arterial system, but an "effective" arterial elastance. The effective arterial elastance changes more with changes in physical arterial resistance than with changes in physical arterial compliance. This was evident in our experimental data, in which arterial resistance was far more important than arterial compliance in determining the stroke volume. This point is discussed in more detail later in this chapter.

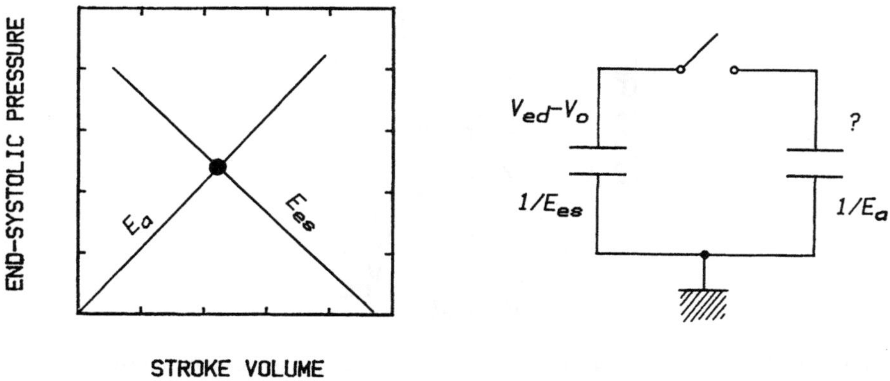

FIGURE 10.8. *Left,* the ventricle and the arterial systems are characterized by the P_{es}-SV (end-systolic pressure to stroke volume)) relationship. Stroke volume is obtained as the abscissa of the intersection between these two lines. *Right,* both ventricle and arterial systems are considered elastic chambers with known volume elastance [E_{es} (end-systolic elastance)] and E_a *(arterial elastance)*]. V_{ed} = end-diastolic volume; V_o = volume axis intercept of end-systolic pressure–volume relationship.(From Sunagawa et al. 1983).

Graphic Analysis of Ventricular/Arterial Coupling

The following examples illustrate how this approach to coupling allows a simple evaluation of the specific contributions of ventricular end-systolic properties and arterial system parameters in determining stroke volume. Although the fundamental concept of ventricular/arterial coupling using the pressure–volume relationship is best explained by discussion of the end-systolic pressure–stroke volume relationship of both the ventricular and arterial system, graphic analysis is easier using the end-systolic pressure–ventricular volume plane. Figure 10.9 shows the analysis on the pressure–volume plane. On this plane the arterial system property is expressed by a diagonal line labeled "A" (arterial end-systolic pressure–volume relationship). The slope of the arterial ESPVR is E_a and the volume axis intercept is V_{ed}. The equilibrium end-systolic pressure–volume point when the ventricle is coupled with the arterial system is determined graphically as the intersection between the arterial and ventricular ESPVR.

Figure 10.10 diagrams the graphic analysis of ventricular/arterial coupling under a variety of conditions. The left side of Figure 10.10 illustrates the effect of changes in arterial resistance. If we increase the arterial resistance while maintaining a constant contractility and preload, the absolute value of the slope of the arterial ESPVR increases (because E_a can

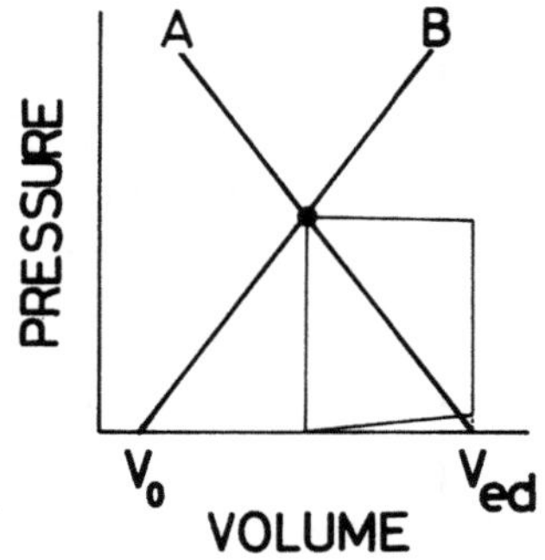

FIGURE 10.9. The slope of the arterial end-systolic pressure–volume relationship (ESPVR) (line A) is E_a with the volume axis intercept of V_{ed}. Line B represents the ventricular ESPVR, with the volume axis intercept of V_o. End-systolic volume is given as the intersection between these two lines. Stroke volume is the difference between the end-diastolic volume (right bottom corner of pressure–volume loop and end-systolic volume (top left corner of pressure–volume loop).

be approximated by R_T/T), which in turn decreases stroke volume by shifting the intersection between the ventricular and arterial ESPVR upward. If we decrease the arterial resistance, the intersection point shifts downward, resulting in an increase in stroke volume.

The middle of Figure 10.10 illustrates the effect on stroke volume of changing preload while maintaining constant contractility and arterial resistance. With an increase in end-diastolic volume, the arterial ESPVR shifts rightward and the stroke volume increases from the control equilibrium value (closed circle) to a new equilibrium value. If the end-diastolic

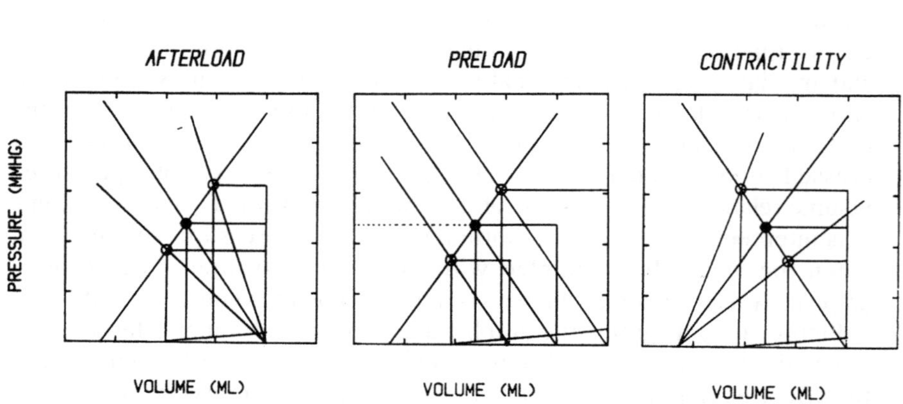

FIGURE 10.10. Effects of changes in arterial resistance *(left)*, end-diastolic volume *(middle)*, and contractility *(right)* on stroke volume. Closed circle represents equilibrium end-systolic pressure–volume point under control conditions. See text for discussion.

volume decreases, the stroke volume will also decrease because the intersection between the ventricular and arterial ESPVR shifts leftward.

If the contractility of the ventricle is augmented (i.e., if E_{es} increases), the slope of the ventricular ESPVR will increase without changes in the volume axis intercept, as shown on the right side of Figure 10.10. The arterial ESPVR remains constant as long as the end-diastolic volume is held constant. Therefore stroke volume will increase from a control equilibrium value (closed circle) to a new equilibrium value. A decrease in contractility results in a decrease in stroke volume.

Pressure Source, Source Resistance, and Arterial Resistance

We have shown that the ESPVR is extremely useful in describing the coupling of the ventricle with the arterial system. The volume elastance concept applied to both the ventricle and the arterial system enables us to predict stroke volume. The goal, however, is not to predict stroke volume, but to predict cardiac output. Therefore we must modify the coupling framework described in order to determine cardiac output directly on the basis of ventricular and arterial system properties.

Since cardiac output (CO) is flow, it can be obtained by dividing stroke volume by cardiac cycle length (T). Therefore, dividing Equations 10.9 and 10.10 by T and approximating E_a by R_T/T yield

$$\mathrm{CO} = \frac{E_{es}(V_{ed} - V_o)}{TE_{es} + R_T} \tag{10.11}$$

and

$$\mathrm{CO} = \frac{E_{es}(V_{ed} - V_o) - P_d}{TE_{es} + R_T} \tag{10.12}$$

respectively. The electrical analog of Equation 10.11 is easily generated (Fig. 10.11). Unlike the model shown in Figure 10.8, the ventricle here consists of a pressure source in series with source resistance. The arterial load in this model framework becomes a simple resistance R_T. The value of the source pressure is $E_{es}(V_{ed} - V_o)$, which is maximal ventricular pressure if the ventricle contracts isovolumically. The source resistance is equivalent to TE_{es}. In other words, an increase in contractility increases the source pressure, but also increases the source resistance. An increase in heart rate, which shortens the cardiac cycle length T, decreases the source resistance.

The advantage of this model over that based on the volume elastance is direct association of ventricular and arterial properties to cardiac output. Therefore we prefer to use this model in later sections dealing with cardiac output.

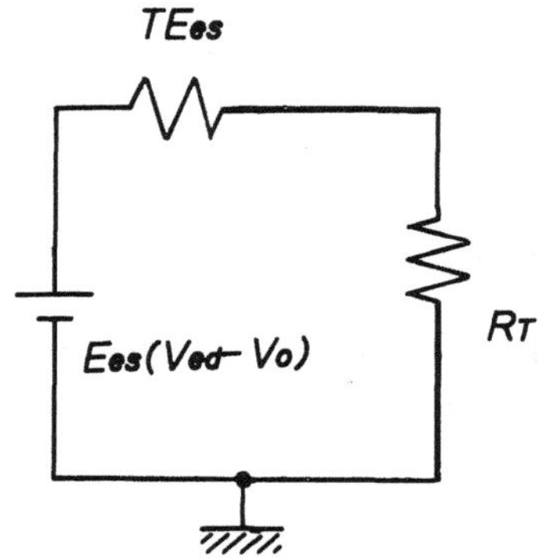

FIGURE 10.11. Electrical analog of the coupled ventricular-arterial system. Ventricular properties are lumped into the source pressure and resistance. The source pressure is equivalent to isovolumic ventricular pressure at the end-diastolic volume V_{ed}. The source resistance is equivalent to TE_{es}. Increasing the heart rate and/or increasing contractility make the source resistance larger. E_{es} = end-systolic elastance; R_T = total arterial resistance; V_o = volume axis intercept of end-systolic pressure-volume relationship.

Stroke Volume and Variations in Input Impedance

In the previous section we considered "arterial resistance" the primary factor determining stroke volume when ventricular properties are constant. In the natural arterial system, however, because the arterial tree comprises elastic tubes and has finite resistance and compliance, and because blood in the vessels has inertia, aortic input impedance is frequently invoked as the proper and correct measure of afterload against ventricular ejection (Milnor 1975, Noble 1979). Thus some investigators have studied the effects on stroke volume of artificially stiffening the canine descending aorta by applying Lucite ferrules (O'Rourke 1967) or by replacing it with a rigid bypass (Urschel et al. 1968). Others (Pouleur, Covell, and Ross 1979; Wilcken et al. 1964) occluded the descending aorta to decrease aortic compliance while increasing peripheral resistance, and compared the effects on stroke volume with those caused solely by a similar increase in resistance caused by phenylephrine (Pouleur, Covell, and Ross 1979).

Despite these studies, the relative quantitative importance for stroke volume of the arterial input impedance in different frequency ranges has not been well defined. Therefore, we investigated the relative importance of input impedance moduli over specific frequency ranges as determinants of stroke volume (Sunagawa et al. 1985a).

Experimental Conditions

Figure 10.12 illustrates on the left the three-element Windkessel model used for this experiment (Westerhof, Elzinga, and Sipkema 1971) and on the right its impedance spectrum. Unlike the ordinary vascular input impedance spectrum, both the ordinate and abscissa are logarithmically scaled. This is the so-called "Bode plot" of the three-element Windkessel model. Three major parts are clearly recognizable in the Bode plot of the impedance spectrum. Under a set of control parameter values [Z_o (is the characteristic impedance) $=0.2$ mm Hg sec/ml, $R=$ 3.0 mm Hg sec/ml, and $C=$ 0.4 ml/mm Hg], the impedance modulus in the extremely low-frequency range (below approximately 0.13 Hz) is large and asymptotically approaches the total resistance value, R_T (i.e., $R_T = R + Z_o$). Similarly, in the high-frequency range (above approximately 2 Hz), the impedance modulus is small and asymptotically approaches the characteristic impedance value, that is, Z_o. Between these low-frequency and high-frequency ranges there is a transitional range. The corner frequency at which the impedance modulus becomes 3 db smaller than the DC modulus (F_{c1} in Fig. 10.12) is known to be $1/(2\pi CR)$. The corner frequency at which the impedance modulus in 3 db is greater than that over the extremely high-frequency range (F_{ch} in Fig. 10.12) can be approximated by $1/(2\pi CZ_o)$.

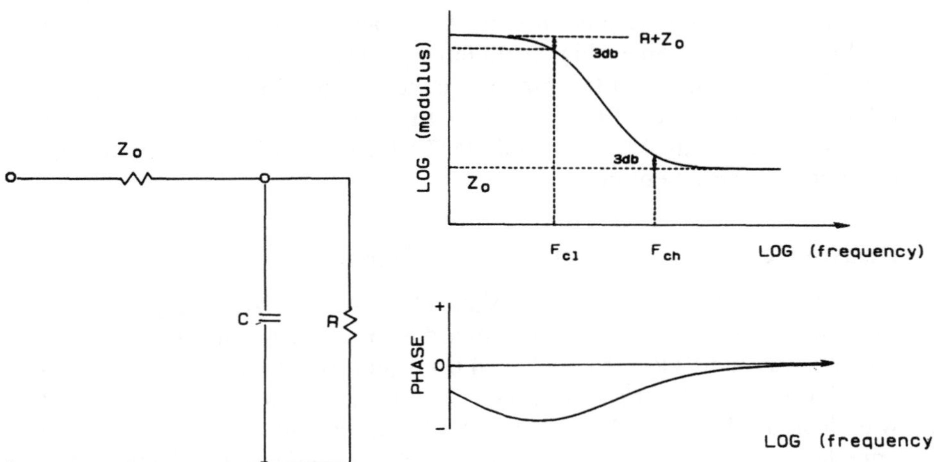

FIGURE 10.12. *Left,* three-element Windkessel model consisting of characteristic impedance *(Z_o)*, arterial resistance *(R)*, and arterial compliance *(C)*. (From Westerhof, Elzinga and Sipkema 1971). *Right,* the impedance spectrum of the three-element model. F_{cl}, corner frequency at which impedance modulus becomes 3db smaller than the DC modulus; F_{ch}, corner frequency at which it becomes 3db greater than the characteristic impedance. (From Sunagwa et al. 1985a).

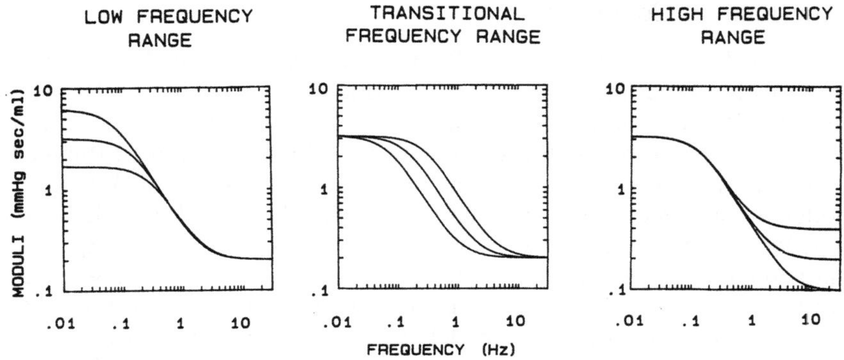

FIGURE 10.13. Bode plots of arterial input impedance for three protocols. *Left,* changes in resistance values vary the modulus of impedance below 0.13 Hz; *middle,* changes in compliance values vary corner frequencies and shift the impedance curve in the transitional range; *right,* changes in characteristic impedance combined with compensatory changes in resistance and compliance vary the modulus of impedance above 2 Hz. (From Sunagawa et al. 1985a).

The frequency range between the lower (F_{cl}) and higher (F_{ch}) corner frequencies is referred to hereafter as the transitional frequency range.

Using an isolated heart preparation, we measured stroke volume while varying the impedance modulus by the impedance loading system over the low-frequency range (Fig. 10.13 left), the transitional frequency range (Fig. 10.13 middle), and the high-frequency range (Fig. 10.13 right) in a total of nine ventricles. Each one of the three impedance components was changed from control to 50% and 200% of control while the rest of the components were kept constant.

Resistance as Determinant of Stroke Volume

Changes in stroke volume in each protocol are summarized in Figure 10.14. The data show mean values and standard error of the mean (SEM). Stroke volume was increased from 11.9±0.6 ml (control) to 16.1±0.7 ml by a 50% decrease in the impedance modulus in the low-frequency range, and was decreased to 7.4±0.5 ml by a 100% increase in the low frequency impedance (left). These changes were highly significant ($P<0.0001$). Selective changes in the impedance modulus over the transitional frequency range resulted in minimal changes in stroke volume, although the measured changes were statistically significant ($P<0.0001$). Halving the corner frequencies increased stroke volume slightly, from 11.9±0.6 ml to 12.3±0.7 ml. Doubling the corner frequencies slightly decreased stroke volume to 11.2±0.6 ml (middle). Selective changes in the high-frequency impedance modulus also made changes in the stroke volume, which while minimal were still statistically significant ($P<0.01$) (right).

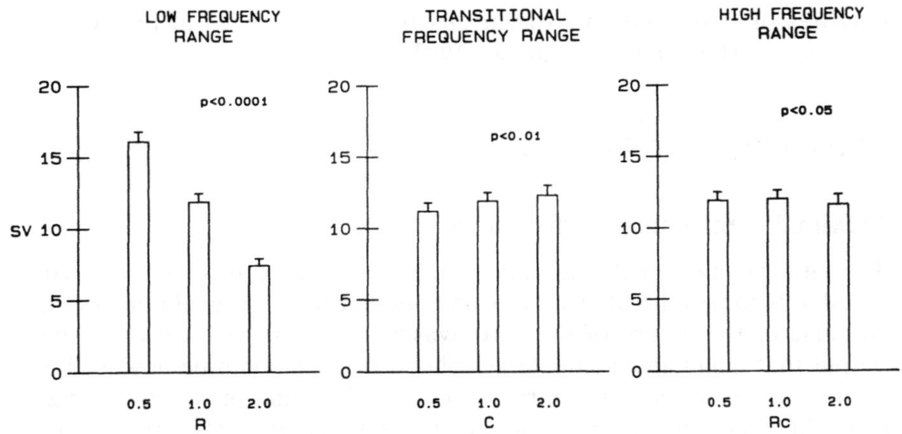

FIGURE 10.14. Stroke volume (SV) is highly sensitive to changes in arterial resistance *(R) left*. Effects of changes in compliance *(C) middle* or characteristic impedance *(Z_o–right)* on stroke volume are statistically significant yet qualitatively small. (From Sunagawa et al. 1985a).

The relative changes in stroke volume produced by selectively halving the impedance modulus over the low-frequency range, transitional frequency range, and high-frequency range were 100%, 8%, and −1%, respectively. Selective doubling of the modulus in these frequency ranges caused relative changes in stroke volume of 100%, 15%, and 9%. Thus the average relative sensitivity of stroke volume to the fourfold changes in the impedance modulus over the low-, transitional, and high-frequency ranges was approximately 100%, 12%, and 4%, respectively.

The mechanism of the observed differential sensitivity of stroke volume to the impedance moduli over different frequency ranges is unclear. Since the interaction of the ventricle with the arterial loading system comprises static and dynamic components, however, the magnitude of interaction between the ventricle and arterial load that determines stroke volume may be indexed by the amounts of external power partitioned by the static and oscillatory terms. It has been shown that the oscillatory component of left ventricular external power is in general less than 20% of the total external power (Milnor 1982, Nichols et al. 1977). About 70% of the oscillatory power is in the fundamental frequency. Interestingly, the external power distribution among the static, fundamental frequency and higher harmonic components is quite similar to the relative sensitivities of stroke volume to different impedance moduli over the low-, transitional, and high-frequency ranges. This parallelism may explain the differential sensitivity of stroke volume to changes in the impedance modulus in different frequency ranges. If this mechanism is significant, interventions that alter the partition of static and oscillatory work would change the relative sensitivities of stroke volume to various frequency ranges of impedance (Sunagawa, Maughan, and Sagawa 1982).

Maximizing Stroke Work

Maximal Ventricular Performance

The vital function of the cardiovascular system is to provide necessary blood flow to peripheral tissues to maintain normal biological functioning. Because of the nature of the cardiovascular system, a level of cardiac output that maintains adequate peripheral perfusion can be achieved by various combinations of ventricular contractility and preloading and afterloading conditions. There should also be a unique combination of ventricular contractility and loading condition at which operation of the cardiovascular system is optimal. It is not clear, however, how optimal functioning of the cardiovascular system should be defined, whether in terms of cardiac output, stroke volume, and/or external work or alternatively, the efficiency of the ventricular ejection. In view of the vital function of the cardiovascular system, performance is more essential than efficiency, although the optimal loading conditions for performance and for efficiency may well be the same. We may, therefore, define the optimal loading condition as that at which the performance of the left ventricle becomes maximal. It has been shown that power output of the normal excised feline left ventricle is maximum when the normal arterial impedance is loaded (Elzinga, Piene, and DeJong 1980). This is also the case in the right ventricular/pulmonary arterial coupling (Elzinga, Piene, and DeJong 1980; Piene and Sund 1982b). However, how changes in end-diastolic volume, ventricular contractility, heart rate, and arterial compliance, which are likely to vary under physiologic conditions, affect the optimal ventricular performance and loading conditions has not been known.

Therefore, we sought a basic relationship between left ventricular contractility, preloading and afterloading conditions under which ventricular performance would be maximized (Sunagawa, Maughan, and Sagawa 1985b). Since the peripheral tissue perfusion of vital organs (i.e., brain, heart, and kidney) is dependent not only on cardiac output or stroke volume but also on blood pressure, the performance of the ventricle can be measured by the stroke work, which accounts for both arterial blood flow and pressure. This allows us to evaluate the energy efficiency of the ventricle as well, since stroke work has the same units as energy. To cover a wide variety of physiologic coupling conditions between the ventricle and the arterial system, we changed, one at a time, the following parameters: left ventricular end-diastolic volume, ventricular contractility, heart rate, and arterial compliance. We then determined the optimal loading condition by varying arterial resistance so that stroke work became maximal in the blood-perfused isolated canine heart preparation.

Optimal Arterial Resistance

Figure 10.15 summarizes the effects of end-diastolic volume (top left), contractility (top right), heart rate (bottom left), and arterial compliance (bottom right) on the stroke work versus arterial resistance relationship. In all experimental conditions, stroke work was smallest (50–60% of maximum) when arterial resistance was large (i.e., 8 mm Hg sec/ml). Stroke work rapidly increased as the arterial resistance decreased. There were obviously optimal arterial resistances at which stroke work became maximal. When the arterial resistance was further decreased, stroke work started decreasing, although the amount of reduction was not as large as that observed in the higher resistance range.

The influence of changes in end-diastolic volume, contractility, heart rate, and arterial compliance on the maximal stroke work and optimal arterial resistance are summarized in Figure 10.16. As can be seen, the stroke work increased significantly ($P<0.0001$) with increases in the end-diastolic volume (A). However, the optimal arterial resistance did not change significantly. Increases in contractility altered the maximal stroke work ($P<0.0001$) with concomitant increases in the optimal arterial resistance ($P<0.0001$) (B of Fig. 10.16). Alterations in heart rate did not change the maximal stroke work. The optimal arterial resistance tended to decrease with increases in heart rate, although this trend was statistically insignificant (C of Fig. 10.16). Increases in arterial compliance did slightly increase ($P<0.01$) the maximal stroke work (D of Fig. 10.16). Although quantitatively small, the optimal resistance significantly increased ($P<0.01$) with increases in the arterial compliance.

Ventricular/Arterial Matching

In our analysis of ventricular/arterial coupling, the arterial system was treated as an elastic chamber with a volume elastance E_a, and the ventricle as an elastic chamber with an end-systolic volume elastance E_{es}. One can predict, analytically or intuitively, that the mechanical energy transferred from the ventricular elastance to the arterial elastance will become maximal when the slopes of the ventricular and arterial P_{es}-SV relationships are equal. Under this optimally loaded condition, an effective ejection fraction as defined by the ratio of stroke volume to the effective end-diastolic volume (V_{ed}-V_o) would be 50%. To test these predictions, we directly measured the optimal slope E_a of the P_{es}-SV relationship of the arterial system in each of four different protocols by calculating the ratio of the arterial end-systolic pressure to the stroke volume when the arterial resistance is optimal (Fig. 10.17). The P_{es} was measured as the arterial pressure at the

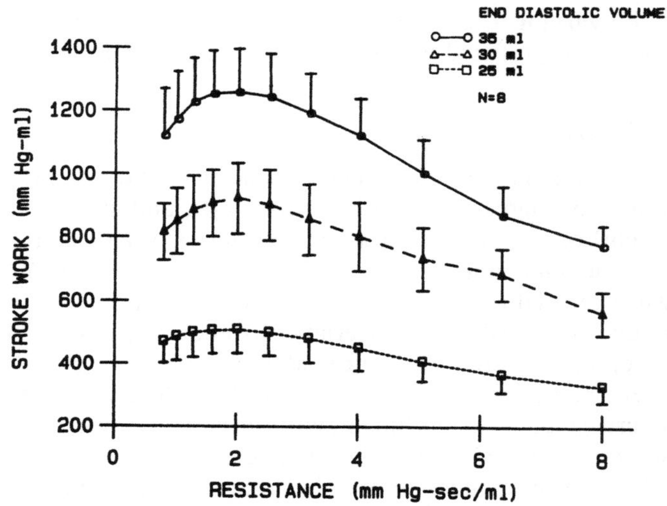

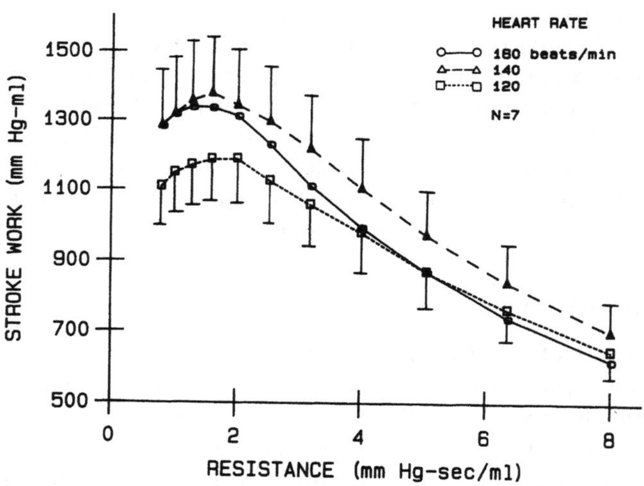

FIGURE 10.15. Effects of changes in arterial resistance on stroke work under various end-diastolic volumes *(top left)*, contractilities *(top right)*, heart rates *(bottom left)*, and compliance *(bottom right)*. Stroke work reached its peak value as arterial resistance decreased. Further reductions of arterial resistance decreased stroke work. (From Sunagawa, Maughan, and Sagawa 1985b; by permission of the American Heart Association, Inc.)

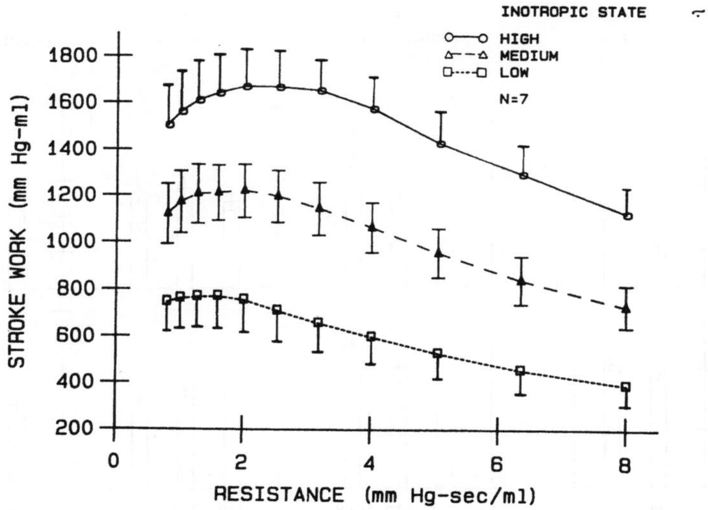

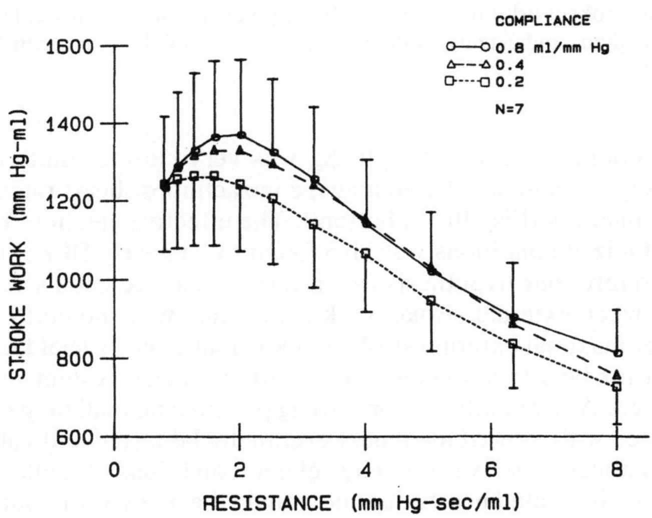

FIGURE 10.15. continued.

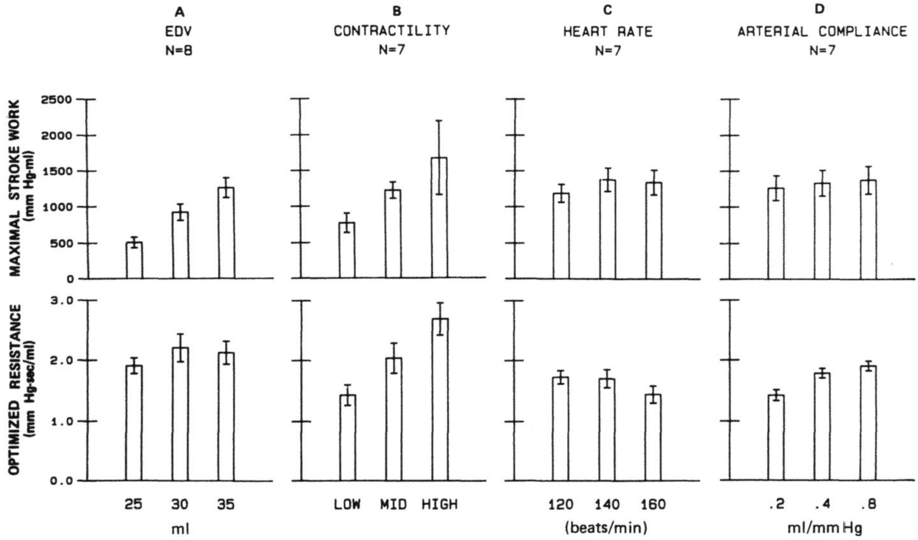

FIGURE 10.16. Effect of changes in end-diastolic volume (EDV) (A), contractility (B), heart rate (C), and arterial compliance (D) on maximal stroke work and optimal resistance. Increases in end-diastolic volume increased the maximal stroke work without changing the optimal arterial resistance. Increases in contractility increased the maximal stroke work with significant increases in the optimal arterial resistance. Changes in heart rate did not alter the maximal stroke work but tended to decrease the optimal arterial resistance. Alterations in arterial compliance did not change the optimal stroke work but increased the optimal resistance a little. (From Sunagawa, Maughan, and Sagawa 1985b; by permission of the American Heart Association, Inc.)

end of ejection. The ratio of E_a to E_{es} was very close to unity despite a wide variety of end-diastolic volumes, contractilities, heart rates, and arterial compliances (Fig. 10.17, bottom). The effective ejection fraction in all the optimized conditions was also found to be nearly 50% (Fig. 10.17, top). Therefore, our hypothesis that arterial elastance equals ventricular elastance when external stroke work is maximal was shown to be valid.

Whether maximal external stroke work is really the crucial factor to be optimized for best functioning of the cardiovascular system remains to be answered. Any hypothesis concerning physiologic matching of cardiac performance with arterial load must eventually be tested and validated in an intact animal under various physiologic conditions. Needless to say, our study with isolated hearts does not mimic the in vivo situation. Nevertheless, the information from the present study attests to the compatibility of our hypothesis with the basic mechanical property of the ventricular pump.

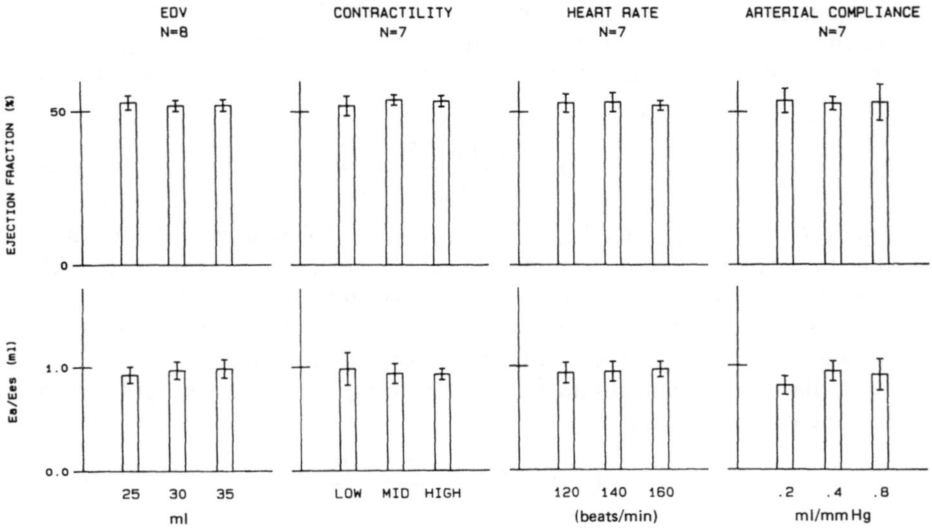

FIGURE 10.17. Effects of changes in end-diastolic volume (EDV) (A), contractility (B), heart rate (C), and arterial compliance (D) on ejection fraction and E_a/E_{es}. Ejection fraction is defined as the ratio of stroke volume to V_{ed}-V_o. As shown in the top panel, ejection fraction remains virtually constant at 50% under the optimal arterial load. The ratio of effective arterial stiffness (E_a) to ventricular end-systolic stiffness (E_{es}) remains constant at the matched condition irrespective of end-diastolic volume, contractility, heart rate and arterial compliance (From Sunagawa et al. 1985b by permission of the American Heart Association, Inc.)

Cardiac Output Surfaces

Mathematical Expression

To extend our model of ventricular/arterial coupling to describe the cardiac output surface, both end-diastolic volume (V_{ed}) and downstream pressure (P_d) in Equation 10.12 must be redefined in terms of left atrial pressure (P_{la}) and right atrial pressure (P_{ra}).

As a first approximation, we assume a linear end-diastolic pressure–volume relationship (EDPVR) for both right and left ventricles. With this approximation, end-diastolic volume is described as

$$V_{\text{ed}} = \frac{P_f}{E_{\text{ed}}} + V_u \tag{10.13}$$

where P_f = filling pressure
E_{ed} = linearized slope of the EDPVR,
V_u = volume axis intercept.

If we approximate the filling pressure for the left ventricle by P_{la} and downstream pressure by P_{ra}, substituting Equation 10.13 into Equation 10.12 yields an analytical expression of the systemic cardiac output surface as

$$\text{CO}_s = \frac{H_l P_{\text{la}} + P_{o,l} - P_{\text{ra}}}{T E_{\text{es},l} + R_{T,s}} \tag{10.14}$$

where $H_l = E_{\text{es},l} 1 E_{\text{ed},l}$ and $P_{o,l} = E_{\text{es},l}(V_{u,l} - V_{o,l})$. "$l$" in subscripts indicates that these parameters represent left ventricular properties. H_l is the left ventricular pressure gain and $P_{o,l}$ is the end-systolic pressure at the diastolic unstressed volume of the ventricle. $R_{T,s}$ denotes the systemic arterial resistance. T is the cycle length of the cardiac contraction.

Similarly, one can obtain an analytical expression of the pulmonary cardiac output surface as

$$\text{CO}_p = \frac{H_r P_{\text{ra}} + P_{o,r} - P_{\text{la}}}{T E_{\text{es},r} + R_{T,p}} \tag{10.15}$$

"r" in subscripts indicates that these parameters denote right ventricular properties. $R_{T,p}$ is the pulmonary arterial resistance.

Electrical Analog

Figure 10.18 shows an electrical circuit analogous to Equations 10.14 and 10.15. The circuit is similar to that shown in Figure 10.11 except that the source pressure is not a function of end-diastolic volume but of filling pressure. For the left ventricle, the source pressure is $H_l P_{la} + P_{o,l}$. The left ventricle behaves as an amplifier with a gain of H_l. From this analog circuit, it is obvious that the pumping performance of the ventricle relies on four major system properties: H_l, $P_{o,l}$, $E_{\text{es},l}$, and $R_{T,s}$.

The Frank-Starling Relationship

Nearly a century ago, Frank (1895) and subsequently Patterson, Piper, and Starling (1914) demonstrated that cardiac performance increases with an increase in preload. The so-called "Frank–Starling relationship" has occupied the attention of many physiologists and cardiologists for years because of its tremendous clinical and physiologic significance. Although it is generally accepted that the Frank–Starling relationship is determined by properties of cardiac muscle, the role of other factors, such as properties of cardiac chambers or the blood vessels, is not well known.

To elucidate determinants of this relationship between cardiac output and filling pressure, we make use of the analytically derived formula (Eq. 10.14) that expresses the relationship between systemic cardiac output and left atrial pressure ($\text{CO}_s - P_{\text{la}}$). We ignore the term p_{ra} because of its minor quantitative contribution to the relation. Equation 10.14, the slope

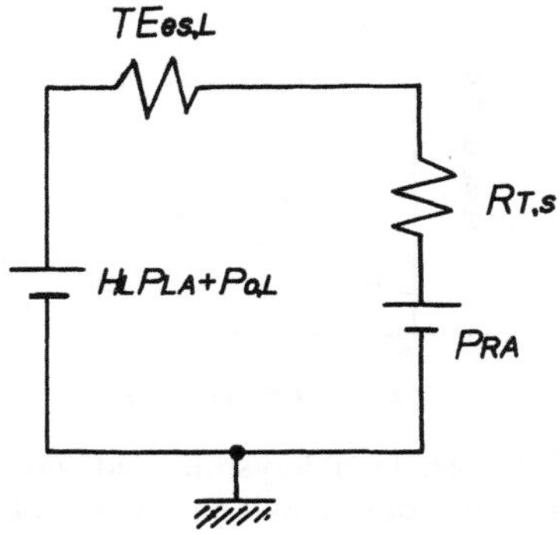

FIGURE 10.18. Electrical analog of the cardiac output surface. There are two pressure sources. One is ventricular source pressure and the other is downstream pressure P_{ra}. The amount of cardiac output is obtained by dividing the pressure gradient between the pressure sources $(H_l P_{la} + P_{o,l} - P_{ra})$ by the total resistance of the circuit $(TE_{es,L} + R_{T,s})$. The ventricle behaves as an analog amplifier with a gain of H_l and source resistance of $TE_{es,l}$. Since the H_l is the ratio of $E_{es,l}$ to $E_{ed,l}$, $E_{ed,l}$ is as important as $E_{es,l}$ in determining the source pressure. See text for definitions.

of the $CO_s - P_{la}$ relationship (S_s) is

$$S_s = \frac{E_{es,l}/E_{ed,l}}{TE_{es,l} + R_{T,s}} \tag{10.16}$$

the CO_s axis intercept $(CO_{s,o})$ is

$$CO_{s,o} = \frac{E_{es,l}(V_{u,l} - V_{o,l})}{TE_{es,l} + R_{T,s}} \tag{10.17}$$

and the P_{la} axis intercept $(P_{la,o})$ is derived as

$$P_{la,o} = -E_{ed,l}(V_{u,l} - V_{o,l}) \tag{10.18}$$

With an increase in $E_{es,l}$, a decrease in $R_{T,s}$, or a shortening in T (an increase in heart rate), the slope of the $CO_s - P_{la}$ relationship increases without changing the P_{la} axis intercept (Fig. 10.19, left). With an increase in $E_{ed,l}$, the slope of the $CO_s - P_{la}$ relationship decreases but the CO_s axis intercept remains unaltered (Fig. 10.19, middle). With a decrease in $V_{u,l}$ or an increase in $V_{o,l}$, which would occur under pathologic conditions such as regional ischemia (Sunagawa, Maughan, and Sagawa 1983), the $CO_s - P_{la}$ relationship shifts rightward without changing its slope (Fig. 10.19, right).

Since the dynamics of the $CO_s - P_{la}$ relationship are revealed predominantly by changes in slope, we sought to determine the effect on the slope of changes in various system properties. From Equation 10.16, doubling $E_{es,l}$, doubling heart rate (therefore halving T), or halving $R_{T,s}$ increases the slope by 33% relative to control, whereas halving $E_{es,l}$, halving heart

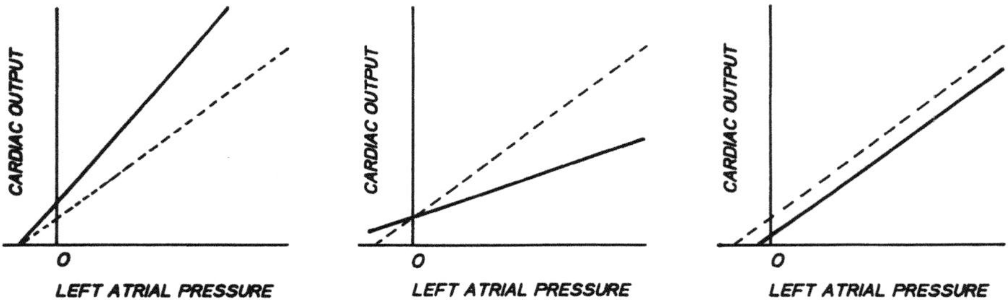

FIGURE 10.19. Effects of changes in end-systolic and end-diastolic elastance ($E_{es,l}$ and $E_{ed,l}$, respectively) and in volume axis intercepts of end-diastolic and end-systolic pressure volume relationship of the left ventricle ($V_{o,l}$ and $V_{u,l}$, respectively) and in total systemic resistance *($R_{T,s}$)* on the cardiac output versus left atrial pressure (CO_s-P_{la}) relationship. Dashed lines represent the control CO_s-P_{la} relationship. With an increase in contractility or heart rate or a decrease in arterial resistance, the slope of the CO_s-P_{la} relationship increases without the P_{la} axis intercept being altered *(left)*. With an increase in $E_{ed,l}$, the slope of the CO_s-P_{la} relationship decreases without the CO_s axis intercept being changed *(middle)*. *If the difference between* $V_{u,l}$ and $V_{o,l}$ decreases, the CO_s-P_{la} relationship shifts rightward *(right)*.

rate, or doubling $R_{T,s}$ decreases the slope by 33%. Therefore, relative sensitivity of all three system parameters are comparable over 400% changes in each parameter. This implies that decreases in the slope of the $CO_s - P_{la}$ relationship due to impaired ventricular systolic function can be restored by adjusting heart rate and/or arterial resistance.

On the contrary, doubling ventricular diastolic stiffness $E_{ed,l}$ decreases the slope by 50%, and halving it increases the slope by 100%. This is to say that the most important determinant of the slope of the $CO_s - P_{la}$ relationship is not systolic stiffness of the ventricle, heart rate, or arterial resistance, but diastolic ventricular stiffness. Unlike systolic ventricular properties, heart rate, or arterial resistance—all of which are closely regulated by neural, humoral, and metabolic control mechanisms—the ventricular diastolic property does not vary easily. However, once it is impaired, the resultant deterioration of pumping ability is serious, as for example in cardiac tamponade.

Venous Return Surface

Venous Return of Single Vascular System

To obtain a mathematical description of the venous return surface, we first derive an analytical equation representing the venous return properties of a single vascular system. We then extend the analysis to include two vascular systems, the systemic and pulmonary systems.

To be general, we model the vascular system using a distributed model.

Suppose that compliance and resistance are serially distributed in the systemic vascular system as shown in Figure 10.20. The compliance distribution $C(x)$ and pressure distribution $P(x)$ are expressed as a function of distance x from the venous outflow port. Stressed systemic blood volume V_s in the vascular system is

$$V_s = \int_0^L P(x)C(x)dx \tag{10.19}$$

where L represents the distance between the arterial and venous ports. If we denote the cumulative resistance over a distance x from the venous port by $R(x)$, the serial pressure distribution can be expressed as

$$P(x) = R(x)\mathrm{CO}_v + P_{\mathrm{ra}} \tag{10.20}$$

in which CO_v represents venous return. Substituting Equation 10.20 into Equation 10.19 yields

$$V_s = \mathrm{CO}_v \int_0^L C(x)R(x)dx + P_{\mathrm{ra}} \int_0^L C(x)dx \tag{10.21}$$

The first integral, which sums systemic compliance weighted by cumulative resistance, is equivalent to Guyton's resistance for systemic venous return R_{vs} multiplied by total systemic compliance C_s (Guyton 1963). Since the second integral equals C_s, Equation 10.21 can be rewritten as

$$V_s = \mathrm{CO}_v C_s R_{\mathrm{vs}} + P_{\mathrm{ra}} C_s \tag{10.22}$$

By the same token, stressed blood volume in the pulmonary vascular system V_p is

$$V_p = \mathrm{CO}_v C_p R_{\mathrm{vp}} + P_{\mathrm{la}} C_p \tag{10.23}$$

where C_p is total pulmonary compliance and R_{vp} the resistance for pulmonary venous return.

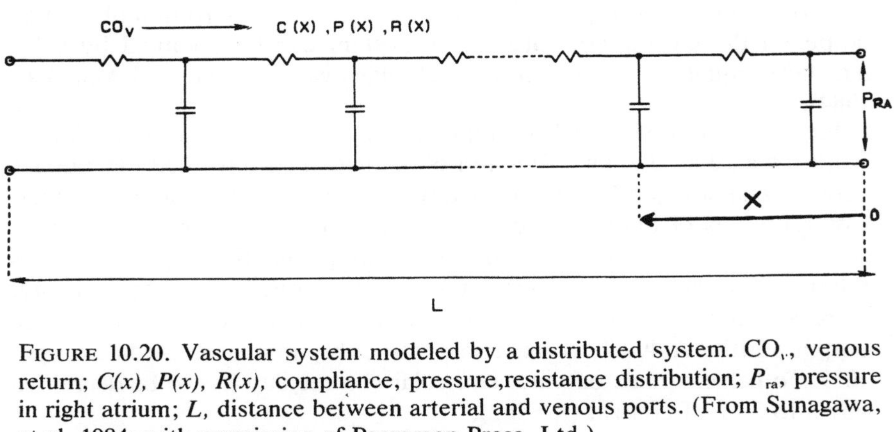

FIGURE 10.20. Vascular system modeled by a distributed system. CO_v, venous return; $C(x)$, $P(x)$, $R(x)$, compliance, pressure, resistance distribution; P_{ra}, pressure in right atrium; L, distance between arterial and venous ports. (From Sunagawa, et al. 1984; with permission of Pergamon Press, Ltd.)

Mathematical Analysis of the Venous Return Surface

The sum of the stressed blood volume in the systemic vascular system and pulmonary vascular system should be constant as long as total stressed blood volume V_T remains constant irrespective of its distribution. Thus, adding Equations 10.22 and 10.23 and rearranging yields

$$CO_v = \frac{V_T - (C_s P_{ra} + C_p P_{la})}{C_p R_{vp} + C_s R_{vs}} \tag{10.24}$$

which represents the venous return surface. The venous return axis intercept $CO_{v,max}$ is obtained by making both left and right atrial pressure zero as

$$CO_{v,max} = \frac{V_T}{C_p R_{vp} + C_s R_{vs}} \tag{10.25}$$

It is clear from Equation 10.25 that the maximal venous return is proportional to total stressed blood volume.

Through the inspection of Equation 10.24 one can obtain the P_{la} axis intercept as V_T/C_p and the P_{ra} axis intercept as V_T/C_s. The former represents the mean pulmonary filling pressure when the entire stressed blood volume is shifted exclusively to the pulmonary vascular system, and the latter represents the mean systemic filling pressure when the entire stressed blood volume is shifted exclusively to the systemic vascular system. The ratio of the slope of the venous return surface along the P_{la} axis to that along the P_{ra} axis equals the ratio of total pulmonary compliance to total systemic compliance (C_p/C_s).

Cardiac Output in the Closed Circulatory Loop

Since all three cardiac output surfaces are described by relatively simple analytical equations (Eqs. 10.14, 10.15, and 10.24), description of cardiac output in the closed-loop circulatory system can be obtained by solving these equations simultaneously (Sunagawa, Sagawa, and Maughan 1984).

Figure 10.21 shows cardiac output calculated by the developed framework for a 20-kg dog as a function of ventricular contractility (K) under various total stressed blood volumes, heart rates, and systemic resistances. Contractilities of both ventricles are simultaneously altered. As can be seen, cardiac output under the normal loading condition (open circles) is relatively insensitive to changes in ventricular contractility. Specifically, increases in contractility above the normal level ($K = 1$) increase cardiac output a little. A 400% increase in contractility increases cardiac output 10%. To halve cardiac output, ventricular contractility has to be decreased down to 20% of control ($K = 0.2$).

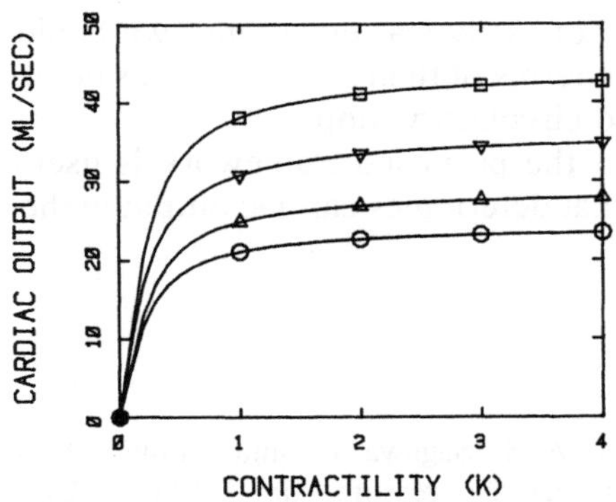

FIGURE 10.21. Analytical estimation of the relationship of cardiac output to contractility during changes in heart rate (triangles), arterial resistance (inverted triangles), and stressed blood volume (squares). Circles represent cardiac output under control loading conditions. (From Sunagawa, Sagawa, and Maughan 1984; with permission of Pergamon Press, Ltd.)

Doubling the heart rate (open triangles in Fig. 10.21) from 80 to 160 beats/min increases cardiac output by 20% at each level of contractility. Halving the systemic resistance increases cardiac output about 50% regardless of ventricular contractility (inverse triangles in Fig. 10.21). Doubling the total stressed blood volume doubles cardiac output irrespective of contractility. Under all these different loading conditions, cardiac output at control contractility is about 90% of the maximal possible value.

These analyses indicate that cardiac output in the closed circulatory system is determined predominantly by extracardiac factors, such as total stressed blood volume and arterial resistance, rather than by ventricular contractility. Ventricular contractility would play a significant role in determining cardiac output only when cardiac output decreases markedly.

Summary

The objective of this investigation was to develop a framework to predict cardiac output of the closed circulatory system on the basis of system properties. We used the end-systolic pressure–volume relationship of ventricles and arterial systems to define their mechanical properties. This definition of system properties made it possible to estimate analytically the cardiac output of the coupled ventricular/arterial system. Thanks to the mathematical simplicity of the definition, we could derive a set of analytical equations that describe systemic cardiac output surface, pulmonary cardiac output surface, and venous return surface. Because of the inherent nature of analytical equations, we could interpret various system properties in terms of their significance as determinants of cardiac output in the closed circulatory loop.

We conclude that the proposed framework is useful in analyzing the global mechanism that determines cardiac output in the closed circulatory system.

Bibliography

Brunner, M. I., Greene, A. S., Sagawa, K., and Shoukas, A. A. 1983. Determinants of systemic zero-flow arterial pressure. *Am. J. Physiol.* 245 (*Heart Circ. Physiol.* 14):H453–H460.

Elzinga, G., Piene, H., and DeJong, J. P. 1980. Left and right ventricular pump function and consequences of having two pumps in one heart: a study on the isolated cat heart. *Circ. Res.* 46:564–574.

Frank, O. 1895. Zur Dynamik des Herzmuskels. Z. Biol. 32:370–447.

Guyton, A. C. 1963. Venous return. In Hamilton, W. F., ed. *Handbook of Physiology,* Sec. 2, Vol. 2., *Circulation*. Washington, DC: American Physiological Society, pp. 1099–1133.

Guyton, A. C., Jones, C. E., and Coleman, T. G. 1973. *Circulatory Physiology: Cardiac Output and its Regulation*. Philadelphia: Saunders.

Herndon, C. W., and Sagawa, K. 1969. Combined effects of aortic and right atrial pressures on aortic flow. *Am. J. Physiol.* 201:102–108.

Maughan, W. L., and Sunagawa, K. 1984. Factors affecting the end-systolic pressure–volume relationship. *Fed. Proc.* 43:2408–2410.

Maughan, W. L., Sunagawa, K., and Sagawa, K. 1984. Effect of arterial impedance changes on the end-systolic pressure–volume relationship. *Circ. Res.* 54:595–602.

Milnor, W. R. 1975. Arterial impedance as ventricular afterload. *Circ. Res.* 36:566–570.

Milnor, W. R. 1982. *Hemodynamics*. Baltimore: Williams & Wilkins, pp. 157–191.

Nichols, W. W., Conti, C. R., Walker, W. E., and Milnor, W. R. 1977. Input impedance of the systemic circulation in man. *Circ. Res.* 40:451–458.

Noble, M. I. M. 1979. Left ventricular load, arterial impedance, and their interrelationship. *Cardiovasc. Res.* 13:183–198.

O'Rourke, M. F. 1967. Steady and pulsatile energy losses in the systemic circulation under normal conditions and in simulated arterial disease. *Cardiovasc. Res.* 1:313–326.

Patterson, S. W., Piper, H., and Starling, E. H. 1914. The regulation of the heart beat. *J. Physiol. (Lond.)* 48:465–513.

Piene, H. 1980. Interaction between the right heart ventricle and its arterial load: a quantitative solution. *Am. J. Physiol.* 238 (*Heart Circ. Physiol.* 7):H932–H937.

Piene, H., and Sund, T. 1982a. Calculation of flow pressure curves from the ventricular pressure–volume–time relationship and load impedance. In Kenner, T., Busse, R., and Hinghofer-Salkay, H., eds. *Cardiovascular System Dynamics: Models and Measurements*. New York: Plenum, pp. 47–56.

Piene, H., and Sund, T. 1982b. Does normal pulmonary impedance constitute the optimum load for the right ventricle? *Am. J. Physiol.* 242 (*Heart Circ. Physiol.* 11):H154–H160.

Pouleur, H., Covell, J. W., and Ross, J. Jr. 1979. Effects of alterations in aortic input impedance on the force–velocity–length relationship in the intact canine heart. *Circ. Res.* 45:126–135.

Sagawa, K. 1967. Analysis of the ventricular pumping capacity as a function of input and output pressure loads. In Reeve, E. B., and Guyton, A. C., eds. *Physical Basis of Circulatory Transport: Regulation and Exchange*. Philadelphia: Saunders, pp. 143–151.

Sagawa, K. 1981. The end-systolic pressure–volume relation of the ventricle: definition, modifications and clinical use. *Circulation* 63:1223–1227.

Sagawa, K., and Eisner, A. 1975. Static pressure–flow relation in the total systemic vascular bed of the dog and its modifications by the baroreceptor reflex. *Circ. Res.* 36:406–413.

Suga, H., and Sagawa, K. 1974. Instantaneous pressure–volume relationships and their ratio in the excised, supported canine left ventricle. *Circ. Res.* 35:117–125.

Suga, H., Sagawa, K., and Shoukas, A. A. 1973. Load independence of the istantaneous pressure–volume ratio of the canine left ventricle and effects of epinephrine and heart rate on the ratio. *Circ. Res.* 32:314–332.

Sunagawa, K., Burkhoff, D., Lim, K., and Sagawa, K. 1982. Impedance loading servo-pump system for excised canine ventricle. *Am. J. Physiol.* 243 (*Heart Circ. Physiol.* 12):H346–H350.

Sunagawa, K., Maughan, W. L., Burkhoff, D., and Sagawa, K. 1983. Left ventricular interaction with arterial load studied in isolated canine ventricle. *Am. J. Physiol.* 245(*Heart Circ. Physiol.* 14):H773–H780).

Sunagawa, K., Maughan, W. L., and Sagawa, K. 1982. Influence of heart rate on the sensitivity of stroke volume to afterload compliance and resistance (abstr). *Circulation* 66 (Suppl. II):304.

Sunagawa, K., Maughan, W. L., and Sagawa, K. 1983. Effect of regional ischemia on the end-systolic pressure–volume relationship. *Circ. Res.* 52:170–178.

Sunagawa, K., Maughan, W. L., and Sagawa K. 1985a. Stroke volume effect of changing arterial input impedance over selected frequency ranges. *Am. J. Physiol.* 248 (*Heart Circ. Physiol.* 17):H477–H484.

Sunagawa, K., Maughan, W. L., and Sagawa, K. 1985b. Optimal arterial resistance for the maximal stroke work studied in isolated canine left ventricle. *Circ. Res.* 56:586–595.

Sunagawa, K., Sagawa, K., and Maughan, W. L. 1984. Ventricular interaction with the loading system. *Ann. Biomed. Eng.* 12:163–189.

Urschel, C. W., Covell, J. W., Sonnenblick, E. H., Ross, J. Jr., and Braunwald, E. 1968. Effects of decreased aortic compliance on performance of the left ventricle. *Am. J. Physiol.* 214:298–304.

Westerhof, N., Elzinga, G., and Sipkema, P. 1971. Artificial arterial system for pumping hearts. *J. Appl. Physiol.* 31:776–781.

Wilcken, D. E. L., Charlier, A. A., Hoffman, J. I. E., and Guz, A. 1964. Effects of alterations in aortic impedance on the performance of the ventricle. *Circ. Res.* 14:283–293.

11 Left Ventricular Systolic Resistance and Its Role in Coupling the Ventricle to the Arterial Circulation

SANJEEV G. SHROFF, JOSEPH S. JANICKI, AND KARL T. WEBER

Introduction

The overall performance of the left ventricle coupled to the arterial circulation is dependent on the intrinsic mechanical properties of the individual subsystems and their mutual interaction. The myocardium of the left ventricle is a viscoelastic material whose mechanical properties are reflected in the behavior of the ventricular chamber (i.e., the relationships among chamber pressure, volume, and flow). The relationship between instantaneous ventricular pressure and volume, analogous to the force–length relation of the cardiac muscle, has been described in terms of a time-varying volume elastance (Sagawa 1978; Suga, Sagawa, and Shoukas 1973). The role of ventricular elastance in determining the overall performance (e.g., stroke volume) of the left ventricle has been clearly established (Sunagawa, Sagawa, and Maughan 1984) and is presented in Chapter 10. Further, it has recently been shown that the left ventricle exhibits a viscouslike behavior that can be described in terms of the ventricular pressure–flow relation and, phenomenologically, can be represented as ventricular resistance (Hunter et al. 1979; Hunter et al. 1983; Ringo et al. 1982; Shroff, Janicki, and Weber 1983, 1985; Suga, Sagawa, and Demar 1980; Vaartjes, van Alste, and Boom 1982). This chapter focuses on the description of ventricular resistance and examines its role in determining the overall performance of the left ventricle coupled to the arterial circulation. The discussion is organized in four sections: (1) definition and experimental evidence of ventricular resistance, (2) quantitative characterization of ventricular resistance, (3) role of ventricular resistance in determining ventricular performance (a simulation study), and (4) ventricular resistance in cardiovascular disease.

Definition and Experimental Evidence of Ventricular Resistance

Instantaneous ventricular pressure, $P(t)$, is assumed to be a function of four independent variables: time, t, during systole measured from the onset of depolarization (i.e., the R wave of the ECG); chamber volume $V(t)$; ventricular flow $\dot{V}(t)$ (i.e., the rate of change in chamber volume); and myocardial contractile state, CS (Hunter et al. 1979; Shroff, Janicki, and Weber 1983, 1985). Accordingly, one can represent $P(t)$ in the following functional form:

$$P(t) = f\,[t,\ V(t), \dot{V}(t), \mathrm{CS}] \tag{11.1}$$

An inertial component, describing the dependence of pressure on volume acceleration, has been omitted, because previous studies (Shroff, Janicki, and Weber 1983; Tallarida, Ryssy, and Longname 1970) have shown it to be negligible in comparison with the elastic and resistive components. For a given t, $V(t)$, and CS, ventricular resistance R is defined as the partial derivative of $P(t)$ with respect to $\dot{V}(t)$:

$$R\,\big|_{t,V(t),\mathrm{CS}} = \left.\frac{\partial P(t)}{\partial \dot{V}(t)}\right|_{t,V(t),\mathrm{CS}} \tag{11.2}$$

This definition emphasizes the fact that, during systole, the left ventricle represents a time-varying system (Hunter and Noordergraaf 1976) and therefore its mechanical properties should be described in the time domain as opposed to the frequency domain. Furthermore, by varying flow and observing the changes in ventricular pressure for a fixed combination of t, $V(t)$, and CS, one can assess the independent effects of flow on pressure.

Utilizing the flow-pulse technique, Hunter and colleagues (1979) provided the experimental evidence of ventricular resistance in the time domain. Using a servo-regulated, isolated canine heart preparation (Janicki et al. 1974) in which the left ventricle was contracting isovolumetrically, a flow pulse was generated by creating small changes in volume (±3 ml) within a short period of time (30–50 msec). The pulse response, $\Delta P/\Delta V$, was defined as the instantaneous pressure difference between the perturbed beat and the control beat (i.e., isovolumetric contraction). Figure 11.1 illustrates three pulse responses resulting from flow pulses during early, middle, and late systole. Since ventricular resistance contributes to the measured pressure difference only when there is flow, restricting the duration of flow to a narrow pulse permits the separation of resistive from elastic (i.e., proportional to volume) properties. From the pulse responses

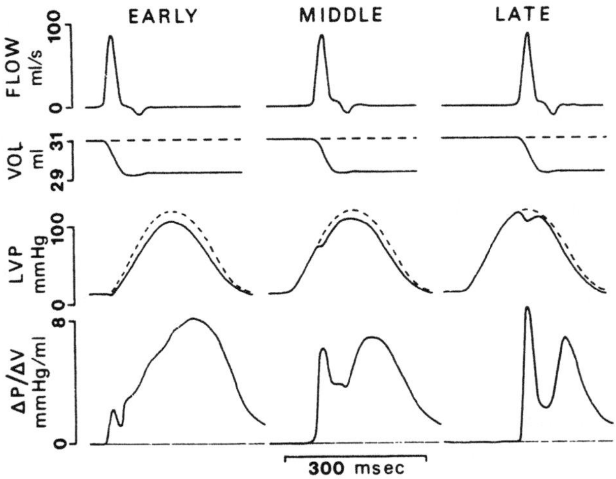

FIGURE 11.1. Examples of the experimentally imposed flow and volume perturbations and the corresponding left ventricular pressure (LVP) responses. The dashed lines show control LVP when chamber volume (VOL) was held constant. The flow–pulse response ($\Delta P/\Delta V$) was calculated by taking the instantaneous difference in pressure produced by a volume step (i.e., the difference between solid and dashed lines in LVP tracings) and dividing it by the difference in volume (ΔV) between the two levels. Steps were imposed at early, middle, and late phases of systole. (From Hunter et al. 1983; by permission of the American Heart Association, Inc.)

presented in Figure 11.1, the resistive effect (i.e., pressure difference in phase with flow) is clearly evident, especially during mid and late systole.

We used the flow-clamp technique (Vaartjes, van Alste, and Boom 1982) to provide additional experimental evidence of ventricular resistance (Shroff, Janicki, and Weber 1985). Figure 11.2 illustrates a typical set of pressure–volume data obtained during a flow-clamp study utilizing the isolated, servo-regulated canine heart preparation. The broken lines represent data corresponding to an isovolumetric beat when ventricular volume was held constant at a specified volume, V_s (20 ml in this example). The remaining three contractions began with successively higher end-diastolic volumes (24, 28, and 32 ml). Beginning at time t_1, ventricular volume was reduced at a constant and successively increasing rate such that at time t_s, all four contractions had the same ventricular volume, V_s. It is evident from Figure 11.2 that even though t_s and V_s were identical, ventricular pressure at t_s and V_s declined progressively as ventricular outflow was increased. This clearly establishes ventricular flow as an independent determinant of ventricular pressure.

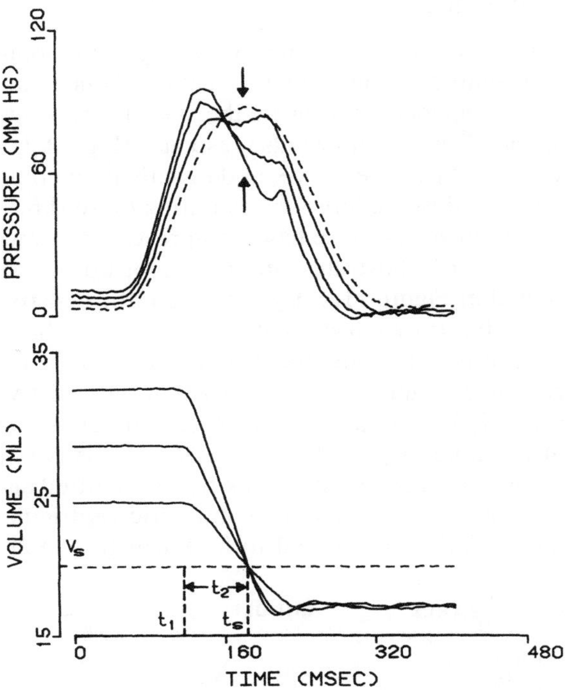

FIGURE 11.2. Typical set of pressure and volume data obtained during a flow-clamp experiment. V_s and t_s represent the specified volume and time, respectively. The volume change was initiated at time t_1, while t_2 was the time from the onset of volume change to the specified time (t_s). Even for a matched time (t_s) and volume (V_s), left ventricular pressure decreased as the flow was increased, indicating that flow was an independent determinant of pressure. (From Shroff, Janicki, and Weber 1985)

Quantitative Characterization of Ventricular Resistance

Two methods for quantitative characterization of ventricular resistance are presented here: the flow-pulse and flow-clamp techniques. The flow-pulse technique represents a general analysis that accounts for the three major components of left ventricular systolic mechanical behavior: (1) elastic, (2) resistive, and (3) deactivation. The flow-clamp technique, on the other hand, provides a better insight into the pressure–flow relation of the ventricle. Results from these two techniques have been found to be in close agreement (vide infra).

Flow-Pulse Technique

The experimental procedure to obtain the flow-pulse response of the left ventricle was presented in the previous section. Hunter and colleagues (1983) utilized this technique to calculate the systolic mechanical properties of the left ventricle. The total flow-pulse response (Fig. 11.1) was assumed to be the sum of the elastic, resistive, and deactivation components. The detailed computational procedure for separating these three components from the total flow-pulse response has been presented elsewhere (Hunter et al. 1983). Figure 11.3 illustrates the decomposition of the flow-pulse response presented in Figure 11.2. By definition, the resistive component exists only during the flow pulse. Once the flow pulse has ceased, only the elastic and deactivation components contribute to the observed pressure difference. From Figure 11.3 it is clear that the resistive component increases in parallel with ventricular pressure. Therefore, ventricular resistance (R) at the time of peak flow (i.e., R = resistive component of flow-pulse response $\times$ pulse volume/peak flow) is plotted against the corresponding (i.e., for the same time) isovolumetric ventricular pressure in Figure 11.4. This relation was found to be linear ($r > .92$) for all cases

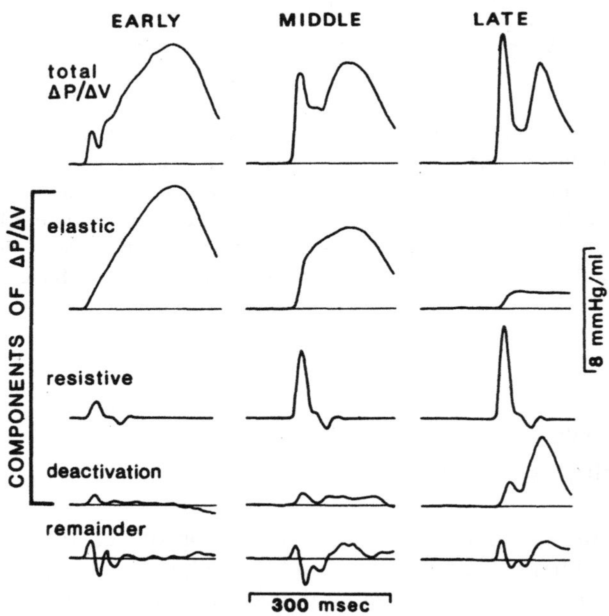

FIGURE 11.3. In each column the total flow-pulse response ($\Delta P/\Delta V$) corresponding to the data shown in Figure 11.1 has been resolved into its elastic, resistive, and deactivation components. As the timing of the flow pulse was progressively delayed from early *(left)* to late *(right)* systole, the resistive component increased in parallel with the left ventricular pressure. (From Hunter et al. 1983; by permission of the American Heart Association, Inc.)

(Hunter et al. 1983), with R increasing at an average rate of 0.107 mm Hg sec/ml per 100 mm Hg (range for all hearts 0.076 to 0.134). The intercept of R at zero pressure was statistically not different from zero (range for all hearts -0.013 to $+0.011$ mm Hg sec/ml). Changes in contractile state did not significantly ($P>0.1$) alter the relationship between R and isovolumetric pressure (Fig. 11.4 *left*). Changing ventricular volume (Fig. 11.4 *right*) slightly altered this relationship such that the slope was reduced as the end-diastolic volume was raised. However, this reduction in slope was statistically significant only for a large (> 13 ml) change in end-diastolic volume. Thus the results from the flow-pulse technique indicate that ventricular resistance in systole can be characterized as a linear function of the isovolumetric pressure and that this relationship is essentially independent of changes in ventricular volume and/or contractile state.

Flow-Clamp Technique

The flow-clamp technique (Vaartjes, van Alste, and Boom 1982) was used to quantify ventricular resistance in an isolated, servo-controlled canine heart preparation (Shroff, Janicki, and Weber 1985). This technique consists of producing various amounts of ventricular flow at a specified time (t_s) and volume (V_s) in systole and for a given contractile state (CS) (Fig. 11.2). By plotting the observed ventricular pressure against flow under

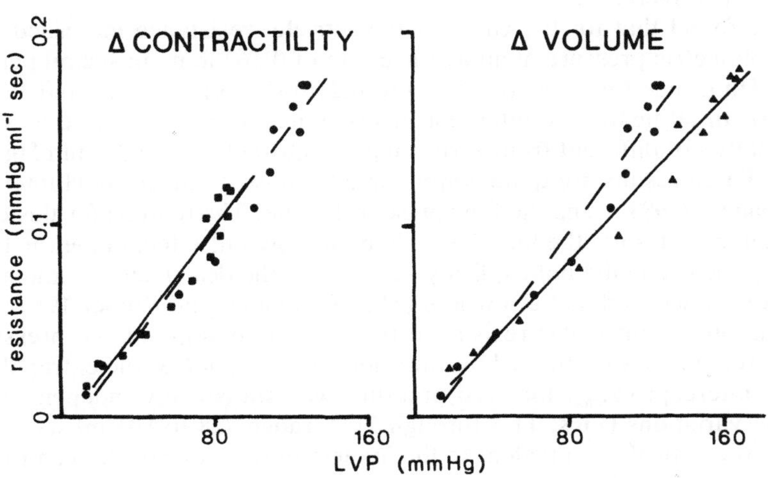

FIGURE 11.4. Influence of changes in contractility or volume on the linear relation between left ventricular isovolumetric pressure (LVP) and resistance. The solid or dashed line is the regression line for each condition. *Left,* beginning from a control contractile state (■), dobutamine was infused directly into the coronary perfusion line of the isolated heart at 6.6 μg/min (●); volume was constant at 36 ml. *Right,* chamber volume was increased from 36 ml (●) to 46 ml (▲), while dobutamine infusion was maintained at 6.6 μg/min. (From Hunter et al. 1983; by permission of the American Heart Association, Inc.)

these circumstances, one can obtain a ventricular pressure–flow relation for a specific combination of t_s, V_s, and CS. Finally, t_s, V_s, and CS can be varied in a systematic manner to evaluate the effects of each of these parameters on the ventricular pressure–flow relation. Figures 11.5 through 11.7 illustrate the individual effects of t_s, V_s, and CS, respectively, on the ventricular pressure–flow relation. It should be noted that for each illustration, only one parameter was changed; the other two were kept constant. Furthermore, for the sake of clarity, only four of eight data points are shown for each pressure–flow relation. However, the regression line was fitted to all eight points in each condition. Ventricular pressure–flow relations were always linear ($r>.97$) within the range of flows examined (0–250 ml/sec). The negative slope of this relation is defined as ventricular resistance. It was noted that ventricular resistance increased with increments in time of observation, t_s (Fig. 11.5); with larger chamber volume, V_s, at the time of observation (Fig. 11.6); and with augmentation of CS (Fig. 11.7). Since ventricular isovolumetric pressure depends on t_s, V_s, and CS in a similar manner, ventricular resistance was plotted against the corresponding isovolumetric pressure (i.e., pressure at zero flow in Figs. 11.5 through 11.7) for 16 combinations of t_s, V_s, and CS obtained in one experiment (Fig. 11.8). Data presented in Figures 11.5 through 11.7 are included in Figure 11.8. The slope and intercept values for the linear regression line ($r=.987$) were 0.00145 sec/ml and -0.0105 mm Hg sec/ml, respectively.

We found that for the canine left ventricle, resistance increased with isovolumetric pressure at an average rate of 0.162 mm Hg sec/ml per 100 mm Hg (range for all hearts 0.110 to 0.210) (Shroff, Janicki, and Weber 1985). In all hearts the intercept of resistance at zero pressure was statistically not different from zero (range -0.0105 to $+0.0112$ mm Hg sec/ml). These results are quite comparable to those obtained by Hunter and colleagues (1983) using the flow-pulse technique. The reasons for this close agreement, despite the fact that unlike the flow-pulse technique the flow-clamp technique did not explicitly account for the deactivation component, are discussed in detail elsewhere (Shroff, Janicki, and Weber 1985).

Although ventricular resistance (i.e., negative slope of the pressure–flow relations) was affected by variations in t_s, V_s, or CS, the extrapolated flow intercept ($\dot{V}_{max}$) for zero pressure was statistically independent of these variations (Figs. 11.5 through 11.7, range 728 to 801 ml/sec). $\dot{V}_{max}$ is mathematically equivalent to the inverse of the slope of the ventricular resistance–isovolumetric pressure relationship (Fig. 11.8) because the intercept of this relationship is zero (Shroff, Janicki, and Weber 1985). The results indicate that augmentation of contractile state with dobutamine infusion does not alter $\dot{V}_{max}$. This is consistent with the concept that catecholamine stimulation improves contractility primarily by altering excitation-contraction coupling (Stull and Mayer 1979) and not by increasing the rate of the individual contractile processes. Therefore we feel that the rate-limiting properties of the contractile apparatus are causally related to the observed resistance–isovolumetric pressure relationship and $\dot{V}_{max}$.

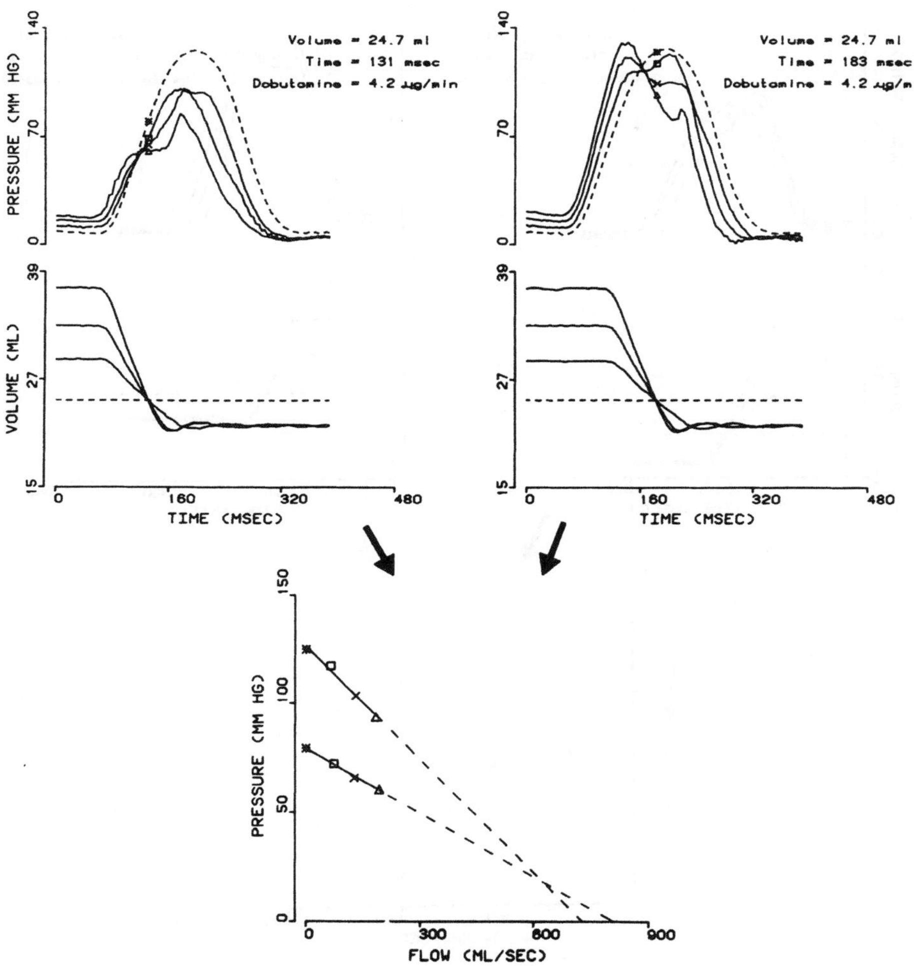

FIGURE 11.5. *Top,* ventricular pressure and volume data when different flows were produced at a fixed specified volume (24.7 ml) and contractile state (dobutamine infusion 4.2 μg/min) and specified time was varied *top left,* 131 msec; *top right,* 183 msec). *Bottom,* corresponding pressure–flow relations. Ventricular resistance (i.e., negative slope of the pressure–flow relation) increased with the increment in specified time without a significant change in the zero pressure flow intercept. (From Shroff, Janicki, and Weber 1985)

This hypothesis is further supported by the results of Daniels and coworkers (1984) regarding sarcomere force–velocity relations in isolated trabeculae carneae from the rat right ventricle. They found that under physiologic conditions, the maximum (i.e., unloaded) velocity of sarcomere shortening was independent of sarcomere length, time in systole, and Ca^{++} concentrations. It should be emphasized, however, that for the intact heart, other factors such as extracellular components (e.g., collagen), shearing

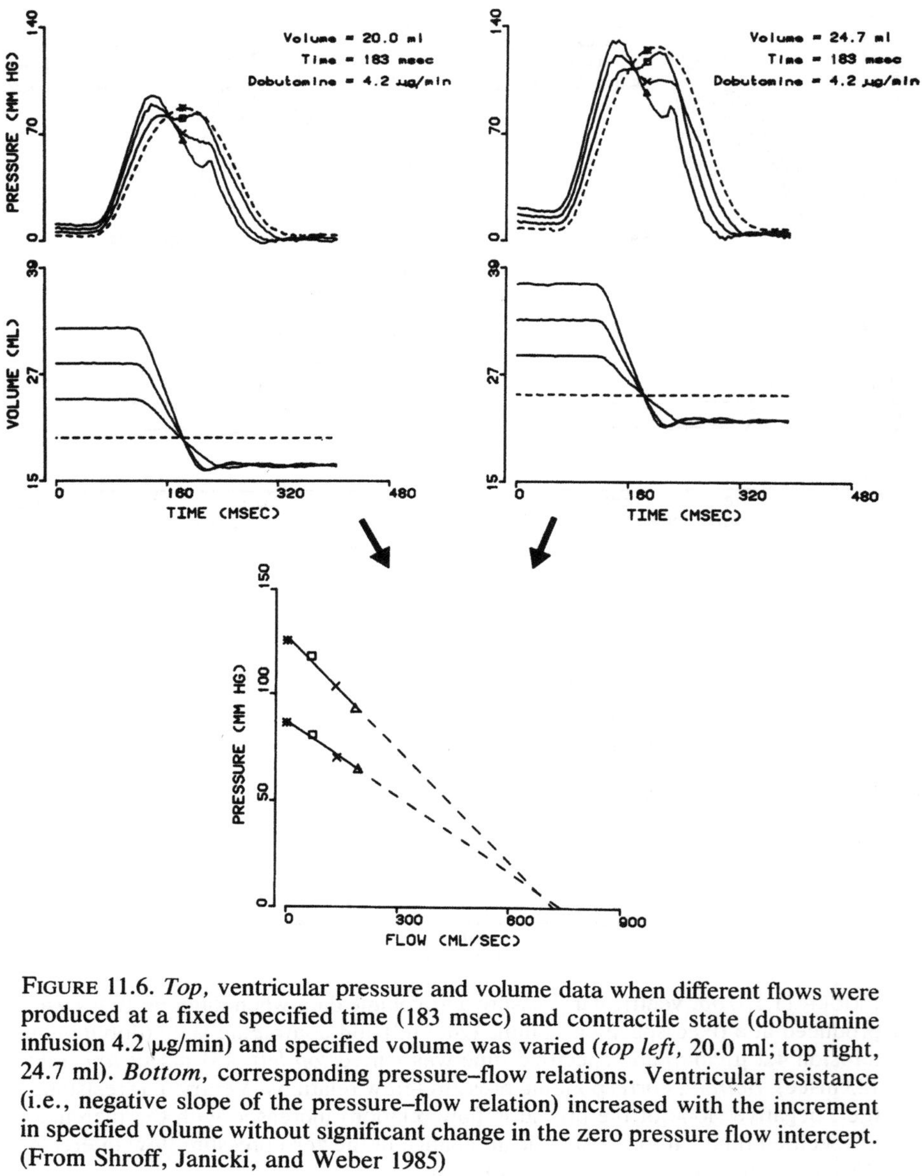

FIGURE 11.6. *Top,* ventricular pressure and volume data when different flows were produced at a fixed specified time (183 msec) and contractile state (dobutamine infusion 4.2 μg/min) and specified volume was varied (*top left,* 20.0 ml; top right, 24.7 ml). *Bottom,* corresponding pressure–flow relations. Ventricular resistance (i.e., negative slope of the pressure–flow relation) increased with the increment in specified volume without significant change in the zero pressure flow intercept. (From Shroff, Janicki, and Weber 1985)

forces between myocardial fibers, and geometric deformation may also contribute to the observed resistive behavior (Shroff, Janicki, and Weber 1985). The relative contribution of each of these factors is presently unknown.

It should be noted that large differences in ventricular chamber volume (e.g., ventricular dilatation in pathologic states or volume differences

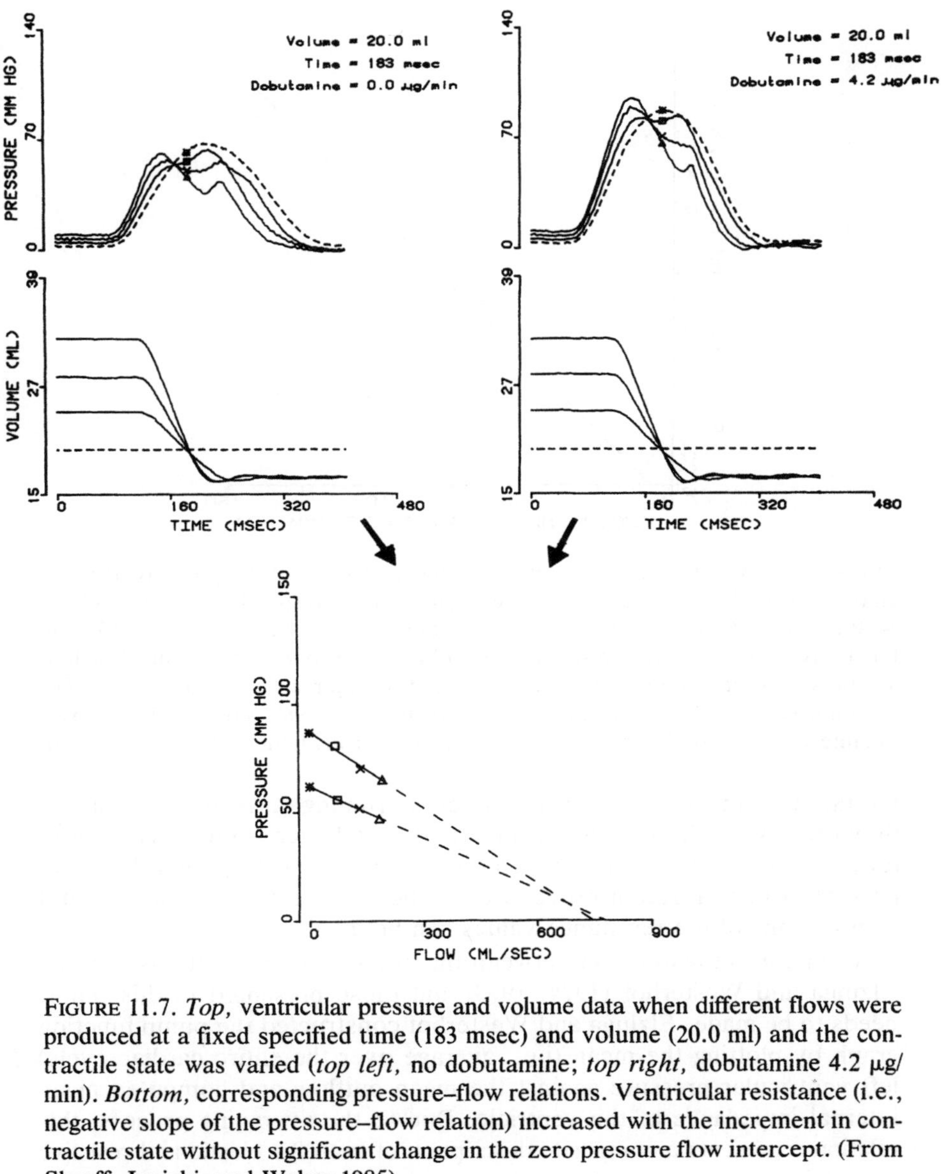

FIGURE 11.7. *Top,* ventricular pressure and volume data when different flows were produced at a fixed specified time (183 msec) and volume (20.0 ml) and the contractile state was varied (*top left,* no dobutamine; *top right,* dobutamine 4.2 μg/min). *Bottom,* corresponding pressure–flow relations. Ventricular resistance (i.e., negative slope of the pressure–flow relation) increased with the increment in contractile state without significant change in the zero pressure flow intercept. (From Shroff, Janicki, and Weber 1985).

across species) could alter the magnitude of ventricular resistance: the higher the chamber volume the lower the resistance. For example, the slope of the left ventricular resistance–isovolumetric pressure relation in dogs is approximately 0.0015 sec/ml (Campbell et al. 1984; Hunter et al. 1983; Shroff, Janicki, and Weber 1983, 1985), while it is 0.0426 sec/ml in rabbits (Vaartjes, Van Alste, and Boom 1982). This difference is due to

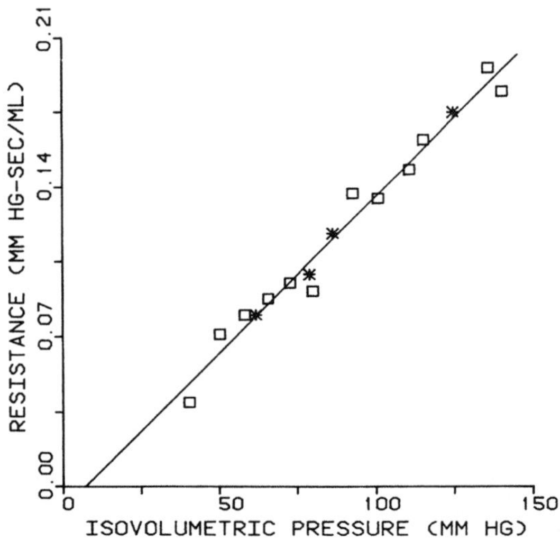

FIGURE 11.8. Ventricular resistance (i.e., negative slope of the pressure–flow relations in Figs. 11.5 through 11.7) plotted against the corresponding (i.e., for volume $= V_s$, time $= t_s$, and same contractile state) isovolumetric pressure for 16 combinations of V_s, t_s, and contractile state obtained in one experiment. The four symbols (*) correspond to the data presented in Figures 11.5 through 11.7. The solid line corresponds to the linear regression line ($r = .987$, slope $= 0.00145$ sec/ml, intercept $= -0.0105$ mm Hg sec/ml). (From Shroff, Janicki, and Weber 1985).

the fact that the same velocity of fiber shortening corresponds to larger flow rates when the ventricle contracts from a larger volume. Therefore, for those circumstances in which large differences in chamber volume are present, a normalization procedure will be necessary before meaningful comparisons of the resistance values can be made.

An important distinction between the pressure–flow relations used by Elzinga and Westerhof (1974, 1978) and those presented in this report needs to be made. Elzinga and Westerhof constructed the pump function graph by plotting the mean (i.e., average over the entire cardiac cycle) left ventricular pressure against the mean outflow and computed mean internal impedance of the ventricle. Due to the averaging process, this relation represents a composite description of both the elastic and resistive properties of the left ventricle. In contrast, the instantaneous pressure–flow relations and resistance presented in this report exclusively describe the flow-dependent properties of the left ventricle.

From the preceding discussion it is clear that ventricular resistance reduces chamber pressure compared with the value that would have existed for zero flow condition. We have shown (Shroff, Janicki, and Weber 1985) that for the canine left ventricle, this reduction in pressure (or resistive component) can be as high as 15% to 20% at the time of peak flow. In addition, the effect of ventricular resistive behavior on the end-systolic pressure–volume relation is of interest because this relation is currently used as an approximation of the peak isovolumetric pressure–volume relation which, in turn; is a sensitive indicator of ventricular contractile state (Sagawa et al. 1977; Suga, Sagawa, and Shoukas 1973; Weber, Janicki, and Hefner 1976). Results from the canine heart experiments (Shroff, Janicki, and Weber 1985) indicate that for those conditions wherein substantial (> 35 ml/sec) end-systolic flow exists (e.g., aortic and/or mitral regurgitation), the end-systolic pressure–volume relation does not approximate the desired peak isovolumetric pressure–volume relation, and therefore caution must be exercised when using the end-systolic relation as an index of contractile state under these conditions.

Role of Ventricular Resistance in Determining Ventricular Performance

To this point we have indicated that the magnitude of canine left ventricular resistance is about 0.15 mm Hg sec/ml at 100 mm Hg. In contrast, the resting vascular resistance of the arterial circulation in normal dogs is 2.5 to 3.3 mm Hg sec/ml. Therefore it would appear that ventricular resistance may not influence the overall ventricular performance (e.g., stroke volume) significantly. One way to examine the effects of ventricular resistance on both the dynamic and static components of ventricular performance is to perform a computer simulation study (Shroff, Janicki, and Weber 1985). The electrical equivalent network representing the left ventricle coupled to its arterial load is depicted in Figure 11.9. The left ventricle and the

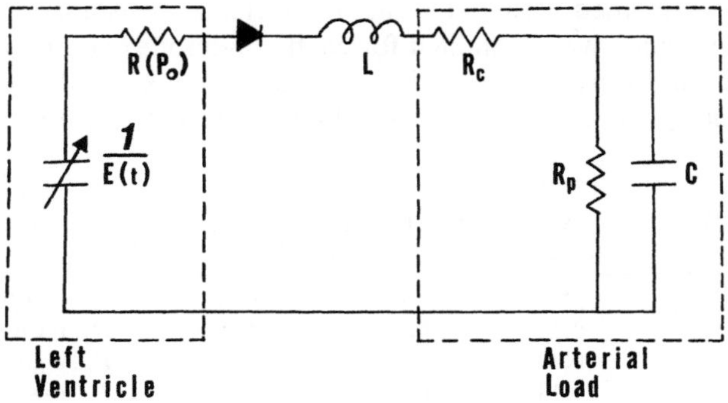

FIGURE 11.9. Model of the left ventricle coupled to the arterial circulation used in the simulation study discussed in the text. The left ventricle was represented by time-varying elastance, $E(t)$, in series with a pressure-dependent resistance, $R(P_o)$. The arterial load consisted of R_p (peripheral resistance), C (lumped arterial compliance), and R_c (characteristic impedance). L corresponded to the inertia of the blood column in the outflow tract and ascending aorta, while the diode simulated the aortic valve. (From Shroff, Janicki, and Weber 1985)

arterial load were represented by a time-varying elastance, $E(t)$, in series with a pressure-dependent resistance, $R(P_o)$, (Campbell et al. 1984; Hunter et al. 1983; Shroff, Janicki, and Weber 1985) and a three-element modified Windkessel (Westerhof, Elzinga, and Sipkema 1971), respectively. The three elements of the Windkessel corresponded to peripheral resistance, R_p, lumped arterial compliance, C, and characteristic impedance, R_c. The fluid inertia between the ventricular cavity and ascending aorta was represented by L, while the diode simulated the aortic valve.

Values for $E(t)$ and $R(P_o)$ were chosen to correspond to those found in the normal canine left ventricle (Hunter et al. 1983; Shroff, Janicki, and Weber 1983; Suga, Sagawa, and Shoukas 1973). $E(t)$ was represented by a third-order polynomial in time (Shroff, Janicki, and Weber 1983) and $R(P_o)$ as a linear function of isovolumetric pressure, P_o (Hunter et al. 1983; Shroff, Janicki, and Weber 1985). The polynomial coefficients for $E(t)$ were chosen such that peak elastance and time to peak elastance were 6.11 mm Hg/ml and 297 msec, respectively. These values are within the normal range for the canine left ventricle beating at a rate of 120/min (Shroff, Janicki, and Weber 1983). The following equations define $E(t)$ and $R(P_o)$:

$$E(t) = 0.2 + 3.75t + 418.87t^2 - 1449.18t^3 \quad (11.3)$$

$$R(P_o) = 0.0015\, P_o(t) \quad (11.4)$$

The normal values chosen for the arterial circulation (i.e, R_p, R_c, and C) were 3.3 mm Hg sec/ml, 0.1 mm Hg sec/ml, and 0.4 ml/mm Hg, respectively. For study of ventricular performance at various arterial loads, R_p and C were varied to yield nine different conditions (Table 11.1). Left ventricular end-diastolic volume, heart rate, and L were held constant at 37 ml, 120 beats/min, and 0.0015 mm Hg sec^2/ml, respectively. The simulation was performed with and without ventricular resistance in the model. Steady-state results were obtained for each condition by numerical solution of the coupled differential equations (Shroff, Janicki, and Weber 1985).

Ventricular pressure, volume, and flow are depicted in Figure 11.10 for

TABLE 11.1. Parameter values used for the computer simulation of the left ventricle coupled to the arterial circulation

Run	R_p	C	R_c	L	Comments
1	3.3	0.40	0.1	0.0015	R_p: normal; C: normal
2	7.0	0.40	0.1	0.0015	R_p: high; C: normal
3	1.6	0.40	0.1	0.0015	R_p: low; C: normal
4	3.3	0.13	0.1	0.0015	R_p: normal; C: low
5	7.0	0.13	0.1	0.0015	R_p: high; C: low
6	1.6	0.13	0.1	0.0015	R_p: low; C: low
7	3.3	1.20	0.1	0.0015	R_p: normal; C: high
8	7.0	1.20	0.1	0.0015	R_p: high; C: high
9	1.6	1.20	0.1	0.0015	R_p: low; C: high

R_p, peripheral resistance (mm Hg sec/ml); C, lumped arterial compliance (ml/mm Hg); R_c, characteristic impedance (mm Hg sec/ml); L, inertia of blood column in outflow tract and aorta (mm Hg sec^2/ml).

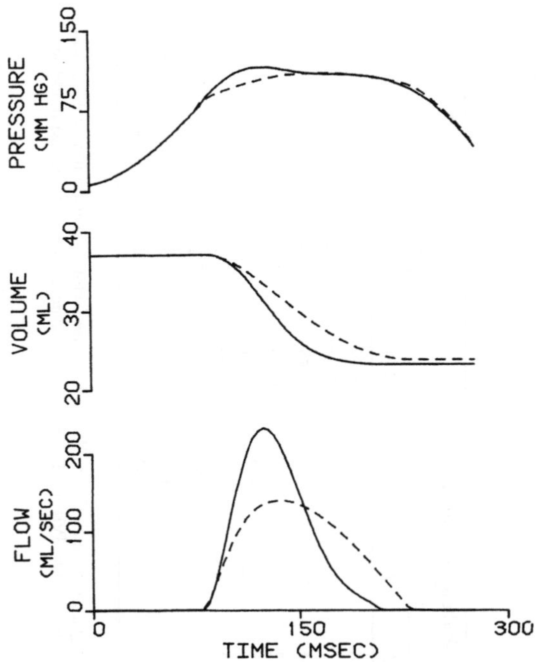

FIGURE 11.10. Instantaneous left ventricular pressure, volume, and flow obtained from the simulation study for the parameter values corresponding to run 1 in Table 11.1. The dashed and solid lines correspond to the simulation results with and without ventricular resistance, respectively. (From Shroff, Janicki, and Weber 1985)

one of the nine combinations of arterial load parameters (run 1 in Table 11.1). The solid lines represent data for simulation without ventricular resistance, while the broken lines correspond to the data where ventricular resistance was included. Table 11.2 lists the mean aortic pressure, stroke volume, duration of ejection, and peak aortic flow for all nine arterial loads, both with and without ventricular resistance. It was observed that mean aortic pressure and stroke volume decreased only slightly (range 2.8% to 7.1% for stroke volume and 3.1% to 11.0% for mean arterial pressure) when ventricular resistance was included in the simulation. However, the decline in peak flow was disproportionately greater; the simulation with resistance had considerably lower peak flow (range of reduction in peak flow 31% to 70%). To quantify the effect of ventricular resistance on the pulsatile nature of flow, a normalized pulse factor (PF) was calculated as the ratio of peak flow and average flow, where average flow was computed as the ratio of stroke volume and cardiac cycle duration. It is evident from the data in Table 11.2 that the inclusion of ventricular resistance in the computer simulation reduced the pulse factor by amounts varying from 26.9% to 62.1%. This reduction was significantly higher than the reduction in stroke volume (range 2.8% to 7.1%).

TABLE 11.2. Computer simulation results

Run	With resistance					Without resistance					ΔSV	ΔM_{ap}	ΔPF
No.[a]	M_{ap}	SV	E_D	Q_{max}	PF	M_{ap}	SV	E_D	Q_{max}	PF	(%)	(%)	(%)
1	92	13.2	152	139	5.06	96	13.8	130	232	8.04	4.5	4.3	58.9
2	121	8.6	126	106	5.93	130	9.1	94	181	9.60	5.8	7.4	61.9
3	64	17.9	178	170	4.55	66	18.4	164	263	6.85	2.8	3.1	50.5
4	84	11.8	168	117	4.74	87	12.2	156	170	6.70	4.1	3.6	41.4
5	115	7.8	136	95	5.81	119	8.1	120	151	8.94	3.8	3.5	53.9
6	57	16.0	210	125	3.75	59	16.5	202	164	4.76	3.1	3.5	26.9
7	91	14.0	154	149	4.99	98	15.0	114	254	8.09	7.1	7.7	62.1
8	118	9.4	130	113	5.77	131	10.0	94	193	9.31	6.4	11.0	61.4
9	68	18.6	170	179	4.61	71	19.2	138	295	7.37	3.2	4.4	59.9

[a]Run numbers correspond to different combinations of arterial load as listed in Table 11.1.

M_{ap}, mean aortic pressure (mm Hg); SV, stroke volume (ml); E_D, duration of ejection (msec); Q_{max}, maximum flow (ml/sec); PF, pulse factor (maximum flow/average flow); ΔM_{ap}, % change in mean aortic pressure; ΔSV, % change in stroke volume; ΔPF, % change in pulse factor. Percent changes are with respect to the condition with resistance.

Similar results were obtained by Campbell and colleagues (1984) who examined the sensitivity of stroke volume to left ventricular (i.e., elastance and resistance) and arterial (i.e., R_p and C) parameters. Therefore it can be concluded that ventricular resistance primarily affects the pulsatile nature of flow, while its effects on mean values of flow and pressure (e.g., stroke volume and mean arterial pressure) are minimal. Moreover, it should be emphasized that ventricular resistance plays an important role in describing and predicting the *dynamic* behavior (i.e., instantaneous pressure and flow) of the left ventricle (Campbell et al. 1984, Ringo et al. 1982).

An alternative approach to analyzing the effects of ventricular resistance on the dynamic behavior of the left ventricle is to examine its power output. There are two components of the mechanical power generated by the left ventricle (Milnor, Bergel, and Bargainer 1966; O'Rourke 1967). The first is the steady power, which is dissipated mainly in the arterioles as the blood flows forward. The second is the pulsatile power, which is lost in the arterial pulsations and which does not contribute to the forward blood flow. The ratio of oscillatory (or pulsatile) power to steady power (an index of relative power wasted due to pulsations) is plotted against ventricular resistance in Figure 11.11. Ventricular resistance was quantified by the slope of the resistance–isovolumetric pressure relation (Eq. 11.4). These data were calculated from the computer simulation study in which the arterial circulation parameters (run 1 in Table 11.1) and elastance function (Eq. 11.3) were held constant. A clear minimum of the ratio of oscillatory power to steady power with respect to ventricular resistance is evident (Fig. 11.11). To the left of this minimum, the pulsatile power decreased more than the steady power as resistance was increased. On the other hand, to the right of this minimum, the steady power decreased more than the pulsatile power, thus producing the observed minimum. It is interesting to note that the resistance value observed in the normal canine left ventricle (slope = 0.0015 sec/ml) corresponds to the region where oscillatory power-to-steady power ratio is minimum. Thus, under normal conditions, the value of ventricular resistance is such that it critically damps the flow pulsations and minimizes the relative power lost due to arterial pulsations.

Ventricular Resistance in Cardiovascular Disease

Although ventricular resistance does not affect *average* performance (e.g., mean arterial pressure, stroke volume) of the left ventricle coupled to the arterial circulation under normal conditions, the magnitude and relative effect of its contribution in various pathologic states remain to be elucidated. Recently, the response in ventricular resistance to the myocardial hypertrophy that accompanies spontaneous hypertension in the rat has been examined (Shroff et al. 1985). For this purpose, 25-week-old, spon-

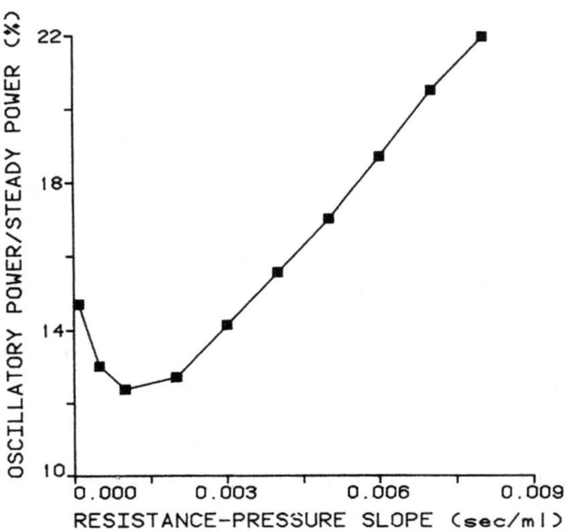

FIGURE 11.11. Ratio of the oscillatory power to steady power is plotted against ventricular resistance described in terms of the slope of the resistance–isovolumetric pressure relationship. Data were obtained from a computer simulation study in which arterial circulation parameters and elastance function were held constant. The ratio of the oscillatory power to steady power exhibited a minimum with respect to ventricular resistance.

taneously hypertensive rats (SHR) and their age- and sex-matched normotensive controls, Wistar-Kyoto (WKY) rats, were studied. This age represents a stage of established hypertension and hypertrophy without signs of overt cardiac failure. Instantaneous left ventricular pressure and aortic flow were recorded in open-chest, anesthetized, artificially ventilated rats. Data were recorded for steady-state ejecting beats and the corresponding isovolumetric beat obtained by transient occlusion of the ascending aorta in diastole (Fig. 11.12). From these data ventricular elastance, $E(t)$, and resistance, $R(P_o)$, were computed using the following model of ventricular systolic dynamics (Campbell et al. 1984; Shroff, Janicki, and Weber 1983):

$$E(t)[V(t) - V_d] - R(P_o)\dot{V}(t) = P(t) \qquad (11.5)$$

where $P(t)$, $V(t)$, $\dot{V}(t)$, and V_d are instantaneous ventricular pressure, volume, flow, and absolute residual volume, respectively. Equation 11.5 can be rewritten in terms of the measured quantities $P(t)$, $\dot{V}(t)$, $P_o(t)$ (isovolumetric pressure), and $V_e(t)$ [$= \int_0^t \dot{V}(t)dt$ = volume ejected up to the time *t*] as follows:

$$AP_o(t)V_e(t) + BP_o(t)[1 - AV_e(t)]\dot{V}(t) = P_o(t) - P(t) \qquad (11.6)$$

$$\text{where } AP_o(t) = E(t)$$

$$A = 1/(\text{end-diastolic volume} - V_d)$$

$$BP_o(t)[1 - AV_e(t)] = R(P_o)$$

From Equation 11.6, *A* and *B*, and consequently $E(t)$ and $R(P_o)$, were calculated using the least squares parameter estimation technique. The results indicated that maximum elastance (E_{max}) was significantly higher in SHR than in WKY rats (1100±163 vs. 577±88 mm Hg/ml). However, most of the difference in E_{max} could be accounted for by the increased left ventricular muscle mass seen in SHRs (916±69 vs. 571±49 mg). Thus the difference in normalized E_{max} (i.e., E_{max}/LV mass) was greatly attenuated (1.19±0.12 mm Hg/ml mg for SHRs vs. 1.01±0.13 mm Hg/ml mg

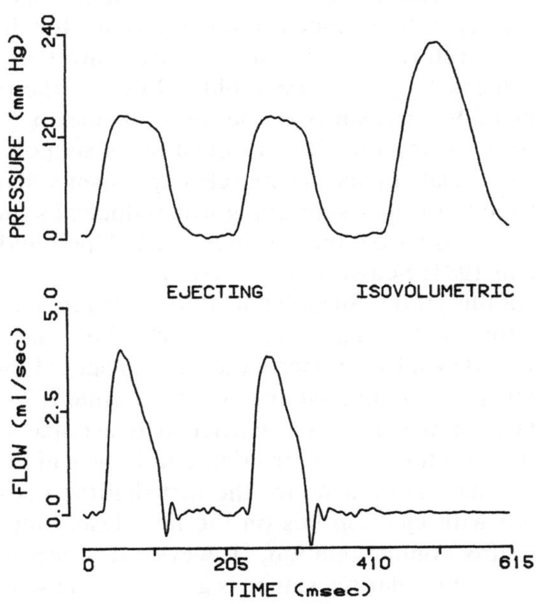

FIGURE 11.12. Typical left ventricular pressure and aortic flow data obtained from an anesthetized, open-chest rat. Two steady state ejecting beats followed by an isovolumetric beat, obtained by transient occlusion of the ascending aorta in diastole, are depicted. Left ventricular elastance and resistance were calculated from these data (see text).

for WKY rats). Ventricular resistance was significantly higher in SHRs such that the slope of the resistance–isovolumetric pressure relation (B) was 0.206±0.034 sec/ml for SHRs and 0.096±0.018 sec/ml for WKY rats. In other words, the extrapolated maximum flow ($\dot{V}_{max}$), which is simply the inverse of B (vide supra), was significantly reduced for SHRs (4.96±0.81 vs. 10.78±2.07 ml/sec). It is interesting to note that the pump function of the left ventricle expressed in terms of cardiac index (237±30 ml/min kg for SHRs vs. 221±17 ml/min kg for WKY rats) was similar between the two groups. We would hypothesize that at this stage of compensated hypertrophy, the opposing effects of increased E_{max} (primarily due to hypertrophy) and reduced $\dot{V}_{max}$ negate each other, and thus the pump function is preserved. If ventricular resistance continued to increase without a proportional increase in E_{max}, cardiac failure could result. The validity of this hypothesis needs to be carefully examined in a longitudinal study.

The major difference between left ventricular myocardium of 25-week-old SHRs and WKY rats was the relative proportions of myosin isozymes (V3 myosin isozyme 25–40% in SHR vs. 8–14% in WKY). The collagen concentration of the myocardium, on the other hand, was similar for the two groups (hydroxyproline concentration 3.6 μg/mg for SHR vs. 3.4 μg/mg for WKY). Therefore it is tempting to speculate that the increased resistance or reduced $\dot{V}_{max}$ in 25-week-old SHRs was the result of an increased amount of V3 myosin isozyme and not due to a structural remodeling of the myocardium. This speculation is supported by the observation from the isolated papillary muscle experiments that the unloaded or maximum velocity of fiber shortening was reduced as the relative proportion of the V3 myosin isozyme was increased (Alpert and Mulieri 1982; Pagani and Julian 1984; Schwartz et al. 1981).

The findings of this study (Shroff et al. 1985) indicate that the responses of the volume- (or length-) dependent property (i.e., elastance) and the flow- (or velocity-) dependent property (i.e., resistance) of the myocardium to a given stimulus can be discordant. For the normal left ventricle (both in dogs and rats), the magnitude of ventricular resistance is such that the limit to its shortening for a given arterial load is provided by the elastic property. This is the main reason for the fact that the end-systolic point for a contraction with ejection lies on the line describing the peak isovolumetric pressure–volume relation. However, it is conceivable that for conditions where ventricular resistance is greatly increased without a proportional increase in E_{max}, the limit to shortening would be provided by the resistive property. Thus the role of ventricular resistance in a variety of disease conditions, in which structural (e.g., collagen concentration) and/or biochemical (e.g., myosin isozyme composition) alterations may occur, needs to be examined. Examples of such pathologic conditions include cardiomyopathy with interstitial fibrosis, chronic myocardial ischemia with local or diffuse fibrosis, and hypertrophy at various stages of pressure and/or volume overload.

Summary

Instantaneous left ventricular flow, along with systolic time, volume, and contractile state, is an independent determinant of instantaneous chamber pressure. Phenomenologically, the ventricular pressure–flow relationship can be represented by the instantaneous ventricular resistance, which can be uniquely quantified as a linear function of the corresponding (i.e., for the same time, volume, and contractile state) isovolumetric pressure. For a given heart, acute variations in systolic time, chamber volume, or contractile state do not affect this linear relationship. On the other hand, large variations in chamber volume (e.g., chamber dilatation in pathologic conditions or volume differences across species) significantly alter this relationship. In the normal canine left ventricle, the contribution of the resistive component to ventricular pressure can be as high as 15% to 20% of the elastic component at the time of peak flow. While the indices of performance of the left ventricle coupled to the arterial circulation in terms of the mean values (e.g., mean arterial pressure, stroke volume) are minimally affected by ventricular resistance, the effects of such resistance on the pulsatile nature of flow are significant. The magnitude of ventricular resistance for the normal canine left ventricle is such that the relative power wasted due to arterial pulsations is minimized. Thus ventricular resistance plays an important role in determining the dynamic component of ventricular performance. Cardiac disease can significantly affect ventricular resistance. Changes in resistance can be related to the structural (e.g., collagen concentration), biochemical (e.g., myosin isozyme distribution), or geometric (e.g., chamber volume) alterations. For example, ventricular resistance is significantly higher in myocardial hypertrophy associated with systemic hypertension (spontaneously hypertensive rats), and this increase appears to be correlated with the increase in the proportion of V3 myosin isozyme. Thus future investigations should be directed toward elucidating the role of ventricular resistance in a variety of cardiac disease conditions.

Bibliography

Alpert, N. R., and Mulieri, L. 1982. Heat, mechanics and myosin ATPase in normal and hypertrophied heart muscle. *Fed. Proc.* 41:192–198.

Campbell, K. B., Ringo, J. A., Neti, C., and Alexander, J. E. 1984. Informational analysis of left ventricle/systemic arterial interaction. *Ann. Biomed. Eng.* 12:209–231.

Daniels, M., Noble, M. I. M., ter Keurs, H. E. D. J., andWohlfart, B. 1984. Velocity of sarcomere shortening in rat cardiac muscle: relationship to force, sarcomere length, calcium and time. *J. Physiol.* 355:367–381.

Elzinga, G., and Westerhof, N. 1974. End-diastolic volume and source impedance of the heart. In Porter, R., and Fitzsimons, D. W., eds. *The Physiologic Basis of Starling's Law of the Heart.* New York: Associated Science, Publishers, pp. 241–255.

Elzinga, G., and Westerhof, N. 1978. The effect of an increase in inotropic state and end-diastolic volume on the pumping ability of th feline left ventricle. *Circ. Res.* 42:620–628.

Hunter, W. C., Janicki, J. S., Weber, K. T., and Noordergraaf, A. 1979. Flow pulse response: a new method for the characterization of ventricular mechanics. *Am. J. Physiol.* 237 (*Heart Circ. Physiol.* 6):H282–H292.

Hunter, W. C., Janicki, J. S., Weber, K. T., and Noordergraaf, A. 1983. Systolic mechanical properties of the left ventricle: effects of volume and contractile state. *Circ. Res.* 52:319–327.

Hunter, W. C., and Noordergraaf, A. 1976. Can impedance characterize the heart? *J. Appl. Physiol.* 40:250–252.

Janicki, J. S., Reeves, R. C., Weber, K. T., Donald, T. C., and Walker, A. A. 1974. Application of a pressure servo system developed to study ventricular dynamics. *J. Appl. Physiol.* 37:736–741.

Milnor, W. R., Bergel, D. H., and Bargainer, J. D. 1966. Hydraulic power associated with pulmonary blood flow and its relation to heart rate. *Circ. Res.* 19:467–480.

O'Rourke, M. F. 1967. Steady and pulsatile energy losses in the systemic circulation under normal conditions and in simulated arterial disease. *Cardiovasc. Res.* 1:313–326.

Pagani, E. D., and Julian, F. J. 1984. Rabbit papillary muscle myosin isozymes and the velocity of muscle shortening. *Circ. Res.* 54:586–594.

Ringo, J. A., Campbell, K. B., Slinker, B. K., and Robinette, J. B. 1982. Internal series resistance: an important LV pump property. *Proc. 35th ACEMB* 24:64.

Sagawa, K. 1978. The ventricular pressure–volume diagram revisited. *Circ. Res.* 43:677–687.

Sagawa, K., Suga, H., Shoukas, a. A., and Bakalar, K. M. 1977. End-systolic pressure–volume ratio: a new index of ventricular contractility. Am. J. Cardiol. 40:748–753.

Schwartz, K., Lecarpentier, Y., Martin, J. L., Lompre, A. M., Mercadier, J. J., and Swynghedauw, B. 1981. Myosin isozymic distribution correlates with speed of myocardial contraction. *J. Molec. Cell. Cardiol.* 13:1071–1075.

Shroff, S. G., Janicki, J. S., and Weber, K. T. 1983. Left ventricular systolic dynamics in terms of its chamber mechanical properties. *Am. J. Physiol.* 245 (*Heart Circ. Physiol.* 14):H110–H124.

Shroff, S. G., Janicki, J. S., and Weber, K. T. 1985. Evidence and quantitation of left ventricular systolic resistance. *Am. J. Physiol.* 249 (*Heart Circ. Physiol.* 18):H358–370.

Shroff, S. G., Motz, W., Janicki, J. S., and Weber, K. T. 1985. Importance of quantifying left ventricular systolic resistance in hypertrophy due to systemic hypertension. *J. Am. Coll. Cardiol.* 5:487.

Stull, J. T., and Mayer, S. E. 1979. Biochemical mechanisms of adrenergic and cholinergic regulation of myocardial contractility. In *Berne, R. M., Sperelakis, N., and Gaiger, S. R.*, eds. *Handbook of Physiology,* Sec. 2, Vol. 1, *Cardiovascular System, The Heart*. Washington, DC: American Physiological Society, pp. 741–774.

Suga, H., Sagawa, K., and Demar, L. 1980. Determination of instantaneous pressure in canine left ventricle: time and volume specification. *Circ. Res.* 46:256–263.

Suga, H., Sugawa, K., and Shoukas, A. A. 1973. Load independence of the instantaneous pressure–volume ratio of the canine left ventricle and effects of epinephrine and heart rate on the ratio. *Circ. Res.* 32:314–322.

Sunagawa, K., Sagawa, K., and Maughan, W. L. 1984. Ventricular interaction with the loading system. *Ann. Biomed. Eng.* 12:163–189.

Tallarida, R. J., Ryssy, B. F., and Longname, M. H. 1970. Left ventricular wall acceleration and the law of Laplace. *Cardiovasc. Res.* 4:217–223.

Vaartjes, S. R., van Alste, J. A., and Boom, H. B. K. 1982. Active resistance during left ventricular contraction. *Proc. 35th ACEMB* 24:142.

Weber, K. T., Janicki, J. S., and Hefner, L. L. 1976. Left ventricular force–length relations of isovolumic and ejecting contractions. *Am. J. Physiol.* 231:337–343.

Westerhof, N., Elzinga, G., and Sipkema, P. 1971. An artificial arterial system for pumping hearts. *J. Appl. Physiol.* 31:776–781.

12
Influence of Extracardiac Forces on the Cardiopulmonary Unit

JOSEPH S. JANICKI, SANJEEV G. SHROFF, AND
KARL T. WEBER

Introduction

The cardiopulmonary unit consists of the heart and lungs. It is responsible for the transport and exchange of oxygen and carbon dioxide between the metabolizing tissues and the atmosphere (Weber et al. 1983). To ensure an adequate supply of O_2 and appropriate elimination of CO_2, the heart and lungs must function in a coordinated fashion. For example, with exercise cardiac output and minute ventilation are simultaneously increased. A defect in either the heart or lungs will comprise the ability of the cardiopulmonary unit to transport and exchange oxygen and carbon dioxide, particularly during increased muscular work when the aerobic requirements of working muscles and the CO_2 they produce are increased.

The functional integration of the heart and lungs is accomplished through their anatomic arrangement, their mechanical interplay, and a complex neurohumoral control system (Weber et al. 1983). This chapter focuses on the forces that act on the surface of the heart and/or the lung owing to their anatomic arrangement and mechanical interplay. These forces, which we have collectively termed "extracardiac forces," influence the filling volumes and ejection pressures of the ventricles and the hemodynamics of the pulmonary circulation. We will consider in particular to what extent the mechanical interplay between the left and right ventricles, the presence of the pericardium, intrathoracic pressure, and abnormal distension of the lungs influence the functional integration of the heart and lungs. We will begin by considering the anatomic aspects of the cardiopulmonary unit.

Anatomic Aspects of the Cardiopulmonary Unit

A schematic representation of the cardiopulmonary unit is depicted in Figure 12.1. As can be seen, the major vessels to and from the heart, the lungs, and thereby the entire pulmonary circulation, are contained within the thorax. The two reservoirs external to the thorax represent the venous and arterial systems, which are conceptualized to be a certain height or pressure above the right and left ventricles, respectively.

For the steady state, the output of the right heart is balanced with the output of the left heart. Balance is accomplished primarily by the in-series arrangement of the right and left sides of the heart, with the output of the right ventricle becoming the input of the left ventricle. In addition, the interventricular septum and the common muscular wall that surrounds each chamber establish a continuous anatomic structure in which the mechanical events in one ventricle influence the behavior of the other, thereby fostering coordination of their activity. The interventricular septum has been shown (Pearlman et al. 1982) to have a muscle fiber alignment morphologically similar to that of the left ventricular free wall, and therefore it could be considered an integral component of the left ventricle. The concentric reduction of left ventricular chamber size during systole can be considered functional evidence in support of this viewpoint. The muscle fiber orientation of the right ventricular free wall and its relation to the septum and left ventricular free wall, particularly at the sites of attachment, remain to be defined. However, there exists physiologic evidence indicating that the right ventricular free wall is drawn toward the septum during systole by the contraction of the septum and left ventricular free wall, to aid right ventricular ejection (Brooks et al. 1977, Starr et al. 1943).

The pericardium also contributes to the integrated function of the heart and lungs. It surrounds the entire heart and plays a role in determining the limits to acute diastolic expansion and therefore the extent to which the Frank–Starling mechanism is utilized (Janicki, Shroff, and Weber 1985). It consists of a thin serous membrane (visceral pericardium) and a thicker fibrous sac (parietal pericardium). The serous membrane covers the epicardium and lines the interior of the fibrous sac. The fibrous portion of the pericardium surrounds the entire heart and is closed by its attachment to the great vessels (Holt 1970).

The pericardium is in contact with and has several attachments to various components of the thorax. It is attached caudally to the central tendon of the diaphragm, ventrally to the sternum, and dorsally to the vertebral column. Laterally the pericardium is surrounded by the lungs. Thus with inspiration the pericardium is subjected to elongating stresses via its connections to the sternum and diaphragm and to a compressive stress resulting from the increase in lung volume. These forces are then transmitted to the heart through the resulting variations in intrapericardial pressure, and are therefore capable of influencing ventricular performance. In addition, the concomitant decrease in intrathoracic pressure with inspiration represents a decrease in the external force acting on the pericardium. This too influences ventricular function, as will be discussed subsequently in greater detail.

Physiologic Implications of Extracardiac Forces

Mechanical determinants of right and left ventricular function include ventricular filling volume and ejection pressure (Weber et al. 1982a). Extracardiac forces that alter one or both of these determinants would therefore modify ventricular function.

For better understanding of the dependence of ventricular function on extracardiac forces, we have found it useful to depict the cardiopulmonary unit and its connections with the venous and arterial circulations, as shown in Figure 12.1. The large-capacity venous circulation, which is located entirely outside the thorax, can be represented by an external venous reservoir that resides at a certain height above the right ventricle. This height represents the filling pressure gradient for venous return to the right heart (normally 5 to 10 mm Hg) and is determined by the difference between extrathoracic and intrathoracic venous pressures. The right ventricle must eject venous blood, with low O_2 and high CO_2 concentrations, into the low-impedance pulmonary circulation. In doing so, the right ventricle effectively pumps blood to another reservoir (pulmonary reservoir in Fig. 12.1), which is located entirely within the thorax and at a certain height or pressure above the right heart; this height represents the level of ejection pressure for the right ventricle. At the same time, the height of this reservoir above the left heart represents the filling pressure gradient for the left heart. Accordingly, blood flow through the lung, or equivalently

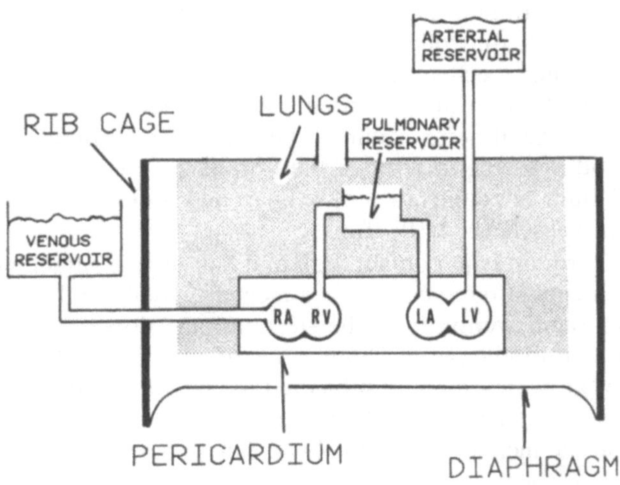

FIGURE 12.1. Schematic representation of the cardiopulmonary unit. The heart is depicted as two pumps connected in series (RA = right atrium, RV = right ventricle, LA = left atrium, LV = left ventricle). The pulmonary, systemic arterial and systemic venous circulations are shown as either intrathoracic or extrathoracic reservoirs.

the filling of the left heart, is sustained by the difference between pulmonary arterial and left atrial pressures. Finally, the left ventricle must eject arterial blood, with increased O_2 and reduced CO_2 concentrations, into the high-impedance systemic arterial circulation. Most of this circulation lies outside the thorax and is depicted accordingly in Figure 12.1. The height of the reservoir with respect to the left ventricle is analogous to the level of ejection pressure for the left ventricle.

Not depicted in Figure 12.1 is the connection between the arterial and venous reservoirs. The mean pressure in the central aorta (90 mm Hg) decreases as blood traverses the systemic circulation. The greatest drop occurs at the smaller resistance vessels, or arterioles. The pressure within the arterioles is typically 40 mm Hg, while the pressure in the venous circulation is less than 15 mm Hg.

From Figure 12.1 it can be seen that the venous return to the right heart is increased by either increasing the height of the venous reservoir with respect to the right heart or by lowering the right heart with respect to the venous reservoir. Both maneuvers result in a larger filling pressure gradient for venous return to the right heart. The filling gradient for the left heart is similarly augmented by raising the height of the pulmonary reservoir with respect to the left heart or lowering the heart with respect to the pulmonary reservoir.

The ejection pressure, which influences the output of each ventricle (Weber et al. 1982a), is altered in a manner similar to the filling of the ventricles. For example, the ejection pressure of the left ventricle will be increased if the height of the arterial reservoir is increased with respect to the left ventricle or if the left ventricle is lowered with respect to the external arterial reservoir.

Extracardiac forces are capable of altering these filling and ejection pressure gradients, thereby producing changes in right and left ventricular function. In the remainder of this chapter we will consider how these gradients in pressure are affected by the mechanical interplay between ventricles, the presence of the pericardium, intrathoracic pressure, and the distension of the lungs under normal and abnormal conditions. In so doing we will be able to evaluate their influence on the functional integration of the heart and lungs.

Ventricular Interaction

Diastolic Interaction

As mentioned earlier, the anatomic structure of the two ventricles results in the ability of one ventricle to mechanically interact with the other ventricle (Bove and Santamore 1981, Janicki and Weber 1980a). An example of ventricular interaction is presented in Figure 12.2. Shown are the responses in left ventricular (LV), developed (i.e., peak isovolumetric pressure minus end-diastolic pressure), and end-diastolic (EDP) pressures to stepwise increments in right ventricular (RV) end-diastolic volume (EDV) when left ventricular end-diastolic volume (LVEDV) is held constant. These data were obtained in an isolated heart preparation following the removal of the pericardium, to demonstrate ventricular interaction resulting solely from the anatomic arrangement of the ventricles. The level of EDP in the left ventricle (where volume is constant) is related to the increment in RVEDV. LVEDP becomes greater with each increment in RVEDV. In the intact heart, in which the volume in each ventricle can vary, acute increments in RVEDV would, via mechanical interaction, result in a reduced filling pressure gradient to the left heart. Similar results are obtained when the influence of variations in LVEDV on RVEDP is investigated. In the absence of the pericardium, Janicki and Weber (1980b) reported the maximum diastolic interaction for either ventricle (i.e., elevation in filling pressure of the ventricle with constant EDV as the volume of the other ventricle was increased) to be 1 to 6 mm Hg.

Distortions in left ventricular geometry have been reported following right ventricular distension. These configurational changes have been

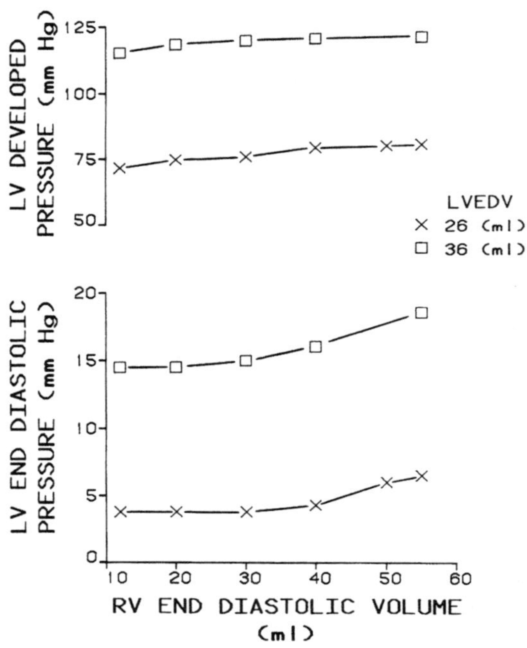

FIGURE 12.2. Example of ventricular interaction. Left ventricular (LV) end-diastolic and developed pressures increase as volume in the right ventricle (RV) is incremented, even though LV volume (LVEDV) is held constant at either 26 or 36 ml. The data were obtained in an isolated, isovolumetrically contracting canine heart preparation following removal of the pericardium.

characterized as a significant decrease in the septal-lateral dimension of the LV (Bemis et al. 1974, Santamore et al. 1976, Stool et al. 1974) and either an equal increase (Bemis et al. 1974) or a disproportionate decrease (Stool et al. 1974) in its anterior-posterior dimension. Likewise, the septal-to-free-wall dimension of the RV has been found to decrease as LV volume is increased (Santamore et al. 1976). Thus the basis for ventricular interaction during diastole is a shift of the interventricular septum from the ventricle that is being distended toward the other ventricle.

Ventricular diastolic interaction has been demonstrated in patients with heart failure (Alderman and Glantz 1976). A substantial downward shift of the instantaneous (i.e., from the end of rapid filling to the end of diastole) left ventricular pressure–volume relation was observed in such patients with nitroprusside administration. Owing to venous pooling and the concomitant reduction in right ventricular volume produced by nitroprusside, there was less ventricular interaction and an increase in left ventricular distensibility.

Systolic Interaction

Mechanical events in one ventricle can influence the systolic function of the other ventricle (Elzinga, Piene, and DeJong 1980; Janicki and Weber 1980b). For example, systolic interaction, consisting of an elevation in the developed isovolumetric pressure of one ventricle, occurs as the filling volume and developed pressure of the opposite ventricle are increased (Fig. 12.2). The maximum systolic interaction has been found to be 3.2 ± 1.2 mm Hg for the right ventricle and 7.4 ± 1.5 mm Hg for the left ventricle (Janicki and Weber 1980b). Here, too, a shift of the septum may be responsible. As volume and systolic pressure are increased in one ventricle, the septum appears stiffer and therefore minimizes the dissipation of pressure generated by the other ventricle.

The role of the septum appears to be particularly important in the case of RV performance. For example, Brooks, Holland, and Al-Sakir (1977) were able to induce a deterioration in RV performance following right ventricular infarction only when the interventricular septum was also damaged. Thus systolic performance in the RV appears to be influenced by the basic interaction between ventricles associated with normal LV contraction and by that occurring as LV volume is raised. This view is consistent with that of Starr and colleagues, who found in their classic studies (1943) that cardiac output was maintained without an elevation in central venous pressure following extensive cauterization of the RV free wall that left the interventricular septum intact. Furthermore, one can infer from these findings that LV systolic performance is not significantly influenced by the normal RV. However, in the patient with severe pulmonary hypertension in whom the right ventricle generates systemic levels of pressure, the motion of the interventricular septum during systole becomes abnormal. That is, the septum shifts toward the left ventricle. The septum may then partially obstruct the left ventricular outflow tract and thereby increase its afterload (Weber et al. 1983).

Presence of the Pericardium

Influence on Diastolic and Systolic Function

The parietal pericardium significantly influences the diastolic pressure (EDP)–volume (EDV) relations of both ventricles (Glantz et al. 1978, Holt 1970, Janicki and Weber 1980b, Maruyama et al. 1982, Shirato et al. 1978). This is demonstrated in Figure 12.3, in which the EDP–EDV relations are presented for the right and left ventricles before and after pericardiectomy. Removal of the pericardium results in a downward, nonparallel shift of the EDP–EDV curve. The amount by which the curve is shifted is a function of ventricular volume. For both ventricles, the greatest decrease occurs at higher volumes. Similar observations have been made for the atria (Maruyama et al. 1982). Therefore, the pericardium limits cardiac size and minimizes acute cardiac distension.

The influence of the pericardium on the developed pressure (DP)–EDV relation is similar to its influence on the EDP–EDV relation (Fig. 12.3). That is, the DP–EDV relation is shifted downward following pericardiec-

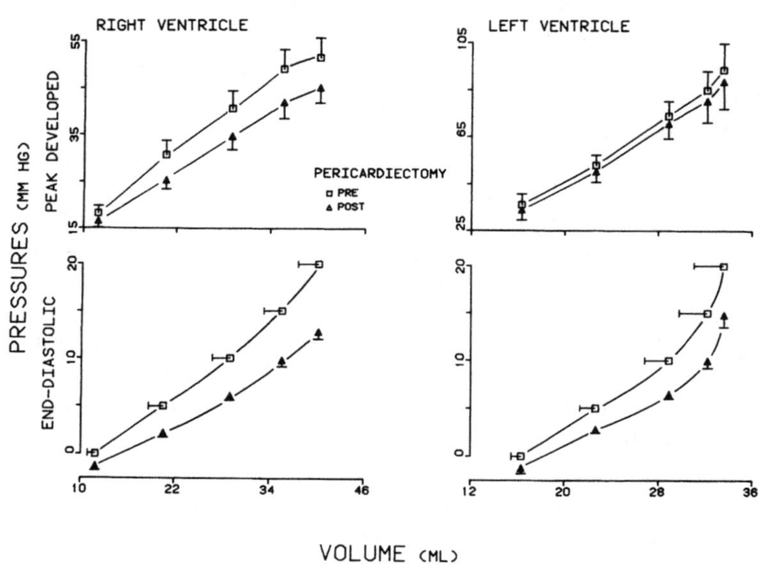

FIGURE 12.3. Effects of pericardiectomy on the developed and diastolic pressure–volume relations of both ventricles. The data were obtained in an isolated, isovolumetrically contracting canine heart preparation. (From Janicki and Weber 1980b).

tomy. The degree to which it shifts is a function of pericardial stiffness; the stiffer the pericardium, the greater the amount of the downward shift (Janicki and Weber 1980b). These results imply that, if the filling pressures were maintained after pericardiectomy by raising EDV, ventricular performance would be essentially unchanged.

Influence on Ventricular Interaction

The pericardium also enhances the mechanical interaction or interplay between ventricles (Glantz et al. 1978, Janicki and Weber 1980b, Shirato et al. 1978). In Figure 12.4, the maximum diastolic (ΔEDP) and systolic (ΔDP) interaction for each ventricle, with and without the pericardium, is presented. The ratio of ventricular volume at a given pressure to that obtained at zero EDP (i.e., unstressed volume, or V/V_o), is used here to compare hearts of different size and distensibility. Let's consider, for example, the total increase in LVEDP (upper right of Fig. 12.4) obtained at an LV volume 1.6 times larger than the unstressed LVEDV, or at a V/V_o value of 1.6. It can be seen that an average increase in LVEDP of 6 mm Hg with the pericardium intact and only 3.4 mm Hg after its removal was obtained when RV volume was varied from its unstressed value to a value that corresponded to an RV end-diastolic pressure of 20 mm Hg.

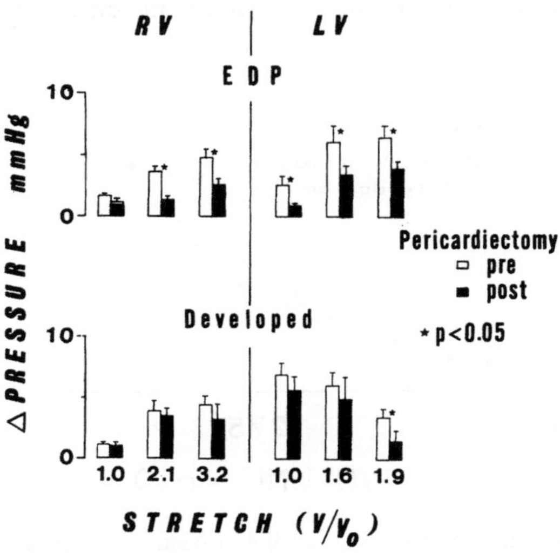

FIGURE 12.4. Diastolic and systolic interaction for the right (RV) and left (LV) ventricles with and without the pericardium. Shown are the increases in end-diastolic pressure (EDP) and peak developed pressure for one ventricle whose volume (V) was held constant at one of the three values of stretch (V/V_o, where V_o is the volume at EDP = 0) as EDP in the other ventricle was raised from 0 to 20 mm Hg. (From Janicki and Weber 1980b).

These reductions following pericardiectomy in ventricular interaction at end diastole (except that for the RV at $V/V_o = 1$) were significant and ranged from 30% to 65%. For either ventricle the maximum amount of diastolic interaction ranged from 4 to 22 mm Hg with the pericardium intact and from 1 to 6 mm Hg following pericardiectomy (Janicki and Weber 1980b).

As can be seen in Figure 12.4 systolic interaction, consisting of an elevation in the developed isovolumetric pressure of one ventricle as the filling volume of the opposite ventricle was increased, occurs with and without the pericardium. The influence of the pericardium on the magnitude of systolic interaction, however, was much less than its influence on diastolic interaction. The maximum systolic interaction before and after pericardiectomy for the right ventricle was 4.4 ± 0.7 and 3.2 ± 1.2 mm Hg, respectively. For the left ventricle the maximum values were 9.2 ± 1.3 mm Hg with and 7.4 ± 1.5 mm Hg without the pericardium. In general, the systolic interaction values before and after pericardiectomy were not significantly different from each other (Janicki and Weber 1980b).

The fact that the diastolic interaction between ventricles is significantly greater with the pericardium intact, particularly when ventricular filling volumes are large, suggests that the extent of this interaction must be related to intrapericardial volume. This relationship has been identified (Janicki and Weber 1980b) and is depicted in Figure 12.5. The solid squares represent the pericardial pressure–volume relation and the circles represent the amount of diastolic interaction in either ventricle as a function of total

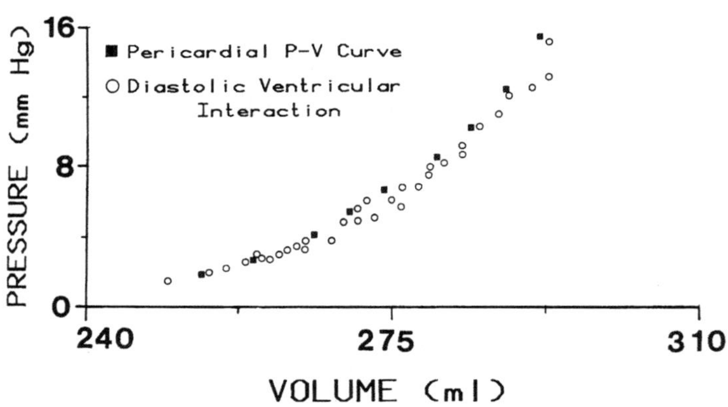

FIGURE 12.5. Comparison of the pericardial pressure–volume curve with a series of diastolic interaction curves from both ventricles. The amount of diastolic interplay between ventricles (i.e., the amount end-diastolic pressure in the ventricle with constant volume increases as the volume in the opposite ventricle is elevated) is seen to be a direct consequence of the pressure–volume (P–V) relation of the pericardium. (From Janicki and Weber 1980b).

cardiac volume (i.e., muscle volume plus total chamber volume). As can be seen, diastolic interaction in the presence of the pericardium is a direct consequence of the pericardial pressure–volume relation. That is, as volume in one ventricle is increased to a point where the pericardium is stretched, the resulting increase in intrapericardial pressure is transmitted to the other ventricle, causing either a rise in its filling pressure and a decrease in the filling pressure gradient, a decrease in its filling volume, or some combination of these two responses.

The data of Shirato and colleagues (1978), together with the analysis of Tyberg and co-workers (1978), support this view. Shirato found that, with the pericardium intact, a significant upward shift of the instantaneous diastolic pressure–segment length relation occurred during volume loading. After pericardiectomy, pressure–segment length data obtained before and after volume loading tended to form a single relation. Tyberg concluded that left ventricular pressure measured with the pericardium intact represented the sum of the intrapericardial pressure plus the pressure necessary to distend the left ventricle in the absence of the pericardium.

Clinical Significance of the Pericardium

For normal cardiac size and ventricular volume, the pericardium imposes little if any physical constraint on the heart (Fig. 12.3). Moreover, when cardiac size increases gradually, as occurs from infancy to adulthood or in the chronically failing heart, the pericardium itself grows (Freeman and LeWinter 1984). As a result, the functional role of the pericardium normally remains minimal. In either the normal-sized heart or the chronically dilated heart, the pericardium becomes influential only during an acute increase in heart size, retarding further filling of the heart. Thus, normally, the functional role of the pericardium appears to be minimal but is difficult to assess, owing to our inability to measure intrapericardial volume and pressure accurately. Inferential evidence to the contrary (Janicki, Shroff, and Weber 1985; Janicki and Weber 1984; Janicki et al. 1985) has been obtained in patients with enlarged and failing hearts by comparing the hemodynamic responses of the left and right heart obtained during upright, incremental treadmill exercise. Details of the exercise protocol and methodology can be found elsewhere (Weber et al. 1982b).

The responses in stroke volume, pulmonary capillary wedge pressure (P_{cw}), and right atrial pressure (P_{ra}) to increasing levels of work or O_2 uptake ($\dot{V}O_2$) are shown for a patient with mild heart failure (i.e., maximum $\dot{V}O_2 > 20$ ml/min/kg) in Figure 12.6. Stroke volume continually increases, with increments being greater at the lower work loads than at the higher work loads. Right atrial pressure, which reflects the filling pressure of the right ventricle, and pulmonary capillary wedge pressure, which reflects the filling pressure of the left ventricle, also increase as the work load is

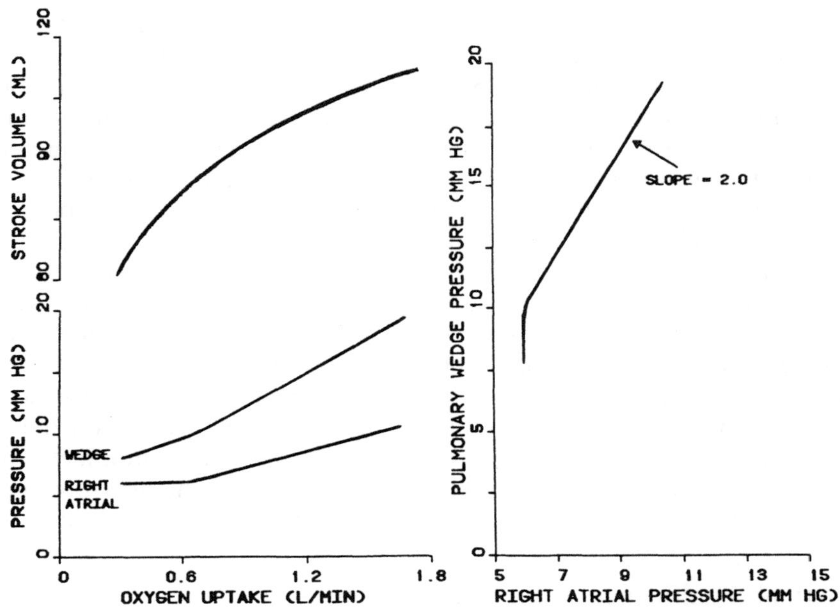

FIGURE 12.6. Hemodynamic response to incremental, upright treadmill exercise for a patient with mild heart failure. See text for discussion.

elevated. However, the rates of rise of these two pressures are not the same, as can be seen when P_{cw} is plotted versus the corresponding P_{ra} (Fig. 12.6, right). At rest and for several of the lighter work loads, P_{cw}, increases with little to no change in P_{ra}. At higher work loads, both pressures increase and the relation between them becomes linear, with the increase in P_{cw} being typically twice that of P_{ra}.

From this response little can be inferred regarding the functional significance of the pericardium. The increments in P_{cw} and P_{ra} were associated with increments in stroke volume and therefore were the result of larger ventricular filling volumes. In addition, P_{cw} should increase by a greater amount than P_{ra} as the filling volumes are increased, because the left ventricle is less distensible than the right (see Fig. 12.3).

The influence of the pericardium is evident when similar hemodynamic responses are obtained in patients with more advanced heart failure. For example, in Figure 12.7, the patient's maximum O_2 uptake and stroke volume are approximately one half those in Figure 12.6. Unlike the previous response in Figure 12.6, stroke volume increases minimally (i.e., <20ml) and becomes essentially invariant at the higher work loads. Nevertheless, increments in P_{cw} and P_{ra} occur throughout, with the rate of increase at the higher work loads being the same for both pressures. The concordant increase in filling pressures and the invariant stroke volume in all likelihood reflect the pericardial constraint to further elevations

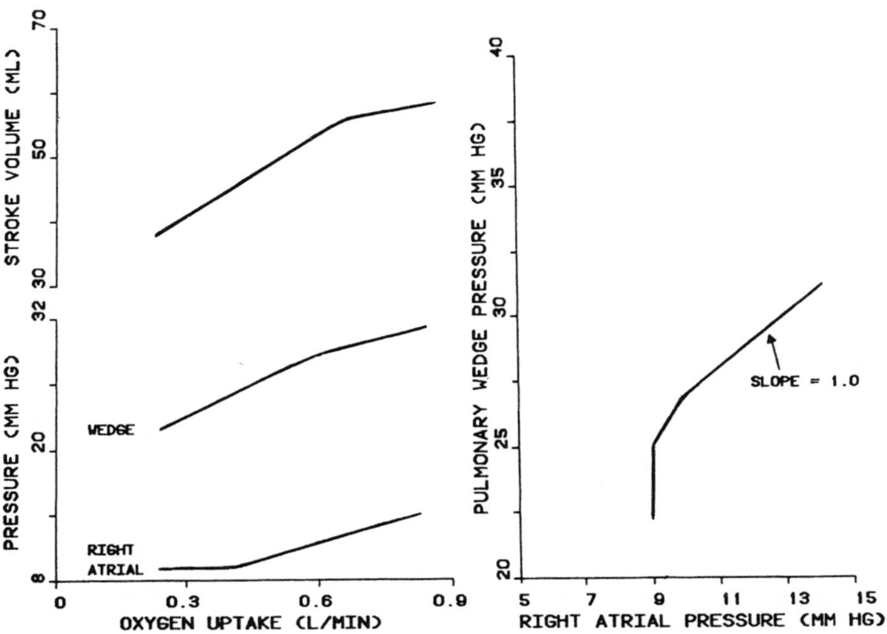

FIGURE 12.7. Hemodynamic response to incremental, upright treadmill exercise for a patient with moderate heart failure. See text for discussion.

in ventricular filling volumes. Thus in humans, the pericardium can play a substantial role in determining limits to diastolic function and systolic performance of the failing, enlarged heart.

Intrathoracic Pressure

In humans, intrathoracic pressure is indirectly determined from measurements of pressure within the esophagus. This measurement is made without surgical intervention and has been shown to be a reliable estimate of intrathoracic pressure (Milic-Emili et al. 1964). Intrathoracic pressure during quiet breathing varies between −3 mm Hg (by convention a negative pressure is below atmospheric pressure) with expiration and −8 mm Hg with inspiration. This range in intrathoracic pressure is quite small when compared with the range associated with a maximum inspiratory or expiratory effort (Fig. 12.8). Accordingly, intrathoracic pressure can vary over a wide range. For example, a single "ahem" can cause a rise in intrathoracic pressure of about 20 mm Hg, while spontaneous coughing can elicit a rise in pressure of 100 mm Hg or more (Sharpey-Schafer 1965). In the study of the acute effects of intrathoracic pressure on the cardiovascular system, the Mueller and Valsalva maneuvers are frequently used

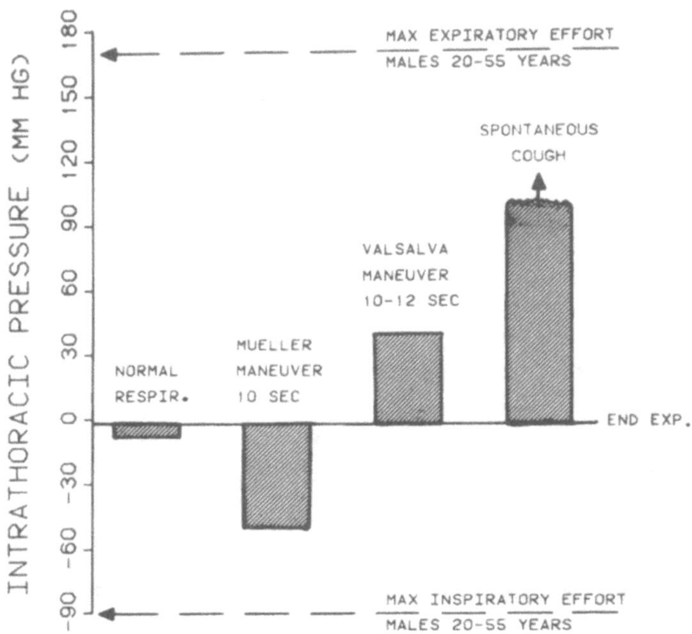

FIGURE 12.8. Intrathoracic pressures associated with normal respiration, Mueller maneuver, Valsalva maneuver, and a spontaneous cough are compared with the pressures obtained by Black and Hyatt (1969) during maximum inspiratory and expiratory efforts. Maximum inspiratory pressure was measured near residual volume after a maximum expiration. Maximum inspiratory pressure was measured near total lung capacity after a maximum inspiration.

to elicit a large decrement or increment, respectively, in intrathoracic pressure. The Mueller maneuver consists of a strong sustained inspiratory effort against a closed glottis. With this maneuver, most subjects can maintain a negative intrathoracic pressure of 50 mm Hg for 10 seconds. The Valsalva maneuver is the reverse of the Mueller maneuver and consists of a forcible exhalation against the closed glottis. Most adults can easily generate and sustain for 12 seconds a positive intrathoracic pressure of 40 mm Hg.

Decreased Intrathoracic Pressure

Variations in intrathoracic pressure can alter the output of both ventricles (Fig. 12.9). With inspiration, the stroke volume of the right ventricle increases 10% to 20%, while left ventricular stroke volume falls (Hoffman et al. 1965). The magnitude of change in stroke volume of each ventricle depends on the degree to which intrathoracic pressure falls. This is particularly true during rapid shallow breathing. In this case stroke volume

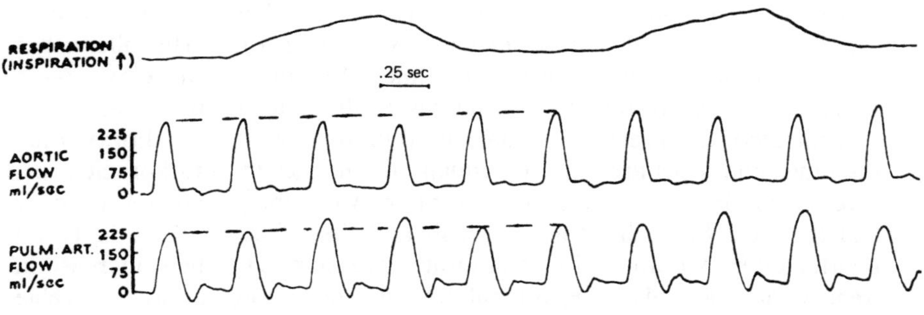

FIGURE 12.9. Variations of aortic and pulmonary arterial flow with respiration. Aortic flow decreases and pulmonary flow increases as intrathoracic pressure becomes more negative (i.e., during inspiration). (From Hoffman et al. 1965).

usually remains constant. If inspiration is deep, left ventricular output falls significantly as right ventricular output significantly increases (Table 12.1).

The increase in right ventricular stroke volume during inspiration is due to the increased filling volume that accompanies the increased filling pressure gradient between the extrathoracic venous circulation and the right heart. That is, a decrease in intrathoracic pressure results in a reduction of the pressure within the chambers of the right heart and effectively "lowers" the heart with respect to the venous reservoir (Fig. 12.1). The increased filling pressure gradient results in a larger right ventricular end-diastolic volume, which then leads to augmentation of the stroke volume.

In considering the influence of intrathoracic pressure on the cardiopulmonary unit, it is helpful to calculate the transmural pressure (i.e., the difference in pressure between the inside and outside of a cardiac chamber or blood vessel). When intrathoracic pressure is not measured, which is the case during most clinical hemodynamic studies, one could consider the intracardiovascular pressures at the end of expiration as transmural, since intrathoracic pressure is very close to the atmospheric level at this

TABLE 12.1. Mean inspiratory changes in cardiopulmonary variables

Variable	Charlier et al. (1974)	Scharf et al. 1979a
Intrathoracic pressure[a]	− 6.3	− 9.6
Right atrial pressure	—	+ 5.0
Mean pulmonary arterial pressure	+ 3.2	+ 2.0
Pulmonary blood flow	+15.0%[b]	+27.0%[c]
Left atrial pressure	+ 2.2	+ 3.0
Mean diastolic aortic pressure	—	+ 4.0
Aortic blood flow	−25.0%[b]	−17.0[c]

[a]All pressures transmural, in millimeters of mercury; difference between end expiration and peak inspiration.
[b]Percent change in blood flow.
[c]Percent change in peak blood flow.

phase of respiration. As can be seen in Table 12.1, Scharf and co-workers found transmural right atrial pressure to increase 5 mm Hg following a decrease in intrathoracic pressure of almost 10 mm Hg. This elevation in transmural pressure reflects the increased filling of the right heart.

If intrathoracic pressure is considered uniform within the thorax, then the lungs and heart are exposed equally to changes in intrathoracic pressure. Accordingly, the pressure gradients between the pulmonary reservoir and the right heart and this reservoir and the left heart (Fig. 12.1) are unaffected by variations in intrathoracic pressure. The slight increase in transmural mean pulmonary arterial pressure seen during inspiration (Table 12.1) is due to the concomitant rise in right ventricular stroke volume, and therefore is indirectly related to the change in intrathoracic pressure.

There is also an increase in transmural left atrial pressure during inspiration (Table 12.1). However, current evidence indicates that this rise is not associated with an increase in left ventricular end-diastolic volume (Robotham 1981). As discussed earlier, there is a complex change in the shape of the left ventricle as right ventricular volume increases. Typically, the anteroposterior dimension becomes larger, while the septal-left ventricular free-wall dimension becomes shorter. That is, the septum is shifted leftward toward the left ventricular chamber, making the left ventricle "stiffer" so that a given volume becomes associated with a higher transmural filling pressure. Thus, because of ventricular interaction, left heart filling during inspiration either remains the same or is slightly reduced (Robotham 1981).

A reduction in left ventricular filling volume during inspiration could account in part for the decrease in stroke volume. The reduced stroke volume is also the result of an increase in the pressure the left ventricle must generate to maintain the same arterial pressure. In fact, this is probably the major cause. The fall in the pressure surrounding the heart during an inspiratory effort effectively "lowers" the left ventricle with respect to the arterial reservoir (Fig. 12.1), or equivalently, the left ventricle must generate higher pressures before blood can leave the thorax. Therefore, since stroke volume is inversely related to the level of ejection pressure (Weber et al. 1982a), a decrease in left ventricular stroke volume occurs as intrathoracic pressure is reduced.

The effects of a sudden decrease in intrathoracic pressure on the cardiopulmonary unit are summarized in Table 12.1 and Figure 12.10. It should be emphasized that these alterations are a function of the magnitude of the decrease in inspiratory intrathoracic pressure; for lesser or greater changes in intrathoracic pressure, the changes in cardiovascular pressures and ventricular outputs will likewise be smaller or larger. However, the increase in venous return is not linearly proportional to the decline in intrathoracic pressure, because the veins are collapsible and negative pressures cannot be transmitted through them. This is analogous to a waterfall (Fig. 12.11), where the amount of water flowing over the waterfall

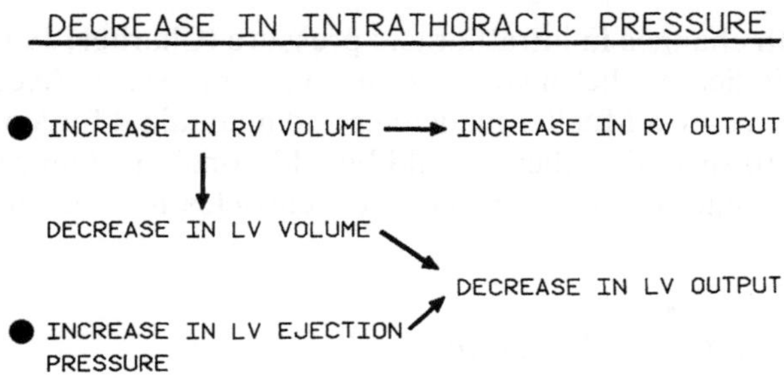

FIGURE 12.10. Summary of effects of a sudden decrease in intrathoracic pressure on the cardiopulmonary unit. RV, right ventricle; LV, left ventricle.

is not influenced by the height of the water level at the bottom of the waterfall (Permutt and Riley 1963). Thus there is a limit to which venous return to the right heart can be increased by lowering intrathoracic pressure. In addition, it is important to remember that respiratory-related hemodynamic alterations are the result of sudden, acute changes in intrathoracic pressure. The fact that the output of the two ventricles becomes

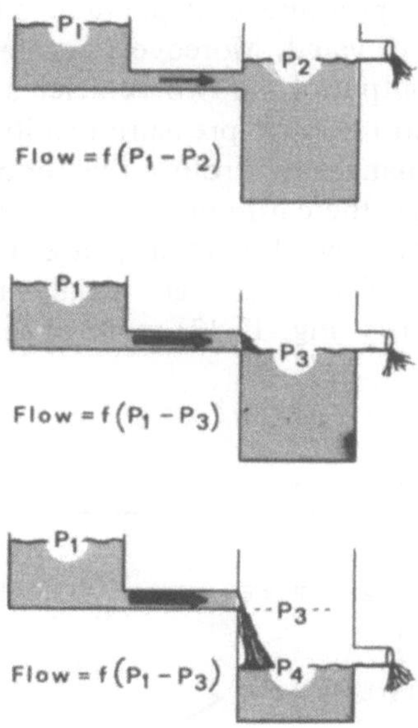

FIGURE 12.11. Diagram depicting the waterfall phenomenon. *Top,* the fluid flow from one reservoir at constant level P_1 to another at constant level P_2 is a function of the difference in heights (i.e., P_1-P_2). *Middle,* fluid flow is a function of P_1-P_3. *Bottom,* fluid flow remains a function of P_1-P_3 despite the marked decrease in the fluid level of the second reservior to P_4.

imbalanced with changes in intrathoracic pressure (Charlier, Jaumin, and Pouleur 1974) indicates the presence of a nonsteady state. Accordingly, if one were to execute a Mueller maneuver and maintain it for a prolonged period (e.g., >10 seconds), there would be additional mechanical and reflexive changes tending to return the two ventricles to a balanced state of blood flow.

Increased Intrathoracic Pressure

The response of the cardiopulmonary unit to sudden elevations in intrathoracic pressure is opposite to its response when intrathoracic pressure falls (Fig. 12.12). An increase in intrathoracic pressure effectively "raises" the right heart with respect to the venous pressure reservoir and the left heart with respect to the arterial reservoir; the relations between the left and right heart and the pulmonary reservoir are unchanged. Accordingly, venous return to the right heart, right ventricular volume, and the ejection pressure of the left ventricle are all reduced. As a result there is a decrease in RV output, an increase in the amount of blood the left ventricle can accommodate for a given filling pressure (i.e., ventricular interaction), and an augmented left ventricular stroke volume.

The magnitude of the response of the cardiopulmonary unit to an acute increase in intrathoracic pressure is dependent on the degree to which intrathoracic pressure is elevated. Moreover, because the response results in an imbalance in the output of the two ventricles, a prolonged increase (i.e., >10 seconds) in intrathoracic pressure results in compensatory mechanical and reflexive changes that return the ventricles to a state of balanced flow. For example, the cardiopulmonary response to an extended Valsalva maneuver can be divided into four phases (Sharpey-Schafer 1965). The first phase consists of a rise in right atrial, pulmonary arterial, left atrial, and aortic pressures (Fig. 12.13) more or less equal to the rise in

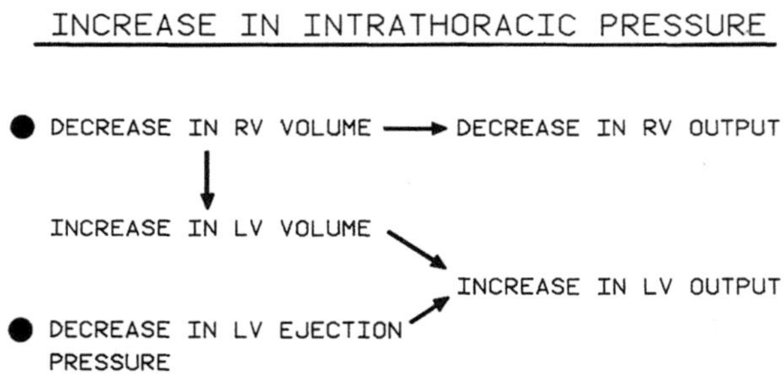

FIGURE 12.12. Summary of effects of a sudden increase in intrathoracic pressure on the cardiopulmonary unit. RV, right ventricle; LV, left ventricle.

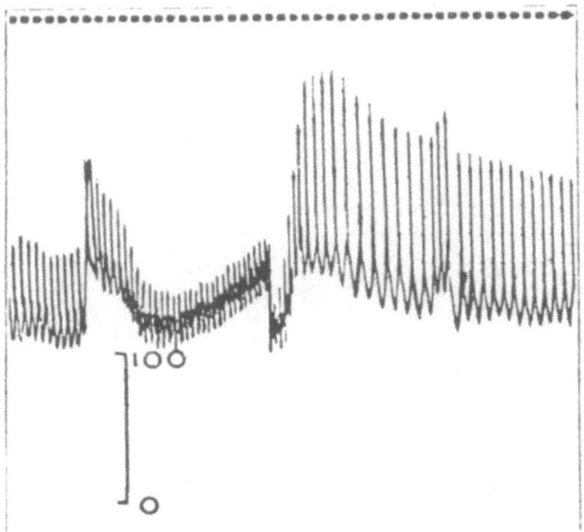

FIGURE 12.13. Continuous arterial pressure record showing the effect of the Valsalva maneuver in a normal subject. Calibration is in millimeters of mercury. From this response four phases can be identified (see text). (From Sharpey-Schafer 1955).

intrathoracic pressure. In the second phase the filling gradient to the right heart is reduced and venous return is throttled. As the elevated intrathoracic pressure is maintained, the stroke volumes of both ventricles decrease, that of the left ventricle always decreasing more slowly (Hoffman 1965). With this reduction in cardiac output there is a parallel reduction in arterial pressure (Fig. 12.13) and a reflexive vasoconstriction. In the third phase the maneuver is terminated; arterial pressure and right ventricular stroke volume begin to increase. Because of the elevated vascular resistance that occurred in phase two, the fourth phase is characterized by an "overshoot" of arterial pressure as stroke volume returns to normal.

Intrathoracic Versus Intrapericardial Pressure

From the preceding discussion it is obvious that extracardiac pressures within the pericardium and thorax are both capable of altering cardiopulmonary hemodynamics. However, the response of each ventricle to increasing intrapericardial pressure and intrathoracic pressure is different. Intrapericardial pressure is the more influential of the two external pressures (Janicki, Shroff, and Weber 1981). An example to illustrate this point is presented in Figure 12.14. As can be seen, right and left ventricular end-diastolic (EDP) pressures (measured relative to atmosphere) increase in a linear fashion as either external pressure is increased. However, the response in LVEDP is dependent on the source of the external pressure. While the slopes of the RVEDP– and LVEDP–intrathoracic pressure re-

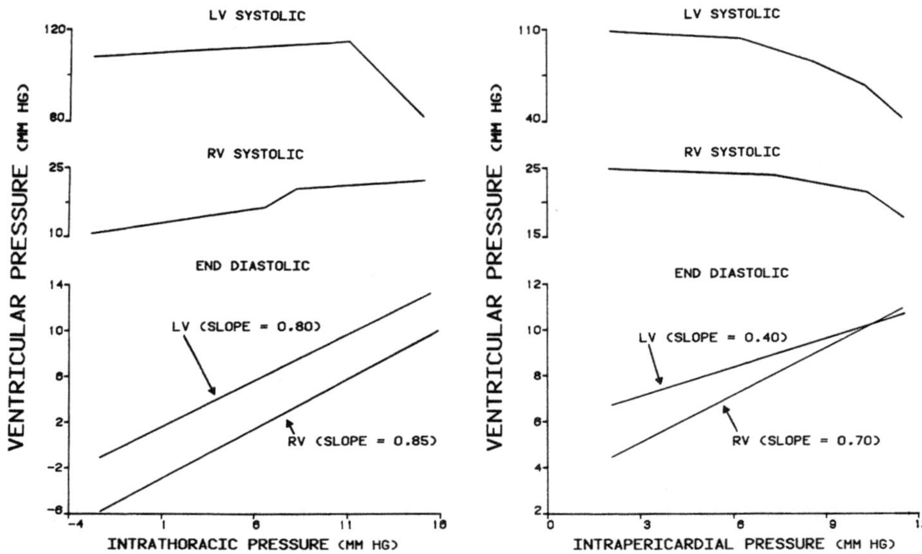

FIGURE 12.14. Example of left (LV) and right (RV) ventricular peak systolic and end-diastolic pressure responses to increases in intrathoracic pressure (*left*) and intrapericardial pressure (*right*). During variation of intrathoracic pressure, the transpulmonary pressure and consequently the lung volume were held constant (closed-chest dog preparation). Responses to intrapericardial pressure were obtained in an open-chest dog preparation.

lations and the RVEDP–pericardial pressure relation are statistically the same, the slope of the LVEDP–pericardial pressure response is significantly lower. This indicates that there is a greater reduction in LV filling volume with increments in pericardial pressure than with increments in intrathoracic pressure.

A comparison of the RV systolic pressure responses also reveals a dependence on the source of external pressure. With increments in intrathoracic pressure, RV systolic pressure continually increases (Fig. 12.14). In contrast, with increases in pericardial pressure, RV systolic pressure either remains constant or decreases. LV systolic pressure declines regardless of the source of increasing external pressure. However, the percent decline is much greater with increments in pericardial pressure.

An explanation for this dependence of ventricular dynamics on the source of external pressure can be obtained from Figures 12.1 and 12.15. As mentioned previously, intrathoracic pressure acts equally on all components of the cardiopulmonary unit, and therefore the filling pressure gradient of the left ventricle is not directly affected by this extracardiac force. Consequently, as intrathoracic pressure is raised, both RV systolic and LVEDP are seen to increase in the example shown in Figure 12.14. The LV filling gradient, however, is indirectly affected if pulmonary arterial

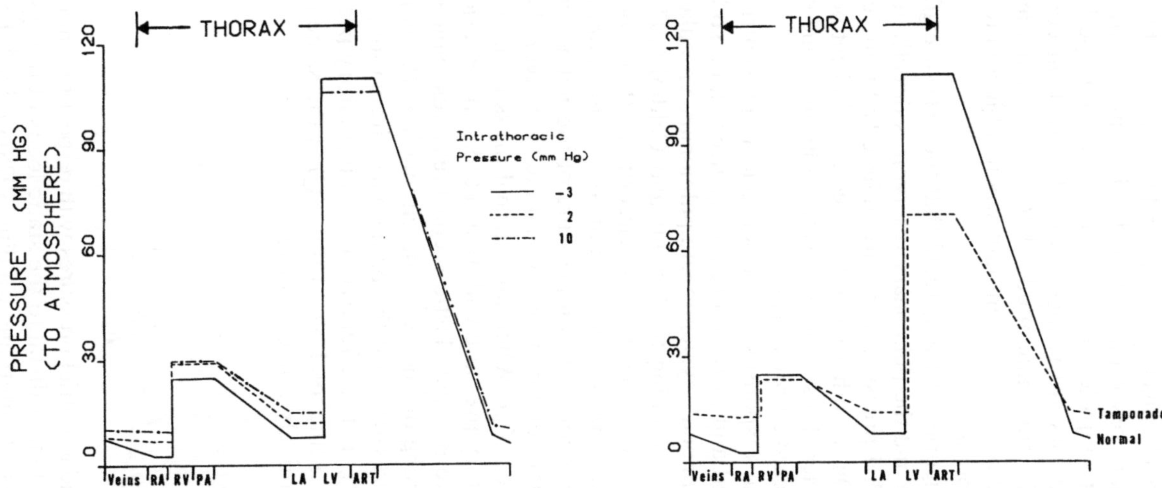

FIGURE 12.15. Conceptual representation of hemodynamic responses to variations in intrathoracic and intrapericardial pressures. RA and RV, right atrium and ventricle, respectively; PA, pulmonary artery; LA and LV, left atrium and ventricle, respectively; ART, aorta. An intrapericardial pressure of 13 mm Hg is assumed for the tamponade condition. With increases in intrapericardial pressure, there is a decrease in the effective pressure generated by the right ventricle (i.e., PA-RA pressures) and in the filling pressure gradient of the left ventricle (i.e., PA-LA pressures). These decreases are not seen with variations in intrathoracic pressure. Consequently, the decline in arterial pressure is much greater with increments in pericardial pressure.

pressure, or the height of the pulmonary reservoir in Figure 12.1, cannot be maintained because of the reduced RV filling and output.

Intrapericardial pressure, on the other hand, acts only on the cardiac chambers (Fig. 12.1). Therefore this pressure directly influences the filling of both ventricles by effectively raising the right and left hearts with respect to the venous and pulmonary reservoirs, respectively. That is, since both of these reservoirs are not subjected to the increments in intrapericardial pressure, both filling gradients are reduced (Fig. 12.15). Consequently, for extracardiac forces of similar magnitude, LV filling volume and LV systolic pressure decrease by a greater amount when subjected to increments in intrapericardial pressure.

It can also be seen in Figures 12.1 and 12.15 that the pressure gradient between the right ventricle and the pulmonary reservoir is directly reduced by increments in pericardial pressure and only indirectly by increments in intrathoracic pressure. On the other hand, the pressure gradient between the left ventricle and the arterial reservoir is directly reduced by increments in either of the two extracardiac pressures.

These differences in the responses of the left and right ventricles to increasing intrapericardial and intrathoracic pressures can also be explained on the basis of "lumped" systemic and pulmonary vascular compliances and resistances. According to Permutt and Caldini (1978), the elastic recoil or mean pressure of the systemic circulation (P_{ms}) is obtained by dividing the systemic volume (V_s) by the lumped systemic compliance (C_s). Similarly, the elastic recoil or mean pressure of the pulmonary circulation (P_{mp}) is equal to the total pulmonary volume (V_p) divided by the total pulmonary compliance (C_p). When an external pressure is present, the effective mean circulatory pressure with respect to atmospheric pressure is the sum of the external and elastic recoil pressures. Since there is no external pressure acting on the systemic circulation, P_{ms} is the effective mean circulatory pressure. For the pulmonary circulation, P_{mp} + intrathoracic pressure (P_{th}) is the effective mean circulatory pressure. When a steady state is present, total blood flow (Q) is the same for the right and left heart. Accordingly one can write

$$Q = \frac{P_{ms} - P_{ra}}{R_s} \tag{12.1}$$

$$Q = \frac{P_{mp} + P_{th} - P_{la}}{R_p} \tag{12.2}$$

here P_{ra} and P_{la} are right and left atrial pressures with respect to atmosphere and R_s and R_p are the resistances distal to the static locus of the systemic and pulmonary circulations. Permutt and Caldini (1978) defined the static locus as the locus of all points in the circulation where the pressure is the elastic recoil pressure both during flow and in the absence of flow.

P_{ra} can be written as the sum of the transmural right atrial pressure,

TMP_{ra}; pericardial pressure, P_p; and P_{th}. Similarly, P_{la} can be written as the sum of the transmural left atrial pressure, TMP_{la}; P_p; and P_{th}. Rewriting Equations 12.1 and 12.2 in terms of these sums one obtains

$$Q = \frac{P_{ms} - (TMP_{ra} + P_p + P_{th})}{R_s} \tag{12.3}$$

$$Q = \frac{P_{mp} + P_{th} - (TMP_{la} + P_p + P_{th})}{R_p} \tag{12.4}$$

To obtain a relation that combines the effects of the systemic and pulmonary circulation on venous return, Equation 12.3 is first multiplied by C_s ($= V_s/P_{ms}$) and Equation 12.4 by C_p ($= V_p/P_{mp}$). The resulting equations are then solved for the volumes (i.e., V_s and V_p).

$$V_s = Q \cdot R_s \cdot C_s + (TMP_{ra} + P_p + P_{th})C_s \tag{12.5}$$

$$V_p = Q \cdot R_p \cdot C_p + (TMP_{la} + P_p)C_p \tag{12.6}$$

The products $R_s \bullet C_s$ and $R_p \bullet C_p$ are the time constants (T_s and T_p, respectively) of drainage of the systemic and pulmonary beds. Finally, by adding Equations 12.5 and 12.6, setting $V_s + V_p = V$, $T_s + T_p = \mathrm{T}$, and solving for Q, one obtains

$$Q = \frac{V - (TMP_{ra} + P_{th} + P_p)C_s - (TMP_{la} + P_p)C_p}{T} \tag{12.7}$$

This relation expresses the venous return (or cardiac output) in terms of total circulatory volume, transmural right and left atrial pressures, intrathoracic and pericardial pressures, total systemic and pulmonary compliances, and the total time constant of drainage of the two circulations. If one assumes a constant total circulatory volume it is possible to compare the effects of varying pericardial and intrathoracic pressures on circulatory flow. That is, Equation 12.7 is differentiated with respect to P_p and P_{th} to yield Equations 12.8 and 12.9.

$$\frac{dQ}{dP_{th}} = \frac{-C_s\,(1 + dTMP_{ra}/dP_{th} - C_p(dTMP_{la}/dP_{th})}{T} \tag{12.8}$$

$$\frac{dQ}{dP_p} = \frac{-C_s\,(1 + dTMP_{ra}/dP_p) - C_p(dTMP_{la}/dP_p + 1)}{T} \tag{12.9}$$

From Equations 12.8 and 12.9 it is clear that the change in blood flow with a change in pericardial pressure (Eq. 12.9) exceeds the change in blood flow due to a change in intrathoracic pressure (Eq. 12.8) by the amount $-C_p/T$. Thus, even if pericardial and intrathoracic pressures had equal effects on the end–diastolic volumes of the ventricles, the change in blood flow due to a change in pericardial pressure would still exceed that due to a change in intrathoracic pressure.

Lung Distension

The lung is in contact with the lateral surfaces of the pericardium. Accordingly, the distended lung is capable of reducing filling of the left ventricle by direct compression. Abnormally high lung volume occurs during continuous positive pressure ventilation with positive end-expiratory pressure (PEEP). Associated with PEEP is a decrease in cardiac output. Various mechanisms for this decrease have been proposed and supported with experimental data, including reduced ventricular volume (Fewell et al. 1980a), ventricular interaction (Robotham and Mitzner 1979, Scharf et al. 1979b), and heart–lung interaction (Cassidy and Ramanathan 1984, Fewell et al. 1980b, Wallis et al. 1983). Here we will focus only on the experimental evidence that demonstrates the influence of lung distension on ventricular function.

Fewell and colleagues (1980b) reported that the decreases in cardiac output, ventricular filling volumes, and transmural ventricular end-diastolic pressures (measured relative to lateral and posterior epicardial surface pressure) that occurred as expiratory pressure was increased were eliminated when the chest was opened and the lungs held away from the heart. Wallis and co-workers (1983) found left and right ventricular pressures at end expiration (arrested heart, closed chest, positive ventilation) to increase with the application of PEEP; even greater increases were obtained during inspiration (Fig. 12.16). This increased ventricular pressure dem-

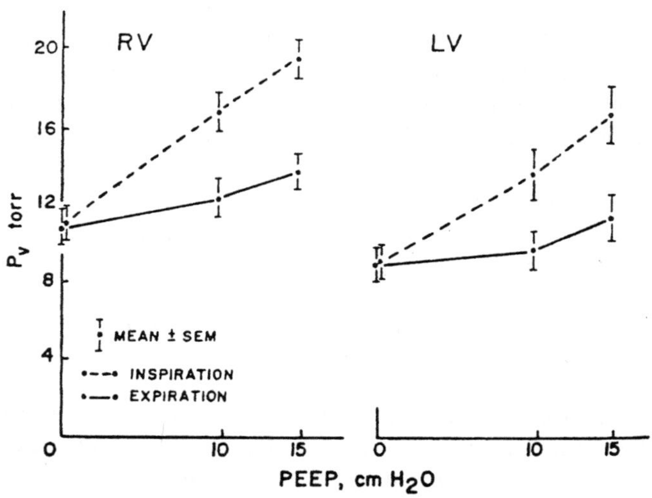

FIGURE 12.16. Inspiratory-expiratory differences in right (RV) and left (LV) ventricular pressures (P_v) during positive end-expiratory pressure ventilation (PEEP) with the pericardium intact and the chest closed (arrested-heart preparation). (From Wallis et al. 1983).

onstrates that a net extracardiac force is applied to the heart by the lungs as lung volume is expanded with PEEP. Recently, Cassidy and Ramanathan (1984) were able to demonstrate radiographically a deformation of the left ventricle with PEEP that could only be due to lung compression. The lateral-septal wall dimension during diastole was reduced with PEEP by an inward displacement of the lateral wall; the septum was not displaced into the left ventricular cavity by PEEP to any greater degree than were the anterior or posterior walls. Thus these findings indicate that marked increases in lung volume result in compressive extracardiac forces of sufficient magnitude to influence diastolic and systolic ventricular function.

Summary

The cardiopulmonary unit is responsible for the transport and exchange of oxygen and carbon dioxide between the metabolizing tissues and the atmosphere. Its functional integration is accomplished through the anatomic arrangement of, and the mechanical interplay between, the components of this unit and a complex neurohumoral control system. In this chapter we have indicated that mechanical interplay between the left and right ventricles, the presence of the pericardium, positive and negative intrathoracic pressures, and abnormal distension of the lungs are associated with extracardiac forces (either compressive or elongating) that are capable of influencing ventricular diastolic and systolic function and hence the functional integration of the cardiopulmonary unit. We emphasized that the net effect of an alteration in one of these extracardiac forces could include the effects of several other extracardiac forces. Finally, we pointed out that the influence of variations in intrapericardial pressure on ventricular function are greater than the influence of variations in intrathoracic pressure.

Bibliography

Alderman, E. L., and Glantz, S. A. 1976. Acute hemodynamic interventions shift the diastolic pressure–volume curve in man. *Circulation* 54:662–671.

Bemis, C. E., Serur, J. R., Borkenhagen, D., Sonnenblick, E. H., and Urschel, C. W. 1974. Influence of right ventricular filling pressure on left ventricular pressure and dimension. *Circ. Res.* 34:498–504.

Black, L. F., and Hyatt, R. E. 1969. Maximal respiratory pressures: normal values and relationship to age and sex. *Am. Rev. Resp. Dis.* 99:696–702.

Bove, A. A., and Santamore, W. P. 1981. Ventricular interdependence. *Prog. Cardiovasc. Dis.* 23:365–387.

Brooks, H., Holland, R., and Al-Sadir, J. 1977. Right ventricular performance during ischemia: an anatomic and hemodynamic analysis. *Am. J. Physiol.* 233:H500–H513.

Cassidy, S. S., and Ramanathan, M. 1984. Dimensional analysis of the left ventricle during PEEP: relative septal and lateral wall displacements. *Am. J. Physiol.* 246:H792–H805.

Charlier, A. A., Jaumin, P. M., and Pouleur, H. 1974. Circulatory effects of deep inspirations, blocked expirations and positive pressure inflations at equal transpulmonary pressures in conscious dogs. *J. Physiol.* 241:589–605.

Elzinga G., Piene, H., and DeJong, J. P. 1980. Left and right ventricular pump function and consequences of having two pumps in one heart: a study on the isolated cat heart. *Circ. Res.* 46:564–574.

Fewell, J. E., Abendschein, D. R., Carlson, C. J., Murray, J. F., and Rapaport, E. 1980a. Continuous positive-pressure ventilation decreases right and left ventricular end-diastolic volumes in the dog. *Circ. Res.* 46:125–132.

Fewell, J. E., Abendschein, D. R., Carlson, C. J., Rapaport, E., and Murray, J. F. 1980b. Mechanism of decreased right and left ventricular end-diastolic volumes during continuous positive-pressure ventilation in dogs. *Circ. Res.* 47:467–472.

Freeman, G. L., and LeWinter, M. M. 1984. Pericardial adaptations during chronic dilation in dogs. *Circ. Res.* 54:294–300.

Glantz, S. A., Misbach, G. A., Moores, W. Y., Mathey, D. G., Lekven, J., Stowe, D. F., Parmley, W. W., and Tyberg, J. V. 1978. The pericardium substantially affects the left ventricular diastolic pressure–volume relationship in the dog. *Circ. Res.* 42:433–441.

Hoffman, J. I. E., Guz, A., Charlier, A. A., and Wilcken, D. E. L. 1965. Stroke volume in conscious dogs: effect of respiration, posture and vascular occlusion. *J. Appl. Physiol.* 20:865–877.

Holt, J. P. 1970. The normal pericardium. *Am. J. Cardiol.* 26:455–465.

Janicki, J. S., Shroff, S., and Weber, K. T. 1981. Extracardiac pressure and ventricular hemodynamics: contrasting effects of pericardial and pleural pressures. *Circulation* 64:IV–54.

Janicki, J. S., Shroff, S. G., and Weber, K. T. 1986. Functional significance of the pericardium. In Sideman, S., and Beyar, R., eds. *Simulation and Control of the Cardiac System.* Boca Raton: Fla.: CRC Press, in press.

Janicki, J. S., and Weber, K. T. 1980a. Factors influencing the diastolic pressure–volume relation of the cardiac ventricles. *Fed. Proc.* 39:133–140.

Janicki, J. S., and Weber, K. T. 1980b. The pericardium and ventricular interaction, distensibility, and function. *Am. J. Physiol.* 238:H494–H503.

Janicki, J. S., and Weber, K. T. 1984. Functional significance of the pericardium in chronic cardiac failure. *Circulation* 70:II–350.

Janicki, J. S., Weber, K. T., Likoff, M. J., and Fishman, A. P. 1985. The pressure–flow response of the pulmonary circulation in heart failure and pulmonary vascular disease. *Circulation,* 72:1270–1278.

Maruyama, Y., Ashikawa, K., Isoyama, S., Kanatsuka, H., Ino-Oka, E., and Takishima, T. 1982. Mechanical interactions between four heart chambers with and without the pericardium in canine hearts. *Circ. Res.* 50:86–100.

Milic-Emili, J., Mead, J., Turner, J. M., and Glauser, E. M. 1964. Improved technique for estimating pleural pressure from esophageal pressure. *J. Appl. Physiol.* 19:207–211.

Pearlman, E. S., Weber, K. T., Janicki, J. S., Pietra, G. G., and Fishman, A. P. 1982. Muscle fiber orientation and connective tissue content in the hypertrophied human heart. *Lab. Invest.* 46:158–164.

Permutt, S., and Caldini, P. 1978. Regulation of cardiac output by the circuit: venous return. In Baan, J., Noordergraaf, A., and Raines, J., eds. *Cardiovascular System Dynamics*. Cambridge, Mass.: MIT Press, pp. 465–477.

Permutt, S., and Riley, R. L. 1963. Hemodynamics of collapsible vessels with tone: the vascular waterfall. *J. Appl. Physiol.* 18:924–932.

Robotham, J. L. 1981. Cardiovascular disturbances in chronic respiratory insufficiency. *Am. J. Cardiol.* 47:941–949.

Robotham, J. L., and Mitzner, W. 1979. A model of the effects of respiration on left ventricular performance. *J. Appl. Physiol.* 46:411–418.

Santamore, W. P., Lynch, P. R., Heckman, J. L., Bove, A. A., and Meier, G. D. 1976. Myocardial interaction between the ventricles. *J. Appl. Physiol.* 41:362–368.

Scharf, S. M., Brown, R., Saunders, N., and Green, L. H. 1979a. Effects of normal and loaded spontaneous inspiration on cardiovascular function. *J. Appl. Physiol.* 47:582–590.

Scharf, S. M., Brown, R., Saunders, N., Green, L. H., and Ingram, R. H. Jr. 1979b. Changes in canine left ventricular size and configuration with positive end-expiratory pressure. *Circ. Res.* 44:672–678.

Sharpey-Schafer, E. P. 1955. Effects of Valsalva's manoeuvre of the normal and failing circulation. *Br. Med. J.* i:693–695.

Sharpey-Schafer, E. P. 1965. Effect of respiratory acts on the circulation. In Hamilton, J., and Dow, P. eds. *Handbook of Physiology,* Sec. 2, *Circulation,* Vol. 3. Washington, D.C.: American Physiological Society, pp. 1875–1886.

Shirato, K., Shabetai, R., Bhargara, V., Franklin, D., and Ross, J. Jr. 1978. Alteration of the left ventricular diastolic pressure–segment length relation produced by the pericardium. *Circulation* 57:1191–1198.

Starr, I., Jeffers, W. A., and Meade, R. H. Jr. 1943. The absence of conspicuous increments of venous pressure after severe damage to the right ventricle of the dog, with a discussion of the relation between clinical congestive failure and heart disease. *Am. Heart J.* 26:291–301.

Stool, E. W., Mullins, C. B., Leshin, S. J., and Mitchell, J. H. 1974. Dimensional changes of the left ventricle during acute pulmonary arterial hypertension in dogs. *Am. J. Cardiol.* 33:868–875.

Tyberg, J. V., Misbach, G. A., Glantz, S. A., Moores, W. Y., and Parmley, W. W. 1978. A mechanism for shifts in the diastolic, left ventricular, pressure–volume curve: the role of the pericardium. *Eur. J. Cardiol.* 7:163–175.

Wallis, T. W., Robotham, J. L., Compean, R., and Kindred, M. K. 1983. Mechanical heart-lung interaction with positive end-expiratory pressure. *J. Appl. Physiol.* 54:1039–1047.

Weber, K. T., Janicki, J. S., Hunter, W. C., Shroff, S., Pearlman, E. S., and Fishman, A. P. 1982a. The contractile behavior of the heart and its functional coupling to the circulation. *Prog. Cardiovasc. Dis.* 24:375–400.

Weber, K. T., Janicki, J. S., Shroff, S. G., and Likoff, M. J. (1983). The cardiopulmonary unit: the body's gas transport system. *Clin. Chest. Med.* 4:101–110.

Weber, K. T., Kinasewitz, G. T., Janicki, J. S., and Fishman, A. P. 1982b. Oxygen utilization and ventilation during exercise in patients with chronic cardiac failure. *Circulation* 65:1213–1223.

13
Sensitivity Analysis of Interaction Between the Left Ventricle and the Systemic Arteries

KENNETH B. CAMPBELL, JOHN A. RINGO,
N. S. PETERSON

Introduction

In an earlier work (Campbell et al. 1984) we proposed that a useful interpretation of interaction between the left ventricle (LV) and systemic arteries (SA) could come from couching the relationship between them in a block diagram form. Such a representation would emphasize the relationship between system components and the flow of information within the system. With the block diagram it was possible to refer to elementary linear system concepts, to understand how a change in LV/SA properties might affect LV/SA behavior. Predictions based on these elementary considerations were compared with more detailed calculations of sensitivities derived from mathematical models of the system, and there was general agreement between the predictions and the calculations. Even though the LV/SA system was nonlinear and time varying, the simple linear block diagram analysis was useful in generating insights and hypotheses about interactions within the system. We felt that this approach would prove fruitful in relating LV/SA interaction to other cardiovascular interactions as more and more aspects of the cardiovascular system were taken into account.

LV/SA Sensitivities at Basal Physiologic State

The variables of primary interest in LV/SA interaction are flow, Q, out of the LV and into the SA, and aortic pressure, P_a, at the interface between these two system components. When Q was considered in the block diagram representation (Fig. 13.1), LV and SA related to one another as LV in a feedforward configuration and SA in a feedback configuration. When P_a was considered, both LV and SA were configured as feedforward components. In engineering, feedback is often employed to reduce the sensitivity of the output variable to changes in the feedforward system component. This led us to predict that Q would be more sensitive to changes

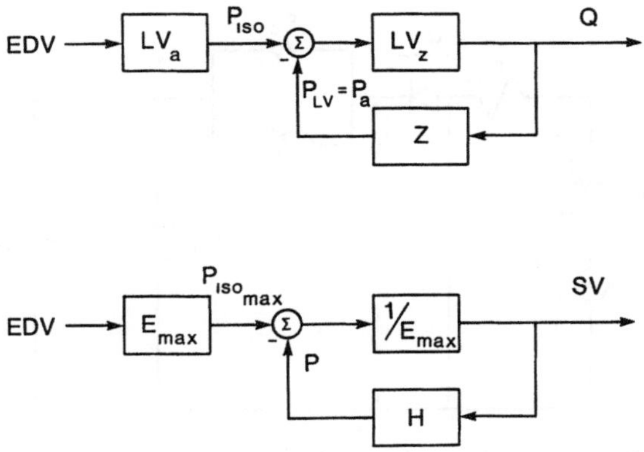

FIGURE 13.1. Block diagram representation of left ventricle (LV)/systemic arterial (SA) interaction. *Top,* generalized diagram splits the LV functions into active functions, LV_a, and loading functions, LV_z. LV_a governs the transformation between end-diastolic volume, EDV, and isovolumic pressure, P_{iso}. LV_z governs the conversion between LV pressure and LV outflow, Q. The pressure that is being converted to flow is the instantaneous difference between P_{iso} and left ventricle pressure, P_{lv}. P_{lv} is not different from aortic pressure, P_a, while the aortic valves are open. P_a is generated as a result of Q encountering the arterial input impedance, Z. *Bottom,* reduced version of generalized diagram in which end state rather than instantaneous values of the respective variables are represented. See text for rationale of this reduced version and definition of symbols.

in SA than to changes in LV, while P_a would be equally sensitive to changes in both LV and SA.

To test these predictions we used an LV/SA model that employed time-varying elastance, $E(t)$, and internal resistance, R_{lv}, as LV pump properties and a lumped second-order model of arterial input impedance that included total peripheral resistance, R, as a model of SA load properties. An electrical analog of this model is shown in Figure 13.2. Parameters for the LV and SA components of the model were estimated separately by using heuristic optimization techniques to fit model predictions to measured data. Two sets of measured data were used for estimating parameters of the LV component. One set was obtained from recordings of LV pressure, P_{lv}, and outflow during a normal ejecting beat, and the second set was obtained from LV pressure recorded during an isovolumic beat, P_{iso}. This technique of LV identification had two advantages: (1) It minimized the number of parameters needed for characterizing $E(t)$ by placing much of the time dependence of $E(t)$ in the isovolumic pressure record according to

$$E(t) = \frac{P_{iso}(t)}{a_1 + a_2 * t} \tag{13.1}$$

where a_1 and a_2 are parameters to be estimated, and (2) it removed the

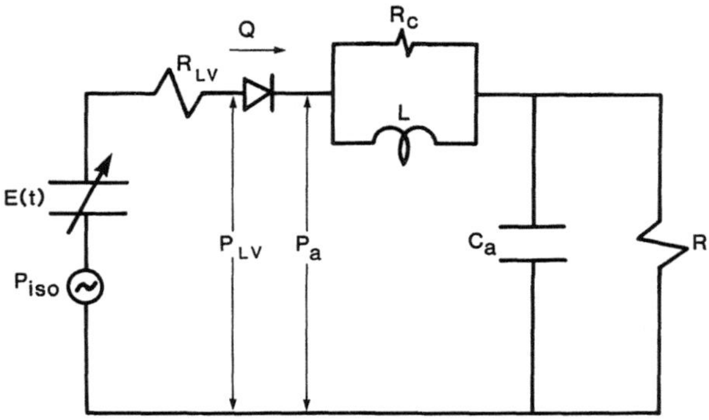

FIGURE 13.2. Electrical analog of block diagram model for studying left ventricle (LV)/systemic artery (SA) interaction. The aortic valves are modeled as a perfect diode. The structure to the left of the diode is the LV component, while the structure to the right is the SA component. The parameters of each component were identified independently and then were merged to predict left ventricular pressure (P_{lv}), aortic pressure (P_a), and left ventricular outflow (Q). $E(t)$, time-varying elastance; R, peripheral resistance; P_{iso}, isovolumic beat; R_{lv}, left ventricular resistance; R_c and L, components of characteristic impedance.

necessity of measuring LV volume for estimation of $E(t)$. This second advantage occurred because the parameterization of $E(t)$ allowed P_{lv} to be expressed as

$$P_{lv} = E(t)_*(a_1 + a_{2*}t - V_e) - R_{lv*}Q \tag{13.2}$$

where V_e is ejected volume. The parameters a_1, a_2, and R_{lv} were adjusted to minimize the sum of squares of the difference between predicted and observed P_{lv}. Data for estimating parameters of the SA component of the model were obtained from aortic root pressure and flow records. Once the basal parameters were estimated for each component, a descriptive test for model validation was conducted. The values of the independently estimated parameters were introduced into equations describing the merged LV/SA model, the equations were solved, and predictions from the merged model were compared with the measured records. The model predictions of LV and aortic pressure and flow closely matched the actual measurements and the model was considered valid insofar as it accounted for those features of the system that are important to LV/SA interaction.

The model was used to calculate sensitivities of flow and pressure in the LV/SA system to changes in $E(t)$ and R. Because sensitivities of Q and P_a are functions of time, the non-time-varying sensitivities of the related time-averaged variables—stroke volume, SV, and mean arterial pressure, P—were calculated. Sensitivities of SV and P to changes in $E(t)$ and R were calculated by a model empirical approach. For instance, the sensitivity of SV to R, S_R^{sv}, was calculated as the percent change in SV for a 10% change in R. The 10% change in R was made by adjusting R

by 5% of its basal value on either side of the basal value. The model was used to predict SV at each value of R. The resulting incremental change in SV was then divided by SV at the basal value of R to get the percent change in SV. Sensitivity was finally calculated as the percent change in SV divided by the percent change in R.

At the nominal values of LV and SA properties found in one open-chest dog, the magnitudes of the sensitivities of SV and P to changes in $E(t)$ and R were ranked as

$$\begin{array}{ccccccc} |S_R^{SV}| & > & |S_R^P| & > & |S_{E(t)}^{SV}| & \sim & |S_{E(t)}^P| \\ 0.51 & & 0.42 & & 0.38 & & 0.36 \end{array}$$

where S_R^{sv} is the magnitude of the sensitivity of SV to R, S_R^P is the magnitude of the sensitivity of P to R, and so on. Of the sensitivities, only S_R^{sv} is negative, that is, an increase in R causes a decrease in SV. This ranking was consistent with the preliminary expectations derived from consideration of the block diagram in Figure 13.1; the output variable SV was more sensitive to properties of the feedback component, R, than to properties of the feedforward component, $E(t)$.

The spread of sensitivity magnitudes in this dog, however, was not very great: $|S_R^{sv}| = 0.51$ and $|S_{E(t)}^{sv}| = 0.38$. Since reporting these values we have found that values and rankings of sensitivities vary from dog to dog. Common to all dogs, however, is the finding that $|S_R^{sv}|$ and $|S_{E(t)}^{sv}|$ always possess values close to one another in the neighborhood of 0.5.

Further analysis of relations in a linear feedback system gave greater insight into the first findings. In the standard linear feedback system with output Y, feedforward component with transfer function G, and feedback component with transfer function H, the sensitivity of Y to G and to H depends on the properties of the system according to

$$S_G^Y = \frac{1}{1 + G*H} \tag{13.3}$$

and

$$S_H^Y = \frac{-G*H}{1 + G*H} \tag{13.4}$$

The $G*H$ product is the prime determinant of sensitivities. For values of $G*H$ much greater than 1, $|S_H^y| > 0.5 > |S_G^y|$; for $G*H$ near 1, $|S_H^y|$ approximates $|S_G^y|$ and both have values near 0.5; and for $G*H$ much less than 1, $|S_H^y| > 0.5 > |S_G^y|$.

To apply this analysis to the LV/SA system, we formulated a reduced LV/SA model based on end-diastolic and end-systolic LV states and average pressure and flow states in the SA. In this model, SV $= G*(P_{iso} - P)$ and $P = H*$SV. We assumed that P very nearly equals end-systolic pressure and that the appropriate end-state value of P_{iso} is its maximal value. With these assumptions, it can be shown that in an elastance-resistance LV model, $P_{iso} = E_{max}*(EDV - V_d)$, where E_{max} is the maximal value of $E(t)$, and V_d is dead volume, and $G = 1/E_{max}$. From these relationships and simple algebra applied to the block diagram in Figure 13.1, the following can be derived:

$$SV = \left(\frac{E_{max}}{E_{max} + H}\right)*EDV = \left(\frac{1}{1 + G*H}\right)*EDV \tag{13.5}$$

and

$$S_G^{SV} = \frac{-\ G*H}{1 + G*H} = S_H^{SV} \tag{13.6}$$

Further, $S_G^{sv} = -S_{Emax}^{sv}$. From Sunagawa and co-workers (1983)

$$P = \frac{R}{t_s + T*[1 - \exp(-\ t_d/T)]}*SV \tag{13.7}$$

where t_s = systolic interval

T = time constant of pressure decay in the aorta during diastole

and t_d = diastolic interval

Therefore

$$H = \frac{R}{t_s + T*[1 - \exp(-\ t_d/T)]} \tag{13.8}$$

and $S_H^{sv} = S_R^{sv}$. Typical values from our data give $t_s = t_d = 0.25$ sec, $T = 0.78$ sec, and $R = 2900$ dyne sec cm^{-5}. Substituting these values into the equation for H gives $H = 6250$ dyne/cm^5. Typical E_{max} was 6500 dyne cm^5. Since in this system

$$G*H = \frac{H}{E_{max}} = \frac{R}{E_{max}*\{t_s + T*[1 - \exp(-\ t_d/T)]\}} \tag{13.9}$$

the $G*H$ product was indeed very nearly 1 and we would predict both S_R^{sv} and $S_{E(t)}^{sv}$ to have magnitudes close to 0.5. This was observed.

Further confirmation came from calculations on data from Sunagawa and colleagues (1983), in which t_s, = 0.16 sec, $t_d = 0.31$ sec, $T = 1.2$ sec, $R = 3.0$ mm Hg sec/ml, and $E_{max} = 5.4$ mm Hg/ml. These gave a $G*H$ product of 1.28 and $|S_R^{sv}| = |S_{Emax}^{sv}| = 0.58$. Calculation from the experimentally derived data of these authors gave $|S_R^{sv}| = 0.46$. There were no data that allowed calculation of S_{Emax}^{sv}. The estimates from our own data and from those of Sunagawa agree well with one another and with our earlier interpretations and conclusions. This further suggests the usefulness of the reduced model and block diagram approach to understanding LV/SA interaction.

LV/SA Sensitivities Depend on Physiologic State

Changes in *E*(*t*) and *R* may change the effective *G*∗*H* product. Thus one would expect strong dependence of the sensitivities on the values of the properties in terms of both absolute magnitudes and relative rankings. The consequence would be that the most effective strategy for bringing about a change in either SV or *P* in one condition (e.g., normal resting state) may be quite different from that in another (e.g., exercise, shock, heart failure, etc.). Such differences must be appreciated for the design of effective cardiovascular therapies and are important to an understanding

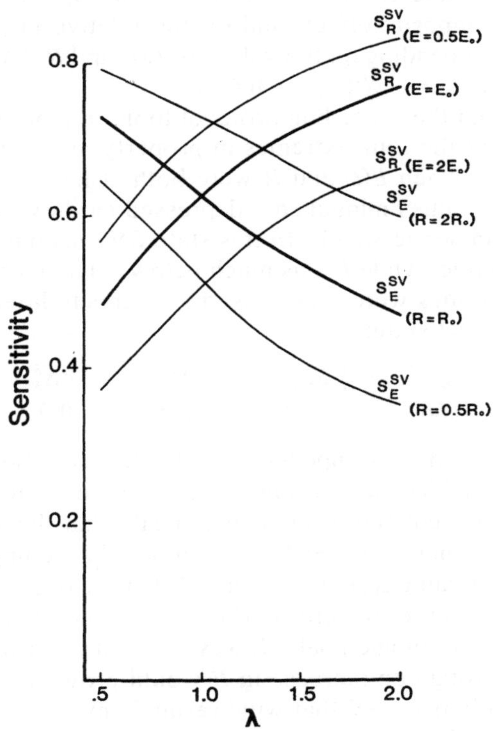

FIGURE 13.3. Dependence of model-derived left ventricle (LV)/systemic artery (SA) stroke volume sensitivities on LV/SA properties. The independent variable, λ, is the multiplying parameter that was used to change the selected property, either *E*(*t*) or *R*. For instance, while *E*(*t*) was held constant at its nominal value, $S\frac{\text{SV}}{R}$ was calculated at 0.1 increments over the range $0.5{*}R_o$ to $2{*}R_o$, where R_o is the nominal value estimated under basal physiologic conditions. This was repeated for *E*(*t*) held constant at 0.5 and 2.0 times its nominal value (E_o). A similar procedure was used in the calculation of $S\frac{\text{SV}}{E(t)}$.

of natural cardiovascular control mechanisms that move the cardiovascular system from one state of operation to another.

Sensitivity calculations were conducted at numerous values of properties over a broad range of physiologic states using the following protocol. $E(t)$ and R were varied independently in 10% increments over the range of one half to two times their basal values. At each increment in the range of values, $S^{sv}_{E(t)}$ $S^{P}_{E(t)}$, S^{sv}_{R}, and S^{P}_{R} were calculated. Results for $S^{sv}_{E(t)}$ and S^{sv}_{R} are shown in Figure 13.3, where several features are demonstrated: (1) There were significant changes in $S^{sv}_{E(t)}$ and S^{sv}_{R} with changes in either $E(t)$ or R; (2) increases in R above its basal value caused increases in both $S^{sv}_{E(t)}$ and S^{sv}_{R} while decreases in R had the opposite effect; (3) increases in $E(t)$ above basal value caused decreases in both $S^{sv}_{E(t)}$ and S^{sv}_{R}, while decreases had the opposite effect; and (4) the relative magnitudes of $S^{sv}_{E(t)}$ and S^{sv}_{R} changed depending on the values of $E(t)$ and R. (At the basal state in this dog, $|S^{sv}_{R}| = |S^{sv}_{E(t)}| = 0.63$.)

The results from the preceding protocol took on more relevance when the sensitivities at the four extremes in property values were examined. One extreme was when $E(t)$ and R were both at half their basal values. These parameter values mimicked a depressed cardiovascular system as might be found in septic shock. In this state SV was only slightly below (95%) its basal value, while P was much below (51%) its basal value. The sensitivity magnitudes were now ranked very much differently than was the case in the basal state:

$$\begin{array}{ccccccc} |S^{P}_{E(t)}| & \sim & |S^{SV}_{E(t)}| & > & |S^{SV}_{R}| & > & |S^{P}_{R}| \\ 0.66 & & 0.64 & & 0.57 & & 0.29 \end{array}$$

Because the sign of $S^{sv}_{E(t)}$ is opposite that of S^{sv}_{R}, restoration of the system by increasing both $E(t)$ and R would have competing effects on changing SV. However, the sensitivity rankings indicate that the improvement in SV realized from increasing $E(t)$ would more than compensate for the decline in SV that came from increasing R. On the other hand, increases in $E(t)$ and R both serve to increase P and do not conflict in achieving an elevated P as a physiologic goal. However, an attempt to achieve an elevated P only through an increase in R would prove disastrous because of the adverse effects on SV that wuld result from an increase in R alone with no compensation from an increase in $E(t)$.

A second extreme was when $E(t)$ and R were twice their basal values. This mimicked a condition of a generalized increase in sympathetic activity. Here, because of the competing effects of $E(t)$ and R, SV was not different from its basal value (99.5%). However, $E(t)$ and R have complementary effects on P, and P thus was elevated to 193% of its basal value. The sensitivity magnitudes were ranked as

$$\begin{array}{ccccccc} |S^{P}_{E(t)}| & \sim & |S^{SV}_{R}| & > & |S^{SV}_{E(t)}| & > & |S^{P}_{R}| \\ 0.68 & & 0.66 & & 0.60 & & 0.23 \end{array}$$

The sensitivity rankings with generalized activation are much the same as in the previous case of generalized depression. Again, consideration must be given to the competing effects of $E(t)$ and R on SV when an optimal strategy is devised for returning P to its normal value.

A third extreme was when $E(t)$ was twice and R was half its basal value. Such changes would occur during exercise. In this condition, SV was increased to twice its basal value but P was maintained nearly equal to (105%) its basal value. The sensitivity magnitudes were ranked

$$\begin{array}{ccccccc} |S_R^P| & > & |S_R^{SV}| & \sim & |S_{E(t)}^P| & \sim & |S_{E(t)}^{SV}| \\ 0.52 & & 0.37 & & 0.36 & & 0.35 \end{array}$$

These rankings, in which the properties $E(t)$ and R have been changed in opposite directions, are quite different from the earlier rankings in generalized activation and generalized depression, in which the properties had been changed in the same direction. Whereas P had not been responsive to changes in R in both generalized activation and generalized depression, it now becomes very responsive to changes in R under conditions that mimic light exercise, even though its value has not changed.

The fourth extreme was when $E(t)$ was half and R was twice its basal value. This distribution of values mimicked that found in congestive heart failure. Here SV was one third and P was two thirds its basal value. The sensitivity magnitude rankings were

$$\begin{array}{ccccccc} |S_R^{SV}| & > & |S_{E(t)}^P| & \sim & |S_{E(t)}^{SV}| & > & |S_R^P| \\ 0.83 & & 0.80 & & 0.79 & & 0.14 \end{array}$$

The high sensitivity of SV to R emphasized once again the improvement in SV that can be realized from reducing R in congestive failure. Further, the improvement in SV comes about without a large decrease in P because P is not very sensitive to R in this case. Of course, more complete restoration of both SV and P requires an increase in $E(t)$ in addition to the decrease in R.

A partial summary of these results is shown in Figure 13.4. S_R^{SV} and $S_{E(t)}^{SV}$ are displayed in histogram form for the basal state and the four states just reviewed. Two generalizations can be made from the information in this figure. First, at any given state over the full range of properties investigated, SV is about equally sensitive to changes in R and in $E(t)$. This can be contrasted to P (data not shown in figure), which exhibited a fivefold difference in sensitivity to $E(t)$ and R in heart failure $S_{E(t)}^P = 0.80 > S_R^P = 0.14$), virtually no difference in the basal state, and a reversal of these rankings in exercise $S_R^P = 0.52 > S_{E(t)}^P = 0.36$). The second generalization is that conditions in which $E(t)$ and R change in the same direction, as in generalized activation or generalized depression, do not bring about an appreciable deviation from the basal value in the sensitivity of SV to either $E(t)$ or R, but conditions in which $E(t)$ and R change in opposite directions,

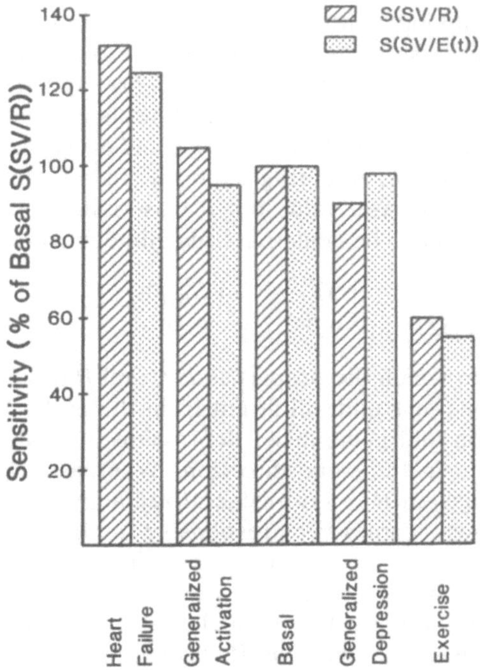

FIGURE 13.4. Comparison of sensitivity (S) of stroke volume (SV) to changes in total peripheral resistance, (R), $S\frac{\text{SV}}{R}$, and to changes in time-varying elastance [$E(t)$], $S\frac{\text{SV}}{E(t)}$, in simulations of different physiologic states. These two sensitivities possessed similar values over a broad range of parameter variations. Changes in sensitivity values were in accordance with predictions from linearized reduced version of the detailed model.

as in heart failure or exercise, cause marked deviations from the basal value in the sensitivity of SV to these properties. In exercise, where SV is elevated, SV is much less sensitive to properties of the LV/SA system than in the basal state, while in heart failure, where SV is depressed, SV is much more sensitive to these properties.

The preceding calculated dependence of sensitivities on physiologic state can be compared with expectations from the block diagram analysis. As can be seen from Equation 13.9, equal proportional changes in $E(t)$ and R would not change the $G{*}H$ product. Therefore we would predict that in generalized activation and generalized depression $S^{\text{sv}}_{E(t)}$ and S^{sv}_{R} would not be different from their basal values nor different from one another. Indeed, this was the case. On the other hand, changes in $E(t)$ and R of different proportionalities and in different directions, as in simulated heart failure and simulated exercise, would be expected to cause $S^{\text{sv}}_{E(t)}$ and S^{sv}_{R} to deviate from their basal values together and by the same amount. As can be seen in Figure 13.4, this was the case. Also, the magnitude of deviation from basal value was approximately that predicted from the block diagram analysis. This gives even further validity to the linear block diagram approximation of the nonlinear time-varying interacting LV/SA system.

LV/SA Sensitivities Change When the LV/SA System Becomes Part of the Heart–Lung/SA System

The sensitivities in the LV/SA model were determined while holding the input into the LV/SA system, that is, end-diastolic pressure, or EDP, constant. However, when the LV/SA system is made part of a larger section of the cardiovascular system, such as the SA-loaded heart–lung blood pump, a change in an LV/SA function will bring about a change in EDP as well as a change in Q and P_a. Thus the SV and P response to a change in $E(t)$ or R will be the result of both a change in property and a change in EDP. The respective sensitivities will reflect these changes. It is expected that the sensitivities of SV and P to changes in $E(t)$ and R will be different in the heart–lung/SA system than in the LV/SA system.

A heart–lung/SA model was constructed that consisted of the following serially connected components: right ventricle, pulmonary circulation, left ventricle, and systemic arterial impedance. An electrical analog version of the model is shown in Figure 13.5. In the heart–lung/SA model, the right atrial pressure served as system input and was held constant. As before, left ventricle SV and P were taken as model outputs. The LV/SA section of the model was identical to that used in the previous studies with the exception that a nonlinear passive elastance was added in series with $E(t)$ as an LV component. This passive elastance was needed to represent realistically changes in LV end-diastolic volume when the LV receives continued inflow from the pulmonary circulation while SV has been decreased due to increased R or decreased left ventricular $E(t)$.

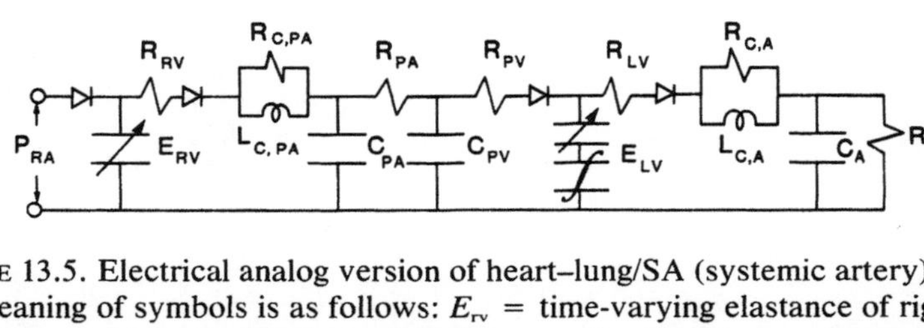

FIGURE 13.5. Electrical analog version of heart–lung/SA (systemic artery) model. The meaning of symbols is as follows: E_{rv} = time-varying elastance of right ventricle; R_{rv} = right ventricle internal resistance; $R_{c,Pa}$ and $L_{c,Pa}$ = components of characteristic impedance of the pulmonary artery; C_{pa} = pulmonary arterial compliance; R_{pa} = resistance of pulmonary arteries; C_{pv} = pulmonary venous compliance; R_{pv} = resistance to flow from pulmonary circuit into LV; E_{lv} = time-varying elastance of LV including a nonlinear passive component; R_{lv} = internal LV resistance; $R_{c,a}$ and $L_{c,a}$ = components of systemic arterial input impedance; C_a = arterial compliance; R = total peripheral resistance.

Validation of the heart–lung blood pump component was performed by simulating the experiment of Herndon and Sagawa (1969) and comparing our model-predicted results with the experimental results reported by these workers. The comparison is shown in Figure 13.6, where mean aortic flow is plotted against right atrial pressure and P over a broad range of these two variables. There is good qualitative agreement between model prediction and experimental observation. The differences between the model-predicted and experimentally observed surfaces are discussed in detail elsewhere (Peterson and Campbell 1985). What is important here is that both model-predicted and experimental surfaces are very shallowly sloped in the region of low P and high right atrial pressure. Further, both surfaces demonstrate a shallow slope of Q versus P for values of P below 150 mm Hg. This agreement between prediction and experiment provides the needed validation and sets the stage for anticipating the results that follow.

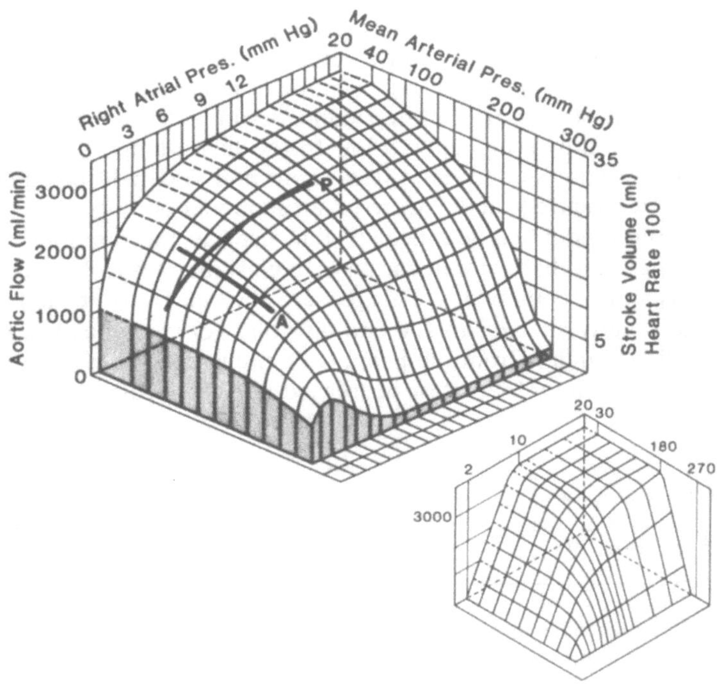

FIGURE 13.6. Results from model experiments and actual experiments (inset) as reported by Herndon and Sagawa (1969). Similarities and difference between results are discussed in the text.

Sensitivities were calculated as before. The values and rankings of the sensitivities were very much different in the heart–lung/SA system than in the LV/SA system. Most dramatically, S_R^P increased by a factor of 2, while S_R^{sv}, $S_{E(t)}^{sv}$, and $S_{E(t)}^P$ all decreased to approximately two thirds to one half their value in the LV/SA system. The sensitivity rankings in the heart–lung/SA system were as follows:

$$| S_R^P | >> | S_R^{SV} | \sim | S_{E(t)}^{SV} | \sim | S_{E(t)}^P |$$

There is a dramatic change in the SV and P sensitivities to changes in $E(t)$ and R as one goes from the LV/SA system to the heart–lung/SA system. Because of the concomitant increase in complexity as larger aspects of the cardiovascular system are being considered, the resultant changes in sensitivities cannot always be anticipated without an analysis of the informational relationships within the system.

A block diagram representation of the interacting heart–lung/SA system is given in Figure 13.7. When Q is taken as the output, there is a feedback path around the LV/SA system placing the entire system in a feedforward position. With the LV/SA system part of a system inside a feedback loop, the sensitivity of SV to changes in any part of the LV/SA subsystem would be expected to be less than it was when the LV/SA system was considered in isolation. On the other hand, when P_a is considered the output, the SA lies outside this circular path and the sensitivity of P_a to a change in SA would not be expected to be reduced.

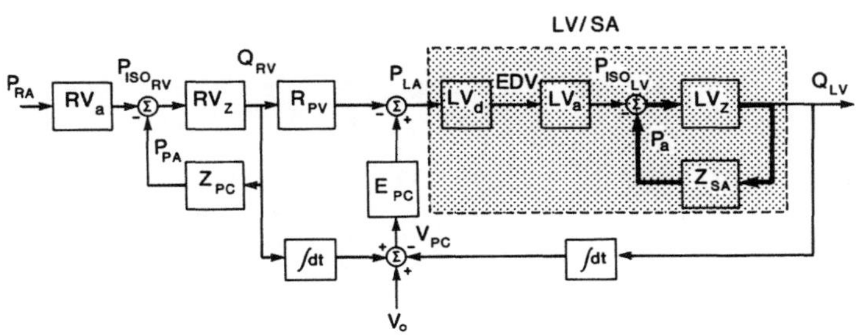

FIGURE 13.7. Block diagram representation of interactions within heart–lung/SA (systemic artery) model. Symbols not identified in earlier figures include RV_a = active function of RV properties, RV_z = loading function of RV properties; Z_{pc} = input impedance of pulmonary circulation; E_{pc} = lumped elastance of pulmonary circulation; V_o = initial volume within pulmonary circulation; and V_{pc} = resident volume in pulmonary circulation. The diagram shows that the entire left ventricle (LV)/SA system is buried within a feedback loop when the LV/SA is considered a component of this larger system.

Conclusion

There is a dramatic change in *P* and SV sensitivities to changes in properties of the LV pump and SA load when the LV/SA system is considered a component of larger aspects of the cardiovascular system. Block diagram representations help to interpret the changes that result, because they elucidate the nature of the complexity, that is, the number of interacting parts, sites where interactions occur, and the amount of feedback. Such representations offer a framework for anticipating how system response will change over a broad range of property values. An additional advantage is that block diagrams point the way for considering "reduced" representations that allow prediction of the behavior of a complex system without recourse to computer models.

Bibliography

Campbell, K. B., Ringo, J. A., Neti, C., and Alexander, J. E. 1984. Informational analysis of left-ventricle/systemic-arterial interaction. *Ann. Biomed. Eng.* 12:209–231.

Sunagawa, K., Maughan, W. L., Burkhoff, D., and Sagawa, K. 1983. Left ventricular interaction with arterial load studied in isolated canine ventricle. *Am. J. Physiol.* 245:H773–H780.

Herndon, C. W., and Sagawa, K. 1969. Combined effects of aortic and right atrial pressure on aortic flow. *Am. J. Physiol.* 217:65–72.

Peterson, N. S., and Campbell, K. B. 1985. Teaching cardiovascular integrations with computer laboratories. *The Physiologist* 28:447–448.

14 Modeling of the Coupling of the Left Ventricle to the Peripheral Circulation

LEWIS K. WALDMAN

Introduction

The interaction of the heart and circulation has been of long-standing interest to physiologists, cardiologists, and researchers in biomechanics. In the absence of neural and chemical controls of the circulation, the mechanical coupling of these systems is an important field in itself that must be studied thoroughly before the fully controlled circulation can be understood clearly. Both theoretical and experimental approaches are needed. This study presents a self-consistent theoretical model of a contracting spherical ventricle along with its accompanying outflow valve and circulatory bed (Waldman 1982). The primary emphasis of the analysis is to account for the aortic valve realistically so that its interaction with a model left ventricle and systemic circulation can be studied during ejection when both the ventricular preload and the circulatory afterload vary with time. To this end, theoretical models of each of the three subsystems are derived. The model ventricle is studied independently during isovolumetric contraction to obtain a realistic pressure preceding ejection. Behavior of the ventricle is studied further by allowing it to contract against a known pressure profile. Then the aortic valve is examined with a specified pressure drop. Finally, coupling of the ventricle, valve, and peripheral bed during ejection is examined.

The next section attempts to bridge the gap between ''isolated'' muscle mechanics and ventricular function. The mechanics of a contracting spherical membrane are discussed. This model represents a left rather than right ventricle because of the greater symmetry and mechanical dominance of the larger ventricle. Stresses generated by contracting muscle fibers are considered a type of membrane stress, so that the well-developed shell theory of continuum mechanics can be used to relate these stresses to external loading, in particular to the pressure developing inside the membrane. These stresses are integrated across the shell thickness, accounting in an average sense for fiber orientation and reducing the problem to a study of the mechanics of the midsurface of the shell, whereby Kirchhoff's hypothesis can be utilized. Then these "active" stresses are related to deformation of the membrane with use of a modified Hill equation (Fung 1970). Frequently this kind of relation has been used to describe the force–length relationship of isolated papillary muscles, but here it is used in an average or "global" sense to account for motion of an entire ventricle. In this way a pressure–displacement or pressure–volume relation can be obtained for the ventricle. In the absence of volume changes, this relation yields the pressure rise during isovolumetric contraction. Various assumptions of this part of the analysis are crude: the small strain hypothesis, surface uniformity of the fiber-winding pattern, and through-wall uniformity of sarcomere lengths. Nevertheless, owing to lack of data regarding many of the parameters, this ventricular model provides a starting point that is consistent with the level of approximation of the remaining coupled subsystems.

Due to the emphasis here on realistic valve modeling to account for coupling during ejection, a discussion of valve mechanics follows. While valve mechanics is a large field in itself, the one-dimensional nonstationary modeling of valve motion is consistent with the ventricular modeling just described. Therefore, in the third section of this chapter, previous work of Lee and Talbot (1979) on valve opening and closure is described and derived in a form more suitable for this study. Because of subsequent difficulty concerning valve closure in the fully coupled model, a study is made of the "isolated" valve closing against a known pressure gradient.

In the final section, a simple model of the peripheral circulation is presented. Then coupling during valve opening and closure is described. A detailed matching of ventricle and valve motion to a branching system of arteries, capillaries, and veins is premature at this time. First, coupling to a simpler model is studied; a Windkessel theory represents the canine peripheral bed. Again, the degree of approximation here is consistent with the other subsystem models. Furthermore, the number of parameters in these initial coupling studies is minimized to permit clearer observation of the effect of parameter variation. Parameters of the peripheral circulation model are varied to show their influence on ventricular and valve motion during valve opening. Valve closure of the coupled model is not as successful, so no parametric studies are reported for this part of systole. Finally, some conclusions are drawn. Possible inadequacies of the valve model and surrounding systems that may contribute to inadequate valve closure and ejection fractions smaller than "physiologic" are discussed, and suggestions for continuing research are made.

Cardiac Mechanics

Starting with the axisymmetric field equations of a spherical membrane, a simplified pressure–flow relationship for a model heart ventricle is derived. This relation, under further restriction, governs the pressure rise during isovolumetric contraction. Once parameters are chosen that yield a physiologic pressure rise preceding ejection, the full relation is utilized to calculate volume change during ejection against a given pressure profile. Later in the chapter this relation serves as one of a system of equations in the coupling analysis. In this analysis, pressure is assumed to be spatially uniform in the ventricle. The discrepancy between the pressure load at the mid surface of a thick-walled shell and the load at the inside surface (e.g., the endocardium of the left ventricle) is neglected. In addition, strains are assumed to be small, and bending moments as well as external forces are ignored. Within these constraints some account is taken of the overall effect of the fiber-winding pattern (Streeter 1979). Here a modified Hill equation (Fung 1970) describes "active" stresses generated by the contractile machinery, while "passive" stresses in the parallel element are ignored.

The modified Hill equation, which relates "active" force or stress in cardiac fibers to ventricular deformation in this analysis, has been used to describe the behavior of papillary muscles (Fung 1970). It is an empirical relation that synthesizes the sliding filament theory (Huxley 1957) with Hill's force–velocity relation (Hill 1938). Abbott and Mommaerts (1959) first suggested the applicability of the Hill model to cardiac muscle mechanics. Previously only the motion of skeletal muscle had been correlated with the Hill model. Since then many have attempted to correlate various versions of the Hill model with the behavior of heart muscle (Brady 1965; Brutsaert, Claes, and Sonnenblick 1971; Edmund and Nilsson 1972; Fung 1970; Pinto and Fung 1973; Sonnenblick 1965).

Originally the model consisted of a contractile component obeying the force–velocity relation of Hill in series with a nonlinear elastic spring. Sonnenblick (1962), in observations of the extrapolated velocity of unloaded shortening (v_o), claimed that v_o was independent of initial muscle length. He proposed the use of v_o as an index of the intrinsic contractile state of the muscle. However, Pollack (1970), on reexamining the data of Sonnenblick and others, declared that v_o was not independent of initial muscle length. Following a number of other authors, Pollack found that a third element, another nonlinear spring, was needed to model the higher resting tension observed in cardiac muscle. This third element can be located either in parallel with the contractile element only (the so-called Voigt model) or in parallel with both of the other elements (the so-called Maxwell model) (Fig. 14.1A). These two versions of the model are equivalent because a choice of parameters for the Voigt model uniquely defines the parameters of the Maxwell model (Fung 1971).

Unlike skeletal muscle, cardiac muscle is difficult to tetanize. Techniques (Blinks and Jewell 1972) used in skeletal muscle experiments to follow the time course of the active state enabled Sonnenblick (1965) and Brady (1965) to show that the active state has a slow onset in cardiac muscle, unlike its rapid onset in skeletal muscle. By allowing muscle segments to contract isometrically and then releasing them against a constant

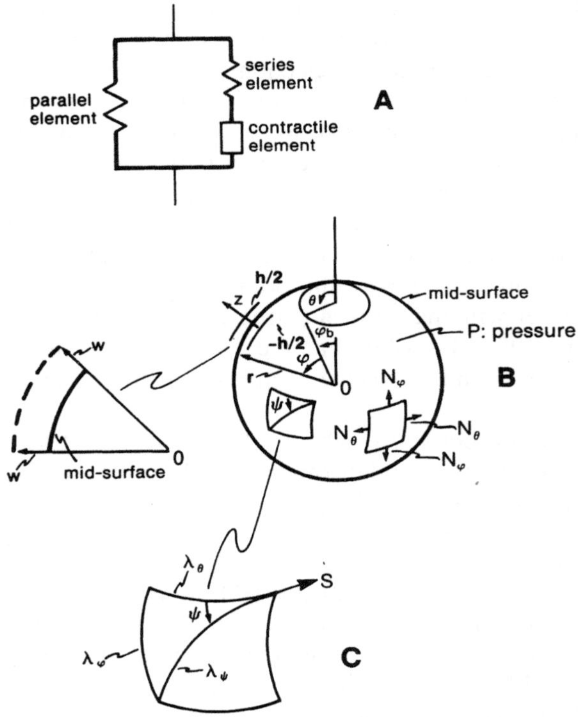

FIGURE 14.1. A, three element model; B, spherical membrane; C, stretch ratios. r, midsurface radius; P, internal pressure; ϕ, latitudinal coordinate; ϕ_b, latitudinal coordinate of valve ring; Θ, longitudinal coordinate; N_φ longitudinal stress resultant; N_Θ, circumferential stress resultant; z, normal coordinate; h, shell thickness; ψ, fiber angle; w, radial displacement; λ_Θ, circumferential stretch ratio; λ_φ, longitudinal stretch ratio; λ_ψ, fiber stretch ratio; S, series element stress.

afterload at a fixed time after stimulation, Edmund and Nilsson (1968) obtained force–velocity curves at a constant level of active state and presumably without a change in length of the contractile element. In this way, the maximum velocity of shortening was obtained at different times during a twitch, and the time dependence of the active state was quantified. Although Sonnenblick (1965), along with Edmund and Nilsson (1968), found a reasonable correlation between these experiments and Hill's model at all muscle lengths, Brady (1965), Noble, Bowen, and Hefner (1969), and Brutsaert, Parmley, and Sonnenblick (1969) found force–velocity curves that were not hyperbolic at muscle lengths below the length at which maximum isometric tension is generated. By introducing the parameter n in the Hill equation, Fung (1970) was able to reconcile the differences between these hyperbolic and nonhyperbolic results. In the present study, n is set equal to 1 as in the original Hill model, although a value of 0.5 could be incorporated to better account for the nonhyperbolic data.

Noble and Else (1972) as well as Pollack, Huntsman, and Verdugo (1972) found that tension–length curves generated in "quick-release" experiments on papillary muscles varied with the time of release. However, McLaughlin and Sonnenblick (1974) claimed that the release velocities were too slow in the preceding experiments, so that the apparent time dependence was an artifact.

Using laser diffraction techniques, Nassar, Manring, and Johnson (1974) were able to follow sarcomere length changes in the central portions of their trabecular (rather than papillary) preparations. In this way, artifacts due to the damaged tissue near the ends of the samples were eliminated. Nevertheless, the traditional concepts of active state and force–velocity relations in striated muscle may be obsolete (Brady 1979). In addition, the classical concept of myofilament overlap used to explain the Frank–Starling law of the heart is clouded by concepts of shortening inactivation and partial activation of the muscle (Endo 1973, Fabiato and Fabiato 1975, Taylor and Rudel 1970). Furthermore, the nonuniformity of sarcomere lengths in multicellular preparations (Nassar, Manring, and Johnson 1974; Pollack and Krueger 1976) has pointed to the importance of attempting to study single-cell preparations. Recent work on single cardiac cells includes research on frog atrial cells (Tarr et al. 1979,1981).

Despite the ever-increasing volume of criticism against classical concepts of muscle mechanics, no better or more comprehensive theory yet exists. Therefore, Fung's modified Hill equation (Fung 1970) is the "active" constitutive relation used in the present study. This theory provides a model with the capability to contract with a changing force or stress. Thus unlike muscle experiments in which either isotonic or isometric conditions prevail, a more general motion, as might be expected in the coupled situation, can be accommodated.

We show that, consistent with the assumptions, stress in the series element is proportional to ventricular pressure. Thus, while isometric muscle experiments may be analogous to the isovolumetric period of systole, ejection of the model cannot be considered isotonic.

A large set of parameters must be selected to integrate the pressure–volume relation. Some of these are based on the geometry of the ventricle, while others represent the stress-generating capability of the muscle. This latter group of parameters is analogous to the set of "physiologic" constants associated with papillary muscle mechanics. Here a considerable assumption must be made in translating the force generated over the cross section of an isolated muscle segment to the stress generated by a given cross section of ventricular fibers. Various parameter sets are chosen, all of which yield a physiologically appropriate pressure rise during the isovolumetric period. Then these sets are used in the calculation of ejection against a specified pressure profile that is initialized as the isovolumetric period ends. Some of the parameter sets result in more realistic ejection fractions than others and are utilized subsequently in the coupling calculation.

Model of a Contracting Spherical Membrane

A spherical heart ventricle is illustrated in Figure 14.1B. Here r, P, ϕ, and θ are the midsurface radius, the internal pressure, the latitudinal coordinate, and the longitudinal coordinate, respectively. The axisymmetric field equations for the stress resultants, N_ϕ and N_θ, in a spherical membrane are (Flugge 1960)

$$N_\phi = \frac{r}{\sin^2\phi}\left[\int P\cos\phi' \sin\phi' d\phi' + C\right] \tag{14.1}$$

and

$$N_\phi + N_\theta = rP \tag{14.2}$$

If external forces, which might include a reaction at the valve ring, and pericardial forces, are ignored, the constant, C, is zero. Further, assuming spatial uniformity of pressure in the ventricle, the resultants are equal, and we have

$$N_\phi = N_\theta = \frac{rP}{2} \tag{14.3}$$

In shell theory these stress resultants are defined by integrating stresses across the shell thickness. As shown in Figure 14.1, the coordinate normal to the shell is z. The shell thickness from endocardium to epicardium is h. In the present study, stresses in cardiac fibers are assumed to be uniaxial stresses generated at any point z across the shell thickness by the three-element model (Figure 14.1A). If stresses in the parallel element are ignored, only the stress S in the series elastic element remains. Furthermore, if a fiber at point z makes an angle ψ with the latitude through that point, and if fiber stresses are transformed to the coordinate directions of the shell, the stress resultants are defined as

$$N_\phi = \int_h S \sin^2\psi \, dz \tag{14.4a}$$

and

$$N_\theta = \int_h S \cos^2\psi \, dz \tag{14.4b}$$

With the additional assumptions that the series element stress, S, does not vary across the thickness and that the fiber angle is a linear function of the through-wall coordinate ($\psi = mz + i$) (Streeter 1979), Equations 14.4a and b can be integrated explicitly:

$$N_\phi = \frac{Sh}{2}(1-F) \tag{14.5a}$$

$$N_\theta = \frac{Sh}{2}(1+F) \tag{14.5b}$$

in which

$$F = \cos 2i \frac{\sin mh}{mh} \tag{14.5c}$$

To be consistent with Equation 14.3,

$$F = 0 \tag{14.6}$$

Therefore

$$N_\phi = N_\theta = \frac{Sh}{2} = \frac{rP}{2}$$

or

$$S = \frac{rP}{h} \tag{14.7}$$

This restriction on F will be examined further when the data contributing to a selection of parameters are reviewed. Here the small-strain hypothesis is implicit in the use of direction cosines.

Length changes of the three-element model are related to series element stress, S, by the following relation, which combines the kinematic relation with the modified Hill equation (Fung 1970):

$$\dot{L} = \left\{\frac{\dot{S}}{\alpha(S+\beta)} - \frac{b[(k_1L+k_2)f-S]}{\gamma(k_1L+k_2)+S}\right\}$$

where α, β, γ, b, k_1, and k_2 are "physiologic" constants and $f(t)$ is a forcing function driving the "active state." The length, L, may be identified with the stretch ratio, λ_ψ shown in Figure 14.1C. This is the stretch ratio in the fiber direction. The stretch ratios in the coordinate directions of the membrane are λ_ϕ and λ_θ (Fig. 14.1C). If, in addition, one assumes that the fiber angle intercept, i, is zero at the midsurface, λ_ψ equals λ_θ. In small strain, λ_θ is related to the midsurface circumferential strain, ε_θ, as follows:

$$\lambda_\theta = 1 + \varepsilon_\theta$$

so that the equation becomes

$$\dot{\varepsilon}_\theta = \left\{\frac{\dot{S}}{\alpha(S+\beta)} - \frac{b\{[k_1(1+\varepsilon_\theta)+k_2]f-S\}}{\gamma[k_1(1+\varepsilon_\theta)+k_2]+S}\right\} \tag{14.8}$$

Substituting Equation 14.7 into 14.8 and again using the small-strain hypothesis yields

$$\kappa_1\dot{\varepsilon}_\theta + \kappa_2\varepsilon_\theta + \kappa_3 = 0 \tag{14.9a}$$

in which

$$\kappa_1 = \alpha\frac{r^2}{h^2}P^2 + \alpha\frac{r}{h}P[\gamma(k_1+k_2)+\beta] + \alpha\beta\gamma(k_1+k_2) \tag{14.9b}$$

$$\kappa_2 = -\frac{r}{h}\dot{P}\gamma k_1 + \alpha\frac{r}{h}Pbfk_1 + \alpha\beta bfk_1 \tag{14.9c}$$

$$\kappa_3 = -\frac{r}{h}[\gamma(k_1+k_2)+\frac{r}{h}P]\dot{P} + \alpha b\frac{r}{h}[f(k_1+k_2)-\beta-\frac{r}{h}P]P \tag{14.9d}$$

$$+\alpha\beta fb(k_1+k_2)$$

In axisymmetry the membrane can undergo displacements in both radial and longitudinal directions, w and u_ϕ, respectively (Fig. 14.1B). The axisymmetric strain–displacement relations in small strain, which relate the circumferential and longitudinal strains, ε_θ and ε_ϕ, respectively, to these displacements are

$$\varepsilon_\theta = \frac{1}{r}(u_\phi \cot\phi + w) \tag{14.10a}$$

$$\varepsilon_\phi = \frac{1}{r}\left(\frac{\partial u_\phi}{\partial \phi} + w\right) \tag{14.10b}$$

In the absence of any longitudinal displacement, u_ϕ, Equations 14.10a and b are redundant, giving

$$\varepsilon_\theta = \varepsilon_\phi = \frac{w}{r} \tag{14.11}$$

This assumption is consistent with the appearance of Equation 14.9a as an ordinary differential equation (ODE). Substituting Equation 14.11 into Equation 14.9a gives

$$\dot{w} + \frac{\kappa_2}{\kappa_1}w + r\frac{\kappa_3}{\kappa_1} = 0 \tag{14.12}$$

This is the equation that governs ventricular ejection. In the absence of any volume change (simulating the isovolumetric period, $w = 0$),

$$\kappa_3 = 0 \tag{14.13}$$

which is a single ODE governing the pressure, P, before ejection.

If the valve ring is located at latitudinal coordinate ϕ_b, and if the model ventricle undergoes a self-similar motion as prescribed, so that the valve ring is still at ϕ_b after deformation, the volume at any time is

$$V = 2\pi\frac{r_i^3}{3}(1+\cos\phi_b)\left(1+\frac{w}{r_i}\right)^3 \tag{14.14}$$

in which r_i is the endocardial radius. Normalizing Equation 14.14 by the end-diastolic volume, V_0, gives

$$\frac{V}{V_0} = \left\{1 + \frac{w}{r_i}\right\}^3 \tag{14.15}$$

For an ejection fraction of 1/2, the value of w/r_i is -0.21. A typical value of V_0 for the canine left ventricle is 50 ml. Therefore a reasonable value for the displacement, w, at end systole of a ventricle of endocardial radius, r_i, 2.3 cm and wall thickness, h, 1 cm, is -0.45 cm. Changes in h are ignored with the small-strain hypothesis.

Parameter Selection

To solve Equation 14.13 for the pressure during isovolumetric contraction, seven parameters as well as the forcing function, $f(t)$, must be measured (or chosen if data are lacking). When correlating his theory with papillary muscle experiments, Fung (1970) chose the following parameters: $\alpha = 40$, $\beta = 0.47$ g, $\gamma = 0.45$, $b = 0.66\ \text{sec}^{-1}$, $k_1 = 5$ g, $k_2 = -3$g. This is the "nominal" parameter set used in the calculations discussed at the end of this section. In addition, the function $f(t)$ was chosen to be a sinusoid, as suggested by Fung (1970), that is, $f = \lambda_a \sin(\omega_1 t)$. Typical values for the canine isovolumetric period and ejection period are 80 msec and 200 msec, respectively. This means that isovolumetric relaxation commences after about 280 msec. Therefore, ω_1 is chosen to be 2.78π rad/sec. In the current theory, muscle fibers act as stress generators rather than force generators so that some average area on which the stress, S, is acting must be chosen. Therefore, β, k_1, and k_2 should be divided by this area. The fibers are assumed to have circular cylindrical cross sections of diameter d.

As mentioned earlier, the parameter F, which indicates the overall variation of the fiber angle through the heart wall, must be zero in this formulation. There is substantial evidence (Streeter 1979) that F is close to zero at many locations in both the left and right ventricles, that is, that the fiber angle ψ varies by π rad from endocardium to epicardium. However, a closer examination of some of the data indicates that F may sometimes be as great as 0.4 (120-degree variation). Furthermore, the fiber angle intercept is not always zero and may be as great as $\pi/8$ rad. Few measurements of fiber angles near the apex or base have been reported. Fiber orientation may be considerably different at these locations; fibers may not lie predominantly in planes parallel to the epicardial tangent plane, making the measurement of angles by serial section difficult. Lastly, fibers at the interface between the intraventricular septum and the left free wall have a different orientation (Torrent-Guasp 1973)—epicardial fibers often pass to the free wall of the right ventricle, while deeper fibers often change direction on entering the septum. Some turn through 90 degrees to a longitudinal orientation. The effect of a nonzero F and possible spatial variation of F will have to await a more sophisticated analysis and a more ambitious program for measuring the fiber angle.

Isovolumetric Contraction

Equation 14.13 is solved numerically for the pressure during isovolumetric contraction. "Active" stress in the membrane, throughout systole, is about three times the ventricular pressure for the thickness-to-radius ratio chosen (Eq. 14.7). Results of the isovolumetric calculation appear in Figures 14.2 and 14.3. The parameter set that results in the solid curve is the "nominal" set just discussed. The fiber diameter, d, is set at 1.4 mm to obtain a physiologic pressure rise without changing any other parameters. In Figure 14.2, the fiber diameter is varied. The accompanying variations in k_1 and k_2 are shown. As expected, the same force acting over a smaller area creates a substantially greater pressure rise. Figure 14.3 shows the effect of varying α. Variation of b yields similar results. This, too, should be expected from the structure of Equation 14.13. Additional parametric studies of this calculation and the one described hereafter appear in Waldman (1982).

Ejection Against a Known Pressure Profile

During ejection, Equation 14.12 must be integrated. In lieu of equations describing a coupled afterload (valve and peripheral circulation), the pres-

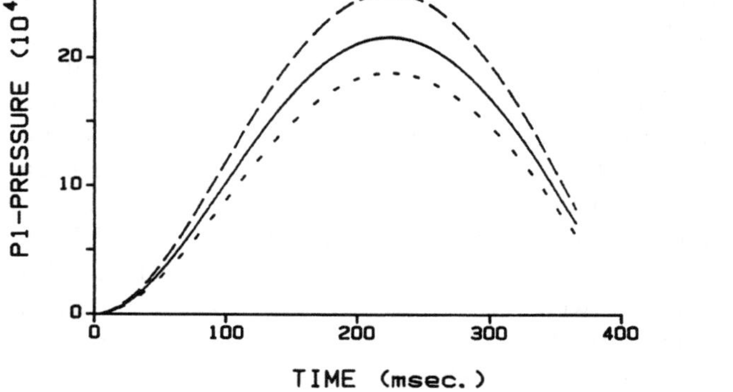

FIGURE 14.2. P_1, ventricular pressure, as a function of time during isovolumetric contraction; k_1 and k_2 varied together. d, Fiber diameter; k_1 and k_2, "physiologic" constants.

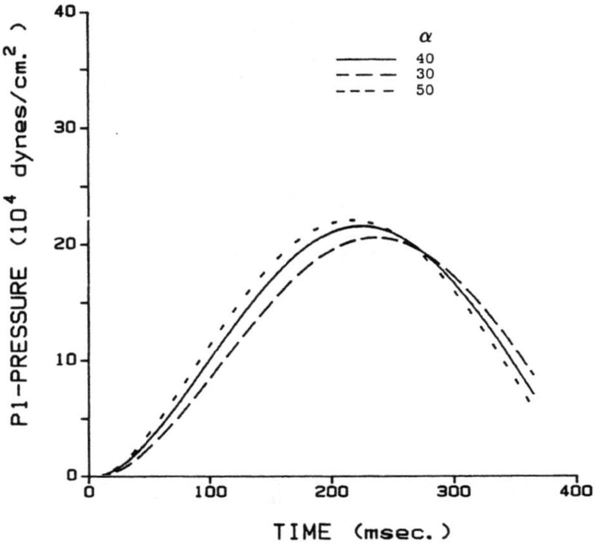

FIGURE 14.3. P_1, ventricular pressure, as a function of time during isovolumetric contraction; α varied. α, "physiologic" constant.

sure profile, $P(t)$, must now be specified. The function is allowed to follow a half sine wave, that is,

$$P = A \sin(\omega_2 t + \phi_2) \qquad t_{eiso} \leq t \leq t_{es} \tag{14.16}$$

in which t_{eiso} and t_{es} signify the end of the isovolumetric period and the end of systole, respectively. The systolic period is 360 msec with a lag of 80 msec, to account for isovolumetric contraction, so that $\omega_2 = 2.78\pi$, $\phi_2 = \pi/4.5$, and $A = 120$ mm Hg (160,000 dyne/cm^2). The calculation is terminated before the pressure is allowed to become negative.

Results of the ejection calculation are given in Figures 14.4 and 14.5. In every case the parameter set chosen resulted in a realistic pressure rise during the isovolumetric calculation. Then the same parameter set was used for the ejection calculation. Figure 14.4 demonstrates that the "nominal" parameter set discussed earlier (solid curve) does not result in a realistic ejection. In fact, the model ventricle is too weak to contract at all. If we simultaneously increase k_1 while lowering α to obtain the correct pressure rise preceding ejection, the "strongest" model ventricle still ejects weakly. A fiber diameter of 1.4 mm is probably much too large. Therefore, in Figure 14.5, the fiber diameter is arbitrarily set to 1.0 mm. The study shown in Figure 14.4 is repeated, but now k_1 is considerably larger while α must be lowered more to obtain the appropriate pressure rise preceding ejection. The "strongest" ejection, here, is fairly realistic. Therefore the parameter set chosen in this case is used later in the coupling analysis.

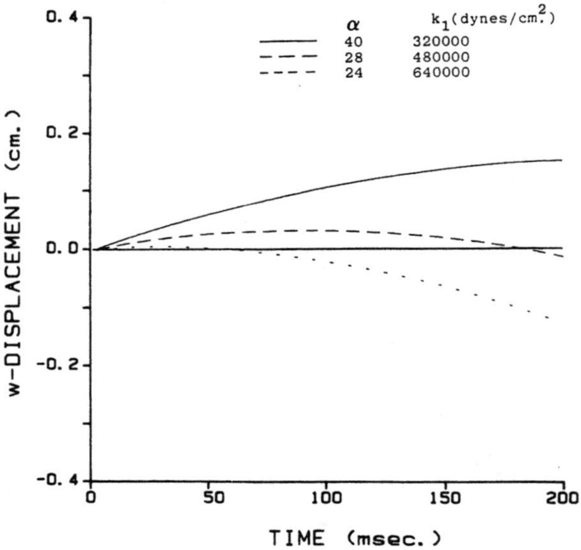

FIGURE 14.4. w, displacement of a spherical ventricle, as a function of time during ejection against a specified ventricular pressure profile; k_1 varied. k_1, "physiologic" constant.

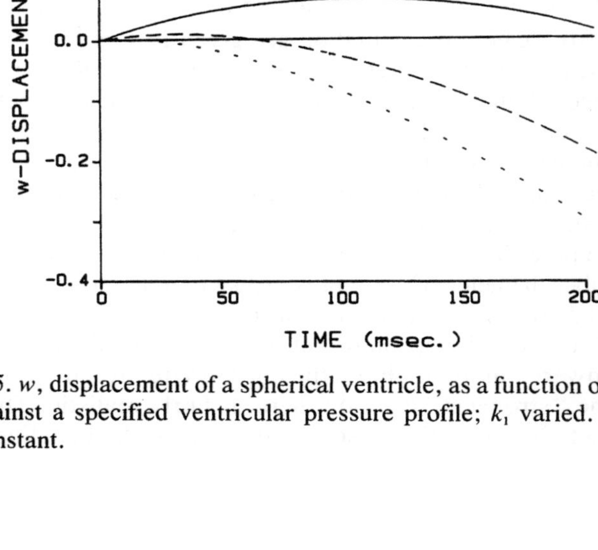

FIGURE 14.5. w, displacement of a spherical ventricle, as a function of time during ejection against a specified ventricular pressure profile; k_1 varied. k_1, "physiological" constant.

Valve Mechanics

Theoretical studies of the motion of heart valves have a wide range of complexity depending on the purpose of the analysis. The simplest representation identifies the valve with a diode (one-way on/off device) and is used by those who model the circulation as an analog circuit. At the other extreme are sophisticated vortex methods such as the work of Peskin (1972), which attempt to account more rigorously for the fluid dynamics of the valves. The latter methods have been predominantly two dimensional and at Reynold's numbers much lower than the physiologic range. Self-consistent with our own approximations of ventricular mechanics on one side of the valve and modified Windkessel models on the other side of the valve are the one-dimensional valve theories. These theories have evolved with the work of Bellhouse and Bellhouse (1969), Bellhouse and Talbot (1969), Snuggs and Aggarwal (1975), Lee (1977), Lee and Talbot (1979), and van Steenhoven and van Dongen (1979) and are based on one-dimensional unsteady fluid dynamics in the form of Euler equations, integral momentum, and integral energy equations. These formulations consider the valve leaflets as passive, rigid objects which respond to the balance of pressure gradient and flow acceleration that surrounds them in the fluid. Two informative reviews that discuss the mechanics of heart valves from both theoretical and experimental viewpoints are the articles of Parmley and Talbot (1979) and Peskin (1982).

The aortic valve consists of three membranous leaflets surrounded by three outpocketings of the aorta, the sinuses of Valsalva. When fully open, the sinuses allow the valve cusps to occupy positions in the aorta that provide a minimum of resistance to blood flow, that is, the leaflets align with the distal aortic lumen and thus are removed from the flow field of the jet of blood issuing from the left ventricle. In the closed position, the valve leaflets are in contact with each other in an axisymmetric configuration: The overlapping surfaces occur at 120-degree intervals around the lumen and provide a tight seal against any back flow while supporting tension in response to the diastolic pressure drop across them. The pulmonic valve has a similar structure so that the mechanics discussed here for the aortic valve probably apply to the other outflow valve as well, although lower dynamic pressures occur on the pulmonary side. Excellent drawings of the heart valves are found in Netter (1969). In addition, a report on the relationship between the structure and function of the aortic valve is available in the histologic study of Sauren and colleagues (1980).

Although Leonardo da Vinci (1513) observed the anatomy of the heart and valves, and even suggested that vortices in the sinuses of Valsalva might influence valve motion, modern understanding commenced with the experiments of Henderson and Johnson (1912). In this early work the ability of heart valves to close during forward flow is explained in terms of the "breaking of a jet." As the jet of blood in motion through the valve cusps dies, surrounding blood fills the gap, closing the valve. In terms of fluid mechanical quantities, the deceleration of the jet is accompanied by an adverse pressure gradient which closes the valve.

In the late 1960s, Bellhouse and Bellhouse (1969) performed measurements on a model valve in a pulsatile water tunnel. They correlated their data with a highly simplified inviscid flow theory in which pressure drop across the valve due to flow deceleration is responsible for closure. In addition, they noted the formation of strong vortices in the model sinuses and suggested that these vortices keep the valve cusps in a stable position when the cusps are fully open. Furthermore, they observed that the valve is three quarters closed by the end of systole. Subsequently, Bellhouse and Talbot (1969) reported additional model experiments and correlated them with a more sophisticated theory. Vorticity in the sinuses is accounted for more explicitly in this theory, in which fluid motion in a sinus is modeled with Hill's hemispherical vortex.

In the 1970s, the theory of Bellhouse and Talbot (1969) was simplified considerably. In the theories of van Steenhoven and van Dongen (1979) and of Lee (1977) and Lee and Talbot (1979), sinus vortices are not represented explicitly. Instead, the assumption that cusp tip pressure is identical to pressure all along the exterior of the cusps is made. Nevertheless, a "hidden" vortex exists in these newer theories, as noted by Peskin (1982). Van Steenhoven and van Dongen (1979) studied motion only in the neighborhood of the fully open position. On the other hand, Lee and Talbot (1977,1979) did not use asymptotics. Instead they solved the one-dimensional equations with the preceding pressure assumption exactly accounting for valve motion throughout most of the time range. This theory is correlated with experiments on model plastic valves (Lee 1977, Lee and Talbot 1979) and with the older experiments of Bellhouse and Bellhouse (1969). Lee and Talbot claimed that the newer theory performs better than the earlier, more complex theory of Bellhouse and Talbot (1969).

In this study the theory of Lee and Talbot (1979) is used. It is found that while their simple model of valve opening is adequate, their closing equations diverge as the closing area approaches zero. This behavior is to be expected, at least at the last instant of closing, because the rigid leaflets cannot support any tension in reaction to the pressure drop across them after they touch. Moreover, under conditions when either the valve closing area or the velocity of blood is small enough, the Reynolds number will be too low to neglect fluid friction.

In the first part of this section the theory of valve opening is discussed. Then the more complicated theory of valve closure is described. Finally, in the last part a study of the "isolated" valve closing against a known pressure drop is presented.

Theory of Valve Opening

Valve opening is handled in a simple but effective way by Lee and Talbot (1979). Their primary assumption is that uniform velocities along the length of the valve are constant. They justify this simplification on the basis that the fluid going through the valve during opening behaves like a jet. The simplest possible valve geometry can be used during opening, that is, the planar, straight valve illustrated on the left in Figure 14.6. Here the model aortic orifice is rectangular in shape. Its width and breadth are $2a$ and b_v, respectively. At the cusp tips the variable width is $2r_v$ while the breadth remains constant. A_1, the area at the valve orifice, is constant; A_2, the area at the cusp tip, is a dependent variable. The velocity U is the spatially uniform velocity of the jet, while l is the length of the cusps. Finally, the control volume, V_v, is contained within the control surface, A_v.

Lee and Talbot discussed only conservation of mass for valve opening, since they were interested only in predicting the opening area from a known velocity profile. Here, their continuity equation for opening is restated and followed with an unsteady Bernoulli equation in order to incorporate the pressure drop needed later for coupling. First, conservation of mass for the valve configuration shown on the left in Figure 14.6 gives

$$U(A_1 - A_2) = \dot{V}_v \tag{14.17}$$

in which

$$V_v = (a + r_v)lb_v \tag{14.18}$$

Combining Equations 14.17 and 14.18 gives

$$\dot{A}_2 = 2(A_1 - A_2)\frac{U}{l} \tag{14.19}$$

The pressures P_1 and P_2 are uniform pressures at the orifice and cusp tip,

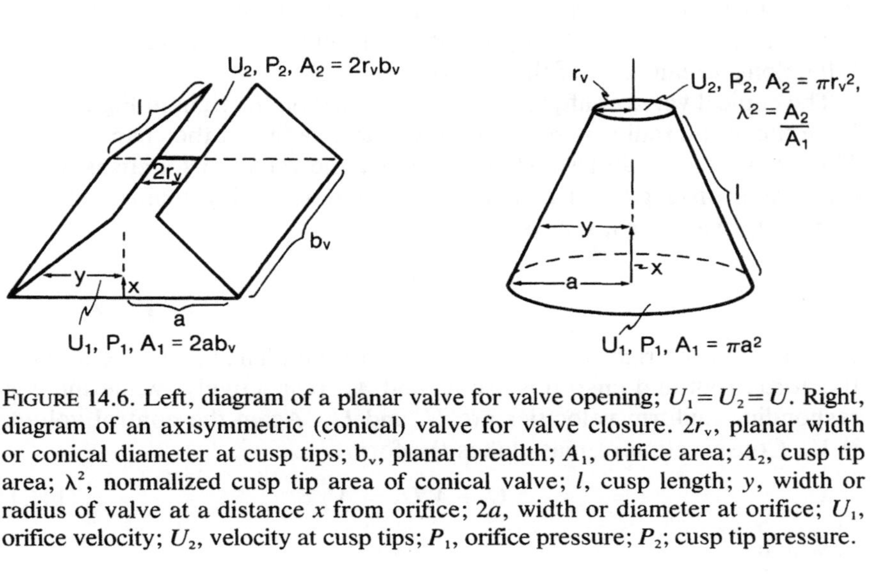

FIGURE 14.6. Left, diagram of a planar valve for valve opening; $U_1 = U_2 = U$. Right, diagram of an axisymmetric (conical) valve for valve closure. $2r_v$, planar width or conical diameter at cusp tips; b_v, planar breadth; A_1, orifice area; A_2, cusp tip area; λ^2, normalized cusp tip area of conical valve; l, cusp length; y, width or radius of valve at a distance x from orifice; $2a$, width or diameter at orifice; U_1, orifice velocity; U_2, velocity at cusp tips; P_1, orifice pressure; P_2; cusp tip pressure.

respectively, while ρ is the density of blood. An unsteady Bernoulli equation along the central streamline is

$$\dot{U} = \frac{P_1 - P_2}{\rho l} \tag{14.20}$$

Because of the assumption that the ventricular pressure is spatially uniform, P_1 is the pressure discussed in the section on cardiac mechanics. A_1, l, and ρ are chosen to be 1.8 cm², 1.0 cm, and 1.0 g/ml, respectively, throughout this study.

Theory of Valve Closure

Valve closing cannot be modeled with the simplicity of valve opening. In fact, an attempt to use for the closing calculation the planar configuration that was used during opening was unsuccessful. The planar closing equations were found to be unrealistically stiff when appropriate parameters were chosen. Therefore, the conical configuration developed by Lee and Talbot (1979) is used here. First, the equations of Lee and Talbot are reiterated in a form more suitable for the purposes here. A primary assumption in this model is that the pressure all along the external length of the valve is equal to the pressure at the cusp tip, P_2. This is justified for the aortic valve by the observation that there is little, if any, "kinetic energy recovery" in the aortic sinuses (Snuggs and Aggarwal 1975). This is interpreted to mean that a sinus vortex exists behind each cusp during closing, and that the external length of the cusp is a streamline of this vortex. This means that the pressure external to the valve at the aortic root is not a stagnation pressure, as assumed more wisely for the inflow valves. Instead, the pressure here is assumed to be the dynamic pressure found at the cusp tip. Both pressure and velocity are assumed to vary little along streamlines of these vortices.

The conical valve configuration is illustrated on the right in Figure 14.6. Here the aortic radius is a while the variable radius at the cusp tips is r_v. The radius at an arbitrary distance, x, along the valve is y for cusps, which are, again, of length l. Lee and Talbot assumed axisymmetric, straight (conical), rigid cusps, that is,

$$y = a - \frac{(a - r_v)}{l} x$$

as shown on the right in Figure 14.6. As in the planar case, the areas at the aortic root and cusp tips are A_1 and A_2, respectively, while the corresponding uniform velocities are U_1 and U_2. Again the control volume is V_v. Conservation of mass takes the form

$$A_1 U_1 - A_2 U_2 = \dot{V}_v \tag{14.21}$$

in which

$$V_v = \int_0^l \pi y^2 dx = \frac{\pi l}{3}(a^2 + ar_v + r_v^2) = \frac{A_1 l}{3}(1 + \lambda + \lambda^2) \qquad (14.22)$$

Here $\lambda^2 = A_2/A_1$. Combining Equations 14.21 and 41.22 gives

$$U_2 = \frac{1}{\lambda^2}\left[U_1 - \frac{(1+2\lambda)\dot{\lambda} l}{3}\right] \qquad (14.23)$$

or at an arbitrary distance along the valve

$$U(x) = \frac{a^2}{y^2}U_1 - \frac{\dot{r}_v}{3y^2}\left[\frac{x^2}{l}\right](a+2y) \qquad (14.24)$$

Lee and Talbot went on and wrote an integral momentum equation for the control volume inside the cusps. Then they wrote a Bernoulli equation along the central streamline. Finally, they combined these two equations to eliminate the pressure drop. The result was a single second-order non-linear ODE that related λ to the velocity U_1. Here these equations are combined in such a way that no derivatives appear on the right side of the first equation. Furthermore, the two second-order equations are written as three first-order equations. First, the integral momentum equation in the x-direction for the control volume inside the control surface, A_v, is restated:

$$F_x = \frac{\partial}{\partial t}\int_{V_v} \rho U dV + \int_{A_v} (\vec{q}\cdot d\vec{A}) \qquad (14.25)$$

Here $\vec{q}$ is the velocity vector at any point. A Bernoulli equation along the central streamline is

$$(P_1 + \rho U_1^2) - (P_2 + \rho U_2^2) = \rho \int_0^l \frac{\partial U}{\partial t} dx \qquad (14.26)$$

Substituting Equation 14.24 into 14.25 and 14.26, the results can be written as

$$\dot{U}_1 = \rho_1(P_1 - P_2) + \rho_2 U_1^2 + \rho_3 \zeta U_1 + \rho_4 \zeta^2 \qquad (14.27)$$

$$\dot{\zeta} = C_1\left[\dot{U}_1 - \frac{\zeta^2 l}{6} - \frac{U_1^2}{l} + \frac{\lambda^2 U_2^2}{l} - \frac{(P_1 - P_2)}{\rho l}\right] \qquad (14.28)$$

$$\dot{\lambda} = \zeta \qquad (14.29)$$

Here the coefficients are defined as follows:

$$\rho_1 = \frac{2+\lambda}{\rho l}$$

$$\rho_2 = \frac{(\lambda^2 - 1)(\lambda^2 + 2\lambda - 1)}{2l\lambda^3}$$

$$\rho_3 = \frac{3\lambda^2 - 1}{3\lambda^3}$$

$$\rho_4 = \frac{l(2\lambda + 1)}{18\lambda^3}$$

$$C_1 = \frac{6}{l(1 + \lambda)}$$

Difficulty in obtaining complete closure of this valve model in the fully coupled situation, discussed in the final section of this chapter, prompted a study of the "isolated" valve. "Isolation" can be accomplished by supplying the valve with a physiologically appropriate pressure drop. Results of this study are reported in the following section.

Valve Closure with a Known Pressure Drop

The valve closure equations (14.27 through 14.29) can be solved independently of the surrounding systems if the pressure drop, $\delta P = P_1 - P_2$, is specified. These equations are initialized at the pressure crossover point of the opening calculation. From Equation 14.20 there is a velocity maximum at this time. Typical values of the opening velocity at this time are between 30 and 35 cm/sec for the parameter sets used. In this study the initial conditions are $U_1 = 32$ cm/sec, $\lambda = 0.99$, and $\zeta = 0$. The pressure drop is chosen to be a linear function of time, that is,

$$\delta P = -M_{\delta P} t$$

The slope, $M_{\delta P}$, is varied from 0 to 12,000 dyne/(cm^2 sec), which means that at the end of 100 msec, δP ranges from 0 to 1200 dyne/cm^2. Note that these are small pressure drops (<1 mm Hg).

The results of this study are presented in Figures 14.7 through 14.9. Figure 14.7 shows the normalized closing area, $\lambda^2 = A_2/A_1$, as a function of time for various slopes, $M_{\delta P}$. The valve remains open as expected for zero pressure drop. On the other hand, increasing gradients tend to close the valve more and more rapidly. In Figure 14.8, the aortic root velocity, U_1, is displayed. It appears that the blood tends to regurgitate substantially before closure is complete. Finally, Figure 14.9 shows the cusp tip velocity, U_2, as a function of time. At small enough gradients, U_2 dips before the calculation diverges. The valve is still wide open as U_1 reverses and U_2 begins to increase following the dip. This result indicates that the divergence is probably not due to the omission of friction at the last instant of the calculation. Rather, the geometry or kinematics of this rigid, straight valve model is in question. Further comments on this difficulty and suggestions for improvement are given in the following section, in which the behavior of the model valve coupled to its surroundings is reported.

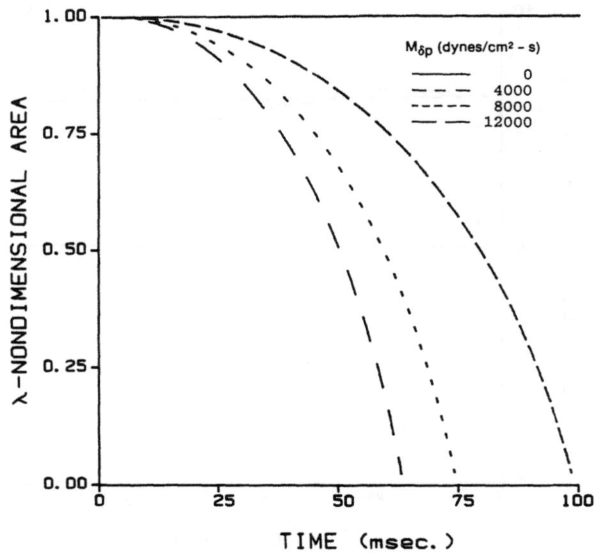

FIGURE 14.7. λ, normalized closing area ($\lambda^2 = A_2/A_1$), as a function of time; pressure *(P)* varied. $M_{\delta P}$, slope of pressure change.

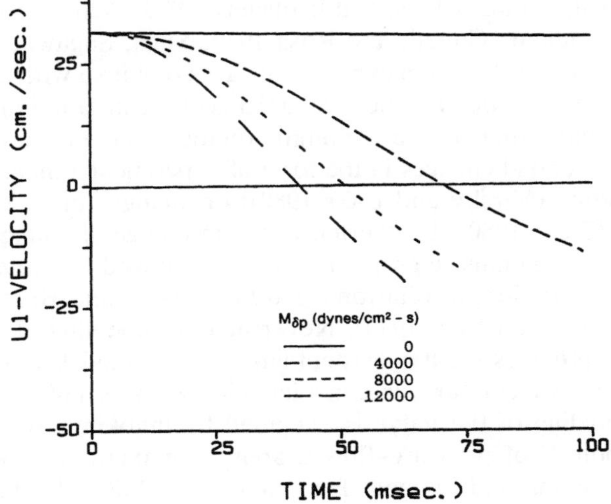

FIGURE 14.8. U_1, aortic root velocity, as a function of time; pressure *(P)* varied. $M_{\delta P}$, slope of pressure change.

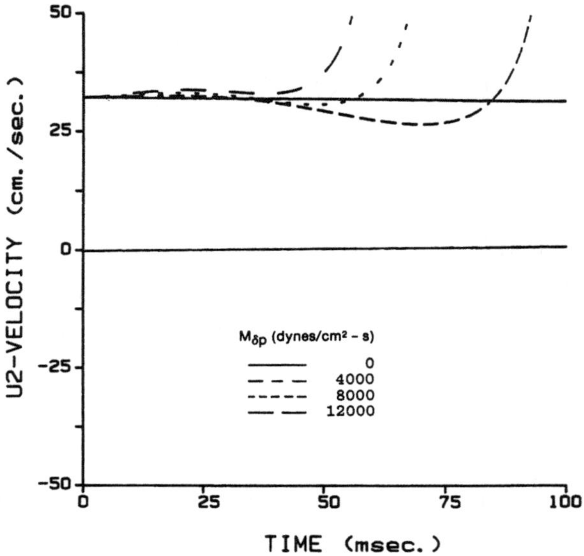

FIGURE 14.9. U_2, cusp tip velocity, as a function of time; pressure *(P)* varied. $M_{\delta P}$, slope of pressure change.

Systolic Coupling of the Heart and Circulation

Attempts have been made to study the coupling of the heart and circulation with both experimental and theoretical approaches (Baan 1978; Mirsky and Laks 1980; Paulus, Claes, and Brutsaert, 1979; Piene 1980; Piene and Sund 1982; Pouleur, Covell, and Ross 1979; Suga, Sagawa, and Demer 1980; Westerhof 1982). Often experiments are combined with analog modeling in an effort to elucidate the mechanisms of the interaction. Statistical changes in ventricular pressure–volume relations have been observed in response to afterload changes in the form of experimental maneuvers [e.g., arterial banding (Mirsky and Laks 1980)] or changes in vascular model parameters (Piene 1980). Furthermore, the influence of changes in aortic input impedance on muscle properties has been studied in terms of changes in force–velocity–length relations (Paulus, Claes, and Brutsaert 1979; Pouleur, Covell, and Ross 1979). Nevertheless, modeling of the outflow valve, which provides a self-consistent link between models of the ventricle and peripheral vasculature, is conspicuously missing (Westerhof 1982). Explicit modeling of the valve is excluded by invoking concepts of the "working point" of pressure–flow graphs or maximum power transfer between the pump and its load (Piene and Sund 1982, Westerhof 1982). In the present study, theoretical modeling of the coupling is attempted, in which modeling of the valve is explicitly included in a self-consistent manner.

Coupling of the two models described previously with a simple model of the canine systemic circulation is considered. During valve opening the coupling problem is represented by a system of five first-order nonlinear ODEs. Coupling during valve closure requires a system of six ODEs. It is found that the opening calculation proceeds smoothly, so that various parametric studies are performed to quantify the interaction between the model heart and circulation during this part of systole. In these studies the constants associated with the model of the peripheral circulation are varied. The closing calculation is not as successful, and no parametric studies are performed for this part of systole. The valve begins to close very rapidly, but the calculation diverges while the valve is still about halfway open. At this time the aortic root velocity, U_1, has reversed substantially, while the cusp tip velocity, U_2, increases without bound as it did in the "isolated" valve study discussed previously and in the study of Bellhouse and Talbot (1969), albeit with a somewhat different theory. The newer valve theory of Lee and Talbot (1979) performs adequately when aortic root velocity data acquired from model experiments (rigid cusps!) is used as a forcing function in their Equation 20. However, no cusp tip velocities or pressures are given. In this study, P_2, U_1, and U_2 behave similarly to the dependent variables in the older spherical vortex theory (Bellhouse and Talbot 1969).

Coupling During Valve Opening

A simple model of the canine peripheral circulation is discussed, which will be coupled with the ventricular and valve models. Here a modified Windkessel theory is proposed. Although wave propagation cannot be accounted for with this model, a more detailed model is unwarranted at this time. A branching model could be included without difficulty by increasing the size of the system of ODEs, but the number of parameters would increase accordingly. A diagram of the analog circuit that is used to derive the pressure–flow relation for this model is shown in Figure 14.10. This relation takes the following form:

$$\dot{P}_2 = L_a\ddot{Q}_H + \left[R_a + \frac{L_a}{(TPR)C_a}\right]\dot{Q}_H + \left(1 + \frac{R_a}{TPR}\right)\frac{Q_H}{C_a} - \frac{P_2}{(TPR)C_a} \quad (14.30)$$

where L_a = aortic inductance,
R_a = aortic resistance,
TPR = total peripheral resistance
C_a = aortic compliance
Q_H = volume flow rate leaving heart

If the first two of these parameters are zero, the model reduces to the classical Windkessel. Here Q_H is the product of two of the dependent variables already discussed ($Q_H = U_2A_2$); therefore it occurs as a nonlinear

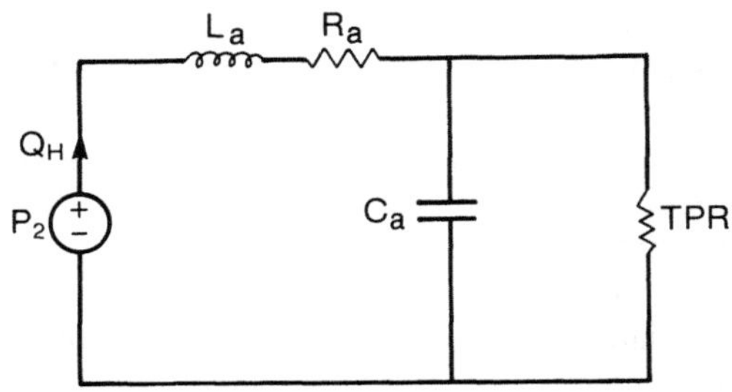

FIGURE 14.10. Circuit diagram for canine peripheral vasculature; modified Windkessel theory. *P*, pressure; Q_H, flow leaving heart; L_a, aortic inductance; R_a, aortic resistance; *TPR*, total peripheral resistance; C_a, aortic compliance.

term in the coupling analysis. Typical values for these constants in the canine systemic circulation are

$$
\begin{aligned}
L_a &= 1 \text{ dyne sec}^2 \text{ cm}^{-5}\\
R_a &= 300 \text{ dyne sec cm}^{-5}\\
TPR &= 4000 \text{ dyne sec cm}^{-5}\\
C_a &= 10^{-4} \text{ cm}^5/\text{dyne}
\end{aligned}
$$

Now, taking the derivative with respect to time of Equation 14.15, one obtains

$$\dot{V} = \frac{3V_o}{r_i}\left(1+\frac{w}{r_i}\right)^2 \dot{w} \tag{14.31}$$

Mass conservation for the ventricle demands that

$$\dot{V} = -A_1U_1 \tag{14.32}$$

Substituting Equation 14.32 into 14.31 and rearranging, one obtains

$$\dot{w} = -\frac{r_iA_1U_1}{3V_o}\frac{1}{(1+w/r_i)^2} \tag{14.33}$$

Equation 14.12 governing ejection can be rewritten as

$$\dot{P} = \frac{\kappa_1\dot{w} + w\alpha bfk_1(P\frac{r}{h}+\beta) + r\alpha b\frac{r}{h}\left[f(k_1+k_2)-\beta-\frac{r}{h}P\right]P + \alpha\beta fb(k_1+k_2)}{w\frac{r}{h}\gamma k_1 + \frac{r^2}{h}\left[\gamma(k_1+k_2)+\frac{r}{h}P\right]} \tag{14.34}$$

Thus, with equation 14.33, Equation 14.34 can be written in a form free of derivatives on the right side. During valve opening, Equations 14.19, 14.20, 14.30, 14.33, and 14.34 are the five ODEs that govern the coupling.

Results of the coupling calculation during valve opening are presented in Figures 14.11 through 14.20. If both the aortic inductance and the aortic

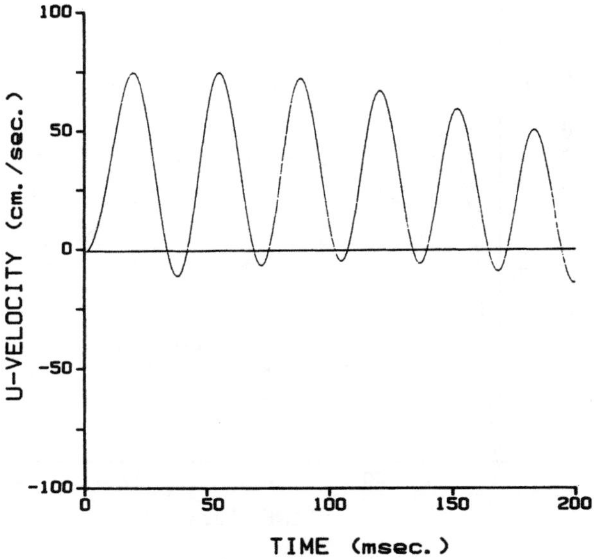

FIGURE 14.11. U, opening velocity, as a function of time; coupling to classical Windkessel during valve opening.

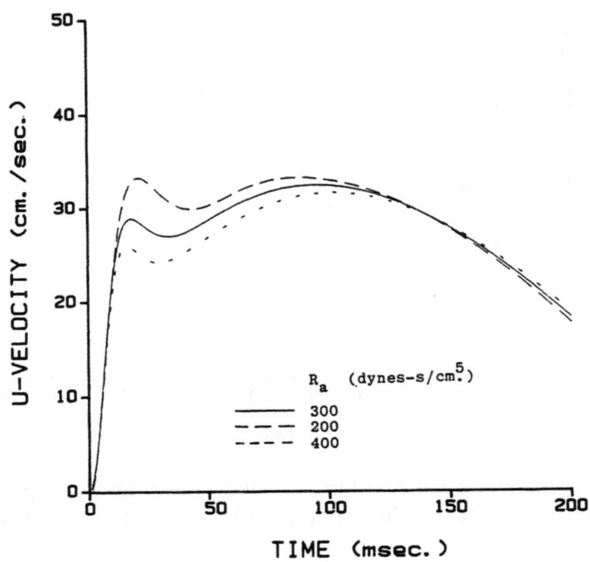

FIGURE 14.12. U, opening velocity, as a function of time; coupling to modified Windkessel during valve opening; aortic resistance (R_a) varied

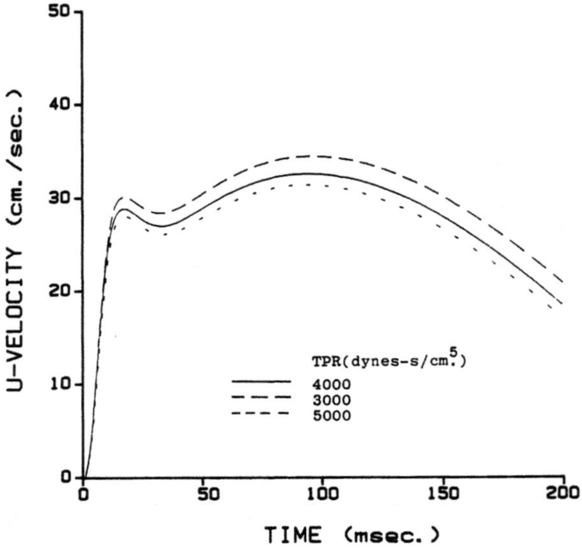

FIGURE 14.13. U, opening velocity, as a function of time; coupling to modified Windkessel during value opening; total peripheral resistance (TPR) varied.

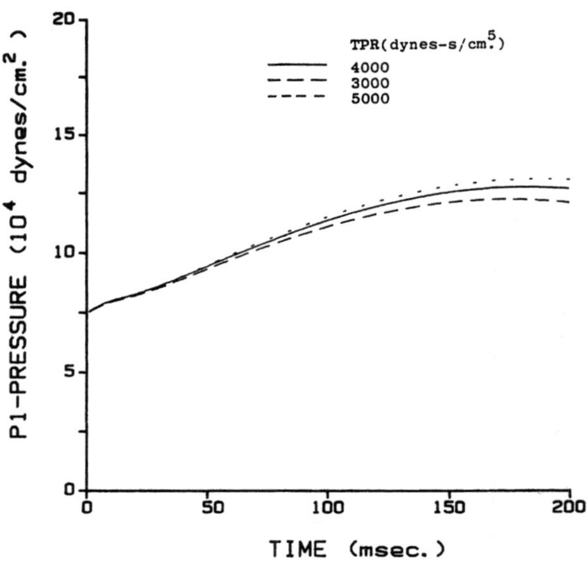

FIGURE 14.14. P_1, ventricular pressure, as a function of time; coupling to modified Windkessel during valve opening; total peripheral resistance (TPR) varied.

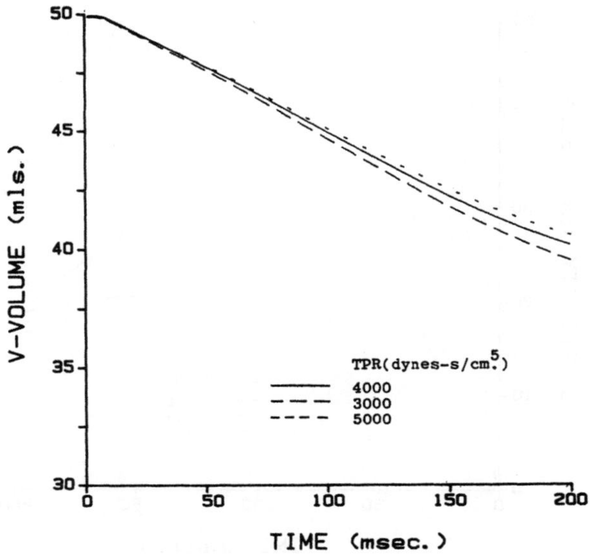

FIGURE 14.15. V, ventricular volume, as a function of time; coupling to modified Windkessel during valve opening; total peripheral resistance (TPR) varied.

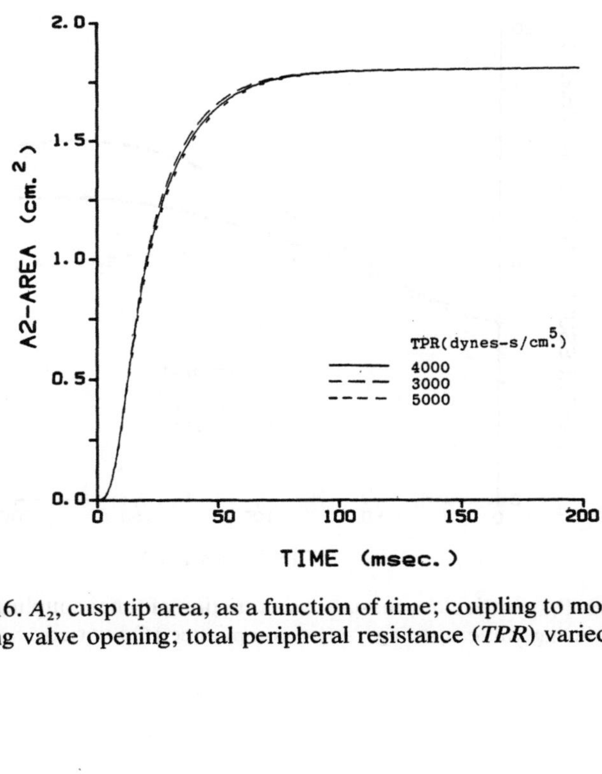

FIGURE 14.16. A_2, cusp tip area, as a function of time; coupling to modified Windkessel during valve opening; total peripheral resistance (TPR) varied.

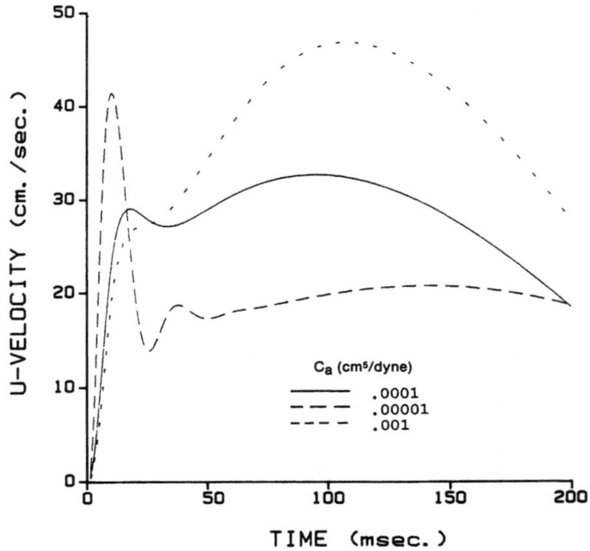

FIGURE 14.17. *U*, opening velocity, as a function of time; coupling to modified Windkessel during valve opening; aortic compliance (C_a) varied.

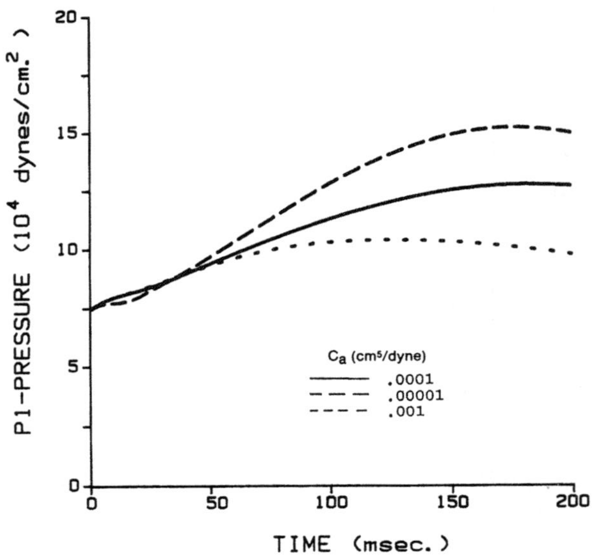

FIGURE 14.18. P_1, ventricular pressure, as a function of time; coupling to modified Windkessel during valve opening; aortic compliance (C_a) varied.

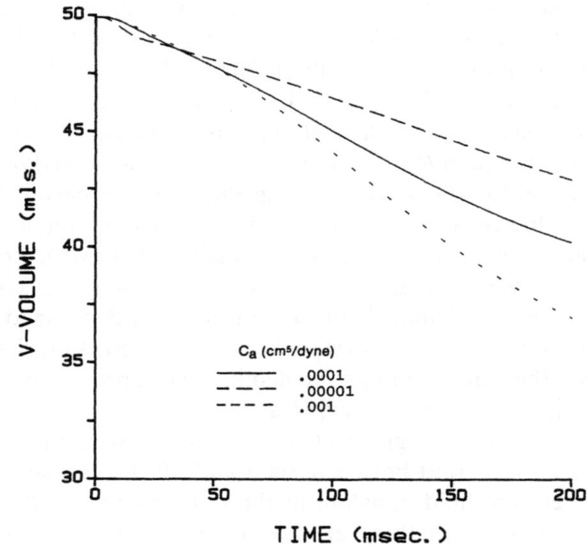

FIGURE 14.19. V, ventricular volume, as a function of time; coupling to modified Windkessel during valve opening; aortic compliance (C_a) varied.

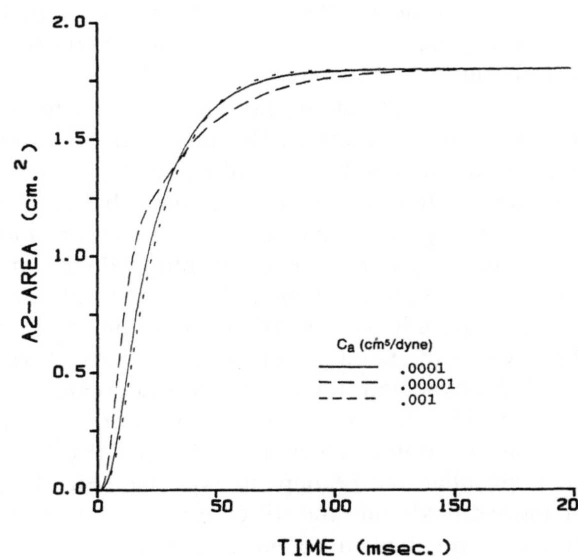

FIGURE 14.20. A_2, cusp tip area, as a function of time; coupling to modified Windkessel during valve opening; aortic compliance (C_a) varied.

resistance are set equal to zero, the peripheral circulation is represented by a classical Windkessel. Under these circumstances, all of the dependent variables exhibit an oscillation at about 30 Hz. In particular, the opening velocity, U, fluctuates wildly (Fig. 14.11). The addition of a resistance in series with the classical Windkessel eliminates the oscillation, as shown in Figure 14.12, in which R_a is varied between 200 and 400 dyne sec cm^{-5}. The initial transient in the velocity is significantly decreased in amplitude at the higher value of R_a. The effect on the other dependent variables of substantial variations in this parameter is small (not shown). The pressures P_1 and P_2 are increased slightly for the higher value of R_a, as expected. Furthermore, the ventricular displacement w has a depressed profile and the valve cusps open at a slower rate when R_a is raised. Ventricular volume always follows the same trend as the displacement profile, because of the algebraic relation between the two (Eq. 14.15).

In the next series of four graphs (Fig. 14.13–14.16), the total peripheral resistance, *TPR*, is varied between 3000 and 5000 dyne sec cm^{-5} while other parameters are held constant at the nominal values discussed previously. The influence of this variation is small also and similar to the effect of variations in R_a for all of the profiles except, again, the opening velocity, U (Fig. 14.13). Here there is a qualitative difference in the results. Not only is the initial transient affected, but the opening velocity is altered throughout the remainder of the time range as well. The cusp tip pressure, P_2, is not shown. Due to the small pressure gradients needed to open or close the model valve, pressures at either end of the control volume are imperceptibly different.

Figures 14.17 through 14.20 show the effect of changes in the aortic compliance, C_a, on all of the profiles. Because decreases in aortic compliance are manifest in atherosclerosis and aging, C_a is varied greatly to emphasize its influence. It should be kept in mind that no compensatory mechanisms are operating in this uncontrolled circulatory model. Figure 14.17 shows the interesting effect on the opening velocity profile. Again this profile manifests the greatest change. The initial transient undergoes a large increase in amplitude for the stiffer aorta, but after only 20 msec or so the profile crosses the nominal curve and remains depressed for the rest of the period. As expected, the heart ejects at greater pressures into the stiffer aorta (Fig. 14.18). Ventricle displacement or volume is considerably less under these circumstances (Fig. 14.19). The effect on the cusp tip area profile is peculiar and perhaps unexpected (Fig. 14.20). At first the cusps open more quickly into the stiffer aorta, but after about 30 msec the profile crosses the nominal curve, indicating that the cusps open more slowly for the remainder of the opening period.

Coupling During Valve Closure

In the closing calculation, Equations 14.19 and 14.20 of the opening model are replaced with Equations 14.27 through 14.29. As mentioned earlier, the coupled closing calculation diverges after about 20 msec. The computation is initialized as the opening velocity U peaks, which occurs after about 100 msec of ejection. During closing the aortic root velocity U_1 differs from the cusp tip velocity U_2. The pressure drop across the valve increases greatly as the calculation diverges. Simultaneously, the ventricular displacement profile reverses slightly at an ejection fraction much smaller than physiologic. The aortic root velocity U_1 also reverses, indicating a substantial regurgitation at large velocity gradients. At the same time, the normalized closing area λ experiences large gradients, but the valve closes only about halfway before the calculation diverges. Finally, the cusp tip velocity U_2 tends to increase without bound, just as in the "isolated" valve study of the previous section. The pressure gradient becomes so large in the coupled situation, however, that U_2 exhibits monotonic behavior, that is, no velocity minimum occurs, as in Figure 14.9, at smaller values of the pressure drop.

Conclusion

An attempt has been made to quantify a large number of events that occur during the systolic period of the cardiac cycle. A model of a contracting spherical membrane is used to calculate the pressure rise during isovolumetric contraction, the ventricular displacement during ejection against a given pressure profile, and finally as part of a coupling calculation during ejection into a valve model and its accompanying model of the circulation. The most tenuous assumptions in the analysis are probably the thin-wall assumption, the small-strain hypothesis, and the symmetry in geometry, deformation, and loading. Left ventricles have thick walls so that the internal pressure load should be applied at the endocardial radius rather than at the substantially greater midsurface radius. Furthermore, left ventricles tend to undergo large displacements during systole, and do so at large strains (Waldman, Fung, and Covell 1985). Moreover, the ventricle is not axially symmetric. Substantial variations in strains at different ventricular sites (Lew and LeWinter 1983, LeWinter et al. 1975) and large transmural variations in both shortening and thickening strains (Waldman, Fung, and Covell 1985) have been observed. Also, the presence of significant shear strains with respect to a cardiac coordinate system, which precludes the use of ordinary membrane theories for the description of ventricular mechanics, has been documented (Waldman, Fung, and Covell 1985). It is possible that the model ventricle is overconstrained, and that by accounting for thick walls and large strains more physiologic ejection fractions might be obtained. If this is the case, it is not clear whether difficulty with the closing calculation might be obviated with a better model of the ventricle.

Although the model valve of Lee and Talbot performs better in the "isolated" valve study, it still manifests many of the problems associated with the coupled closing calculation. The velocity of blood at the distal end of the valve model tends to increase without bound long before the valve is closed. The valve equations diverge at the last instant of closing, so that any expectation of achieving complete closure is perhaps unrealistic. Continuing research on valve mechanics should address the problem of accounting for the deformation of the valve leaflets. One approach is to alter the kinematics of the valve so that the leaflets undergo an increasing change in curvature as closure proceeds. This can be attempted without actually giving the leaflets any load-carrying capability, that is, they are still rigid, but curved. An initial attempt at such a study with a planar valve model has been unsuccessful. At this time it is believed that the new valve model (results of which are not reported here) deforms too much. The next step will be incorporation of a parameter that allows alteration of the amount the leaflets curve as they approach each other. In this way the kinematics of closure might be handled more realistically. However, even if this attempt is successful, the valve equations will probably still diverge at the last instant. To eliminate this mathematical anomaly, a more rigorous attempt at accounting for the true deformation of the leaflets, including their tension-carrying capability, must be initiated. Furthermore, it may be necessary to relax the assumption of potential flow through the valve at the last instant of closure. When the cusp tip area and/or the fluid velocity become small enough, the flow will certainly be a creeping flow in which viscous effects are significant.

It would be of considerable interest to couple a more detailed model to the upstream subsystems. It has been shown here that a classical Windkessel must be modified to avoid oscillations in the dependent variables in the coupled situation. Perhaps a branching model would perform even better. In any case, wave propagation and nonuniformity in the circulatory model can be accounted for only with a more sophisticated analysis. It is not clear whether such a model would alleviate any of the problems of the coupled closing calculation.

A set of studies of the coupling calculation during valve opening has been presented. These studies provide an attempt at quantifying the interaction between ventricle and circulation during ejection. The results indicate that the model is rather insensitive to alterations in the downstream parameters. It is not clear whether this insensitivity is an artifact of the constraints placed on the model ventricle, valve, and peripheral bed or whether this behavior is realistic. It will be necessary to relax some of the constraints and to attempt to correlate these results with experimental data. Lastly, an attempt to couple the model subsystems together during valve closure is only partly successful. Improvements in the modeling of all three subsystems may be needed to obviate difficulties with closure.

Bibliography

Abbott, B. C., and Mommaerts, W. F. H. M. 1959. A study of inotropic mechanisms in the papillary muscle preparation. J. Gen. Physiol. 42:533–551.

Baan, J. 1978. Model of the left ventricle based on an electromagnetic contractile analog of cardiac muscle. In Baan, J., Noordergraaf, A., and Raines, J., eds. *Cardiovascular System Dynamics*. Cambridge, Mass.: MIT Press.

Bellhouse, B., and Bellhouse, F. 1969. Fluid mechanics of model normal and stenosed aortic valves. *Circ. Res.* 25:693–704.

Bellhouse, B. J., and Talbot, L. 1969. The fluid mechanics of the aortic valve. *J. Fluid Mech.* 35(4):721–735.

Blinks, J. R., and Jewell, B. R. 1972. The meaning and measurement of myocardial contractility. In Bergel, D. H. ed. *Cardiovascular Fluid Dynamics*. London: Academic.

Brady, A. J. 1965. Time and displacement dependence of cardiac contractility: problems in defining the active state and force–velocity relations. *Fed. Proc.* 24:1410–1420.

Brady, A. J. 1979. Mechanical properties of cardiac fibers. In Berne, R. M., Sperelakis, N. and Geiger, S. R., eds. *Handbook of Physiology*. Sec. 2, *The Cardiovascular System,* Vol. 1. Bethesda, Md. American Physiological Society.

Brutsaert, D. L., Claes, V. A., and Sonnenblick, E. H. 1971. Effects of abrupt load alterations on force–velocity–length and time relations during isotonic contractions of heart muscle: load clamping. *J. Physiol.* (Lond.) 216:319–330.

Brutsaert, D. L., Parmley, W. W., and Sonnenblick, E. H. 1969. Force–velocity–length–time relations of the contractile elements in heart muscle of the cat. *Circ. Res.* 24:137–149.

Edmund, K. A. P., and Nilsson, E. 1968. Mechanical parameters of myocardial contraction studied at a constant length of the contractile element. *Acta Physiol. Scand.* 72:205.

Edmund, K. A. P., and Nilsson, E. 1972. Relationship between force and velocity of shortening in rabbit papillary muscle. *Acta. Physiol. Scand.* 85:488–500.

Endo, M. 1973. Length dependence of activation of skinned muscle fibers by calcium. *Cold Spring Harbor Symp. Quant. Biol.* 37:505–510.

Fabiato, A., and Fabiato, F. 1975. Dependence of the contractile activation of skinned cardiac cells on the sarcomere length. *Nature* 256:54–56.

Flugge, W. 1960. *Stresses in Shells*. Berlin: Springer-Verlag.

Fung, Y. C. 1970. Mathematical representation of the mechanical properties of heart muscle. *J. Biomech.* 3:381–404.

Fung, Y. C. 1971. Comparison of different models of the heart muscle. *J. Biomech.* 4:289–295.

Henderson, Y., and Johnson, F. E. 1912. Two modes of closure of heart valves. *Heart* 4:69–82.

Hill, A. V. 1938. The heat of shortening and the dynamic constants of muscle. *Proc. Roy. Soc. Lond.* (Series B) 216:136–195.

Huxley, A. F. 1957. Muscle structure and theories of contraction. In Butler, J. A. V., and Katz, B., eds. *Progress in Biophysics and Biophysical Chemistry,* Vol. 7. Oxford: Pergamon Press, pp. 225–318.

Lee, C. S. F. 1977. Fluid mechanical study of mitral valve motion. Ph.D. Thesis, University of California, Berkeley.

Lee, C. S. F., and Talbot, L. 1979. A fluid-mechanical study of the closure of heart valves. *J. Fluid Mech.* 91:41–63.

Leonardo da Vinci. 1513. *Quaderni d'Anatomica* II, 9.

Lew, W., and LeWinter, M. M. 1983. Regional circumferential lengthening patterns in the canine left ventricle. *Am. J. Physiol.* 245:H741–H748.

LeWinter, M. M., Kent, R. S., Kroener, J. M., Carew, T. E., and Covell, J. W. 1975. Regional differences in myocardial performance in the left ventricle of the dog. *Circ. Res.* 37:191–199.

McLaughlin, R. J., and Sonnenblick, E. H. 1974. Time behavior of series elasticity in cardiac muscle. *Circ. Res.* 34:798–811.

Mirsky, I., and Laks, M. M. 1980. Time courses of changes in the mechanical properties of the canine right and left ventricles during hypertrophy caused by pressure overload. *Circ. Res.* 46:530–542.

Nassar, R., Manring, A., and Johnson, E. A. 1974. Light diffraction of cardiac muscle: sarcomere motion during contraction. In Porter, R., and Fitzsimmons, D. W., eds. *The Physiological Basis of Starling's Law of the Heart.* Amsterdam: Elsevier, pp. 57–82.

Netter, F. H. 1969. *The Heart.* In *The Ciba Collection of Medical Illustrations.* Summit, N.J.: Ciba Pharmaceutical Co.

Noble, M. I. M., Bowen, T. E., and Hefner, L. L. 1969. Force–velocity relationship of cat cardiac muscle: studies by isotonic and quick-release techniques. *Circ. Res.* 24:821–833.

Noble, M. I. M., and Else, W. 1972. Reexamination of the applicability of the Hill model of muscle to cat myocardium. *Circ. Res.* 31:580–589.

Parmley, W. W., and Talbot, L. 1979. Heart as a pump. In Berne, R. M., Sperelakis, N. and Geiger, S. R., eds. *Handbook of Physiology,* Sec. 2, *The Cardiovascular System,* Vol. 1, Bethesda, Md.: American Physiological Society.

Paulus, W. J., Claes, V. A., and Brutsaert, D. L. 1979. Physiological loading of isolated feline cardiac muscle. *Circ. Res.* 44:491–497.

Peskin, C. S. 1972. Flow patterns around heart valves. Ph.D. Thesis, Albert Einstein College, Yeshiva University, New York.

Peskin, C. S. 1982. The fluid dynamics of heart valves: experimental, theoretical, and computational methods. *Ann. Rev. Fluid Mech.* 14:235–259.

Piene, H. 1980. Interaction between the right heart ventricle and its arterial load: a quantitative solution. *Am. J. Physiol.* 238:H932–H937.

Piene, H., and Sund, T. 1982. Does normal pulmonary impedance constitute the optimum load for the right ventricle? *Am. J. Physiol.* 242:H154–H160.

Pinto, J. G., and Fung, Y. C. 1973. Mechanical properties of the stimulated papillary muscle in quick-release experiments. *J. Biomech.* 6:617–630.

Pollack, G. H. 1970. Maximum velocity as an index of contractility in cardiac muscle. *Circ. Res.* 26:111–127.

Pollack, G. H., Huntsman, L. L., and Verdugo, P. 1972. Cardiac muscle models: an overextension of series elasticity? *Circ. Res.* 31:569–579.

Pollack, G. H., and Krueger, J. W. 1976. Sarcomere dynamics in intact cardiac muscle. *Eur. J. Cardiol.* 4 (Suppl):53–65.

Pouleur, H., Covell, J. W., and Ross, J. 1979. Effects of alterations in aortic input impedance on the force–velocity–length relationship in the intact canine heart. *Circ. Res.* 45:126–135.

Sauren, A. A. H. J., Kuijpers, W., van Steenhoven, A. A., and Veldpaus, F. E. 1980. Aortic valve histology and its relation with mechanics: preliminary report. *J. Biomech.* 13:97–104.

Snuggs, T. A., and Aggarwal, J. K. 1975. Steady and unsteady flow in a semicylindrical hollow. *Oxf. Univ. Eng. Sci. Lab. Rep.* 1118/75.

Sonnenblick, E. H. 1962. Force–velocity relations in mammalian heart muscle. *Am. J. Physiol.* 202:931.

Sonnenblick, E. H. 1965. Determinants of active state in heart muscle: force, velocity, instantaneous muscle length, time. *Fed. Proc.* 24:1396–1409.
Streeter, D. 1979. Gross morphology and fiber geometry of the heart. In Berne, R. M., Sperelakis, N. and Geiger, S. R., eds. *Handbook of Physiology*, Sec. 2 *The Cardiovascular System,* Vol. 1, Bethesda, Md.: American Physiological Society.
Suga, H., Sagawa, K., and Demer, L. 1980. Determinants of instantaneous pressure in canine left ventricle. *Circ. Res.* 46:256–263.
Tarr, M., Trank, J. W., Goertz, K. K., and Leiffer, P. 1981. Effect of initial sarcomere length on sarcomere kinetics and force development in single frog atrial cardiac cells. *Circ. Res.* 49:767–774.
Tarr, M., Trank, J. W., Leiffer, P., and Shepherd, N. 1979. Sarcomere length–resting tension relation in single frog atrial cardiac cells. *Circ. Res.* 45:554–559.
Taylor, S. R., and Rudel, R. 1970. Striated muscle fibers: inactivation of contraction induced by shortening. *Science* 167:882–884.
Torrent-Guasp, F. 1973. *The Cardiac Muscle.* Barcelona: The Juan March Foundation.
van Steenhoven, A. A., and van Dongen, M. E. H. 1979. Model studies of the closing behavior of the aortic valve. *J. Fluid Mech.* 90(1):21–32.
Waldman, L. K. 1982. On the mechanical coupling of the heart to the circulation. Ph.D. Thesis, University of California, San Diego.
Waldman, L. K., Fung, Y. C., and Covell, J. W. 1985. Transmural myocardial deformation in the canine left ventricle: normal in vivo three-dimensional finite strains. *Circ. Res.* 57:152–163.
Westerhof, N. 1982. Pressure–flow relations of arterial system and heart. In Huiskes, R., Van Campen, D. and De Wijn, J., eds. *Biomechanics: Principles and Applications.* The Hague: Martinus Nijhoff.

15 The Effects of Finite Wave Velocity and Discrete Reflections on Ventricular Loading

TERRY W. LATSON, FRANK C. P. YIN, AND WILLIAM C. HUNTER

Introduction

This chapter examines the nature of ventricular interactions with the *intact* vascular system. Most previous studies on ventricular/vascular interactions have employed modified Windkessel models of the vascular system. Such models differ from the real arterial system in that they do not reproduce the effects of finite wave velocity and discrete reflections. In the first portion of this chapter we examine the differences between the ventricular-loading properties of the three-element Windkessel model and the real arterial system. Methods are then described by which we can study how the ventricle interacts with these complex, in vivo vascular properties. The concluding sections discuss some results from simulations of ventricular/vascular interactions, using measured canine vascular properties, and explore some of the possible clinical implications of these results.

As previously shown by Westerhof, Elzinga, and van den Bos (1973), the three-element Windkessel model does simulate the major ventricular-loading properties of the arterial tree. There are some significant differences between the two, however. One way to illustrate these differences is shown in Figure 15.1. The heavy lines are aortic root pressure recordings taken from chronically instrumented dogs. Although not shown, simultaneous measurements of aortic flow were taken during each pressure tracing. From the recorded pressure and flow, the three-element Windkessel parameters—characteristic resistance, R_c; peripheral resistance, R_p; and compliance, C—were calculated to model the vascular properties for each condition. (The methods used to calculate these parameters are explained later in the chapter.) The dotted lines in each panel are the pressures that would result from putting the measured flow into the corresponding three-element Windkessel models.

In Figure 15.1A, the vascular pressure is initially less than the Windkessel pressure, but later in diastole exceeds the Windkessel pressure. The mean pressures of the two are identical. This pattern of pressure differences is seen in healthy dogs with normal arterial wave velocity.

The initial deficit in pressure occurs because the volume ejected into the aorta is not instantaneously distributed, as it is in the Windkessel model, but sets up traveling flow and pressure "waves." Initially these waves are propagated away from the central aorta and have the effect of decreasing aortic root pressure (relative to an ideal Windkessel). They are then partially reflected at distal reflection sites, and travel back toward the aortic root. The pressure augmentation seen in diastole is due to the return of these "discrete reflections." As these reflections dampen out and disperse, the pressures of the two systems behave similarly. The benefits conferred by a healthy vascular system with normal wave velocity are thus somewhat similar to those obtained with a properly timed intraaortic balloon pump—decreases in systolic and increases in diastolic pressures (O'Rourke 1982). The decrease in systolic pressures leads to decrease in myocardial oxygen demand and a possible increase in stroke volume. The higher diastolic pressures act to increase diastolic coronary perfusion pressure.

Figure 15.1B illustrates a similar comparison, but here the vascular pressure is from an old dog with increased wave velocity along the arterial tree. This increase in wave velocity decreases the transit time for pressure waves to travel from the central aorta to distal reflection sites and back to the aortic root. Thus discrete vascular reflections, which were prominent in diastole in Figure 15.1A, now return earlier and increase systolic pressures. Some of the diastolic pressure augmentation is also lost. In contrast to the situation with normal wave velocity, the vascular system with increased wave velocity may be detrimental to cardiac function (similar to an improperly timed intraaortic balloon pump). The increase in systolic pressures acts to increase myocardial oxygen demand and potentially to decrease stroke volume. These detrimental effects are at least partially reversible with vasodilator therapy, as illustrated in Figure 15.1C. The tracings here are from the same dog as in Figure 15.1B, but during an infusion of nitroprusside. Owing to the decrease in mean pressure, and possibly to direct effects of nitroprusside on vascular smooth muscle, wave velocity has decreased. As a result, the return of prominent vascular reflections is once again delayed into diastole.

The differences between the Windkessel model and the vascular system that give rise to the pressure differences in these examples can be more systematically examined using what is termed the "impulse response function." In general terms, the impulse response function is the *time*-based counterpart of the *frequency*-based impedance that has been described in earlier chapters. As an introduction to the impulse response function, we first examine the impulse response of the three-element Windkessel model. To illustrate the effects of finite wave velocity and discrete reflections, this Windkessel impulse response function is then compared with that of the canine vascular system.

Using the vascular impulse response function, we also perform some simulations that provide more quantitative information about the effects

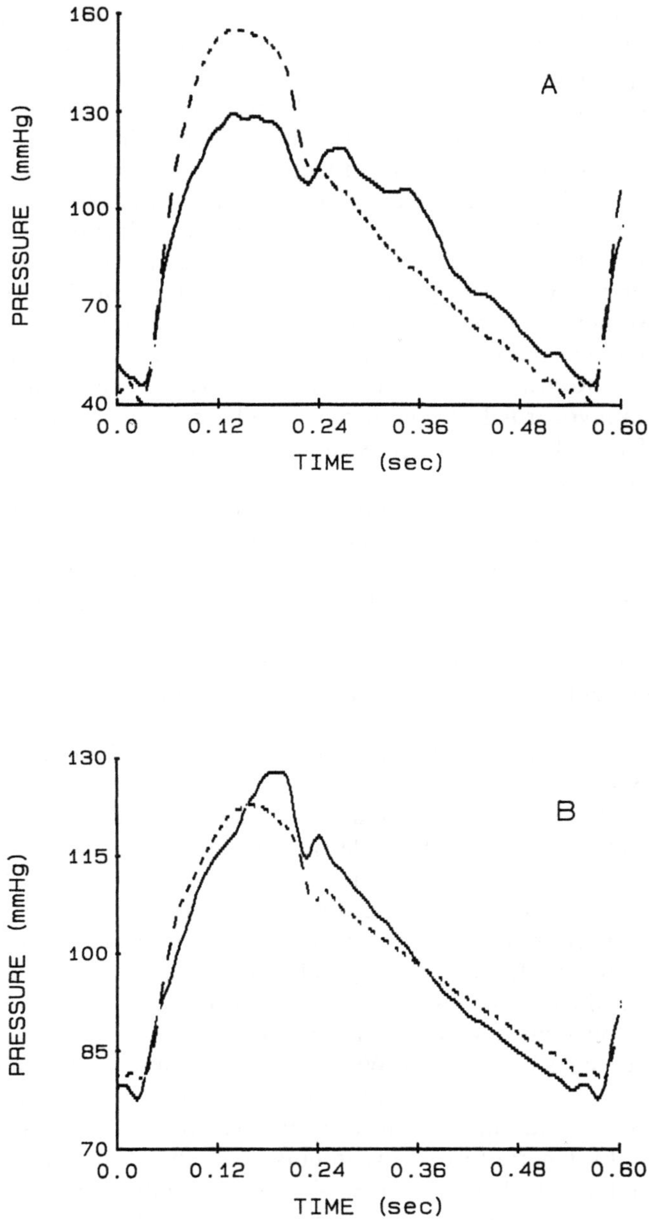

FIGURE 15.1 Comparisons between measured aortic pressure (solid line) and the pressure produced by applying the same flow into a three-element Windkessel model (dashed line) of the measured vascular properties. Recordings were made in chronically instrumented dogs. (A) healthy dogs with normal arterial wave velocity; (B) old dog with increased wave velocity; (C) same dog as in (B), but during infusion of nitroprusside.

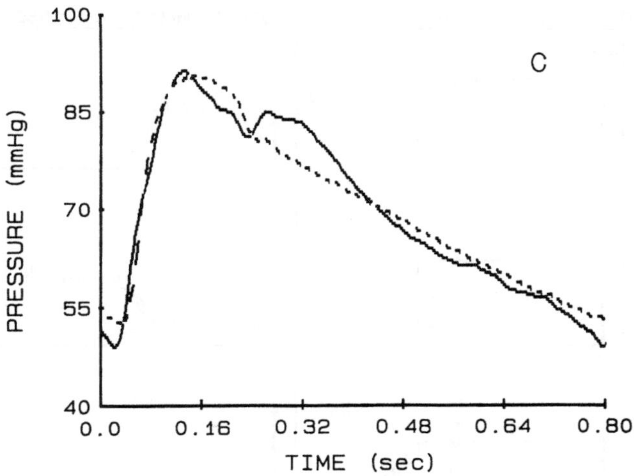

FIGURE 15.1 (Continued)

of vascular reflections. Although the examples just discussed illustrate some of the differences between the Windkessel and the vascular system, they are somewhat "nonphysiologic" in that the same flow is put through both the Windkessel and the vascular system. If a heart were confronted with these different loading systems, the resulting flows would not be identical. To analyze the effects of finite wave velocity and discrete reflections on both pressure *and* flow, we need to "couple" independent descriptions of vascular and ventricular properties. The vascular impulse response function provides a time-based description of vascular properties, which can be used in this coupling process.

The Vascular Impulse Response Function

"Ideal" and "Derived" Impulse Response Functions of the Three-Element Windkessel

Conceptually, the impulse response of any linear, time-invariant system (which the vascular system approximates; see the review in Milnor 1982) may be thought of as the output that would result from the input of an infinitely narrow pulse of unit area, called an "impulse." This is illustrated diagrammatically in Figure 15.2 for the electrical analog of a three-element Windkessel system. For this example the input is an impulse of current and the output is the resulting voltage measured at the location indicated. In the application of this analogy to the circulation, flow would be equivalent to the current, $I(t)$, and pressure would be equivalent to the measured voltage, $V(t)$.

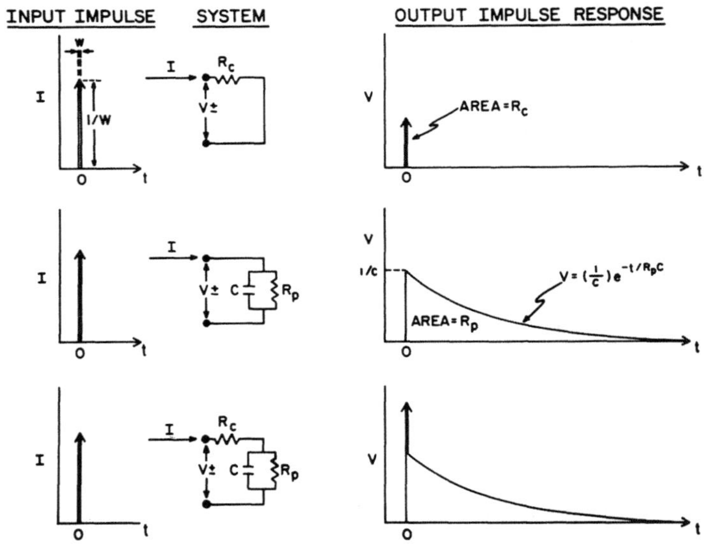

FIGURE 15.2. Impulse response function of the three-element Windkessel model. The single impulse response (SIR) for each circuit is the output voltage V produced by the input of an ideal impulse of current I. Note that the SIR of the three-element Windkessel is the sum of the SIRs of the series resistance R_c and the parallel resistance-capacitance combination R_p-C. C, capacitance; t, time; w, width of the input unit impulse.

The impulse response of the three-element Windkessel is the sum of the impulse responses of (1) the series resistance, R_c, and (2) the parallel resistance-capacitance combination R_p-C. Since the voltage across a series resistance is directly proportional to the input current, the impulse response, $V(T)$, of the series resistance is of the same form as the input impulse, $I(t)$. The area under the curve of $V(t)$ is equal to the proportionality constant between current and voltage, that is, it is equal to the value of the resistance. The impulse response of the parallel resistance-capacitance combination R_p-C has the form of an exponentially decaying function. The impulse of current $I(t)$ acts to "instantaneously" charge the capacitor to a voltage equal to $1/C$, and then the capacitor slowly discharges this current through the resistor R_p. The area under this decay curve is equal to the value of the resistor R_p, and the time constant of the decay is equal to the product of $R_p \times C$. The impulse response of the three-element Windkessel is simply the algebraic sum of the impulse responses of these two components.

In dealing with the vascular system, one way to obtain the vascular impulse response is by inverse Fourier transformation of the vascular impedance spectrum (Laxminarayan, Sipkema, and Westerhof 1978; Sipkema, Westerhof, and Randall 1980). The term "derived impulse response" will be used to denote an impulse response obtained in this manner. The derived response is identical to the ideal response only if a continuous impedance spectrum (known for all frequencies) is used in the inverse Fourier transformation. If only a limited number of impedance values are known, the derived impulse response may deviate considerably from the ideal impulse response.

When calculating vascular impedance by Fourier series methods, the spectrum that we obtain is restricted to a limited number of impedance values. For example, if we were recording steady-state pressure and flow signals from a dog with a heart rate of 120 beats/min, the impedance spectrum that we obtain would consist of terms equally spaced at integer multiples of the primary frequency (2 Hz). A Fourier series analysis would not provide information about impedance values in the interval frequencies between these harmonics. An impedance spectrum in which only certain harmonic values are known is termed a "discrete" impedance spectrum (in contrast to a "continuous" spectrum). Furthermore, because of deteriorating signal-to-noise ratios at higher frequencies, out data might be limited to, for example, the first 10 harmonics. This results in what is termed "high-frequency truncation," that is, our spectrum is truncated after the last known harmonic (the "cutoff frequency"). Digital filtering is sometimes employed to modify the effects of high-frequency truncation (Brigham 1974).

The effects of discrete frequency sampling, high-frequency truncation, and digital filtering on the impedance spectra and impulse response functions of a three-element Windkessel are illustrated in Figure 15.3. Beginning in the first column, A1 depicts a continuous impedance spectrum for a three-element Windkessel model. This impedance spectrum describes the impulse response produced by the ideal impulse shown in A2. Inverse transformation of this spectrum thus yields the ideal impulse response shown in A3. This type of impulse response function is also called the single impulse response (SIR), since it represents the output produced by a single input impulse.

Moving to the second column, B1 depicts the effects of discrete frequency sampling on the impedance spectrum. This spectrum is limited to impedance values at integer multiples of the primary harmonic. This discrete spectrum describes the impulse response produced by the periodic series of impulses shown in B2. The interval *T* between input impulses is equal to the period of the primary harmonic of the discrete spectrum. This change in our effective input impulse function is due to the fact that discrete sampling in the frequency domain produces a periodic function in the time domain (Brigham 1974). Inverse transformation of this discrete spectrum produces the impulse response function shown in B3. This impulse response is composed of an infinite series of a function similar in form to the ideal response of A3, but repeats itself after each period *T*. This period *T* of the derived response corresponds to the repetition interval between the input impulses in B2. This type of impulse response is termed an "impulse train response" (ITR), since it describes the output produced by a

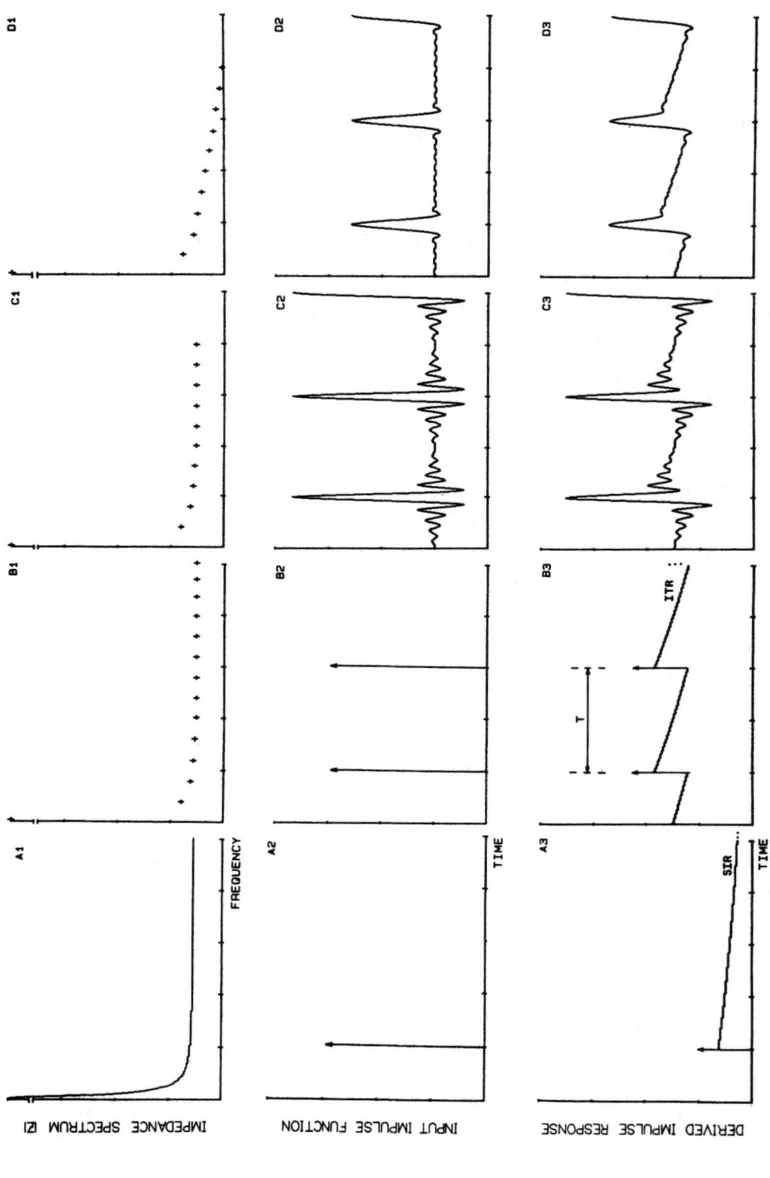

FIGURE 15.3 Effects of discrete frequency sampling (column B), high-frequency truncation (column C), and digital filtering (column D) on the impedance spectrum, effective input impulse function, and derived impulse response function. See text for explanation.

train of impulses. If the period T between successive impulses in the input function is less than the total duration of the single impulse response (A3), the impulse train response will not have time to decay to zero between successive input impulses. This is commonly seen in the vascular impulse response derived by Fourier series impedance methods. The amount that each point in the train response is offset from the corresponding point in the single response (between time $= 0$ to T) depends on the behavior of the single response beyond time $= T$. If the general behavior of the single response beyond time $= T$ is known (e.g., a known monoexponential decay), the single response can be approximated from the train response using known analytical relationships. This fact becomes important later when we take up the topic of convolution.

The effects of high-frequency truncation are illustrated in the third column of Figure 15.3. As depicted in C1, all impedance values after the cutoff frequency are now lost. By limiting the frequency range in our impedance spectrum, we introduce a special type of "low-pass" filtering. This filtering changes our effective input from the train of idealized impulses shown in B2 to the train of modified pulses illustrated in C2. These pulses differ from an impulse in two important ways. First, the main pulse is much broader and rounder than an impulse (although it still does retain the required unit area). The width of each pulse is equal to the period of the highest harmonic below the cutoff frequency. When this more rounded pulse is input into the Windkessel, the output will also have a more rounded contour. This is especially noticeable for the series resistance component R_c. Whereas the width of the R_c component in the ideal impulse response was infinitely narrow (i.e., the width of an ideal impulse), the width now is equal to that of the modified input pulses.

The second important difference between the input pulses modified by high-frequency truncation, shown in C2, and an ideal impulse is that there are "artifactual" oscillations (side lobes) on each side of the main pulse. Although these oscillations rapidly decay, they are of sufficient magnitude to produce corresponding oscillations in the resulting output (i.e., in our derived impulse train response). Such oscillations are evident in the derived impulse train response shown in C3.

The effects of digital filtering are illustrated in the fourth column. With respect to the impedance spectrum, digital filtering acts to "taper" the impedance modulus, $|Z|$, toward zero as the impedance frequency approaches the cutoff frequency. This has the effect of "smoothing out" the artifactual oscillations that result from high-frequency truncation. This smoothing effect is evident in both our effective input function (D2) and the derived impulse response function (D3). However, in addition to lessening the amplitude of the artifactual oscillations, this smoothing process also broadens the effective input pulses. This broadening of the input pulses acts in turn to further lessen the fidelity of our derived impulse response function (i.e., rounds out any sharp peaks or sudden changes in amplitude). When examining the properties of the vascular system, however, it is sometimes necessary to use digital filtering to help distinguish between "true" oscillations caused by discrete vascular reflections and the artifactual oscillations caused by high-frequency truncation (Laxminarayan, Sipkema, and Westerhof 1978).

In subsequent sections of this chapter we make comparisons between the derived impulse response of the vascular system and that of the three-element Windkessel. In making such comparisons, it is important that both responses be derived in the same manner. For example, a vascular impulse response derived from a discrete impedance spectrum of 10 harmonics with a primary frequency of 2 Hz should only be compared with a Windkessel impulse response derived from a similar discrete impedance spectrum. In this way the effects of discrete frequency sampling and high-frequency truncation will be similar for both responses. Likewise, if the vascular response is subjected to digital filtering prior to such a comparison, the Windkessel response with which it is compared should have undergone the same type of digital filtering.

Impulse Response of the Canine Vascular System

Simultaneously recorded pressure and flow traces from the aortic root of a chronically instrumented canine preparation (Yin, Weisfeldt, and Milnor 1981) are shown in Figure 15.4. Fourier series analysis of these signals, followed by division of pressure moduli by flow moduli and subtraction of respective phases, produces the discrete impedance spectra shown by the +'s in Figure 15.5A. As previously described by Laxminarayan, Sipkema, and Westerhof (1978), inverse Fourier transformation of this

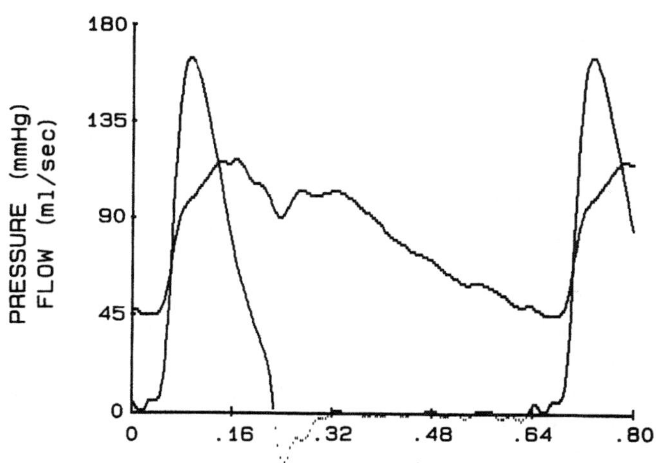

FIGURE 15.4. Aortic root pressure and flow signals recorded in a chronically instrumented canine preparation.

impedance spectrum will produce the vascular impulse response function illustrated by the solid line in Figure 15.5B. As explained previously, this derived impulse response function will have been subjected to the effects of discrete frequency sampling and high-frequency truncation. Discrete frequency sampling converts the derived response into an impulse train response (ITR). High-frequency truncation both lessens the fidelity of the derived response and produces artifactual oscillations in the derived response. Digital filtering may be employed to decrease the magnitude of these latter oscillations, as depicted by the solid line in Figure 15.5C.

To examine the differences between this vascular ITR and that of a corresponding three-element Windkessel model, we can calculate the appropriate values of the Windkessel parameters from the original pressure–flow data and impedance spectrum. The series resistance R_c for the Windkessel model is set equal to the value of characteristic impedance, Z_c, where Z_c is determined by averaging the moduli of the high-frequency terms in the impedance spectra (for these examples Z_c was the average of moduli above 5 Hz up to the cutoff frequency, which was usually between 12 and 15 Hz). The parallel resistance R_p is then set equal to the zero frequency modulus (average pressure divided by average flow) minus R_c (the zero frequency modulus is equal to the total resistance in the Windkessel model). The capacitance C is determined from the value of R_p and the time constant of pressure decay in the latter part of a prolonged diastole. Although there are other methods for determining these parameters, we have chosen this method because it has been employed in the past (Westerhof, Elzinga, and Van den Bos 1973), is straightforward, and provides reasonable results for illustrating the effects of finite wave velocity and discrete reflections in the vascular system.

From these values of the Windkessel parameters we may calculate a discrete impedance spectrum for the Windkessel model corresponding to our vascular spectrum, using the same primary frequency and cutoff frequency. This Windkessel spectrum is depicted by the circles in Figure 15.5A. The Windkessel spectrum faithfully reproduces the major features of the vascular spectrum (high initial amplitude, which then tapers to the value of Z_c), but does not show any of the oscillations that are present in the vascular spectrum. Inverse transformation of this discrete Windkessel spectrum produces the ITR shown by the dashed line in Figure 15.5B. This ITR also shows the effects of high-frequency truncation and discrete frequency sampling. The filtered form of this ITR is shown by the dashed line in Figure 15.5C.

It should be noted that in Figure 15.5C the Windkessel ITR does reproduce the major features of the vascular ITR. Both ITRs show a prominent initial pressure peak followed by a prolonged pressure decay. There are two significant differences, however. First, the initial peak in the vascular ITR is less than that in the Windkessel ITR. As will be explained, this is primarily an effect of finite wave velocity in the arterial system. The second difference is the presence of discrete reflection peaks in the

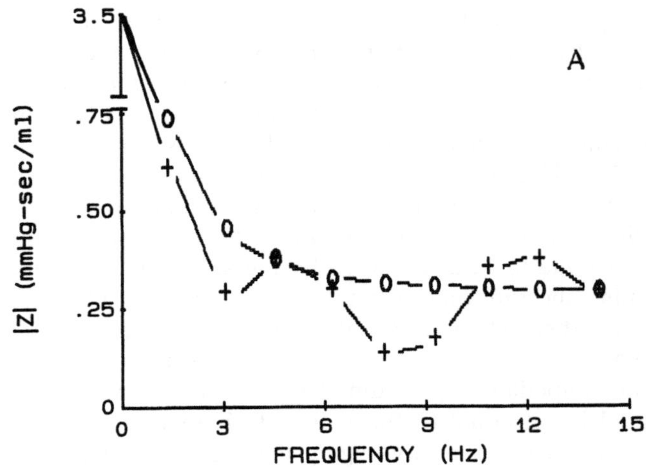

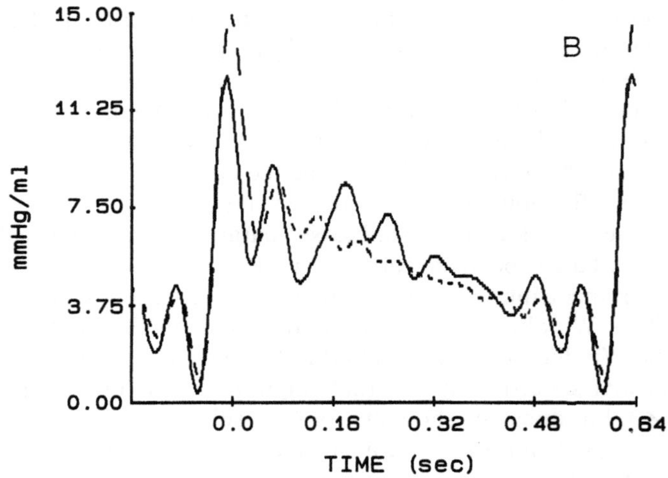

FIGURE 15.5. Impedance and impulse response functions derived from the pressure and flow of Figure 15.4. (A) measured vascular impedance (+) compared with the impedance for a corresponding three-element Windkessel model (o); (B) unfiltered impulse train responses (ITR) derived from the impedance spectra of A (solid line, vascular ITR; dashed line, Windkessel ITR); (C) filtered forms of the ITRs. Arrows indicate discrete reflection peaks. See text for explanation.

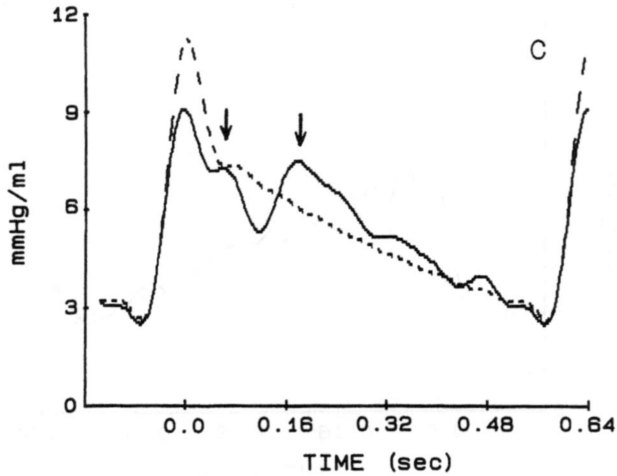

FIGURE 15.5 (Continued)

vascular ITR (marked by the arrows). These additional pressure peaks represent the effects of vascular reflections returning from the upper and lower body regions.

Let us first examine why the heights of the initial peaks in the two ITRs are different. As previously illustrated in Figure 15.2, the initial peak in the Windkessel ITR is due to the effects of the series resistance R_c superimposed on the rise associated with the parallel resistance-capacitance combination R_p-C. This peak in the vascular ITR is due to the effects of characteristic impedance Z_c superimposed on the pressure rise associated with vascular compliance and vascular reflections. Over the applicable frequency range, Z_c behaves similarly to R_c; that is, for the central aorta of the dog, Z_c can be approximated as a series resistance (Westerhof, Elzinga, and Sipkema 1971), and the form of its impulse response representation is thus very similar to that of the series resistance R_c. Furthermore, in constructing our Windkessel model we set R_c equal to Z_c. Z_c and R_c thus have equivalent representations in these two ITRs. The different heights of the initial peaks in the ITRs must therefore arise because of differences in system properties other than Z_c and R_c. To more closely examine the origin of these differences, we may remove the common effects of Z_c and R_c from the vascular ITR and Windkessel ITR, respectively, and examine the resultant "modified" ITRs. These modified ITRs are shown in Figure 15.6 (the analytical method used to remove the Z_c peak is discussed later).

From a knowledge of the design of the three-element Windkessel circuit (Fig. 15.2), it is easily appreciated that the modified Windkessel ITR is simply the derived ITR of the parallel resistance-capacitance combination R_p-C. What the modified vascular ITR represents is a little more complex.

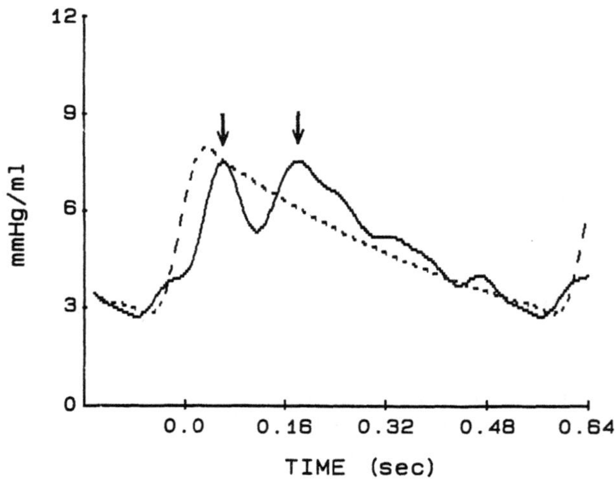

FIGURE 15.6. Modified impulse train response (ITR) functions from Figure 15.5. The R_c (series resistance) component of the Windkessel ITR (dashed line) and the Z_c (characteristic impedance) component of the vascular ITR (solid line) have been removed. Arrows indicate discrete reflection peaks. See text for explanation.

Characteristic impedance, by definition, would represent the impedance of the aorta if it were a tube of infinite length without any reflections (Milnor 1982). Since we have removed the effects of characteristic impedance, the resulting modified vascular ITR of Figure 15.6 may be thought of as that part of the vascular ITR that results from the effects of vascular reflections and/or volume retention within the finite arterial tree.

The initial pressure rise (relative to time = 0) in the modified vascular ITR is less than that in the modified Windkessel ITR. This is why the initial peak of the vascular ITR is less than the initial peak in the Windkessel ITR. Stated another way, the initial peaks in the original ITRs of Figure 15.5C represent the effects of R_c and Z_c *superimposed* on the initial portions of these modified ITRs. Since R_c and Z_c have equivalent impulse response representations, and the initial segment of the modified vascular ITR is less than the corresponding portion of the modified Windkessel ITR, the initial peak in the original vascular ITR is less than that of the Windkessel ITR.

The reason for these differences in the initial segments of the modified ITRs is as follows: As flow enters an ideal Windkessel system, it is instantaneously distributed throughout that system. The subsequent pressure rise is directly proportional to the change of volume in the system (and system compliance)—inflow minus outflow. As flow enters the vascular system, however, it sets up traveling pressure and flow waves. These initially travel away from the central aorta, and only later do they return after being partially reflected "downstream" in the arterial tree. Their egress away from the central aorta causes the initial pressure in the aorta to be less than the corresponding initial pressure of the Windkessel model. Only when the reflected waves return do they augment aortic pressure. As the reflected waves travel back and forth between vascular reflection sites, they eventually attenuate and disperse (the extent of such "rereflections" is debated). As this happens, the volume from the flow pulse becomes "effectively" distributed throughout the major compliance vessels (similar to what happened instantaneously in the Windkessel model), and vascular pressure now behaves like the Windkessel pressure. If wave velocity in the arterial tree were infinite, these pressure and flow waves would "immediately" dampen out (assuming incomplete reflection and wave attenuation/dispersion with travel still existed). Inflowing volume would thus be "instantaneously" distributed, as in the ideal Windkessel, and these differences between the modified vascular ITR and modified Windkessel ITR would theoretically disappear. As has been pointed out by Laxminarayan, Sipkema, and Westerhof (1978), when wave velocity in the dog is increased during hypertension, the properties of the vascular system do more closely match those of the ideal Windkessel.

Before we proceed, a few comments on terminology may be needed. The term "Windkessel" is most often used to refer to a compliant chamber with a variable outlet resistance (equivalent to the parallel resistance-capacitance combination R_p-C). For such a chamber, the pressure within is solely determined by chamber compliance and instantaneous volume. Following a period of rapid inflow (i.e., ventricular ejection), the chamber pressure initially rises owing to the increase in chamber volume, and then decays in a smooth monoexponential pattern as volume is expelled through the outlet resistance. It has been eloquently debated whether the "Windkessel-like" pressure of the arterial system, particularly in diastole, arises in a similar fashion from the compliance characteristics of the major vessels (i.e., phasic inflow, partial volume retention, and gradual egress of volume therefrom) or from the dampening and merging of vascular reflections and rereflections. Probably both factors play a role, with the dominance of either factor varying in different situations (e.g., vasoconstriction versus vasodilation) and at different times during each systolic-diastolic pressure cycle. In discussing vascular reflections and their contribution to both the similarities and differences between the arterial system and a Windkessel, we will use the terms "diffuse" and "discrete" reflections [the terms "diffuse" and "distinct" were previously used by Westerhof's group (Sipkema, Westerhof, and Randall 1980)]. Vascular reflections occur along the entire length of the arterial tree anywhere a change in "impedance properties" exists (e.g., change in vessel diameter, wall stiffness, or vessel branching). These multiple reflection sites give rise to diffuse reflections, which play an important role in wave dispersion, effective volume distribution, and vascular Windkessel-like properties, as discussed previously. Vascular pressure tracings and vascular ITRs also exhibit identifiable pressure changes, or "peaks," not characteristic of a Windkessel system.

The term "discrete reflections" will be used to identify these pressure peaks. It is these discrete reflections that give rise to the oscillations of moduli and phase in the vascular impedance spectra. As will be explained, these discrete reflections "behave" as if they originate from two discrete reflection sites in the upper and lower parts of the body.

The two small peaks marked by arrows in Figures 15.5C and 15.6 are due to these discrete vascular reflections. In the original vascular ITR of Figure 15.5, these additional peaks follow the initial larger peak representing the effects of Z_c. These additional peaks are not present in either the original or modified Windkessel ITR, as the ideal Windkessel is a reflectionless system. Prior work by Westerhof's group has demonstrated that the two reflection peaks usually seen in the canine vascular ITR represent pressure reflections from two "effective" reflection sites. When simultaneous measurements of pulse wave velocity were made, the timing of these reflection peaks indicated that one effective reflection site was located in the upper body distal to the origin of the subclavian and brachiocephalic vessels and the other in the lower body distal to the descending aorta (Sipkema, Westerhof, and Randall 1980). These findings corroborate those of other authors who previously postulated the existence and location of two major reflection sites from the position of frequency minima in the impedance spectra (McDonald, 1968, O'Rourke, 1967). It is important to realize that there is no single anatomic structure or change located at these effective reflection sites, but rather they represent the summation of effects of multiple distal reflections (present evidence suggests that these larger vascular reflections probably originate at terminal arterioles). The effective upper body (more proximal) site gives rise to the first reflection peak following the Z_c peak, and the lower body site gives rise to the second reflection peak. In most cases the second reflection peak is of greater apparent magnitude. As will be shown, however, when wave velocity along the descending thoracic and abdominal aorta is increased, these two reflection peaks may merge and appear as one prominent reflection peak.

Relationship Between Aortic Root Pressure and the Vascular ITR

Once the impulse response function of any linear, time-invariant system is known, the output for any given input may be determined using the principle of superimposition. This principle dictates that if we can describe our input signal as the summation of a group of appropriately timed and scaled unit impulses, our output signal is simply the summation of a corresponding group of appropriately timed and scaled impulse responses. This process is analogous to impedance studies in which the input flow and output pressure can be described as the sum of appropriately timed and scaled sinusoidal waves.

This relationship between input flow impulses and output pressure is illustrated diagrammatically in Figure 15.7. Each impulse of flow into the aortic root produces a pressure waveform described by the impulse response function. When all of the corresponding output impulse responses are added together, the resulting sum predicts the pressure produced by the given input flow. By adding together many such flow impulses, we can reproduce the pattern of ventricular flow; by then adding together the corresponding impulse response functions, we can predict the resultant aortic pressure. In practice, this process of describing the flow signal as a sum of input impulses and then summing up the corresponding impulse responses is done using a mathematical technique termed "convolution."

Knowing this relationship, we can reexamine the vascular versus Windkessel pressure comparisons of Figure 15.1 and relate the observed pressure differences to differences between the vascular and Windkessel impulse response functions. To simplify these comparisons, we will modify the impulse response functions in the same manner as was done in Figure 15.6. That is, since Z_c may be reasonably approximated as a series resistance over the applicable frequency range of flow harmonics, we will remove the equivalent representations of Z_c and R_c from the impulse re-

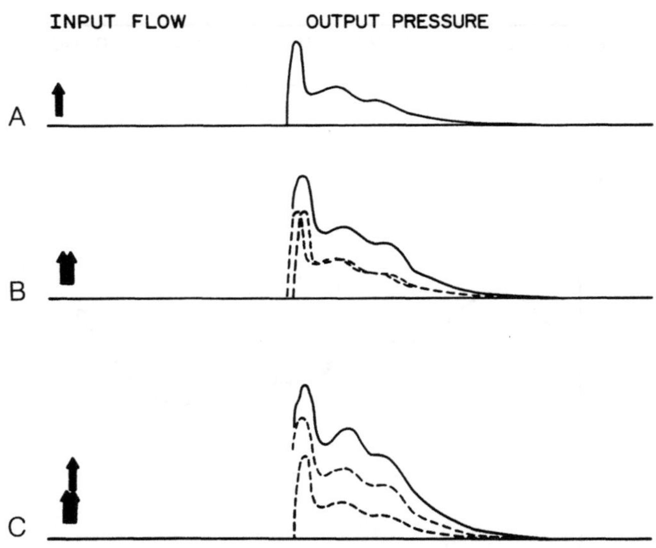

FIGURE 15.7. Relationship between flow and pressure in terms of the impulse response function. (A) single impulse of flow produces a pressure described by the impulse response function; (B) two flow impulses separated in time produce a pressure described by the summation of two impulse responses correspondingly separated in time; (C) the pressure produced by three flow impulses is equal to the sum of the pressure responses shown in B (produced by the first two impulses) plus an additional impulse response corresponding to the third flow impulse.

sponses. These equivalent components do not contribute to the differences between the impulse response functions.

The modified impulse response functions corresponding to the aortic root pressure waveforms in the three parts of Figure 15.1 are shown in the three parts of Figure 15.8. Looking first at Figures 15.1A and 15.8A, we note that the pattern of differences in aortic root pressures in 15.1 (i.e., the recorded vascular pressure is initially less than the simulated

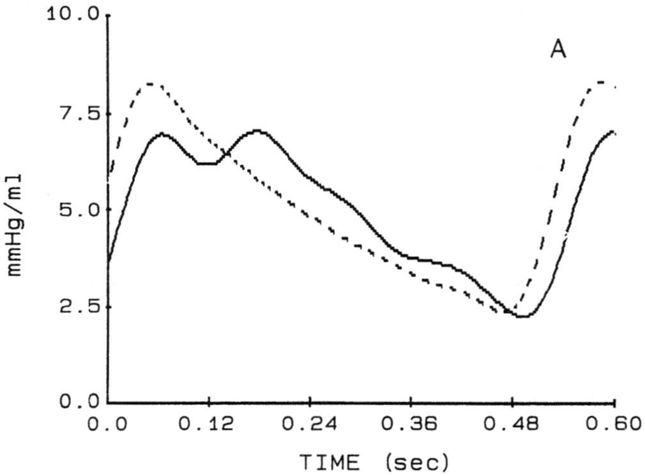

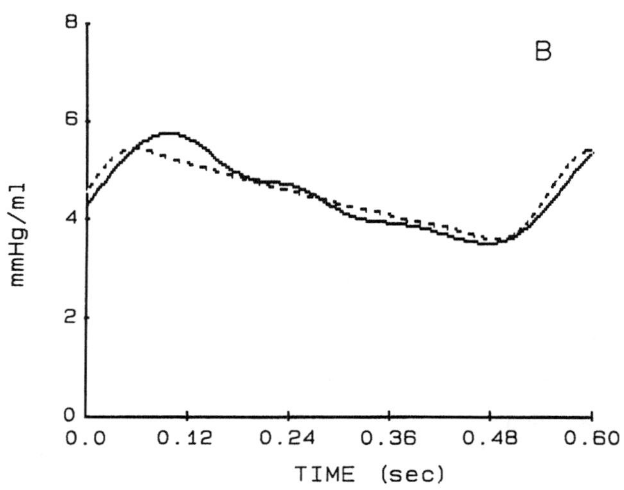

FIGURE 15.8. A-C. Modified vascular (solid line) and Windkessel (dashed line) impulse train responses (ITRs) corresponding to the pressure waveforms in Figure 15.1. See text for explanation.

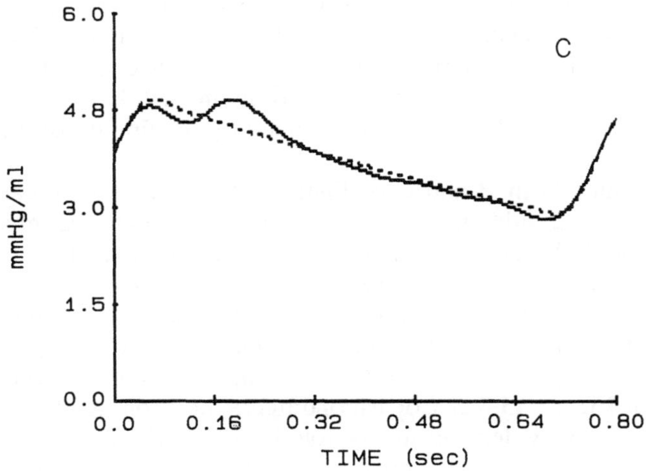

FIGURE 15.8 (Continued)

Windkessel pressure, and then transiently exceeds the Windkessel pressure with the return of discrete vascular reflections) is mirrored by the differences in the impulse response functions shown in 15.8. As already discussed, the slower initial rise in the vascular ITR (and hence in the aortic root pressure) is an effect of finite wave velocity in the vascular system. By comparing the ITRs in Figure 15.8A, we see that with the return of the first discrete reflection peak, the vascular ITR approaches that of the Windkessel. As the second reflection peak returns, it adds to this first peak (and to the pressure from diffuse reflections), and the vascular ITR now transiently exceeds the Windkessel ITR. Similarly, the vascular pressure in Figure 15.1A transiently exceeds the Windkessel model pressure in the early part of diastole, due to the return of these vascular reflections to the aortic root.

Now let's examine the aortic root pressure waveforms and ITRs in Figures 15.1B and 15.8B. Although the general pattern of the differences in the ITRs of Figure 15.8B is similar to that in Figure 15.8A (initial decrease in the vascular ITR relative to the windkessel, followed by a transient increase, then a common smooth exponential decay), the reflection peaks in Figure 15.8B return much earlier than those in Figure 15.8A. As a result, the vascular ITR exceeds the Windkessel ITR at an earlier point in time. This earlier return of reflections is echoed in the aortic pressure trace of Figure 15.1B by the occurrence of a prominent reflection peak during systole rather than diastole.

Also, in the vascular ITR of Figure 15.8B, the first reflection peak is much less distinct than that in Figure 15.8A. With atherosclerotic changes and/or hypertension, wave velocity often tends to increase more in the descending thoracic and abdominal aorta than in the ascending aorta (see

1982 review by O'Rourke). As a result, the second peak returns earlier relative to the first peak, and the two peaks appear to merge. This "merging" of reflections augments the resulting pressure rise. It should be pointed out, however, that our ability to distinguish between these two peaks in the vascular ITR is also influenced by the fidelity of the derived ITR.

By examining both B and C of Figures 15.1 and 15.8 we can see the effects of nitroprusside infusion on arterial wave velocity. Comparing the vascular ITRs in Figure 15.8B and C (ITRs from the same dog), it is easily appreciated that the reflection peaks in C (with nitroprusside infusion) return significantly later than those in B. Whereas the peak reflection in Figure 15.8B occurs at abut 0.11 sec, the dominant reflection peak in Figure 15.8C occurs at approximately 0.20 sec. Also, two distinct reflection peaks reappear in Figure 15.8C. Correspondingly, the prominent vascular reflection that was evident during systole in the aortic pressure tracing of Figure 15.1B has been delayed into diastole in the pressure tracing of Figure 15.1C.

Importance of Flow Patterns

As was evident in our comparisons of vascular pressures and vascular ITRs, a certain amount of information about vascular reflections may be ascertained by examining the central aortic pressure contour. It must be remembered, however, that aortic pressure is determined not only by vascular properties but also by aortic flow. The vascular ITR represents a pressure response to a simple, uniform, and *defined* flow pattern, whereas the aortic root pressure is a pressure response to a highly *variable* flow pattern. Because of this, attempted interpretation of changes in vascular reflection properties solely on the basis of aortic pressure contours is fraught with difficulty.

This concept is illustrated in Figure 15.9, where pressure waveforms from four different simulations are shown. On the basis of the wide differences in pressure contours, it might appear that these pressure traces were derived from several different vascular models. However, all were in fact derived using the same vascular impulse response function. The different pressure contours are solely a function of different flow patterns used in the modeling process. As a simple example, in some of the pressure traces the dominant vascular reflection is in systole, while in others it is clearly in diastole. In contrast to the previous examples, this relative systolic-diastolic positioning is here a function of changes in ejection times rather than changes in wave velocity. With prolonged ejection time, the major vascular reflection is more likely to return while ejection is still proceeding. With a shorter ejection time the same reflection may appear in diastole.

The apparent intensity of reflections in the aortic pressure trace is

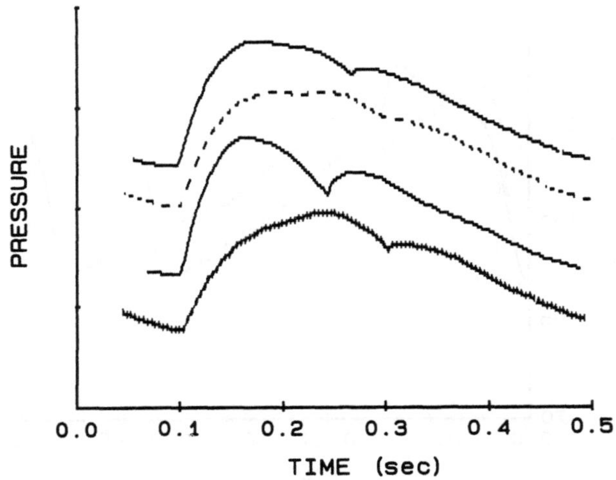

FIGURE 15.9. Effects of flow patterns on the contour of the aortic pressure trace. The differences in pressure contours are due solely to changes in flow patterns and not to differences in vascular properties.

equally as sensitive to small changes in flow patterns as is the relative systolic-diastolic position of the reflections (Figures 15.10 and 15.11). The solid line in each figure represents the control condition and the dashed line the variable following infusion of a low dose of nitroprusside. The control beat appears to have a much stronger systolic reflection peak than the nitroprusside beat. It would be a mistake to infer from these pressure tracings, however, that the nitroprusside had, in this instance, significantly altered the characteristics of discrete reflections. The modified vascular ITRs corresponding to these pressure and flow signals are shown in Figure 15.11A. There is very little difference between them with respect to the timing and magnitude of the major reflection peaks (arrows). The absolute height of each peak is different owing to both the change in heart rate and a decrease in total peripheral resistance, but the height and timing of each peak relative to its own baseline are very similar. The corresponding impedance spectra are illustrated in Figure 15.11B and C. The two spectra are very similar in both magnitude and phase following the first harmonic, again implying little change in reflection dynamics at this dose of nitroprusside.

The apparent stronger reflection in the aortic pressure tracing of the control beat is primarily due to the difference in flow contours and not to changes in reflection magnitude or timing. The broader, more rounded flow contour in the control beat produces a more rounded aortic pressure tracing that is still plateauing at the time the major discrete vascular reflection is returning to the proximal aorta. As a result, the reflection pressure is added to this stable aortic pressure and appears as a prominent

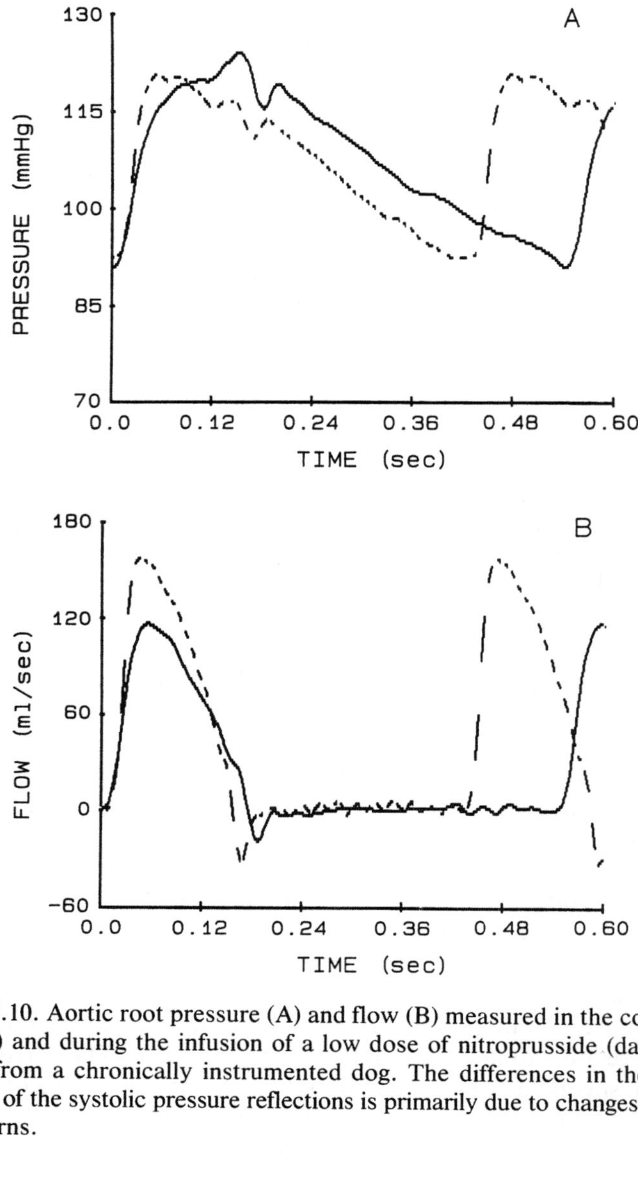

FIGURE 15.10. Aortic root pressure (A) and flow (B) measured in the control state (solid line) and during the infusion of a low dose of nitroprusside (dashed line). Data are from a chronically instrumented dog. The differences in the apparent magnitude of the systolic pressure reflections is primarily due to changes in ejection flow patterns.

peak. In contrast, the flow during nitroprusside administration has a more narrow contour with an early peak and more rapid decline. At the time the major vascular reflection returns to the aorta, the aortic pressure is already noticeably declining owing to the more rapid decline in flow. The reflection pressure is thus added to an already declining ejection pressure, and has an apparent lesser magnitude in the resultant aortic pressure recording.

Murgo and colleagues (1975 and Ch. 6) have shown that certain patterns of pressure tracings recorded at the aortic root correlate with patterns in the impedance spectra. Patients with strong, early reflection peaks in the central aortic pressure (type A beat) usually exhibited a more oscillatory impedance spectra, often with a large peak in impedance modulus around the sixth harmonic. Those individuals exhibiting a less intense and apparently "later" discrete reflection peak (as judged by its relationship to peak aortic pressure) had a much flatter impedance spectra (type C beats). Although this remains a valid observation when applied to a group of patients, the influence of flow contour on the appearance of aortic reflection peaks limits the use of pressure waveform classifications alone for categorizing vascular reflection properties in individual patients. In the example in Figures 15.10 and 15.11, the control beat would have been classified as a type A beat and the nitroprusside beat as a type C beat. Yet, as shown by the impulse response functions, there were only minor changes in the timing and magnitude of discrete vascular reflections (and only minor changes in the impedance spectra after the first harmonic). The major factor accounting for the transition from a type A beat to a type C beat was the change in ejection flow contour and not a change in vascular reflection properties.

Since the impulse response function represents the pressure due to a uniform, defined flow input, any changes in reflection timing and magnitude apparent in the vascular ITR are due to changes in vascular properties. Changes in aortic root pressure contours may, however, reflect changes in vascular properties *and/or* ejection flow contours.

Relationships Between the Vascular Impulse Response Function and the Vascular Impedance Spectrum

As is evident by the way we have derived the vascular impulse response function, there must be a direct correspondence between the information in the vascular impedance spectra and that in the impulse response. In point of fact, they are simply two different ways of presenting the same information. Both measurements describe the pressure produced by a prescribed unit of flow. The difference is that impedance describes this pressure as a function of frequency harmonics (i.e., the relationships between sinusoidal oscillations of pressure and flow), whereas the impulse response function describes this pressure as a function of time.

These two descriptions should be viewed as complementary rather than exclusive. By definition, any information on vascular properties contained in the impedance spectra must likewise be contained in the impulse response function, and vice versa. However, one or the other description may allow that information to be better utilized for a specific purpose. This will be seen, for example, when we take up the topic of ventricular

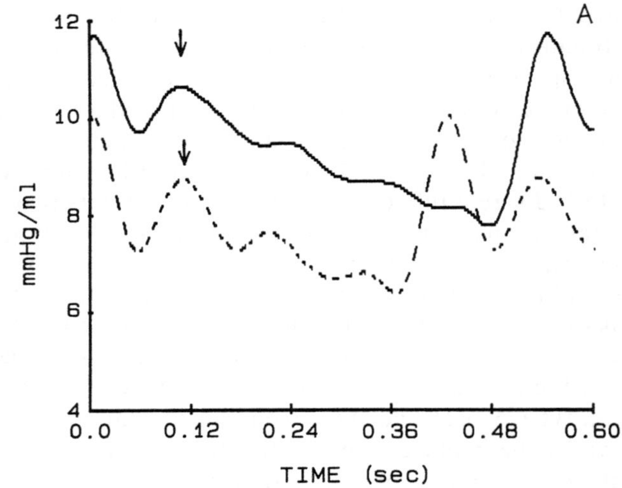

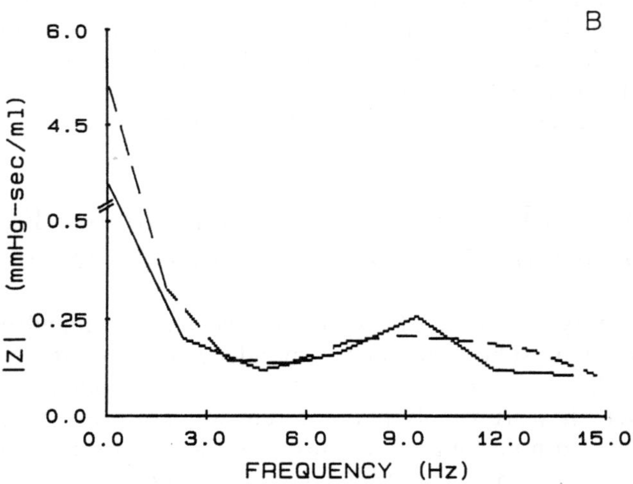

FIGURES 15.11. Impulse train response functions (A) and impedance spectra (B and C) derived from the pressure and flow waveforms of Figure 15.10. Control state, solid line; with nitroprusside infusion, dashed line. Arrows show major reflection peaks.

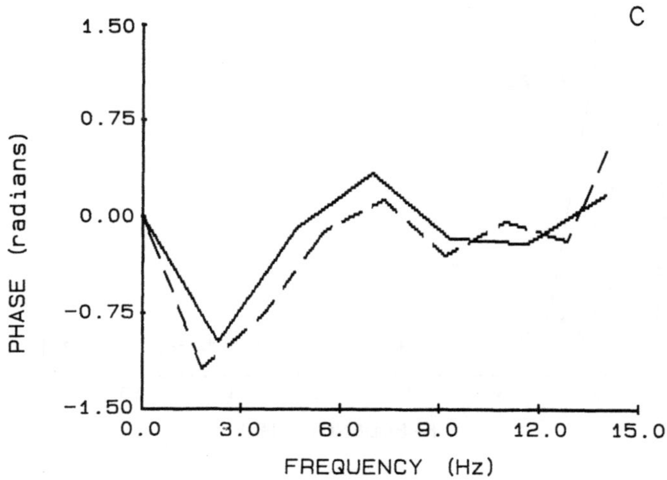

FIGURE 15.11 (Continued)

vascular coupling. Such coupling is more easily performed using the impulse response, which is a function of time, because ventricular function is more easily expressed as a function of time.

For persons not familiar with thinking in the "frequency domain," the impulse response offers the advantage of depicting vascular properties as a function of the magnitude and timing of pressures produced by a known flow pulse. For those familiar with impedance, a discussion of some of the interrelationships between impedance and the time-based impulse response may offer some additional insights. Since most of the previous examples have been limited to the impulse train response (ITR) (rather than the single impulse response SIR), this discussion will be limited to the ITR.

The time period of one cycle in the ITR corresponds to the period of the fundamental harmonic in the discrete impedance spectra. For impedance spectra obtained in the conventional manner (by Fourier series analysis of pressure and flow at a fixed heart rate), this corresponds to the duration of one cardiac cycle. The magnitude of moduli in the impedance spectra often translate into areas in the ITR. As depicted in Figure 15.12, the magnitude of the zero frequency impedance modulus (Z_0) translates into the total area under one cycle of the ITR. In clinical terms, this area corresponds to measured total peripheral resistance (average pressure divided by average flow). The average magnitude of the higher frequency impedance components (i.e., the conventional method for estimating characteristic impedance) correlates with the area of the Z_c peak in the ITR (shown by the hatched area in Figure 15.12B). This peak describes the pressure developed coincident with the input flow pulse, prior to the return of any significant pressure from peripheral reflections. This inter-

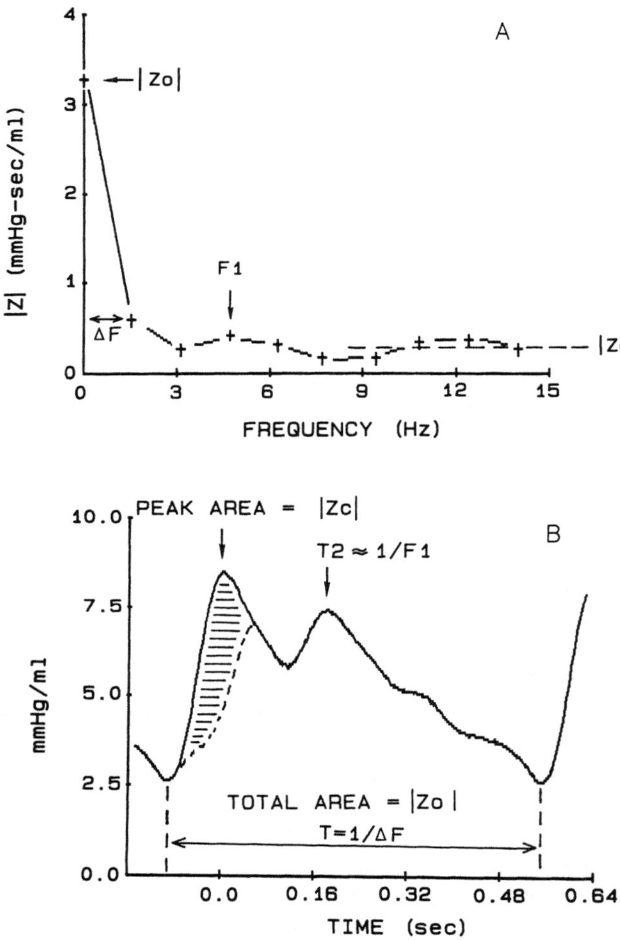

FIGURE 15.12. Relationships between the discrete vascular impedance spectrum (A) and the vascular impulse train response (ITR) function (B). Z_o, zero frequency impedance modulus; Z_c, characteristic impedance; F_1, frequency of first impedance maximum; T, one period of the ITR. See text for discussion.

pretation is consistent with the more general definition of characteristic impedance as the impedance of the system in the absence of reflections.

The timing and magnitude of reflection peaks in the ITR roughly correlate with the frequencies and magnitude of maxima in impedance moduli. As discussed earlier, a typical canine vascular ITR has two discrete reflection peaks, one from the upper body region and one from the lower body region. The peak from the lower body region is usually the more prominent of the two. A canine vascular impedance spectrum typically demonstrates two corresponding impedance maxima. The first maximum correlates with the second reflection peak in the ITR. The inverse of the frequency at which this *first* maximum (F_1) occurs provides a good estimate of the timing of the *second* reflection peak in the ITR. The second impedance maximum correlates with the first reflection peak in the ITR. Although the inverse of the frequency at which this second maximum occurs does provide some information about the timing of the first ITR peak, the apparent timing and magnitude of this peak are also heavily influenced by early "diffuse" reflections and the Windkessel-like properties of the vascular tree. These latter properties, which are most clearly revealed by the exponential decay of the latter portion of the ITR, are described by the very-low-frequency components in the impedance spectrum.

Ventricular/Vascular Coupling: Analytical Methods

Previous studies investigating the effects of different vascular properties on ventricular function can be divided into two groups: (1) analysis of ventricular "power output" and "work" and how the pulsatile and steady components are influenced by vascular properties (see 1979 review by Noble and 1982 review by O'Rourke) and (2) isolated heart studies using artificial ventricular-loading systems. Although studies in the first group provide descriptive information, they do not allow prediction of how time-based variables (e.g., stroke volume and indices of myocardial oxygen demand) are influenced by different vascular properties. Studies in the second group have allowed prediction of such hemodynamic variables, but have been limited to modified Windkessel models of the arterial system (Campbell et al. 1984; Elzinga and Westerhof 1973; Ishide, Shimizu, and Takishima 1980; Piene (1980, 1984); Piene and Sund 1979; Shroff, Janicki, and Weber 1985; Suga 1971; Sunagawa, Sagawa, and Maughan 1984; Sunagawa et al. 1982, 1983; van den Horn, Westerhof, and Elzinga 1984; Westerhof, Elzinga, and Sipkema 1971). As just illustrated, such Windkessel models do not reproduce the effects of finite wave velocity and discrete vascular reflections.

More sophisticated models of the arterial tree will simulate these properties (Noordergraaf, Verdouw, and Boom 1963; O'Rourke and Avolio 1980; Taylor 1966) but the large number of variables that must be determined in these models makes them impractical for simulating ventricular vascular interactions over a wide range of experimental conditions. Although impedance measurements provide a means of describing these vascular properties, as a frequency-based description they cannot be directly coupled to time-based models of ventricular function (Milnor 1975) [and ventricular function cannot be modeled by impedance measurements (Hunter and Noordergraaf 1976)]. By transforming impedance measurements to time-based impulse response functions, however, we can "couple" the resulting descriptions of vascular properties to our ventricular models. This technique thus provides a very powerful analytical method for examing the ventricular-loading effects of complex vascular properties, and how drug-induced changes in these properties influence ventricular pressure and flow.

The analytical methods involved in this technique will be presented in this section. Some familiarity with Fourier transforms and convolution is assumed. An in-depth understanding of the material in this section is *not* required for a qualitative understanding of the material presented in the next section, which describes results from simulations using measured vascular impulse response functions (readers who are not mathematically inclined may wish to proceed directly to the next section).

This approach to ventricular vascular coupling entails three separate steps. The first step is development of a model of ventricular function that is independent of loading conditions. The second step involves transformation of measured impedance data into an impulse response function of a form suitable for "coupling" with the ventricular model, and the last step is performance of this coupling. The first step will be illustrated using data obtained from an isolated canine left ventricle ejecting into a simulated arterial loading system. The second step will be described using impedance data measured in a chronically instrumented canine preparation. The third step will be explained by means of a graphic analysis of the simultaneous equations involved in the coupling process. These methods, as well as some experimental results predicting pressures and flows in the isolated heart, have been presented in more detail elsewhere (Latson et al. 1984).

Ventricular Function

As discussed in Chapters 8, 10, 12, and 13, global ventricular function may be described in terms of a time-varying ventricular elastance $E(t)$:

$$E(t) = \frac{P(t)}{V(t) - V_d}$$

$$\begin{aligned} \text{where } P(t) &= \text{measured ventricular pressure} \\ V(t) &= \text{volume} \qquad\qquad (15.1) \\ V_d &= \text{volume axis intercept extrapolated} \\ &\quad\ \text{from several pressure–volume loops} \end{aligned}$$

Although $E(t)$ defined in this manner has been shown to be relatively insensitive to altered loading conditions (Suga, Kitabatake, and Sagawa 1979; Suga, Sagawa, and Shoukas 1973; Suga and Yamakoshi 1977), there are small flow-related differences between ejecting and isovolumic pressures (at similar volumes and times after onset of systole) which are not accounted for by this conventional $E(t)$ relationship (Hunter et al. 1979; Shroff, Janicki, and Weber 1983,1985; Suga, Sagawa, and Demer 1980). These flow-related differences in pressure have been described by the "internal resistance" concept discussed in Chapter 11. On the basis of this concept, the ejecting pressure $P(t)$ is reduced from the corresponding isovolumetric pressure $P_i(t)$ by a resistance R multiplied by instantaneous flow:

$$P(t) = P_i(t) - [F(t) \times R] \tag{15.2}$$

Direct measurements have indicated that this internal resistance R has a variable magnitude proportional to isovolumetric pressure (Hunter et al. 1979; Shroff, Janicki, and Weber 1985; Suga, Sagawa, and Demer 1980). This time-variant pressure-dependent resistance may be described by

$$R = P_i(t) \times RF \tag{15.3}$$

where RF is a constant called the "resistance factor." The resultant expression relating ejecting pressures to corresponding isovolumetric pressures is now given by

$$P(t) = P_i(t) - \{F(t) \times [P_I(t) \times \mathrm{RF}]\} \tag{15.4}$$

rearranging gives

$$P(t) = P_i(t) \times \{1 - [F(t) \times \mathrm{RF}]\} \tag{15.5}$$

$$P_i(t) = \frac{P(t)}{1 - [F(t) \times \mathrm{RF}]} \tag{15.6}$$

On the basis of results of prior studies using the isolated canine left ventricle, an assumed value of 0.0015 sec/ml was used for RF (Shroff, Janicki, and Weber 1983,1985; Suga, Sagawa, and Demer 1980).

By substituting $P_i(t)$ for $P(t)$ in the numerator of Equation 15.1, we can derive a description of ventricular properties that includes both elastance and resistance effects. The result provides a flow-independent description of ventricular pumping ability, to be denoted $E^*(t)$:

$$E^*(t) = \frac{P_i(t)}{V(t) - V_d^*} \tag{15.7}$$

Here V_d^* is the volume-axis intercept for the $E^*(t)$ model, which is determined as described in the following. By combining Equations 15.6 and 15.7, $E^*(t)$ may be expressed in terms of the measured ejecting pressure $P(t)$:

$$E^*(t) = \frac{P(t)/\{1 - [F(t) \times \mathrm{RF}]\}}{V(t) - V_d^*} \tag{15.8}$$

$E^*(t)$ thus may be determined solely from ejecting beats, without reliance on isovolumic data.

$E^*(t)$ was used to describe the pumping ability of the isolated canine left ventricle ejecting into a computer-simulated three-element Windkessel loading system (Sunagawa et al. 1982). An average $E^*(t)$ curve was determined for each heart as follows: During periods of presumed constant contractility, data were collected over a wide range of loading conditions. For example, the 15 pressure–volume loops measured for one ventricle are shown in Figure 15.13A. $P_i(t)$ was then calculated from Equation 15.6 and the measured variables $P(t)$ and $V(t)$ [where $F(t) = -dV(t)/dt$]. *Pres-*

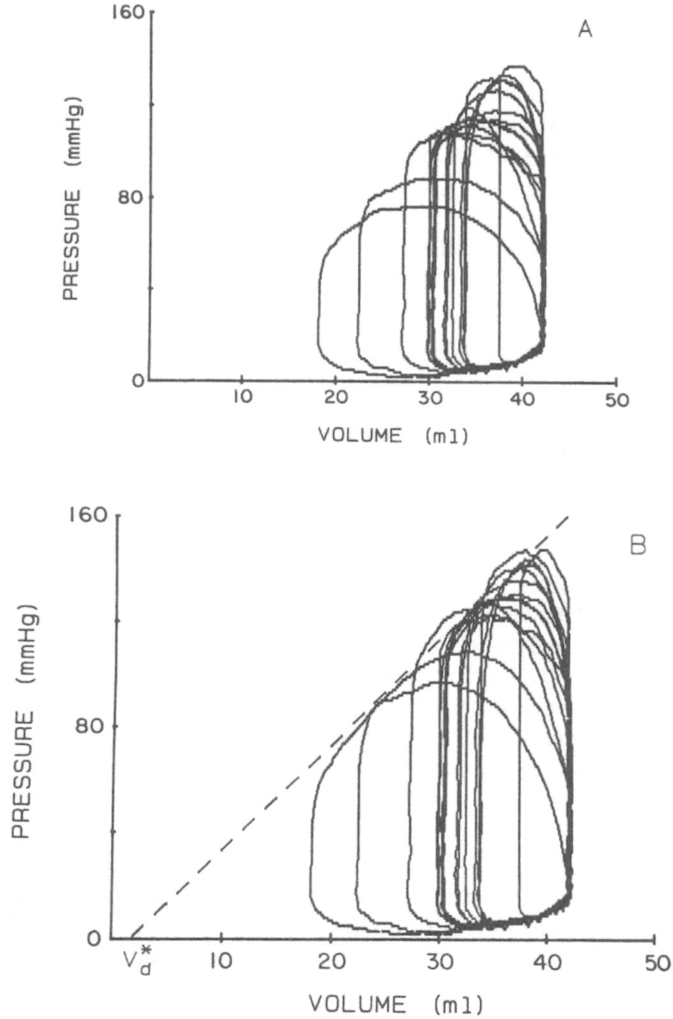

FIGURES 15.13. Measured pressure–volume loops (A) and corresponding $P_i(t)$–volume loops (B) recorded for an isolated canine left ventricle ejecting into a computer-simulated three-element Windkessel load. Data from 15 different loading conditions are shown. Calculations are described in the text. $V_d{}^*$ is the volume–axis intercept of a regression line drawn through the points of maximal elastance.

sure–volume loops of $P_i(t)$ versus $V(t)$ were then constructed, as shown in Figure 15.13B.

$V_d{}^*$ was calculated from a least squares fit regression line drawn through the points of maximal elastance. Zero is used as the initial estimate of $V_d{}^*$ in the equation of $E^*(t)$ to localize these maximal elastance points. A regression line is then constructed through these points, and the projected volume–axis intercept provides a new estimate of V_d* to be used in the next calculation cycle. This process is continued until a stable value for V_d* is obtained.

Figure 15.14A depicts the $E^*(t)$ curves corresponding to the $P_i(t)$ versus $V(t)$ loops in Figure 15.13B. For the 15 curves shown, 12 of the $E^*(t)$ curves essentially superimpose up to the time of maximal $E^*(t)$. Over this same time frame, the remaining three curves shown only small deviations from these 12. More than 90% of ventricular ejection usually occurs during this rising portion of the $E^*(t)$ curve between the onset of contraction and the time of maximal elastance. These $E^*(t)$ curves should be compared with the corresponding $E(t)$ curves constructed using the conventional relationship $E(t) = P(t)/[V(t) - V_d]$ (Fig. 15.14B). Flow-related pressure reductions in $P(t)$ result in corresponding reductions in the derived $E(t)$ curves. Since these reductions depend on the magnitude and timing of $F(t)$, they vary with alterations in loading conditions. These effects are easily appreciated in Figure 15.14C, where $E(t)$ curves are compared with the average curve of $E^*(t)$ from the data in Figure 15.14A (for this comparison, $E(t)$ curves were calculated using V_d*). The $E(t)$ curves superimpose on the $E^*t)$ curve until ejection begins, at which time a distinct change in curve contour is noted. The $E(t)$ curves then begin to fall below the $E^*(t)$ curve owing to the flow-related pressure reductions associated with ejection.

For subsequent computer modeling, the $E^*(t)$ curves constructed from all conditions for each heart can be pooled and an average curve obtained. A numerical description of this average curve (one curve for each heart) can then be used in subsequent modeling calculations.

Loading System Impulse Response

An example canine vascular ITR similar to that derived earlier in this chapter is depicted in Figure 15.15A. Both the unfiltered (dashed line) and filtered (solid line) forms are shown. This derived ITR is subjected to two analytical "modifications" to facilitate the process of "coupling" this impulse response description with $E^*(t)$. These modifications are used to adjust for the alterations in the impulse response function introduced by high-frequency truncation and discrete frequency sampling (discussed earlier).

The first modification adjusts for the effects of high-frequency truncation on the impulse response representation of characteristic impedance. As previously discussed, the characteristic impedance Z_c of the canine aorta may be approximated as a series resistance (Westerhof, Elzinga, and Sipkema 1971; Westerhof, Elzinga, and van den Bos 1973). As such, the IDEAL impulse response representation of Z_c has the same form as that of the series resistance R_c previously illustrated in Figure 15.2, that is, a very narrow impulse of area equal to its resistance value and a width less than or equal to the sampling interval. The initial peak in the derived ITR

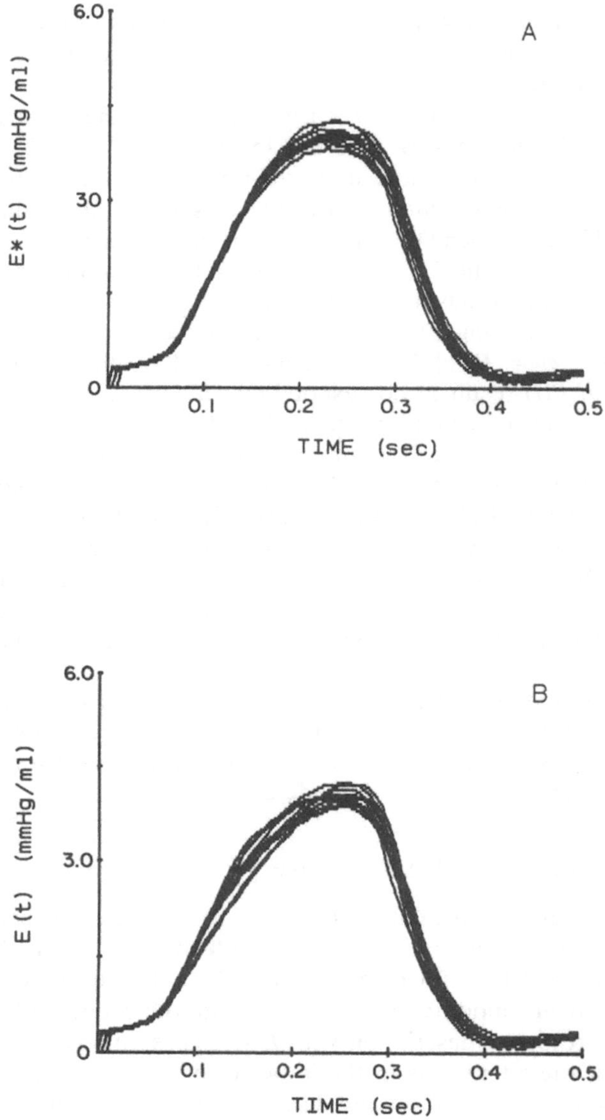

FIGURE 15.14. Elastance [$E(t)$, $E^*(t)$] curves derived from the data of Figure 15.13. (A), $E^*(t)$ curves calculated from [$P_i(t)$] and volume [$V(t)$] for the 15 different loading conditions; (B) $E(t)$ curves calculated from pressure [$P(t)$ and $V(t)$]; (C) average $E^*(t)$ curve from A (dotted line) compared with $E(t)$ curves. Note that the $E(t)$ curves fall below the $E^*(t)$ curve beginning with the onset of ejection. See text for definitions of $E(t)$ and $E^*(t)$ and discussion.

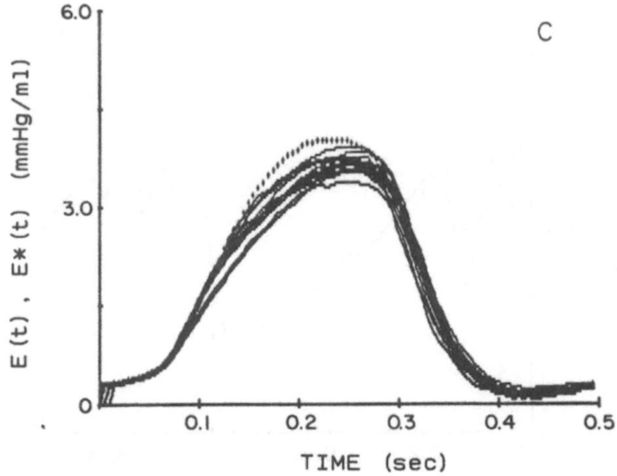

FIGURE 15.14 (Continued)

retains the appropriate area (Bracewell 1965) but is of a much greater width than its idealized representation because of the effects of high-frequency truncation. This width, equal to the period of the highest impedance harmonic, imposes limitations on the minimum timing interval that can be used between sequential calculations in the "coupling" process. The limitation arises from the fact that the pressure associated with this resistive property should be directly proportional to instantaneous flow, and hence the area of the peak (equal to its impedance value) should be ideally represented within one timing interval. One approach to this problem is to calculate the value of Z_c by averaging the high-frequency impedance moduli and then replace its broad-based (frequency-limited) representation in the derived ITR by an appropriate idealized representation of its effects.

Removal of the broad-based Z_c peak in the derived ITR is accomplished by subtracting out the impedance representation of Z_c from the total vascular impedance, and then performing an inverse Fourier transform on this new spectrum. If we approximate Z_c as a series resistance, then the impedance spectrum for Z_c is simply a series of harmonic values with modulus equal to $| Z_c |$ and a phase of zero. Subtraction of this spectrum from the total vascular impedance (using complex numbers to account for phase and magnitude relationships) yields the impedance description of the loading system minus the effects of Z_c. Inverse transformation of this impedance description yields the desired modified ITR (Fig. 15.15B) (several such modified ITRs were also previously shown and discussed). During sequential convolution, the combination of this type of modified impulse response function *plus* a separate numerical representation of Z_c is mathematically equivalent to the original impulse response function. This numerical representation for Z_c is simply a constant, equal to the magnitude of Z_c, which is multiplied by instantaneous flow. For purposes of expla-

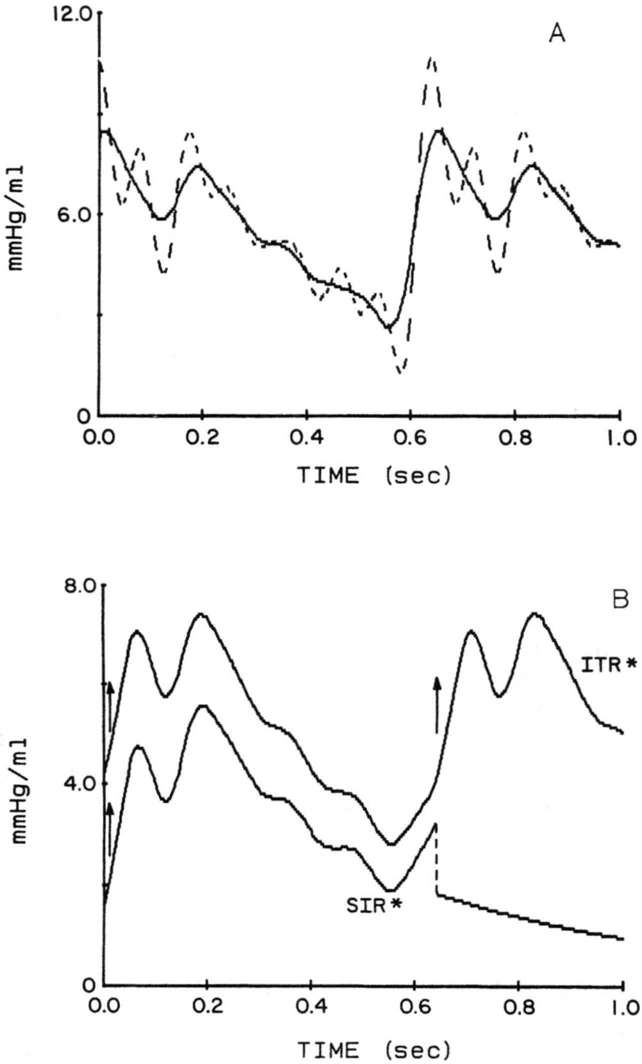

FIGURE 15.15. Modifications made to the vascular impulse train response (ITR) to enable coupling with a ventricular model by sequential convolution. (A) original filtered (solid line) and unfiltered (dashed line) ITR; (B) ITR* derived from the ITR of A by removing the broad-based characteristic impedance (Z_c) peak (the arrow signifies the ideal impulse that replaces the removed Z_c peak during convolution). SIR* is the estimated single impulse response derived from ITR*. See text for explanation.

nation, an asterisk (*) will be used in the equations that follow to denote when an impulse response function has been modified in this manner (e.g., ITR*).

The second modification to the derived impulse train response compensates for the lack of impedance information between harmonics of the heart rate. As illustrated in Figure 15.3, inverse transformation of a discrete impedance spectrum (known only at certain interval values) yields the steady-state response to a train of impulses. For sequential convolutional analysis, it is more desirable to have the response to a single impulse. If the general behavior of the latter portion of the single impulse response (SIR) is known, then the SIR may be approximated from the ITR.

For the intact vascular system, this general behavior of the latter part of the SIR has the form of a first-order exponential decay. Such a decay may be described by a parallel resistance R_p and capacitance C, analogous to that which appears in the three-element Windkessel model. The appropriate values for R_p and C may be calculated from the time constant of decay of the latter portion of the train response (or alternatively from the slope of diastolic pressure decay). Using these parameter values, in conjunction with known analytical equations describing the impulse response functions for a paralleled R_p-C network, we can approximate the SIR from the known ITR.

Conceptually, the steady-state train response ITR(t) between $t = 0$ and T (where T is the repetition period) is equal to the value of SIR(t) plus a variable offset value. This offset value is determined by (1) the interval between flow impulses and (2) the latter part of the single impulse response. For an R_p-C model, this offset may be determined from the following equations describing the impulse response functions for a parallel R_p-C network.

Single response

$$\mathrm{SIR_{RpC}}(t) = \frac{1}{C} \times e^{-t/R_pC} \tag{15.9}$$

Train response (for $t = 0$ to T)

$$\mathrm{ITR_{RpC}}(t) = K \times \mathrm{SIR_{RpC}}(t) \tag{15.10}$$

where

$$K = \frac{1}{1 - e^{-T/R_pC}}$$

Offset of train from single response

$$\begin{aligned} \mathrm{OFFSET_{RpC}}(t) &= \mathrm{ITR_{RpC}}(t) - \mathrm{SIR_{RpC}}(t) \qquad (15.11)\\ &= \mathrm{SIR_{RpC}}(t) \times (K - 1) \end{aligned}$$

To approximate the single response of the canine vascular system from the train response, we divide the single response into two portions on the basis of the value of t relative to T: one portion for $t = 0$ to T and the second portion for $t > T$. The first portion may contain additional components other than that modeled by the R_p-C network. For example, this initial portion of the canine vascular impulse response function also contains the effects of characteristic impedance, finite wave velocity, and discrete reflections. This portion of the single response may be approximated from the train response by subtracting the variable offset (Eq. 15.11) from the train response:

FOR $t = 0$ to T

$$SIR^*(t) = \mathrm{ITR}^*(t) - \mathrm{OFFSET}_{\mathrm{RpC}}(t) \tag{15.12}$$

It should be noted that we have used the modified form (ITR*) of the derived ITR in this equation (although the original derived ITR could be used, this would require extending t backward from 0 to accommodate the increased width of the Z_c peak).

We have assumed that the latter portion of the single impulse response function may be modeled solely by a first-order exponential decay. The second portion of SIR*(t) (for $t > T$) is thus given by substituting the determined values of R_p and C in Equation 15.9 for $\mathrm{SIR}_{\mathrm{RpC}}(t)$:

FOR $t > T$

$$\mathrm{SIR}^*(t) = \mathrm{SIR}_{\mathrm{RpC}}(t) \tag{15.13}$$

This step is illustrated in Figure 15.15B. The single response, as estimated from ITR*, is labeled SIR*. The small upswing in the SIR* function that occurs as t approaches T represents the smoothing effect of high-frequency truncation at points of discontinuity (and is made worse by digital filtering). This upswing will introduce a small error when this SIR* is used in a model operating at a heart rate other than $1/T$. The effect will be minimized by using a high cutoff frequency, and by using the unfiltered form of SIR* in the convolution algorithm. Alternatively, the line described by $\mathrm{SIR}_{\mathrm{RpC}}(t)$ (Eqs. 15.9 and 15.13) can be extended backward for $t < T$ until it intersects with SIR*(t) given by Equation 15.12. This approach also introduces a small error, however, related to the symmetry of the "smoothing" in the ITR* function about $t = 0$ and $t = T$.

This second modification (deriving SIR* from ITR*) assumes that (1) the magnitude of discrete reflections in the loading system (artificial or in vivo) is insignificant outside the period T and (2) the major "time-delaying" storage properties of the system (i.e., those properties of significant duration to cause the train response to not return to baseline between repetitive impulses) may be approximated by a first-order decay. Initial results obtained using the canine vascular ITR support the feasibility of these modeling assumptions for in vivo work, but additional studies are needed.

"Coupling" of Ventricular and Vascular Properties

Sequential convolution may now be used to "couple" these independent, time-based descriptions of ventricular and vascular properties. Conceptually, this involves describing the pressures "opposing" ejection as consisting of two parts: (1) pressure due to all prior flow into the vascular system and (2) pressure generated by flow during the present instant. The first component includes the effects of vascular compliance, finite wave velocity, and discrete vascular reflections. The second component is determined largely by Z_c. $E^*(t)$ describes how the ventricle interacts with these developed pressures in terms of resultant changes in systolic ventricular volumes.

In analytical terms, this method involves sequential calculation of ventricular pressure $P(t_i)$, ventricular volume $V(t_i)$, flow $F(t_i)$, and arterial pressure $P_a(t_i)$ assuming that values for the preceding time point (t_{i-1}) are known. "t_i" is the current value of time in the calculation process, and Δt is the timing interval between successive calculations.

$P_a(t)$ is determined by convolving the arterial system impulse response function $\mathrm{SIR}(t)$ with $F(t)$. This convolution is of the form (for time $= t_i$)

$$P_a(t_i) = \int_{u=0}^{u=+\infty} \mathrm{SIR}(u)\, F(t_i - u)\, du \qquad (15.14)$$

where u is the convolution variable that accounts for the history of flow. We can replace SIR in this integral by the combination of SIR* and Z_c in the following manner:

$$P_a(t_i) = [F(t_i) \times |Z_c|] + \int_{u=0}^{u=+\infty} \mathrm{SIR}^*(u)\, F(t_i - u)\, du \qquad (15.15)$$

The integral in this equation may be divided into two parts on the basis of the value of u: (1) $u = 0$ to Δt and (2) $u = \Delta t$ to $+\infty$. The first part, in combination with the Z_c term, relates pressure and flow over the current time interval between t_{i-1} and t_i. The second part describes pressure due to the prior history of flow (i.e, all $F(t)$ up to time $= t_{i-1}$). Since flow up until time $= t_{i-1}$ is known, this second part of the integral may be solved directly. Equation 15.15 may then be rewritten as

$$P_a(t_i) = [F(t_i) \times |Z_c|] + \left[\{ \int_{u=0}^{u=\Delta t} \mathrm{SIR}^*(u)\, F(t_i - u)\, du \right] + B_i \qquad (15.16)$$

where the known value B_i has now replaced the second part of the original integral. To interact ventricular function with this equation, $F(t)$ (for small

Δt) may be expressed in terms of ventricular volume. The average flow over the time interval between time $= t_{i-1}$ and t_i is given by

$$F_{\text{avg}} = \frac{V(t_{i-1}) - V(t_i)}{\Delta t} \tag{15.17}$$

From this relationship and the known value for $F(t_{i-1})$, we can describe $F(t_i)$ as a function of $V(t_i)$. By using this relationship to make the appropriate substitutions in Equation 15.16, and after performing the indicated numerical integration, we can reduce Equation 15.16 to a linear equation relating arterial pressure and ventricular volume:

$$P_a(t_i) = B_i - [m \times V(t_i)] \tag{15.18}$$

This equation describes a line in the pressure–volume plane, as depicted in Figure 15.16 (the line labeled P_a). For purposes of graphic explanation, this equation may be expanded by adding and subtracting the term $[m \times V(t_{i-1})]$ to the right side of the equation, resulting in:

$$P_a(t_i) = \{B_i - [m \times V(t_{i-1})]\} + \{m \times [V(t_{i-1}) - V(t)_i)]\} \tag{15.19}$$

The bracketed term on the right side of this expanded equation gives the pressure that is due to all prior flow into the arterial system; that is, if there were no additional flow between t_i and t_{i-1}, $P_a(t_i)$ would be equal to this term alone. This term will vary in successive calculations depending on both the nature of the vascular impulse response function and the prior history of flow. In Figure 15.16 this pressure is represented by the heavy dot located at the intersection of the P_a line with the line extended upward from $V(t_{i-1})$. For any change in ventricular volume that does occur in this time interval [i.e., $V(t_i) < V(t_{i-1})$ and hence $F(t_i) > 0$], P_a will increase from this point according to the slope m appearing in the second bracketed term on the right side. The slope m, which is related to the magnitude of Z_c and the initial value of SIR*(t), does not vary between successive calculations. When multiplied by the change in ventricular volume, m describes the pressure generated by flow during the present instant between t_{i-1} and t_i.

By rearranging Equation 15.8 we can arrive at a corresponding equation for ventricular pressure as a function of t_i, $V(t_i)$, $E^*(t_i)$, and $F(t_i)$:

$$P(t_i) = \frac{E^*(t_i)\,[V(t_i) - V_d^*]}{1 - [F(t_i) \times \text{RF}]} \tag{15.20}$$

After making a similar substitution for $F(t_i)$ (based on Eq. 15.17), this equation may be reduced to the form

$$P(t_i) = f[E^*(t_i), V(t_i)] \tag{15.21}$$

$E^*(t_i)$ is obtained from the average $E^*(t)$ curve for this ventricle, determined as previously described. Hence the only unknowns in this equation (for

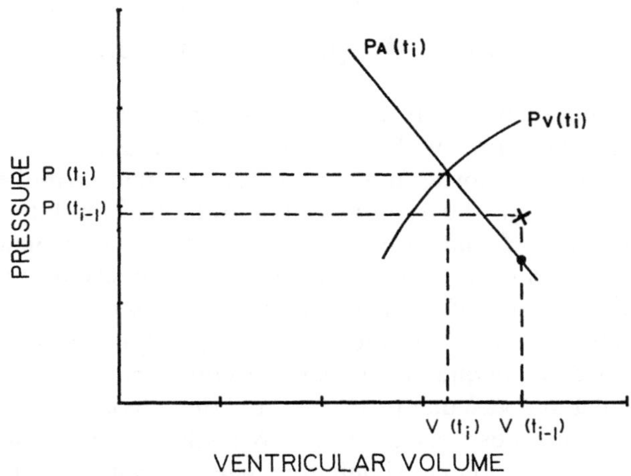

FIGURE 15.16. Graphic analysis of the calculations involved in sequential convolution. x represents the pressure and volume calculated from the preceding step of convolution at time $= t_{i-1}$. The line labeled $P_a(t_i)$ depicts the arterial pressure as a function of ventricular volume at time $= t_i$. The heavy dot is the pressure that would occur if no flow occurred between t_{i-1} and t_i. $P_v(t_i)$ depicts ventricular pressure as a function of ventricular volume at time $= t_i$. The intersection of lines P_a and P_v represents the simultaneous solution predicting both pressures and volume at time $= t_i$.

time $= t_i$) are $P(t_i)$ and $V(t_i)$. As shown in Figure 15.16, this relationship describes a second line (labeled P_v) in the pressure–volume plane which is slightly concave toward the volume axis.

Equations 15.18 and 15.21 contain three unknown variables: $P_v(t_i)$, $P_a(t_i)$, and $V(t_i)$. By adding the constraint that during ejection $P_a(t_i)$ is equal to $P_v(t_i)$, we have three equations with three unknowns that can be solved simultaneously for the three unknowns. The solution is depicted graphically in Figure 15.16 as the intersection of the two lines describing $P_a(t_i)$ and $P_v(t_i)$ as a function of $V(t_i)$. After solving for these values, the present t_i becomes t_{i-1}, and the value of t_i is incremented by Δt. The calculation cycle is then repeated to determine the values of $V(t_i)$, $P_a(t_i)$, and $P_v(t_i)$ for this new t_i. During periods when the ventricle is not ejecting [i.e., $P_v(t) < P_a(t)$ and $F(t) = 0$], the two systems do not interact and can be solved separately.

Once the computer is given a numerical description of $E^*(t)$ and the desired vascular SIR, this calculation cycle is automatically entered. For modeling steady-state conditions, the computer can be made to repeat this cycle until the "arterial" end-diastolic pressure between two successive diastoles is within a given limit (e.g., 1 mm Hg).

Simulations Using a Ventricular Model "Coupled" to the Vascular Impulse Response Function

In the beginning of this chapter, some examples were given comparing recorded aortic pressures with pressures that would have been produced by applying the same aortic flow into a three-element Windkessel model (Fig. 15.1). Although the Windkessel adequately simulated the major properties of the vascular system, the observed discrepancies between these two pressures illustrated that there were some significant differences between the properties of the vascular system (as seen at the aortic root) and the Windkessel model. In the intervening sections of this chapter we have developed techniques by which we could independently describe both ventricular and vascular properties and then "couple" them together to predict resultant pressures and flows. We will now use these techniques to perform more physiologic simulations wherein both pressure *and* flow are allowed to vary.

Figure 15.17 illustrates some sample results comparing the ventricular-loading properties of the real canine arterial system and a corresponding Windkessel model. For the simulations shown here, the vascular ITR and corresponding Windkessel ITR previously illustrated in Figure 15.5C were coupled to the *same* ventricular model. These ITRs were first modified, as explained previously. A typical $E^*(t)$ function (based on experiments such as those described in this chapter) was then choosen to describe the properties of the ventricular model. These independent descriptions of ventricular and vascular properties were then "coupled" by sequential convolution, also described previously. The resultant aortic pressure and pressure–volume loops are shown in the two parts of this figure. The solid line depicts the results using the vascular ITR, and the dashed line shows the corresponding results using the Windkessel ITR.

The aortic pressure from the vascular model is initially less than that from the Windkessel model. This decrease in pressure, which is a manifestation of finite wave velocity in the arterial tree, is qualitatively the same as was observed in Figure 15.1A. In this simulation, however, we can also quantitate an improvement in stroke volume. The timing of the discrete reflections, relative to the duration of ventricular ejection, is such that the pressure augmentation from the dominant reflection (the second reflection peak) does not appear until after the aortic valve has closed. As such, this reflection provides a beneficial effect by augmenting diastolic coronary perfusion pressure.

Let's compare these results with those illustrated in Figure 15.18. The vascular ITR used in these simulations was measured in an old dog with increased arterial wave velocity, and thus was similar to the vascular ITR shown in Figure 15.8B. When this ITR and a corresponding Windkessel ITR were coupled to the same ventricular model, the results were as illustrated here. Although the vascular model pressure (solid line) is again

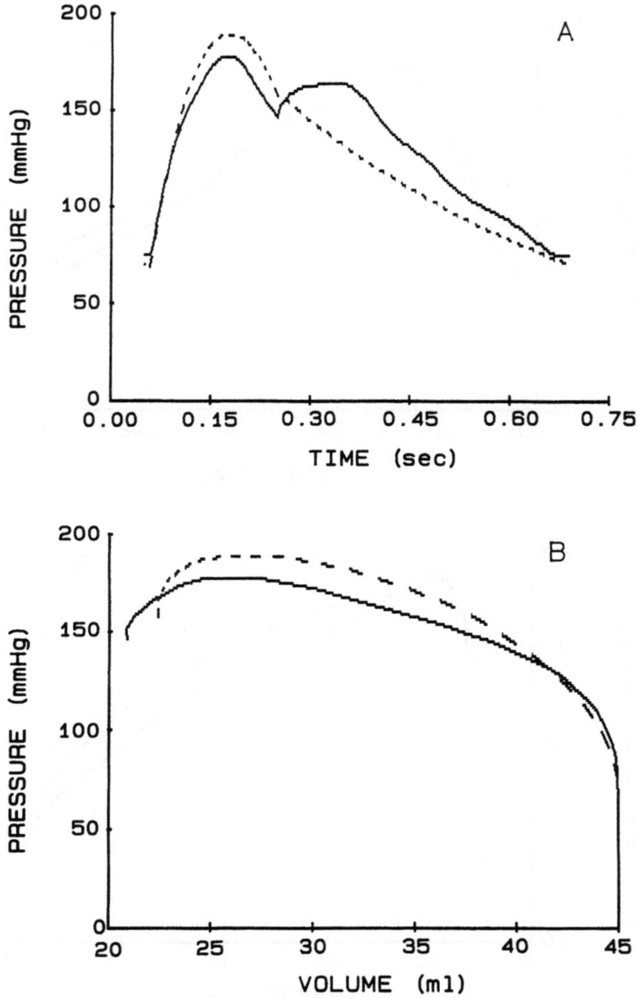

FIGURE 15.17. Simulated aortic pressure (A) and ventricular pressure–volume loops (B). Solid line, simulation using a canine vascular impulse train response (ITR) with normal wave velocity. Dashed line, simulation using the corresponding Windkessel ITR.

initially less than the Windkessel pressure (dashed line), the early returning discrete reflections act to increase the vascular model pressure above that of the Windkessel model much earlier than in Figure 15.17. This resulted in an increase in late systolic pressure, which in turn acted to decrease stroke volume (relative to the Windkessel simulation). Vascular reflections thus produced a detrimental effect in this situation. The diastolic pressure

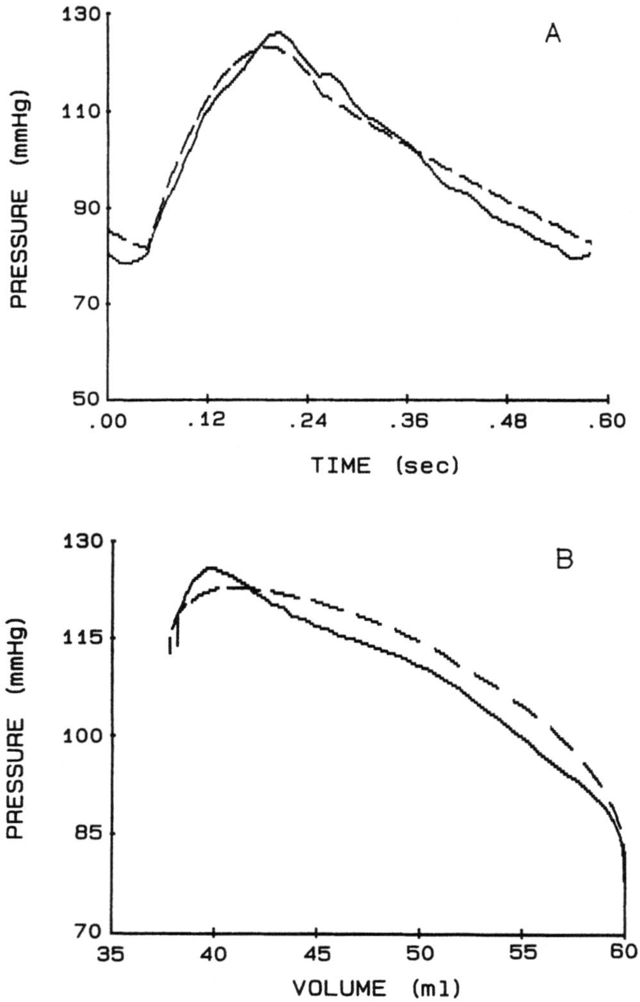

FIGURE 15.18. Simulated aortic pressure (A) and ventricular pressure–volume loops (B). Solid line, simulation using a canine vascular impulse train response (ITR) with increased wave velocity. Dashed line, simulation using the corresponding Windkessel ITR. Compare these results with those in Figure 15.17.

augmentation from returning discrete reflections, which was very evident in Figure 15.17, is almost absent here.

Figure 15.19 illustrates the effects of changing the timing of reflections with a vasodilator. The vascular ITRs for this comparison were measured in the same dog in the control state and after the infusion of a small dose of nitroprusside. In this animal, this dose of nitroprusside was associated with a significant delay in the return of peripheral reflections with only

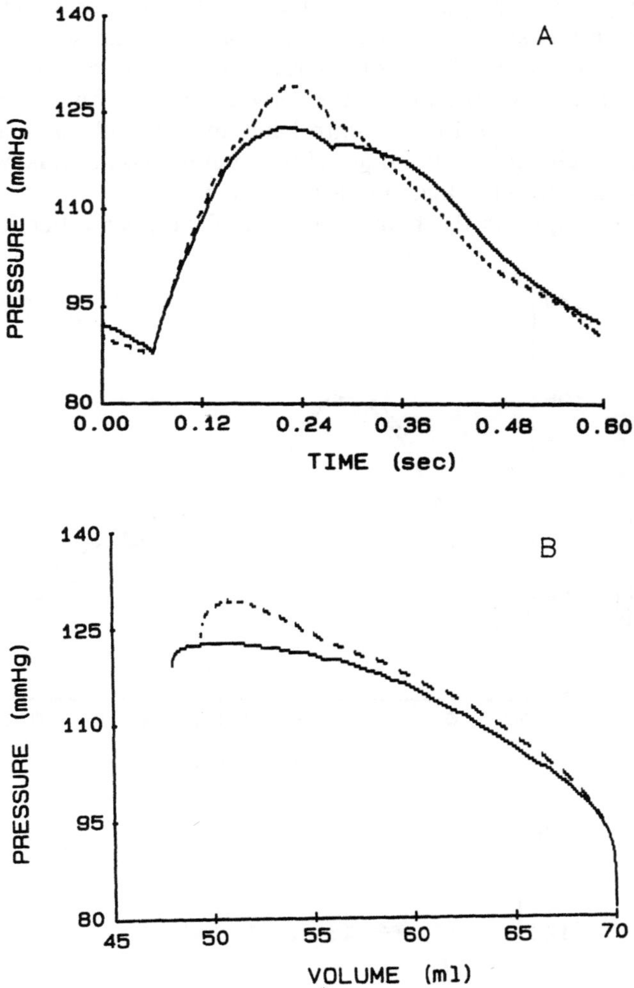

FIGURE 15.19. Aortic pressure (A) and ventricular pressure–volume loops (B) from simulations using vascular ITRs measured in the same dog during control (dashed line) and nitroprusside infusion (solid line).

minimal changes in the values of other vascular parameters (characteristic impedance, total peripheral resistance, effective compliance of the Windkessel model). The simulated aortic pressure traces and pressure–volume loops shown in this figure were obtained by coupling these two ITRs to the same ventricular model (dashed line, control; solid line, nitroprusside). By delaying the return of a prominent vascular reflection, there has been a significant increase in stroke volume, decrease in systolic pressure, and increase in diastolic pressure. If these two situations are simulated instead by coupling the corresponding Windkessel ITRs to this same ventricle, the results are as shown in Figure 15.20. This Windkessel simulation effectively removes the effects of discrete vascular reflections. In this instance there is very little difference between the resulting pressures and pressure–volume loops for the control and nitroprusside simulations. Examining only changes in Windkessel parameters would thus have meant missing an important change in vascular properties.

A final example on the hemodynamics of vascular reflections will il-

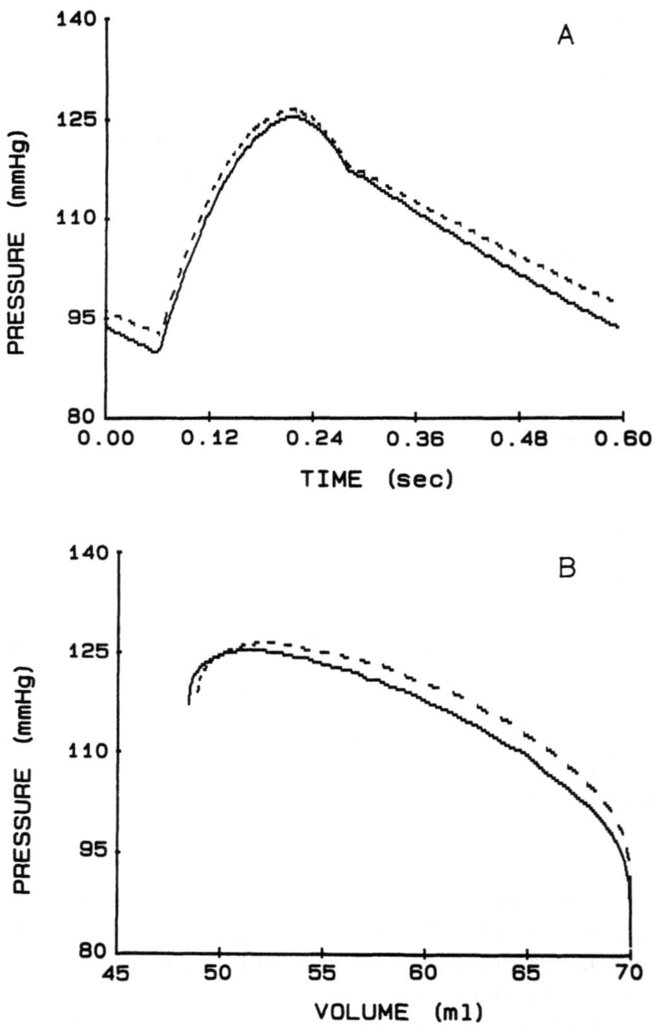

FIGURE 15.20. Aortic pressure (A) and ventricular pressure–volume loops (B) from simulations using the Windkessel impulse train responses (ITRs) corresponding to the vascular ITRs in Figure 15.19.

lustrate the influence of ventricular ejection times. For this example, a vascular ITR and corresponding Windkessel ITR were coupled to two ventricular models. The ventricular models differed only in the parameter T_{max}, the time of maximal ventricular elastance. For the first model, T_{max} was set at 0.14 second (solid line), and for the second it was set at 0.22 second (dashed line). When these two ventricular models were coupled to the Windkessel ITR, the results shown in Figure 15.21 were obtained.

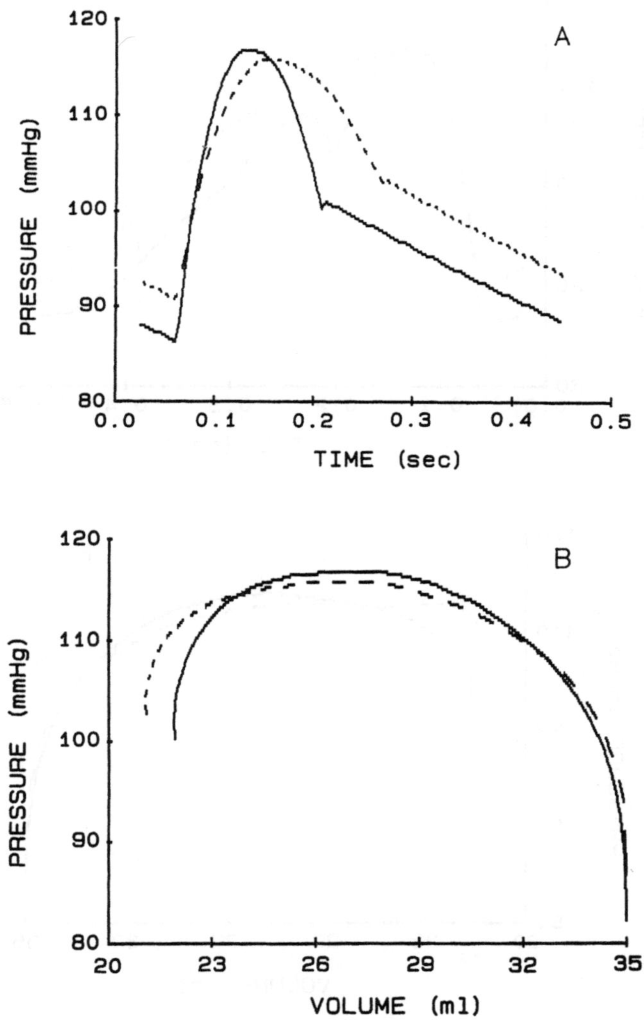

FIGURE 15.21. Aortic pressure (A) and ventricular pressure–volume loops (B) from simulations coupling the same Windkessel impulse train response (ITR) with two ventricular models with different durations of contraction. Dashed line, ventricular model T_{max} = 0.14 sec; solid line, T_{max} = 0.22 sec.

The increase in T_{max} resulted in a significant increase in stroke volume. Figure 15.22 depicts the results when the ventricular models were coupled with the vascular ITR. In contrast to the results shown in Figure 15.21, there was very little change in stroke volume. As T_{max}, and hence ejection time, increased, the peripheral vascular reflection that initially was confined to diastole now returns during systole. Here it acts to augment sys-

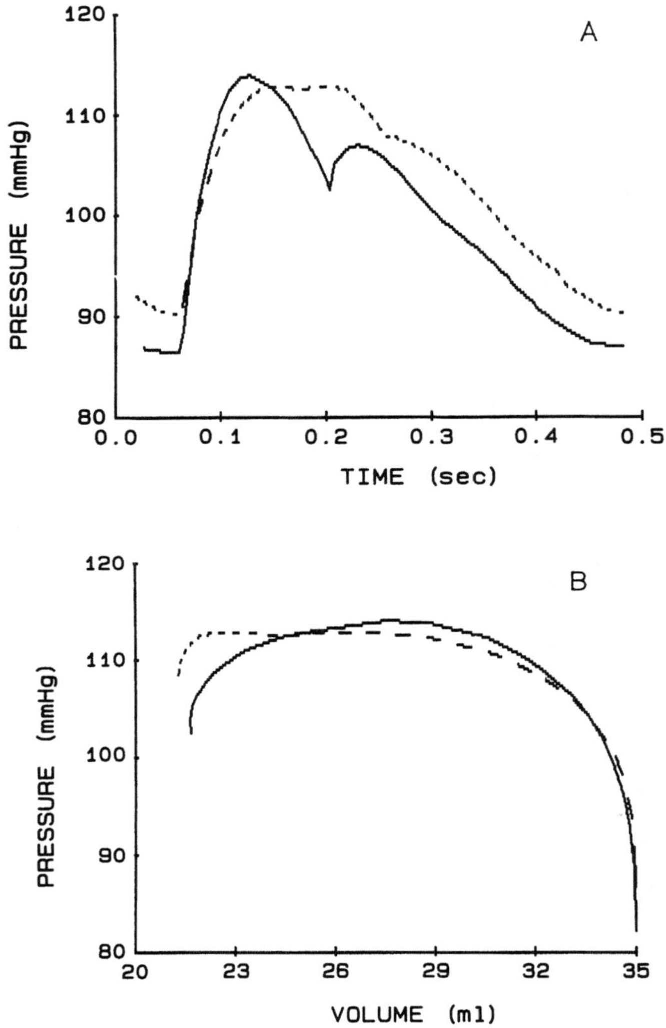

FIGURE 15.22. Aortic pressure (A) and ventricular pressure–volume loops (B) from simulations coupling the same vascular impulse train response (ITR) with two ventricular models with different durations of contraction. Dashed line, ventricular model T_{max} = 0.14 sec; solid line, T_{max} = 0.22 sec. Compare results with those in Figure 15.21.

tolic pressure and reduce any potential increase in stroke volume from the increase in ejection time. In the clinical situation, T_{max} tends to increase both with bradycardia and with heart failure. It is possible that prominent vascular reflections may be more influential in limiting stroke volumes in these instances where ventricular contraction patterns are more prolonged.

Clinical Applications

As has been emphasized throughout this chapter and elsewhere in this book, when we change from a "healthy" arterial system with normal wave velocity to one with increased wave velocity, the effects of wave velocity and discrete reflections change from being beneficial to detrimental. Although many of our comparisons have concerned the vascular system relative to the Windkessel system, the more appropriate comparison for clinical purposes is the absolute change between the beneficial and the detrimental states. O'Rourke and others (see 1982 review by O'Rourke) have repeatedly emphasized that increases in arterial wave velocity and the magnitude of discrete reflections are common with aging and hypertension. Pepine, Nichols, and Conti (1978) have demonstrated that patients with heart failure also show a more oscillatory impedance pattern, suggestive of stronger vascular reflections in this patient group as well.

The possible importance of changes in the magnitude and timing of vascular reflections induced by vasodilators has been the subject of several recent investigations. O'Rourke, Yaginuma, and Avolio (1984) have presented data that such changes may contribute to the beneficial effects of nitroglycerin in the treatment of myocardial ischemia. Yin and colleagues (1983) and Brin and Yin (1984) have studied alterations in vascular properties with infusion of low doses of nitroprusside in heart failure patients, and have suggested that changes in vascular reflections may be involved in the improvements in stroke volumes seen in these patients. Pepine and co-workers (1979) have also presented data suggesting that the combination of nitroprusside and neosynephrine may beneficially affect complex vascular properties in heart failure patients.

None of these studies, however, has offered any quantitative data (e.g., changes in stroke volume or indices of myocardial oxygen demand) on the effects of vascular reflections separate from the other concomitant effects of vasodilators (e.g., changes in total peripheral resistance, characteristic impedance, and/or vascular compliance). Studies using isolated hearts ejecting into simulated physiologic loading systems have been restricted to three-element Windkessel models of the circulation, and so likewise have offered no quantitative information on the effects of changes in wave velocity and discrete vascular reflections.

The analytical methods presented in this chapter were designed to investigate the separate effects of these complex changes in vascular properties in a more quantitative manner. Computer simulations using the impulse response functions of both a three-element Windkessel model and the intact canine vascular system have been presented and compared. From these simulations, some quantitative estimates of the the effects of finite wave velocity and discrete reflections can be made. Initial data from these canine models would suggest that the combined effects of increased wave velocity and a prominent systolic reflection can reduce stroke volume by around 10% to 15% (compared with the situation with normal wave velocity). This same change in arterial properties would also increase myocardial oxygen demand by a similar proportion [as judged by changes in the pressure–volume area index (Suga et al. 1983)]. Shifting of the major vascular reflection into systole also significantly reduces the early diastolic pressures, thus potentially reducing diastolic myocardial perfusion. Although the magnitude of these individual changes is not that large, their combined effects could be important in the setting of congestive heart failure and/or myocardial ischemia. Furthermore, examination of central aortic pressure traces from human studies (see Ch. 4) shows that discrete vascular reflections can be considerably more prominent in humans than in dogs. These data thus add quantitative support to the idea that changes in the timing and magnitude of discrete vascular reflections may play a significant role in the clinical response to vasodilators.

A final important issue requiring further study is whether different vasodilators might have different effects on vascular reflection dynamics. Examples have been given here showing that nitroprusside can reduce wave velocity with resultant beneficial effects on the timing of vascular reflections. Although some of this reduction in wave velocity is due simply to the effects of lowering mean blood pressure, a contribution from direct pharmacologic effects on vascular wall properties cannot be ruled out. Vasodilators are currently evaluated almost solely on their ability to lower total peripheral resistance. On the basis of their differing pharmacologic actions, however (e.g., different types of smooth-muscle relaxants, centrally acting antiadrenergic agents, alpha blockers), discrepant effects on wave velocity and peripheral reflections may well exist. Gundel, Cherry, and Schultz (1981) compared the effects of nitroprusside and isuprel in patients undergoing cardiac catheterization. Although they concluded that there were no significant differences between the effects of these two agents on vascular properties, their study design did not control for mean blood pressure. Furthermore, their results are analyzed by averaging impedance spectra from a group of patients, rather than by looking for differences in individual patients. Differences in reflection dynamics, when impedance spectra are averaged, may cause significant effects in individual patients to be overlooked (O'Rourke 1967). Figure 15.23 shows two vascular ITRs from the same dog during separate infusions of nitroprusside (dashed line) and isuprel (solid line), with a similar mean blood pressure achieved by both infusions. Note that the discrete vascular reflections are significantly more delayed with the nitroprusside. Additional studies comparing the effects of different vasodilators in individual subjects at

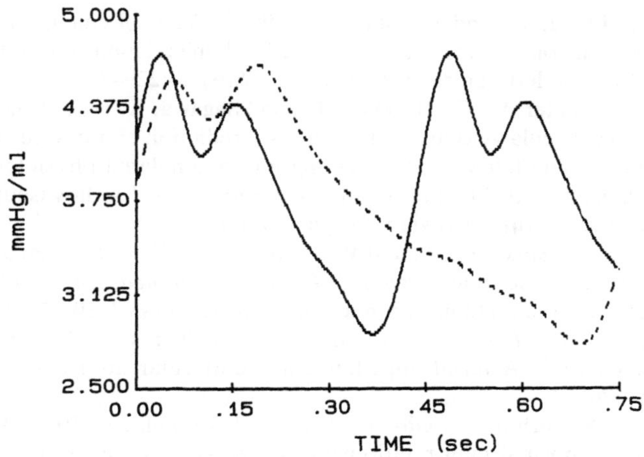

FIGURE 15.23. Impulse train responses (ITRs) from the same dog during infusions of nitroprusside (dashed line) and isuprel (solid line). Note that the discrete vascular reflections are significantly more delayed with nitroprusside than with isuprel.

equivalent mean pressures are needed. Possible different effects by different drugs could have significant implications in the acute and chronic treatment of hypertension, congestive heart failure, and myocardial ischemia.

Bibliography

Bracewell, R. 1965. *The Fourier Transform and Its Applications*. New York: McGraw-Hill.

Brigham, E. O. 1974. *The Fast Fourier Transform*. Englewood Cliffs, N.J.: Prentice-Hall.

Brin, K. P., and Yin, F. C. P. 1984. Effect of nitroprusside on wave reflections in patients with heart failure. *Ann. Biomed. Eng.* 12(2):135–150.

Campbell, K. B., Ringo, J. A., Neti, C., and Alexander, J. E. 1984. Informational analysis of left-ventricle/systemic-arterial interaction. *Ann. Biomed. Eng.* 12(2):209–229.

Campbell, K. B., Ringo, J. A., Wakoo, Y., Klavano, P. A., and Alexander, J. E. 1982. Internal capacitance and resistance allow prediction of right ventricle outflow. *Am. J. Physiol.* 243:H99–H112.

Elzinga, G., and Westerhof, N. 1973. Pressure and flow generated by the left ventricle against different impedances. *Circ. Res.* 32:178–186.

Gundel, W., Cherry, G., and Schultz, D. 1981. Aortic input impedance in man: acute response to vasodilator drugs. *Circulation* 63:(6):1305.

Hunter, W. C., Janicki, J. S., Weber, K. T., and Noordergraaf, A. 1979. Flow-pulse response: a new method for the characterization of ventricular mechanics. *Am. J. Physiol.* 237(3):H282–H292.

Hunter, W., and Noordergraaf, A. 1976. Can impedance characterize the heart? *J. Appl. Physiol.* 40(2):250–252.

Ishide, N., Shimizu, Y., and Takishima, T. 1980. Effects of changes in the aortic input impedance on systolic pressure-ejected volume relationships in the isolated supported canine left ventricle. *Cardiovasc. Res.* 14:229–243.

Latson, T. W., Hunter, W. C., Burkhoff, D., and Sagawa, K. 1984. Time sequential analysis of ventricular-vascular interactions: prediction of pressure and flow for the isolated canine left ventricle ejecting into a simulated physiologic loading system. Presented at 7th International Conference of Cardiovascular System Dynamics Society (under review for publication).

Laxminarayan, S., Sipkema, P., and Westerhof, N. 1978. Characterization of the arterial system in the time domain. *IEEE Trans. Biomed. Eng.* 25(2):177–184.

McDonald, D. A. 1968. Hemodynamics. *Ann. Rev. Physiol.* 30:525–556.

Milnor, W. R. 1982. *Hemodynamics*. Baltimore: Williams and Wilkins.

Milnor, W. R. 1975. Arterial impedance as ventricular afterload. *Circulation* 36(5):565–570.

Murgo, J. P., Westerhof, N., Giolma, J. P., and Altobelli, A. 1980. Aortic input impedance in normal man-relationship to pressure wave forms. *Circulation* 62(1):105–116.

Noble, M. I. 1979. Left ventricular load, arterial impedance and their interrelationship. *Cardiovasc. Res.* 13:183–198.

Noordergraff, A., Verdouw, P. D., and Boom, H. B. K. 1963. The use of an analog computer in a circulation model. *Prog. Cardiovasc. Dis.* 5(5):419–437.

O'Rourke, M. F. 1967. Pressure and flow waves in systemic arteries and the anatomical design of the arterial system. *J. Appl. Physiol.* 23:139–149.

O'Rourke, M. F. 1970. Arterial hemodynamics in hypertension. *Circ. Res.* 26(Suppl. 2):123–133.

O'Rourke, M. F. 1982. *Arterial Function in Health and Disease*. New York: Churchill Livingstone.

O'Rourke, M. F.,and Avolio, A. P. 1980. Pulsatile flow and pressure in human systemic arteries. *Circ. Res.* 46:363–372.

O'Rourke, M. F., and Taylor, M. G. 1967. Input impedance of the systemic circulation. *Circ. Res.* 20(4):365.

O'Rourke, M. F., Yaginuma, T., and Avolio, A. P. 1984. Physiological and pathophysiological implications of ventricular/vascular coupling. *Ann. Biomed. Eng.* 12(2):119–134.

Pepine, C. J., Nichols, W. W., and Conti, C. R. 1978. Aortic input impedance in heart failure. *Circulation* 58(3):460.

Pepine, C. J., Nichols, W. W., Cury, R. C., and Conti, R. C. 1979. Aortic input impedance during nitroprusside infusion. *J. Clin. Invest.* 64:643–654.

Piene, H. 1980. Interaction between the right heart ventricle and its arterial load: a quantitative solution. *Am. J. Physiol.* 238:H932–H937.

Piene, H. 1984. Impedance matching between ventricle and load. *Ann. Biomed. Eng.* 12(2):191–208.

Piene, H., and Sund, T. 1979. Flow and power output of right ventricle facing load with variable input impedance. *Am. J. Physiol.* 237(2):H125–H130.

Pouleur, H., Lefevre, J., and Jaumin, P. M. 1978. Significance of pulmonary input impedance in right ventricle performance. *Cardiovasc. Res.* 12:617.

Shroff, S. G., Janicki, J. S., and Weber, K. T. 1983. Left ventricular systolic dynamics in terms of its chamber mechanical properties. *Am. J. Physiol.* 245:H110–H124.

Shroff, S. G., Janicki, J. S., and Weber, K. T. 1985. Evidence and quantitation of left ventricular systolic resistance.
Am. J. Physiol. 249:H358–H370.
Sipkema, P., Westerhof, N., and Randall, O. S. 1980. The arterial system characterized in the time domain. *Cardiovasc. Res.* 14:270–279.
Suga, H. 1971. Theoretical analysis of a left-ventricular pumping model based on the systolic time-varying pressure/volume ratio. *IEEETrans. Biomed. Eng.* 18:pp. 47–55.
Suga, H., Hisano, R., Goto, Yo., Yamada, O., and Igarashi, Y. 1983a. Effect of positive inotropic agents on the relation between oxygen consumption and systolic pressure volume area in canine left ventricle. *Circ. Res.* 53:306–318.
Suga, H., Hisano, R., Hirata, S., Hayashi, T., Yamada, O., and Ninomiya, I. 1983b. Heart rate-independent energetics and systolic pressure–volume area in dog heart. *Am. J. Physiol.* 244:H206–H214.
Suga, H., Kitabatake, A., and Sagawa, K. 1979. End-systolic pressure determines stroke volume from fixed end-diastolic volume in the isolated canine left ventricle under a constant contractile state. *Circ. REs.* 44:238–249.
Suga, H., Sagawa, K., and Demer, L. 1980. Determinants of instantaneous pressure in canine left ventricle: time and volume specification. *Circ. Res.* 46:256–263.
Suga, H., Sagawa, K., and Shoukas, A. A. 1973. Load independence of the instantaneous pressure–volume ratio of the canine left ventricle and effects of epinephrine and heart rate on the ratio. *Circ. Res.* 32:314–332.
Suga, H., and Yamakoshi, K. 1977. Effects of stroke volume and velocity of ejection on end-systolic pressure of canine left ventricle: end-systolic volume clamping. *Circ. Res.* 40(5):445–450.
Sunagawa, K., Burkhoff, D., Lim, K. O., and Sagawa, K. 1982. Impedance loading servo pump system for excised canine ventricle. *Am. J. Physiol.* 243:H346–H350.
Sunagawa, K., Maughan, W. L., Burkhoff, D., and Sagawa, K. 1983. Left ventricular interaction with arterial load studied in isolated canine ventricle. *Am. J. Physiol.* 245:H773–H778.
Sunagawa, K., Sagawa, K., and Maughan, W. L. 1984. Ventricular interaction with the loading system. *Ann. Biomed. Eng.* 12(2):163–190.
Taylor, M. G. 1966. The input impedance of an assembly of randomly branching elastic tubes. *Biophys. J.* 6:29–51.
van den Horn, C. J., Westerhof, N., and Elzinga, G. 1984. Interaction of heart and arterial system. *Ann. Biomed. Eng.* 12(2):151–162.
Westerhof, N., Elzinga, G., and Sipkema, P. 1971. An artificial arterial system for pumping hearts. *J. Appl. Physiol.* 31(5):776.
Westerhof, N., Elzinga, G., and van den Bos, G. C. 1973. Infuence of central and peripheral changes on the hydraulic input impedance of the systemic arterial tree. *med. Biolog. Eng.* 11:pp. 710–722.
Yin, F. C. P., Guzman, P. A., Brin, K. P., Maughan, W. L., Brinker, J. A., Trail, T. A., Weiss, J. L., and Weisfeldt, M. L. 1983. Effects of nitroprusside on hydraulic vascular loads of the right and left ventricle of patients with heart failure. *Circulation* 67(6):1330–1339.
Yin, F. C. P., Weisfeldt, M. L., and Milnor, W. R. 1981. Role of aortic input impedance in the decreased cardiovascular response to exercise with aging in dogs. *J. Clin. Invest.* 68:28–38.

16
Estimation of Arterial Compliance

FRANK C. P. YIN, ZHAORONG LIU, AND
KENNETH P. BRIN

Introduction

As has been demonstrated in several previous chapters, the Windkessel model of the arterial system, despite its simplicity and attendant limitations, continues to be extremely useful both for understanding vascular physiology and for investigating ventricular/vascular coupling. The important parameters of this model are the compliance and the peripheral resistance. Whereas calculation of total peripheral resistance is straightforward and estimates are easily obtainable in vivo, estimation of total arterial compliance (the increment in volume produced by an increment of pressure) is very difficult. Direct measurement of overall arterial compliance is complicated by the need to measure accurately total arterial blood volume, and has been accomplished only in experimental animals (Shoukas and Sagawa 1973). Therefore in most instances, estimates of compliance are made indirectly by assuming that the aortic diastolic pressure decays exponentially with a time constant that is the product of total arterial resistance and compliance (Bourgeois et al. 1974; Conroy 1969; Defares and Van der Waal 1969, 1973; Deswysen, Charlier, and Gevers 1980; Iriuchijima, Kumazawa, and Kawakami 1971; Levy, Birkui, and Saumont 1978; Levy et al. 1985; Messerli, Frohlich, and Nutura 1985; Randall, Van den Bos, and Westerhof 1984; Simon et al. 1979; Simon, Levenson, and Safar 1985; Ventura et al. 1984). Not only has this assumption not been critically examined, but reflections and other perturbations often distort the diastolic aortic pressure, rendering it difficult to discern the properties of the underlying curve. Thus there is some question as to the reliability of existing estimates of arterial compliance.

In this chapter we propose an alternative method for estimating arterial compliance. This method is still based on a Windkessel model of the vasculature, but utilizes the area under the diastolic pressure waveform rather than depending on a strictly exponential shape of the pressure decay to calculate the time constant of diastolic pressure decay. We will show how estimates of compliance can be obtained both for the simple case of a linear arterial pressure–volume relationship and for three special cases of nonlinear pressure–volume relationships. The different estimates of compliance obtained by existing methods and by the proposed method are compared, using data obtained from patients with congestive heart failure and hypertension. It is shown that compliance estimated with our proposed method is less affected than that estimated with existing methods by deviations of the pressure waveform from an exact exponential.

Governing Equations for a Two-Element Windkessel

For simplicity we will consider only the case where the arterial system is modeled as a two-element Windkessel consisting of a resistance and a compliance in parallel. The same general principles would apply to more complex models. If the volume of the arterial system is denoted by V, the continuity equation is

$$\frac{dV}{dt} + Q_{out} = Q_{in} \tag{16.1}$$

where Q_{in} and Q_{out} denote the rate of flow into the system from the ventricle and from the arterial system into the venous system, respectively. The total arterial compliance is defined as

$$C = \frac{dV}{dP} \tag{16.2}$$

In the most general case of a nonlinear arterial pressure–volume relationship, the compliance is not a constant but depends on the level of the pressure. Only if the pressure–volume relationship is linear will C be a constant. We will derive explicit expressions for compliance assuming both linear and nonlinear pressure–volume relationships.

We assume that the outflow of blood from the arterial system into the venous system is proportional to the pressure difference, that is,

$$Q_{out} = \frac{P - P_o}{R} \tag{16.3}$$

where R is the total peripheral resistance and P_o represents the effective downstream pressure. Usually this is assumed to be the mean venous pressure. Strong evidence suggests, however, that there is a critical opening pressure (a nonzero mean circulatory pressure) at zero flow in the arterial system (Read, Johnson, and Kuida 1957; Sagawa and Eisner 1975; Sylvester et al. 1981), in which case P_o would represent this pressure.

Combining Equations 16.1 through 16.3 we obtain

$$\frac{CdP}{dt} + \frac{P - P_o}{R} = Q_{in} \tag{16.4}$$

Since there is flow into the arterial system only during ventricular systole, Q_{in} in Equation 16.4 can be written

$$Q_{in} = \begin{cases} Q_{in}(t) & 0 < t < t_s^* \\ 0 & t_s^* < t < T \end{cases} \tag{16.5}$$

where t_s* is the time of end of ventricular ejection and T is the total duration of the cardiac cycle as defined in Figure 16.1.

Therefore from Equations 16.4 and 16.5 we obtain the governing equation for the system during any time interval in diastole as

$$\frac{CdP}{dt} + \frac{P-P_o}{\mathrm{R}} = 0 \qquad t_s^* < t < T \tag{16.6}$$

Derivation for a Linear Pressure–Volume Relationship

If we define $P = P_1$ at $t = t_1$ and assume that C is a constant, the solution to Equation 16.6 is

$$P-P_o = (P_1-P_o)\, \mathrm{e}^{(t-t_1)/\mathrm{RC}} \tag{16.7}$$

Therefore in theory, if the pressure decay in diastole from $t = t_1$ onward is an exponential, the compliance C can be calculated from the resistance R and the time constant RC. Usually the further assumption is made that P_o in Equation 16.7 is small enough compared with the arterial pressure so that Equation 16.7 can be simplified to

$$P = P_1\, \mathrm{e}^{(t-t_1)/\mathrm{RC}} \tag{16.8}$$

Most current estimates of the time constant are based on this equation (Bourgeois, et al. 1974; Conroy 1969; Defares and Van Der Waal 1969, 1973; Deswysen, Charlier, and Gevers 1980; Iriuchijima, Kumazawa, and Kawakami 1971; Levy, Birkui, and Saumont 1978; Levy et al. 1985; Messerli, Frohlich, and Nutura 1985; Randall, Van den Bos, and Westerhof 1984; Simon et al. 1979a; Simon, Levenson, and Safar 1985).

If the pressure decay is not an exact exponential or is distorted from a true exponential by reflections or other causes, estimations of the time constant can be affected greatly and, depending on the amount of distortion from a true exponential, can well be erroneous. It is this dependence on strict exponentiality of this method of estimating the time constant that we propose to circumvent. Instead of explicitly solving Equation 16.6, if we integrate directly over a diastolic time interval t_1 to t_2 we obtain

$$C \int_{t_1}^{t_2} \frac{dP}{dt} dt + \frac{1}{R} \int_{t_1}^{t_2} (P-P_o) dt = 0 \tag{16.9}$$

for $t_s^* < t_1 < t_2 < T$, where the various times are defined in Figure 16.1. In Equation 16.9

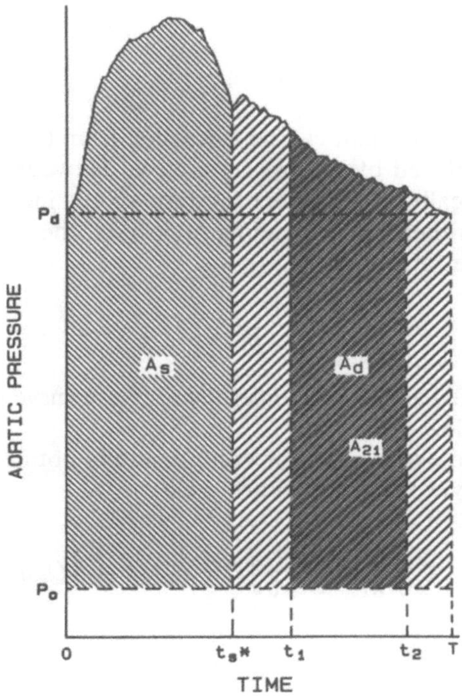

FIGURE 16.1. Schematic drawing of an aortic pressure wave illustrating our nomenclature for the various times and corresponding pressures. The specific areas under the curve are defined by the hatched regions: systolic area, A_s, ; total diastolic area, A_d, ; diastolic area between times t_1 and t_2, and A_{21}, P_o, downstream pressure; P_d, diastolic pressure; T, total duration of cardiac cycle.

$$\int_{t_1}^{t_2} \frac{dP}{dt} dt = \int_{P_1}^{P_2} dP = P_2 - P_1 \tag{16.10}$$

where P_1 and P_2 denote the pressures corresponding to the times t_1 and t_2, respectively. We define the quantity

$$A_{21} = \int_{t_1}^{t_2} (P - P_o)\, dt \tag{16.11}$$

as the area under the pressure curve (above the asymptote P_o) during the time interval t_1 to t_2. Using these definitions, Equation 16.9 can be rewritten

$$C\,(P_2 - P_1) + \frac{A_{21}}{R} = 0 \tag{16.12}$$

from which it follows that

$$\mathrm{RC} = \frac{A_{21}}{P_1 - P_2} \tag{16.13}$$

Therefore the time constant can be calculated from the pressure difference and the area enclosed by the pressure curve between any two points in diastole. Specifically, if we select the time of the incisura, t_s*, and end-diastole, T, with their corresponding pressures, the expression for compliance takes the form

$$C = \frac{A_d}{R * (P_s^* - P_d)} \tag{16.14}$$

where A_d is defined as the area under the pressure curve above the pressure P_o during diastole.

An alternative expression for compliance is obtained by integrating Equation 16.9 over the entire heart cycle

$$C \int_0^T \frac{dP}{dt} dt + \frac{1}{R} \int_0^T (P - P_o) dt = \int_0^T Q_{\text{in}} dt \tag{16.15}$$

Note that

$$\int_0^T \frac{dP}{dt} dt = 0$$

$$\int_0^T Q_{\text{in}}\, dt = \int_0^{t^*_s} Q_{\text{in}}\, dt = \mathrm{SV} \tag{16.16}$$

$$\int_0^T (P - P_o)\, dt = A_s + A_d \tag{16.17}$$

where SV is the stroke volume volume and $A_s + A_d$ is the total area under the aortic pressure curve. Since the total area is equal to the mean pressure times the entire interval of the heart beat, and mean flow is equal to stroke volume times this time interval, resistance can be expressed as

$$R = \frac{A_s + A_d}{\mathrm{SV}} \tag{16.18}$$

Thus, using Equation 16.18, Equation 16.14 can be rewritten

$$C = \frac{\mathrm{SV}}{K * (P_s^* - P_d)} \tag{16.19}$$

where $K = (A_s + A_d)/A_d$ is an area index expressing the ratio of the total area under the aortic pressure curve divided by the diastolic area.

If the diastolic pressure decay is an exact exponential, the expressions for compliance become even simpler. From Equations 16.7 and 16.13 it follows that for the time interval t_1 to t_2

$$A_{21} = \frac{(P_1 - P_2)\,(t_2 - t_1)}{\ln\,[(P_1 - P_o)/(P_2 - P_o)]} \tag{16.20}$$

In this case the time constant for the time interval t_1 to t_2 in diastole is

$$\mathrm{RC} = \frac{(t_2 - t_1)}{\ln\,(P_1 - P_o) - \ln(P_2 - P_o)} \tag{16.21}$$

If one restricts attention to the onset and end of diastole, defined by the times t_s^* and T, with their corresponding pressures P_s^* and P_d, respectively, one obtains

$$\mathrm{RC} = \frac{T_d}{\ln\,(P_s^* - P_o) - \ln(P_d - P_o)} \tag{16.22}$$

where T_d is the total duration of diastole. If P_o is zero, we obtain the simpler expression for the time constant

$$\mathrm{RC} = \frac{T_d}{\ln P_s^* - \ln P_d} \tag{16.23}$$

This last expression was that proposed previously to estimate the time constant (Defares and van der Waal 1969, 1973).

Derivation for Nonlinear Arterial Pressure–Volume Relationships

The derivation of the expression for compliance for nonlinear arterial pressure–volume relationships requires generalization of some of the previously derived expressions. From Equations 16.1 and 16.3 we obtain for the diastolic period

$$\frac{dV}{dt} + \frac{P - P_o}{R} = 0 \tag{16.24}$$

Integrating this equation over the diastolic interval $t_s^* < t < T$ we obtain

$$\int_{t_s^*}^{T} dV + \int_{t_s^*}^{T} \frac{P - P_o}{R}\, dt = 0 \tag{16.25}$$

which can be rewritten as

$$V(P_s^*) - V(P_d) = \frac{A_d}{R} \tag{16.26}$$

where $V(P_s^*)$ and $V(P_d)$ are the arterial volumes corresponding to the pressures P_s^* and P_d, at the times t_s^* and T, respectively. Substituting Equation 16.18 into Equation 16.26 we obtain

$$V(P_s^*) - V(P_d) = \frac{\mathrm{SV}}{K} \tag{16.27}$$

where K is the area ratio defined earlier. To obtain explicit expressions for compliance, we need expressions for the pressure–volume relationships. We will examine three specific cases where, for simplicity, we use the entire diastolic period for estimating the compliance.

Exponential Relationship

If the arterial volume is an exponential function of pressure with the form

$$V = a\,\mathrm{e}^{bP} + k \tag{16.28}$$

where a, b, and k are coefficients to be determined, then

$$\mathrm{V}(P_s^*) - V(P_d) = a[\mathrm{e}^{bP_s^*} - \mathrm{e}^{bP_d}] \tag{16.29}$$

From Equations 16.27 and 16.29 we have

$$a = \frac{\mathrm{SV}}{K} \frac{1}{\mathrm{e}^{bP_s^*} - \mathrm{e}^{bP_d}} \tag{16.30}$$

Thus for this case the expression for compliance is

$$C = ab\mathrm{e}^{bP} = \frac{\mathrm{SV}}{K} \frac{b\mathrm{e}^{bP}}{\mathrm{e}^{bP_s^*} - \mathrm{e}^{bP_d}} \tag{16.31}$$

Logarithmic Pressure–Volume Relationship

If the volume is a logarithmic function of pressure of the form

$$V = A \ln(P) + B \tag{16.32}$$

then

$$V(\mathrm{P}_s^*) - V(P_d) = A \ln \frac{P_s^*}{P_d} \tag{16.33}$$

Substituting this expression into Equation 16.27 we obtain for the constant A

$$A = \frac{\mathrm{SV}}{K} \frac{1}{\ln(P_s^*/P_d)} \tag{16.34}$$

with the resulting expression for the compliance

$$C = \frac{\mathrm{SV}}{Kp} \frac{1}{\ln(P_s^*/P_d)} \tag{16.35}$$

Parabolic Pressure–Volume Relationship

In a previous study, Cope (1960) demonstrated that the arterial pressure–volume relationship could be approximated by a segment of a parabola for $P < P_*$ and by a straight line for $P > P_*$, that is

$$V = \begin{cases} a_2P^2 + a_1P + a_o & \text{for } P < P_* \\ a_2P_*^2 + a_1P_* + a_o & \text{for } P \geqslant P_* \end{cases} \tag{16.36}$$

where P_* is a critical pressure determined experimentally and a_2, a_1, and a_o are constants to be determined that have a linear relationship among themselves, that is,

$$a_2 = ma_1 + n \tag{16.37}$$

From Equation 16.36 we obtain

$$V(P_s^*) - V(P_d) = \begin{cases} a_2[(P_s^*)^2 - P_d^2] + a_1(P_s^* - P_d) & P_s^* < P_* \\ a_2[(P_s^*)^2 - P_d^2] + a_1(P_s^* - P_d) & P_s^* \geqslant P_* \end{cases} \tag{16.38}$$

Substituting Equation 16.27 into Equation 16.38 we obtain, after some algebraic manipulations,

$$a_1 = \begin{cases} \dfrac{(\mathrm{SV}/K) - n[(P_s^*)^2 - P_d^2]}{m[(P_s^*)^2 - P_d^2] + (P_s^* - P_d)} & \text{for } P_s^* < P_* \\[2ex] \dfrac{(\mathrm{SV}/K) - n[(P_*^2 - P_d^2]}{[m(P_*^2 - P_d^2) + (P_* - P_d)]} & \text{for } P_s^* \geqslant P_* \end{cases} \tag{16.39}$$

The expression for compliance for this case is

$$C = \begin{cases} 2(ma_1 + n)P + a_1 & \text{for } P_s^* < P_* \\ 0 & \text{for } P_s^* \geqslant P_* \end{cases} \tag{16.40}$$

Comparison of Values of Compliance Obtained for the Various Methods

Values of compliance using each of the above methods are compared using aortic pressure data obtained from patients with either congestive heart failure or hypertension who were undergoing diagnostic cardiac catheterization. Medication was withheld from each patient for at least 48 hours prior to catheterization. After the diagnostic portion of the procedure, a micromanometer-tipped combination pressure and flow velocity catheter

(Millar Instruments, Houston, Tex.) was introduced into the ascending aorta. To stabilize the flow transducer in the center of the aortic stream, the distal tip of the catheter was positioned in the left ventricle across the aortic valve. The flow velocity was converted to volume flow by multiplying the velocity signal by the cross-sectional area of the aorta obtained from echocardiograms of the ascending aorta. Procedural details such as transducer calibration, minimization of artifacts, data recording, and analog-to-digital conversion have all been discussed in detail previously (Yin et al. 1983). The data used for this illustration consist of a representative beat of the aortic pressure and flow velocity signals during baseline resting conditions prior to any drug infusion for each patient. All calculations were performed on a Data General MV-8000 minicomputer.

Table 16.1 lists some clinical and hemodynamic data for the patients along with the compliance estimates based on a linear arterial pressure–volume relationship obtained from (1) the area method proposed herein (Eq. 16.14), (2) the simplified method, assuming that the pressure decay is an exact exponential from the onset to the end of diastole (Eq. 16.22), and (3) the semilogarithmic regression method. For the latter calculation we used the diastolic period beginning 30 milliseconds after the incisura until 12 milliseconds prior to the next systole, as proposed by Simon and colleagues (1979a). We also calculated the compliances for this portion of diastole using Equations 16.13 and 16.21 and denote these as C_1' and C_2', respectively. All the comparisons in this table were made assuming that P_o equals zero.

The values for compliance estimated by Equations 16.13 and 16.21 utilizing either the entire diastolic period (C_1 and C_2) or the last portion of the diastolic interval (C_1' and C_2') are quite close to one another, even though the absolute values of compliance for the two different intervals differ. In contrast, the values obtained from the semilogarithmic regression using the same diastolic interval may differ (compare C_3 with C_1 and C_2). The differences are sometimes large—30% or more.

The finding that the compliances calculated by using Equation 16.21 depend on the time interval selected is de facto evidence that the pressure decay is not a true exponential function. Since all existing methods for estimating arterial compliance assume a strictly exponential time decay of aortic pressure, there is a strong possibility that these estimates may be erroneous. The method proposed here utilizes the area under the diastolic pressure waveform rather than depending on the waveform itself. While this method is still based on a Windkessel model of the vasculature, and carries the implicit assumption that the underlying pressure decay should be exponential, it has the major advantage that the value of the compliance is not so critically dependent on strict exponentiality of the time course of pressure decay as it is with existing methods. The reason is that distortion of the diastolic waveform affects the area under the curve relatively less than it affects the parameters derived on the basis of the shape of the waveform. This relative insensitivity of the area method to

TABLE 16.1 Clinical profile and hemodynamic data for patients studied

Patient	Diagnosis	SV	P_s	P_d	P_m	R	C_1	C_2	C_1'	C_2'	C_3
1	CHF	33.8	86.6	60.5	72.5	1.34	1.72	1.66	1.16	1.17	1.06
2	CHF	40.3	98.4	67.8	81.1	0.89	1.34	1.37	0.99	0.99	1.03
3	CHF	61.4	106.5	63.6	81.5	1.03	1.53	1.52	1.09	1.31	1.27
4	CHF & HPT	32.4	147.5	77.3	103.5	1.99	0.50	0.52	0.56	0.47	0.49
5	HPT	87.9	169.3	113.5	136.5	1.03	1.86	1.72	1.27	1.33	1.20
6	HPT	104.1	165.3	100.3	128.5	0.98	1.53	1.55	1.35	1.39	1.20
7	HPT	47.7	134.2	92.1	92.1	1.68	1.22	1.17	0.88	0.88	0.79

CHF, congestive heart failure; HPT, hypertension; SV, stroke volume (ml); P_s, peak systolic aortic pressure (mm Hg); P_d, end-diastolic aortic pressure (mm Hg); P_m, mean pressure (mm Hg); R, peripheral resistance (mm Hg sec/ml); C_1 and C_2, compliances calculated using area method of Eqs. 16.14 and 16.22, respectively, for all of diastole (ml/mm Hg); C_1' and C_2', compliances calculated from same equations but using only portion of diastole beginning 36 ms after incisure to 12 ms prior to next beat; C_3, compliance calculated from linear regression of the semilogarithm of pressure versus time for same portion of diastole as for C_1' and C_2'.

nonexponentiality is evidenced by the very similar estimates of compliances C_1 and C_2 and the larger differences between the estimates C_3 and C_2' in Table 16.1. For these reasons we believe that the values of compliance obtained with this method are more likely to match directly measured ones. Unfortunately, since we cannot make this measurement in humans, no experimental data exist with which to compare these estimates.

The finding that aortic pressure decay in heart failure is not exponential verifies previous predictions. The presumption is that the vasodilatation that accompanies failure decreases pulse wave velocity, causing reflected waves to distort the diastolic rather than the systolic portion of the wave. In contrast, it was felt that the increased pulse wave velocity accompanying hypertension should cause earlier appearance of the reflected wave, producing less distortion of the diastolic aortic waveform and rendering this portion closer to a true exponential (Ch. 1). Our data do not support this prediction. We found that the pressure decay also is not exponential in hypertensive patients.

Despite the nonexponentiality of the aortic pressure decay, one could, to a first approximation, assume that it is nearly exponential and obtain useful data about the vasculature in terms of a Windkessel model. Even within the limitations of such an approximation, however, existing methods of obtaining compliance have pitfalls. First, it is clear that compliance estimates obtained with the semilogarithmic method depend on the level of the mean circulatory pressure. This can be seen from examination of Equations 16.6 and 16.7. Calculation of the time constant from the regression of the logarithm of P versus t implicitly assumes that the pressure asymptote is zero. If it is nonzero, the slope of the regression depends on the absolute value of P_o. If one knows what P_o is, one should perform the regression of (P-P_o) versus t. These data, however, are not obtainable in humans. Consequently, we believe that values of compliance obtained with this method are of questionable validity.

Furthermore, there is no agreement as to the most suitable interval during diastole in which to perform the regression. For example, Bourgeois and co-workers (1974) suggested using the interval beginning at 30 milliseconds after the incisura to 20 milliseconds before the onset of the next wave. The rationale for this is to minimize disturbances caused by the opening and closing of the aortic valve. Simon and colleagues (1979a) used the last two thirds of the brachial arterial pressure wave to obtain their estimates of the time constant. They suggested that this would avoid the high-frequency harmonics that occur near the incisura. Rather than using these arbitrarily selected portions of diastole, we propose that the well-defined incisura and end of diastole be used. As the data in the table indicate, using the onset and end of diastole and assuming a simple exponential decay with a zero aymptote, as originally suggested by Defares and van der Waal (1969, 1973), yields results similar to those obtained with our area method. This simpler method obviates the need to calculate the area under the pressure curve. The agreement between the area method and the simpler method, however, is not so good when a subinterval of diastole, such as that suggested by Simon and co-workers (1979a) is used (compare C_1' and C_2'). The probable reason for this can be found from examination of Equation 16.22. Because the differences of the logarithms of the pressures appear in the denominator, any distortions of the pressure waveform will be magnified. Thus, if one wishes to use the simpler method, we suggest that the entire diastolic period be used, since the time points are well defined and would facilitate comparisons among different investigators. It should be emphasized, however, that to properly account for a nonzero mean circulatory pressure, one should use Equation 16.22 and not Equation 16.23.

The expressions for compliance for the exponential and parabolic pressure–volume relationships (Eqs. 16.31 and 16.40) require experimentally determined coefficients. Since there is no feasible method of obtaining such a coefficient for the entire arterial system in vivo, we decided to use values obtained previously in studies in isolated segments of aortas and large arteries (Liu, unpublished data). Segments of aortas and several large arteries (carotid, brachial, femoral) were obtained at autopsy and subjected to cyclic pressurization. Fitting the data to an exponential of the form of Equation 16.28, we found that, remarkably, despite markedly differing pressure–volume relationships for the vessels from different sites, the coefficient b was nearly invariate, averaging -0.013 for all the vessel segments. Therefore we used this value of b in these calculations of compliance.

To obtain the constants needed for the parabolic case, we fitted the same vessel segment data to a parabola. Again for all of the arteries, the constants m and n in Equations 16.39 and 16.40 were nearly independent of location, so we used their average values for all of the segments. These values are compared in Table 16.2 with those previously obtained by Cope (1960), who utilized the pressure–volume data from an earlier study (Remington et al. 1945).

The compliance values obtained for the three nonlinear pressure–volume relationships are listed in Table 16.3. For these nonlinear cases, since the compliances depend on the pressure level the results listed are for the compliances corresponding to peak systolic, mean, and diastolic pressures. In general, the peak systolic compliance is less and the diastolic compliance

TABLE 16.2. Comparison of constants m and n (Eqs. 16.37–16.40) obtained by fitting pressure–volume data for various arterial segments to a parabola with the constants obtained by Cope (1960) utilizing the pressure–volume data of Remington et al. (1945)

Study	m	n
Yin, Liu, and Brin	-2.703×10^{-3}	6.445×10^{-5}
Cope	-3.0×10^{-3}	9.0×10^{-4}

TABLE 16.3. Estimated compliances as function of pressure, assuming three different nonlinear pressure–volume relations

	Exponential			Logarithmic			Parabolic		
Patient	C_s	C_m	C_d	C_s	C_m	C_d	C_s	C_m	C_d
1	1.42	1.70	1.99	1.41	1.69	2.01	1.53	1.75	1.93
2	1.00	1.25	1.49	1.03	1.25	1.49	1.07	1.28	1.43
3	1.25	1.73	2.12	1.33	1.73	2.13	1.35	1.77	2.03
4	0.25	0.44	0.63	0.32	0.45	0.61	0.22	0.46	0.60
5	1.05	1.61	2.18	1.36	1.69	2.05	0.52	1.55	2.26
6	0.84	1.35	1.96	1.10	1.42	1.81	0.48	1.33	1.99
7	0.81	1.06	1.10	0.93	1.10	1.36	0.76	1.06	1.38

C_s, C_m, C_d, compliances corresponding to peak systolic, mean, and diastolic pressures, respectively.

is greater than the corresponding compliances obtained assuming a linear pressure–volume relationship.

Because of the difficulty in measuring the pressure–volume relationship of the arterial system, most studies that require a numerical value for compliance have implicitly assumed that it is a constant in the physiologic range of blood pressure. This is tantamount to assuming that the arterial pressure–volume relationship is linear. Most evidence, however, indicates that the arterial pressure–volume relationship is nonlinear, meaning that compliance must vary according to the level of pressure. Cope (1960, 1961), attempted to take this nonlinear pressure–volume behavior into account in his estimates of arterial compliance by using in vitro pressure–volume data obtained in segments of arteries (Remington et al. 1945). The applicability of these in vitro data to the in vivo setting is not known. The method we propose here can be extended to nonlinear vascular pressure–volume relationships provided we have some idea of these relationships. Our data indicate, moreover, that the exact form of the pressure–volume relationships need not be known. What is required is the exponential coefficient b, in the case of an exponential relationship, and the values m and n, in the case of a parabolic pressure–volume relationship. The isolated vessel segment data indicate that these parameters do not depend on the vessel segment examined. Hence we assume that the value pertains to the entire arterial system. In the case of a logarithmic pressure–volume relationship, the situation is even simpler, since no explicit parameter values are needed. The data indicate that in the physiologic range of pressures, the values of compliance for the logarithmic case are actually closer to those of the exponential case than for the parabolic case.

Summary

We have proposed an improved method for estimating arterial compliance from the aortic pressure waveform. The method is based on a two-element Windkessel model of the arterial system. Unlike existing methods, which depend on a strictly exponential pressure decay, our method depends on the diastolic area under the pressure curve. Using data from patients with congestive heart failure and hypertension, we demonstrated the following: (1) The pressure decay in diastole is not a true monoexponential. (2) Even if one approximates the pressure decay as an exponential, existing methods based on semilogarithmic regression of pressure versus time have several pitfalls. (3) Because the proposed method does not depend on the exact form of the pressure wave, it is less susceptible than existing methods to deviations from a true exponential. (4) Comparisons of the estimated compliance obtained from our method and those of existing methods can differ considerably.

Bibliography

Bourgeois, M. J., Gilbert, B. K., Donald, D. E., and Wood, E. H. 1974. Characteristics of aortic diastolic pressure decay with application to the continuous monitoring of changes in peripheral vascular resistance. *Circ. Res.* 35:56–66.

Brunner, M. J., Greene, A. S., Sagawa, K., and Shoukas, A. A. 1983. Determinants of systemic zero-flow arterial pressure. *Am. J. Physiol.* 245:H453–H460.

Conroy, M. F. 1969. Estimation of aortic distensibility and instantaneous left ventricular volume in living man. *Bull. Math. Biophys.* 31:93–104.

Cope, F. W. 1960. An elastic reservoir theory of the human systemic arterial system using current data on aortic elasticity. *Bull. Math. Biophys.* 22:19–40.

Cope, F. W. 1961. A method for the computation of aortic distensibility in the living human patient and its use for the determination of the aortic effects of aging, drugs and exercise. *Bull. Math. Biophys.* 23:337–353.

Defares, J. G., and van der Waal, H. J. 1969. A method for the determination of systemic arterial compliance in man. *Acta Physiol. Pharmacol. Neerl.* 15:329–343.

Defares, J. G., and van der Waal, H. J. 1973. Theory of the measurement of arterial compliance in humans. *Bull. Math. Biol.* 35:237–244.

Deswysen, B., Charlier, A. A., and Gevers, M. 1980. Quantitative evaluation of the systemic arterial bed by parameter estimation of a simple model. *Med. Biol. Eng. Comput.* 18:153–166.

Iriuchijima, J., Kumazawa, A., and Kawakami, K. 1971. Measurement of aortic compliance in vivo. *Jap. Heart J.* 12:486–493.

Levy, B., Birkui, P., and Saumont, E. 1978. Elasticity modulus of the ascending aorta and systemic arterial compliance in dog. *Colloq. Insti. Nat. Sante Rech. Medi. (INSERM)* 78:141–152.

Levy, B. I., Benessiano, J., Poitevin, P., Lukin, L., and Safar, M. E. 1985. Systemic arterial compliance in normotensive and hypertensive rats. *J. Cardiovasc. Pharmacol.* 7:S28–S32.

Messerli, F. H., Frohlich, E. D., and Nutura, H. O. 1985. Arterial compliance in essential hypertension. *J. Cardiovasc. Pharmacol.* 7:S33–S35.

O'Rourke, M. F. 1982. *Arterial Function in Health and Disease,* Melbourne, Australia: Churchill Livingstone, pp. 79–80.

Randall, O. T., van den Bos, G. C., and Westerhof, N. 1984. Systemic compliance: does it play a role in the genesis of essential hypertension? *Cardiovasc. Res.* 18:455–462.

Read, R. C., Johnson, J. A., and Kuida, H. 1957. Systemic and coronary pressure–flow relationships during total body perfusion in the dog. *Am. J. Physiol.* 190:49–53.

Remington, J. W., Noback, C. B., Hamilton, W. F., and Gold, J. J. 1945. Volume elasticity characteristics of the human aorta and the prediction of the stroke volume from the pressure pulse. *Am. J. Physiol.* 153:298–305.

Sagawa, K., and Eisner, A. 1975. Static pressure–flow relation in the total systemic vascular bed of the dog and its modification by the baroreceptor reflex. *Circ. Res.* 36:406–413.

Shoukas, A. A., and Sagawa, K. 1973. Control of total systemic vascular capacity by the carotid sinus baroreceptor reflex. *Circ. Res.* 33:22–32.

Simon, A. C., Levenson, J. A., and Safar, M. E. 1985. Hemodynamic mechanisms of and therapeutic approach to systolic hypertension. *J. Cardiovasc. Pharmacol.* 7:523–527.

Simon, A. C., Safar, M. E., Levenson, J. A., London, G. M., Levy, B. I., and Chau, N. P. 1979a. An evaluation of large arteries compliance in man. *Am. J. Physiol.* 237:H550–H554.

Simon, A. C., Safar, M. E., Levenson, J. A., and Levy, B. I. 1979b. Systolic hypertension: hemodynamic mechanism and choice of antihypertensive treatment. *Am. J. Cardiol.* 44:505–511.

Sylvester, J., Gilbert, R., Traystman, R., and Permutt, S. 1981. Effects of hypoxia on the closing pressure of the canine systemic arterial circulation. *Circ. Res.* 49:980–987.

Ventura, H., Messerli, F. H., Oigman, W., Suarex, D. H., Dreslinski, G. R., Dunn, F. G., Reisin, E., and Frohlich, E. D. 1984. Impaired systemic arterial compliance in borderline hypertension. *Am. Heart J.* 108:132–136.

Yin, F. C. P., Guzman, P. A., Brin, K. P., Maughan, W. L., Brinker, J. A., Traill, T. A., Weiss, J. L., and Weisfeldt, M. L. 1983. Effect of nitroprusside on hydraulic vascular load on the right and left ventricle of patients with heart failure. *Circulation* 67:1330–1339.

Index

MIX
Papier aus verantwortungsvollen Quellen
Paper from responsible sources
FSC® C105338

If you have any concerns about our products,
you can contact us on
ProductSafety@springernature.com

In case Publisher is established outside the EU,
the EU authorized representative is:
Springer Nature Customer Service Center GmbH
Europaplatz 3, 69115 Heidelberg, Germany

Printed by Libri Plureos GmbH
in Hamburg, Germany